Education

Presentations at the 2008 AIChE Annual Meeting

Philadelphia, Pennsylvania
16 - 21 November 2008

ISBN: 978-1-61567-219-6

Printed from e-media with permission by:

Curran Associates, Inc.
57 Morehouse Lane
Red Hook, NY 12571
www.proceedings.com

Some format issues inherent in the e-media version may also appear in this print version.

Printed by Curran Associates, Inc. (2009)

AIChE
3 Park Avenue
New York, NY 10016-5991

Phone: (203) 702-7660
Fax: (203) 775-5177

www.aiche.org

TABLE OF CONTENTS

2a a Method for the Consistent Creation and Quantitative Testing of Postoperative Pelvic Adhesions in a Porcine Model

Maureen E. Cheung[1], Bradford Fenton[2], M Michelle Evancho-Chapman[3], Mark Kovacik[4], Don Noe[4], Nicholas Ree[5], James Fanning[2], and Stephanie T. Lopina[6]. (1) Chemical and Biomolecular Engineering / Surgery, The University of Akron / Summa Health Systems, 511 E. Buchtel Ave / 525 E. Market Street, Akron, OH 44304, (2) Ob/gyn, Summa Health Systems, 525 E. Market St., Akron, OH 44304, (3) Surgery; Division of Surgical Research, Summa Health Systems, 525 E. Market St., Akron, OH 44304, (4) Orthepedics; Musculoskeletal Research Laboratory, Summa Health Systems, 525 E. Market St., Akron, OH 44304, (5) Pathology, Summa Health Systems, 525 E. Market St., Akron, OH 44304, (6) Department of Chemical & Biomolecular Engineering, The University of Akron, Akron, OH 44325

Introduction: Adhesions following surgery represent a significant problem often resulting in pain, disability, and additional surgeries. There are compounds available for the prevention of post-operative adhesions, but their effectiveness is difficult to assess. The assessment of methods and compounds for the prevention of postoperative pelvic adhesions is hampered by the lack of a quantifiable adhesion measurement technique. Currently accepted models of adhesion comparison are limited to qualitative methods with much potential bias and little comparability.

Objective: The purpose of this study was to develop a standard injury technique and quantitative assessment method to evaluate adhesion prevention by novel compounds.

Development: The primary focus of this research was the creation of an adhesion complex that was suitable to quantitative testing using the Material Testing System (MTS) and histological assessment. This led to several improvements on the initial idea that included improved stabilization technique, testing area, and injury precision.

Methods: A technique was developed to generate consistent and significant adhesions in a swine model that are suitable for quantitative assessment using a Material Testing System (MTS) machine platform. In this method, the uterine horns are cannulated and then attached to the pelvic sidewall using loose silk sutures. The underlying sidewall and adjacent uterine serosa are injured with electrocautery. Following a two week survival the entire complex of uterine horn and sidewall are excised en bloc and prepared for MTS testing.

Result: The MTS measures force by displacement, generating a quantitative assessment of adhesion strength, which can be analyzed and interpreted, while the histology confirmed that adhesion was present and being quantified as well as providing a comprehensive description of the vascularity, density, collagen content, and organization of the adhesion. Our assessments showed that previous visual grading methods were neither consistent nor accurate. This model is unique and advantageous in its ability to quantitatively assess the strength of the adhesion complex.

Discussion: This technique represents a novel quantitative method for adhesion assessment in an animal model. This quantitative technique may then be applied to accurately assess a range of adhesion prevention techniques, producing results which can be standardized for comparison.

2b Bandage Contact Lenses for Anti-Inflammatory Therapeutics

Jeney Zhang and Mark E. Byrne. Biomimetic & Biohybrid Materials, Biomedical Devices, and Drug Delivery Laboratories, Department of Chemical Engineering, Auburn University, Auburn, AL 36849-5127

The US ophthalmic drug market is valued at approximately $4.5 billion and conventional topical ocular therapies control approximately 90% of that market, with inflammation treatments accounting for 55% or $2.5 billion. However, there is still an unmet need in the ocular therapeutic market for effective drug delivery. A major obstacle facing the traditional topical methods of ocular drug delivery is poor bioavailability with only 1-7% of the applied drug absorbed while the rest is either removed by the lacrimal and tear turnover of the blinking mechanism or enters into systemic circulation. We have rationally designed and synthesized novel contact lenses capable of tailorable loading and extended release of anti-inflammatory ocular medication.

This novel drug delivery system will deliver ocular medication via contact lenses in which the drug is released from the hydrophilic, weakly cross-linked polymers of the lens directly onto the surface of the eye. These recognitive polymers are synthesized via a biomimetic imprinting technique, which allows for the formation of pre-polymerized complexes between the drug molecule and functional monomers that are chosen with specific structures designed to interact with the drug molecule via non-covalent chemistry. The polymers were synthesized with a cocktail of FDA-approved methacrylate and acrylate monomers that provided similar side-chain functionalities to amino acids present in the binding pockets of the biomolecules as found in nature. Hydrogels were synthesized with macromolecular "memory" site or reservoirs for the non-steroidal anti-inflammatory drug diclofenac salt.

These biomimetic contact lenses exhibited the ability for biomolecule recognition and memory and demonstrated advantage over hydrogels created without biomimesis by displaying a two-fold increase in drug loading. Further binding studies were conducted with the addition of iniferter to the polymer formulation to reversibly terminate the UV free radical polymerization. The addition of iniferter to the recognitive system created a two-fold increase in loading over the recognitive system without iniferter due to formation of more ordered and uniform crosslinked polymer network. Equilibrium swelling studies indicated that 40% of the swollen gel is water, indicating the adequate comfort of wearing and oxygen permeability of these gels. Dynamic drug release profiles demonstrated that release rates for the biomolecule in the recognitive systems can be tailored to provide a sustained release in artificial lacrimal fluid. The diffusion coefficient calculated for the recognitive system was an order of magnitude lower than that of the control system, demonstrating that imprinting extends the release rate with greater functionality leading to greater delay in release. Further diffusion analysis revealed that while the control system was exhibiting Fickian release profile, driven by pure diffusion, both the recognitive and iniferter systems exhibited release profiles closer to zero-order, constant release. Parametric studies to optimize loading and extend release were performed by synthesizing hydrogels with varying monomer to template ratios. Furthermore, these hydrogels presented the same ocular clarity and mechanical functionalities as conventional contact lenses and could also offer the same comfort and re-epithelialization as hydrophilic bandage contact lenses (HBCLs), which provide a mechanical barrier against the shearing forces of eye lids and can act as a hydrating reservoir.

Designing Scaffolds with Optimal Chemical and Physical Properties for Tissue Engineering

Courteney Cannon, Anna Gromadzka, Lenny Slutsky, Lourdes Collazo, and Miriam Rafailovich

Department of Materials Science and Engineering,
Stony Brook University, Stony Brook, New York 11794-2275

1. Abstract

The design of scaffolds that resemble natural organs is becoming increasingly important for tissue engineering. Cells ability to proliferate and function depends strongly on the mechanical and chemical properties of the scaffold. In this experiment we studied the effects of surface chemistry on cell morphology and proliferation. Polydimethylsiloxane (PDMS) was utilized as a scaffold due to its wide range of medical applications and the results were compared with those of cells cultured on tissue culture plates. The mechanical properties of the PDMS substrate were varied by adding different amount of a cross-linking agent. For each cross linking ratio, sets of samples containing 0, 0.5, 1.0, and 2.0 percent of Na clay were prepared. We observed the mechanical characteristics of the polymer samples using Instron tensile tester and Nicolet Spectra FTIR instrument. We measured the tensile modulus and extension at yield and break as a function of the degree of cross-linking. Cross-linked spherical plug molds were then made of each sample and placed in a 24 well plate where fibroblast-like cells (CF-29) were cultured. Cell counts were obtained after 24 hours and three days. The cells were also stained and imaged under a confocal microscope (Leica-SP). The results indicate that cell proliferation is hindered on PDMS substrates relative to tissue culture plastic (TCP) substrates. Addition of clays in the PDMS susbstrates increased the cell counts to values that were comparable to those obtained on TCP.

2. Introduction

It has been shown that cells can sense the physical and chemical characteristics of their microenvironment. Cells measure the stiffness of the material by adhering to the surface and pulling on it. If the substrate is relatively soft, the cells are shown to express nerve-like protein, and if the substrate if relatively hard, they express bone protein[1]. Polymeric scaffolds can be used to mimic various tissues so the cells can proliferate and function normally.

Fibroblast cells play important role in production of extracellular matrix and are crucial for wound healing. To replicate the mechanical modulus of connective tissue, we cross-linked polydimethylsiloxane (PDMS) and then added clay nanoparticles. PDMS is widely used in the medical industry ranging from joint replacements to implants. PDMS hydrogel can be solidified by cross-linking using a curing agent. The hardness of the solid material depends on the cross-linking density[2]. The PDMS to cross-linker ratios were 20:1, 10:1 and 5:1.

Cross-linked PDMS presents a hydrophobic surface. To modify that surface chemically, we introduced various amounts of sodium Cloisite Na+ clay. Clays are layered compounds that can be exfoliated into particles of nanoscale dimensions. They are used extensively in producing many consumer materials ranging from car parts to beauty supplies. Cloisite® Na^+ is a natural montmorillonite often added to polymers in order to reinforce the polymeric matrix and make it impermeable to gases[3]. Each cross-linked PDMS was modified with 0, 0.5, 1 and 2% clay.

3. Methods

3.1 Materials

Sylgard® 184 Silicone Elastomer Base and Curing Agent were purchased from Dow Corning Corporation.
Cloisite NA^+ clay was purchased from Southern Clay Products, Inc.
Mechanical properties were analyzed using a Nicolet Spectra Fourier Transform Infrared Spectroscopy instrument and an Instron Tensile Tester.
Cell counting was performed under an Olympus inverted phase contrast microscope. Cell imaging was performed with a Leica-SP Confocal microscope.

3.2 Synthesis and Mechanical Analysis of PDMS

PDMS samples were made with varying degrees of cross-linking by mixing Sylgard 184 in 20:1, 10:1, and 5:1 ratios of base to curing agent.
Thin films of the samples were made for FTIR analysis by smearing gel onto Kaptom film and setting between two glass microscope slides. Spectra of Sylgard 184 base, curing agent and mixtures of 20:1, 10:1 and 5:1 were taken.
For each PDMS ratio, samples containing 0, 0.5, 1.0, and 2.0 percent Na clay were prepared. The samples were molded for Instron using appropriate steel plate molds. All

samples were put under vacuum for one half hour followed by two hours in the oven at 65°C to complete curing as per manufacturer's specifications.

3.3 Cell Culture, Counting, Staining, and Imaging

Samples were incubated in 1% fibronectin solution for 24 hrs in a 5% CO_2, 37 °C, 100% humidity. Samples were washed once with PBS and human dermal fibroblast cells (CF-29) were plated with density of 2400 cells/cm^2 in DMEM (low glucose) 20% Fetal Bovine Serum (FBS) 1% Penicillin streptomycin (PS) media.

Cells were counted from three separate dishes of each sample on day 1 and day 3 of incubation in standard conditions of the human body. Media was aspirated out of the dish and the substrate was washed once gently with phosphate buffer saline (PBS) solution. 300μL of trypsin EDTA was added to the substrate to lift the adhered cells and incubated for five minutes. Trypsin solution was then transferred to glass test tubes. Cells were counted under an Olympus inverted phase contrast microscope using a hemocytometer.

Cells were fixed for staining with 3.7% formaldehyde. The cells were then stained using Alexa-fluor 488 Phallodin for the F-Actin protein and Propidium Iodide for the nucleus. In order to determine cell morphology, the stained cells were imaged under the confocal microscope.

4. Results and Discussion

4.1 FTIR

Infrared spectra of our samples were taken in order to quantify the cross-linking density. In order to obtain legible FTIR spectra, very thin films had to be prepared and the instrument dehumidified.

Table 4.1 summarizes the observed peaks from Figures 4.1a – c.

Peaks, cm^{-1}						
Literature value [4]	***Base***	***Curing Agent***	***20:1***	***10:1***	***5:1***	***Assignment*** [4]
2850-3000	2962.72	2963.26 2904.05	2960 2900	2960	2962.87 2904.58	CH_3
2100-2360	-	2162.06	2160	2160	2159.79	Si-H
1900-2000	-	-	1950	1950	1943.86	C=C
1350-1470	-	1410.07	1410	1410	1411.85	CH_2
1250±10	1261.11	1261.04	1260	1260	1259.14	$Si\text{-}CH_3$

Table 4.1. Infrared Spectrum of PDMS

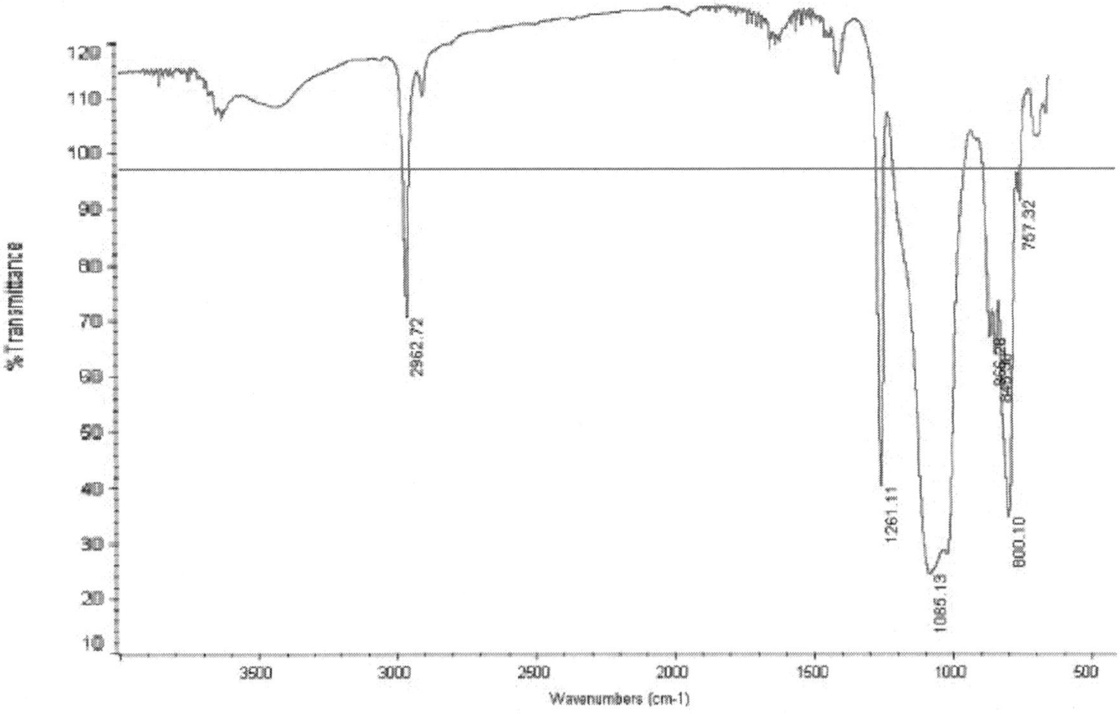

Figure 4.1a. Sylgard 184 PDMS Base

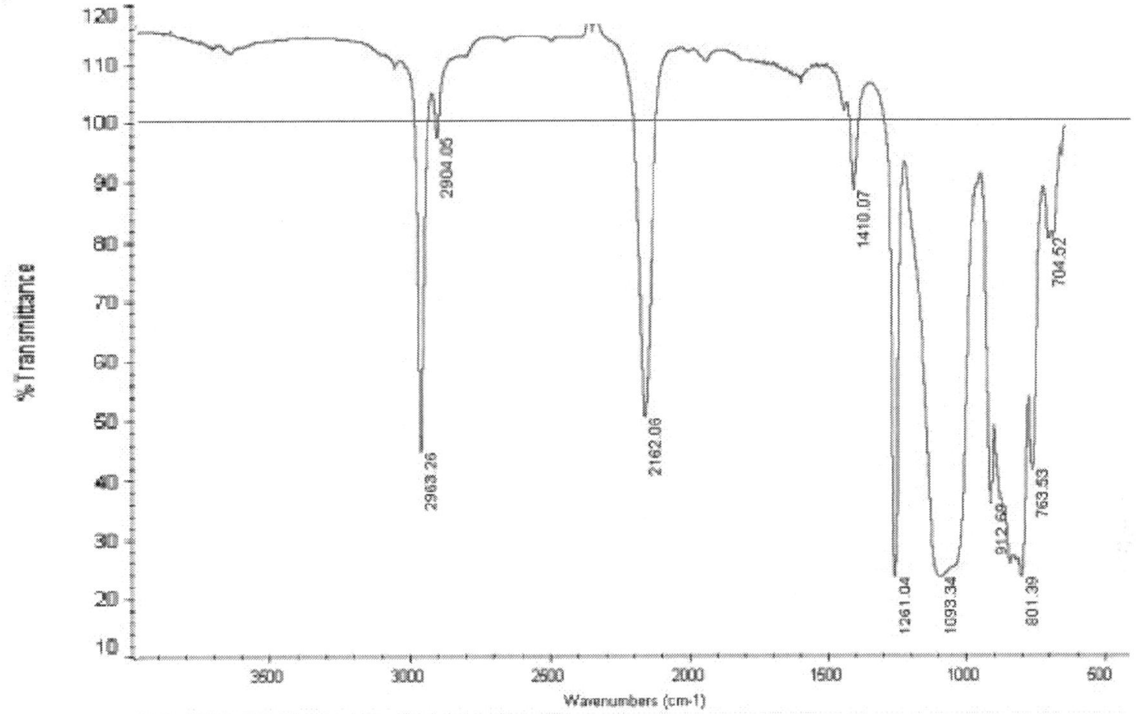

Figure 4.1b. Sylgard 184 Curing Agent

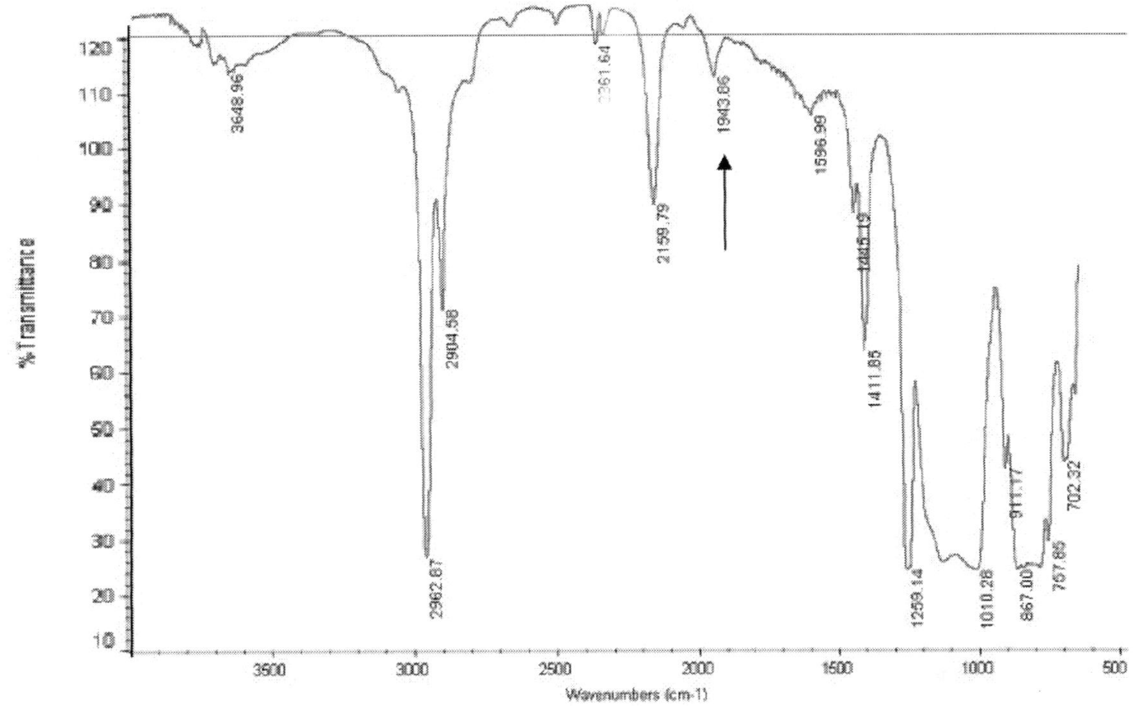

Figure 4.1c. PDMS, 5:1 Cross-linked

From Table 4.1 and Figures 4.1a - 4.1c, it can be recognized that a carbon-carbon double bond (C=C) forms after cross-linking. The peak at 1950 cm^{-1} does not appear on the base or the curing agent's spectrum. To quantify the cross-linking density, the peak area was measured by triangulation. The area of the C=C peak was divided by the area of the CH_3 peak, yielding the transmittance ratio. The results are summarized in Figure 4.3d.

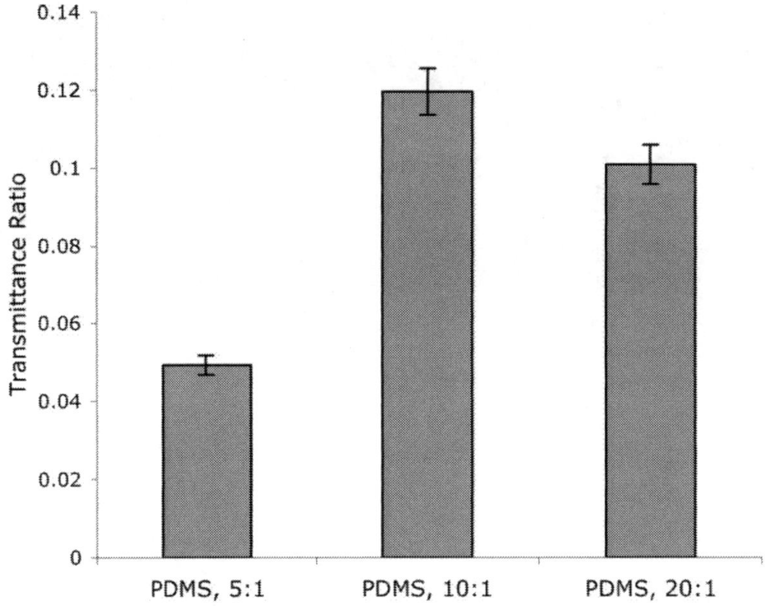

Figure 4.1d. Cross-linked PDMS versus Transmittance Ratio.

From Figure 4.1d, we can observe that 10:1 polymer-to-cross-linker ratio is the optimal one. The transmittance ratio doubles from 20:1 to 10:1, and it decreases for 5:1. The decrease for 5:1 symbolizes oversaturation of PDMS with the curing agent, which consequently inhibits cross-linking.

4.2 Tensile Test

Samples of varying cross-linker were analyzed with an Instron Tensile Tester. Young's Modulus, defined as the slope of the stress-strain curve at the yield point, is calculated and plotted as a function of cross-linking density in Figure 4.2a. The extension of the sample at yield is plotted in Figure 4.2b.

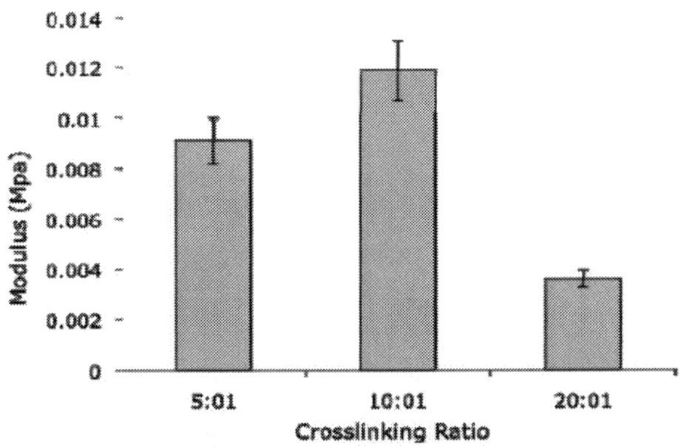

Figure 4.2a. Young's Modulus in PDMS Samples of 5:1, 10:1, and 20:1 Cross-linking Ratios

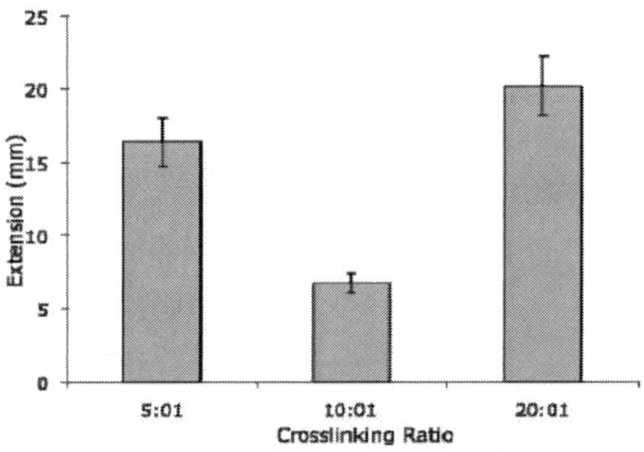

Figure 4.2b. Extension at Breaking Point in PDMS Samples of 5:1, 10:1, and 20:1 Cross-linking Ratios

This data suggests that PDMS with cross-linking density of 10:1 has the relatively highest modulus of 0.012 MPa compared to 5:1 and 20:1. The 10:1 sample had the shortest extension of 20.168 mm before yielding.

The 10:1 was chosen due to its highest modulus for mechanical analysis of varying clay concentrations. Figure 4.2c-d illustrates the Young's Modulus and extension at breaking point in PDMS samples of 0.5, 1, and 2 % clay concentrations. In the samples measured increased clay concentration resulted in a higher modulus material. All samples with nanocomposites had a higher modulus than those without.

In 10:1 samples, 1% clay exhibited the highest extension at breaking point, while 2% displayed the lowest. This suggests clay in amounts larger than 1% make PDMS more brittle than other samples, however still more elastic then samples without clay.

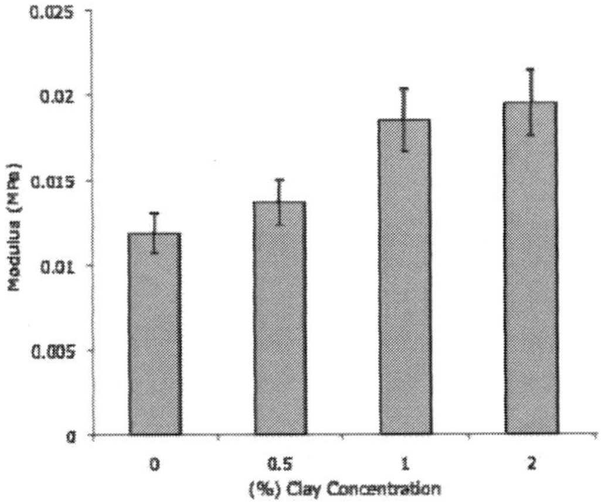

Figure 4.2c. Young's Modulus in 10:1 cross-linked PDMS Samples of 05, 1, 2% clay

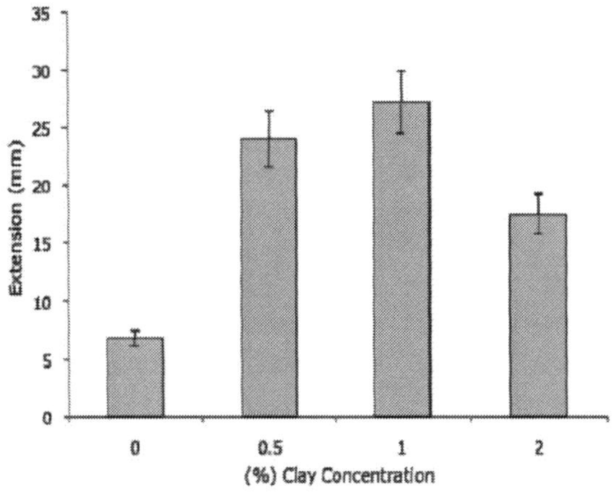

Figure 4.2d. Extension at Breaking Point of 10:1 cross-linked PDMS Samples of 0.5, 1, 2% clay

4.3 Cell Count and Imaging

Cells were counted on Day 1 and Day 3 of incubation at 37°C. Three dishes were counted separately on each day to produce the growth curve in Figure 4.3a. The cell plating density was 2400 cells per cm^2. On Day 1, the 1% clay sample showed the most cell growth with 4350, followed by the 0.5% sample with 3950 cells. On Day 3, the 0.5% showed the most cell growth with 30950 cells, followed by the 1% sample with 29200.

Cells grew the least on pure PDMS with no clay, indicating the polymer itself is not beneficial for cell growth. The polystyrene dishes experienced higher cell growth than the PDMS alone.

All samples containing clays exhibited higher cell proliferation by day three than the PDMS alone and the cell culture dish, indicating scaffolds containing clays nanocomposites are not only cytofriendly but encourage cell proliferation.

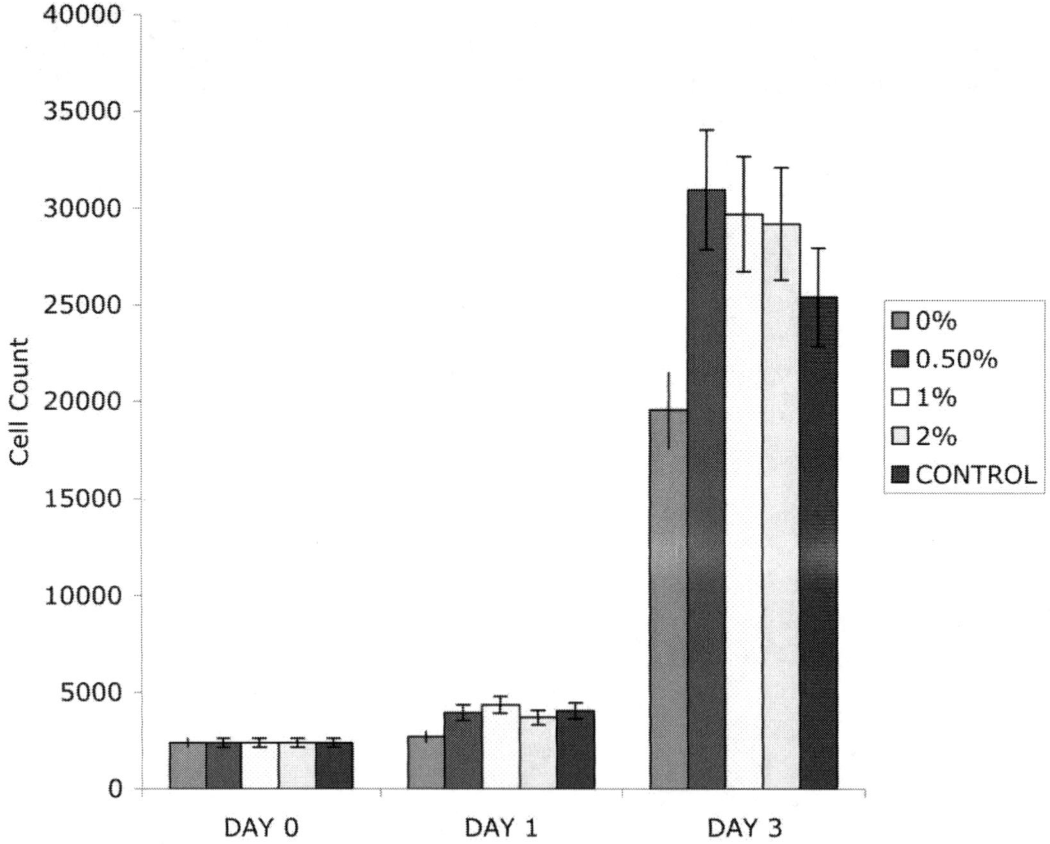

Figure 4.3a. Cell Cytotoxicity for Varying Concentrations of Sodium (Na) Clays in 10:1 Crosslinked PDMS

To determine the cell morphology the samples were examined under a Leica Confocal Microscope using a water lens at 20X magnification. These images may indicate the fibroblast have more preference for .5% clay based on the cell growth after three days.

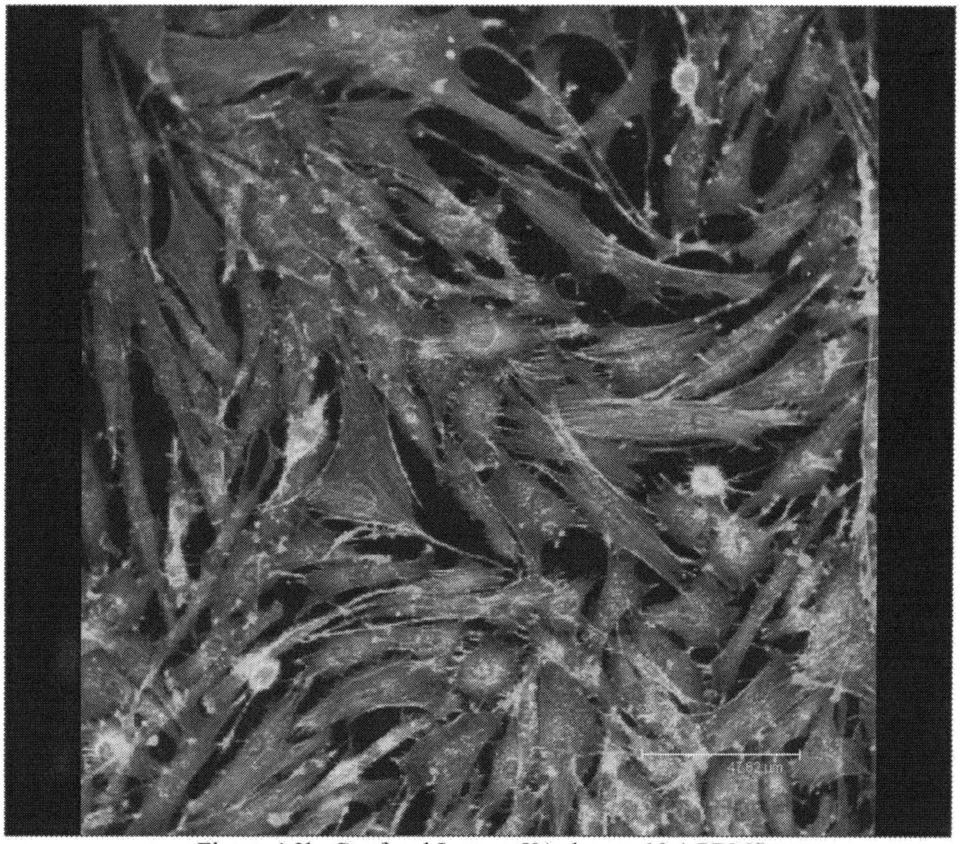

Figure 4.3b. Confocal Image, 5% clay on 10:1 PDMS

5. Conclusions

The FTIR spectra suggest that 10:1 cross-linking ratio yields the optimal cross-linking density for PDMS. The cross-link density increases up to 10:1 PDMS ratio, and then decreases for ratios larger than 10:1. A carbon-carbon double bond appears after the cross-linking reaction. This indicates that the base-curing agent bond contains a C=C bond.

Samples prepared with 10:1 cross-linking ratio had highest Young's Modulus of al10:1 cross-linked PDMS increased the Young's Modulus samples analyzed, whereas 20:1 had the lowest. Adding sodium Cloisite clay nanoparticles to PDMS revealed an increase in modulus in all samples.

Experiment with cell culture reveals sodium clay nanocomposites are not only cytofriendly but encourage cell growth. Cell imagining on confocal microscope discloses normal cell morphology in the presence of clay.

This data may be valuable in tissue engineering when designing scaffolds with more favorable mechanical properties for cell proliferation. Future research will reveal more information on the cell morphology in reaction to varying crosslinking ratios and types of clay nanocomposites, in addition to more detailed mechanical analysis of the materials.

6. References

1. Cloisite Na^+: Typical Physical Properties Bulletin. Southern Clay Products, Inc.: Subsidiary of Rockwood Specialties Inc.

2. Discher, Dennis E., Paul Janmey , and Yu-Li Wang . "Tissue Cells Feel and Respond to the Stiffness of Their Substrate." Science 310 (2005): 1139-1143. 28 Mar. 2008 <www.sciencemag.org>.

3. "Polydimethylsiloxane." Wiki in the Department of Chemistry. 7 Dec. 2006. Imperial College London. 28 Mar. 2008 <http://www.ch.ic.ac.uk/wiki/index.php/It:Polydimethylsiloxane>

4. Reusch, William. "Infrared Spectroscopy." Organic Chemistry On Line. 10 Aug. 2007. Michigan State U. 28 Mar. 2008 <http://www.cem.msu.edu/~reusch/VirtualText/Spectrpy/spectro.htm#contnt>.

Extended Abstract: Improvement of Fiber Reinforced Composites

Danny Kress

Composites represent an important new class of materials made by imbedding fibers or other reinforcements into a polymeric resin. They are often lighter and stronger than other materials of equivalent volume, and are being used in the production of new generations of airplanes, military protective gear and sporting goods, among others. Their properties depend not only on the bulk properties of the resin and the fibers, but also on the adhesion between the fiber surfaces and the matrix, and arrangement of the fibers within the matrix. Surface treatment of fibers to improve adhesion is a common method of improving composite strength. Our research has been exploring different surface treatments of glass fibers, specifically dendrimer coatings, that help improve adhesion and potentially decrease unwanted inter-fiber effects.

Background

While there are many types of composites, the type we studied was a fiber reinforced resin material. Typical composites of this type consist of a matrix with a large number of fibers running parallel through it. To create the composite, several layers of this type are combined, with the fibers in each layer running perpendicular to those in the adjacent layers. This intermediate is known as the prepreg. From there an autoclave compresses the layers into a single lightweight, strong material.

The objective of our research is to investigate whether the use of an appropriate primer coating on the fibers can improve the mechanical properties of fiber-reinforced composites. We wish to accomplish this by getting better adhesion between the fiber and matrix and getting the fibers completely wetter by the matrix. The fibers can be more wetted by the matrix if we reduce fiber-fiber touching and reduce inter-fiber effects.

One of the primary issues with fibers in composites is that the interactions between fibers lead to effects that weaken the mechanical strength of the composite. Inter-fiber effects can be seen in a few as 20 fibers grouped together. Some of the common interactions are fiber-fiber touching, air gaps between fibers, and twisting of the fibers within the matrix. We believe that some of these inter-fiber effects can be reduced through treatment of the fibers before creation of the composite.

Methodology

The system that we used to test different treatment of fibers is a Poly(vinyl butyral) (PVB) matrix with E-glass fibers. The PVB polymer is common in crash-proof windshields and the E-glass fibers are very common in composites. The test we used in this study is known as the fiber fragmentation test. This is a common test for measuring the adhesion between the fiber and the matrix and can be used for both single fibers and bundles of fibers.

In the fiber fragmentation test a single fiber or fiber bundle is suspended in a dog-bone shaped sample and pulled from the ends at a constant rate of 1 mm per minute using an Instron tensile tester. As the matrix stretches, it transfers stress to the fiber through adhesion until the amount of stress over the length of fiber is enough to break it. The fiber breaks repeatedly until it won't break anymore. The lengths of the fiber segments at this point have reached the critical length. The critical length of the fiber in the matrix is the point at which the force of the matrix pulling on the segment is not enough to break it, and is a measure of the level of adhesion.

The critical length will be used to calculate the interfacial shear strength (IFSS), which represents the amount of adhesion between the fiber and matrix. This is calculated by Equation 1.

$$\tau_{\max} = \frac{\sigma_f * d}{2 * l_c} \qquad (1)$$

The IFSS is represented by τ_{max}, l_c is the critical length, σ_f is the tensile strength of the fiber, and d is the diameter of the fiber. The diameter of the E-glass fibers that we used is 17 μm, which is about one quarter of the diameter of the human hair. The critical length, l_c, is a factor of a Weibull distribution based on the failures of the imperfections in the fiber. For simplification we assumed that this length was the sum of all the length segments divided by the total number of segments. The tensile strength of the fiber, σ_f, was used as a value supplied by the manufacturer of 1057 MPa.

Once the sample of fibers within the matrix has been fractured, the critical length can be measured optically using a microscope and counting the number of breaks. The breaking region is shown by a dark area where the fiber has split and a de-bonding zone around each break. One issue with this method arises when trying to count breaks in large bundles of fibers when all fibers cannot be seen. This led to studies looking at measuring breaks with large bundles of fiber through use of acoustic techniques to register each break in the sample. This technique was confirmed in samples where the result could be verified optically. The problem with this technique is that it is very labor intensive and sensitive to picking up cracks in the matrix and other non-fiber breaks. Through this process, however, it was determined that the fibers broke in zones. With all the fibers breaking together, the critical length can be determined optically. The technique of counting all the breaks optically was then re-verified using the acoustic method.

Surface Treatments

As suggested in the objectives, a primer applied to the fibers can increase the adhesion and reduce inter-fiber effects. Primers for fibers in composites are common, but we tried a different type of primer, specifically a dendrimer, that coats all the fibers and shows promise of increasing adhesion and decreasing fiber-fiber touching. A dendrimer is a molecule that is repeatedly branched and is characterized by its structural perfection. The particles can become very large, although the type we used is a 2-generation, amino-based dendrimer. They can become multiply bonded and have a unique modulus that soaks up stress. The dendrimer, however, cannot bond directly to the fiber so a silane must be added to bridge the gap. A silane functional group with an epoxy group attached to the glass fiber in a manner that allows the epoxy group free to react with the dendrimer.

The dendrimer is designed to affect the fiber in two main ways. The first is to reach out into the epoxy and provide a type of mechanical interlocking with the resin. The second utilizes the large size of the dendrimer by providing spacing between the fibers so that there is no fiber-fiber interaction and so that the fibers may be completely wetted by the matrix. Initial results have shown proof of principle by increasing the interfacial shear strength about 30%. From here there are many further areas that may be explored such as the application of larger dendrimers or verification of complete coating of the fibers by the dendrimer.

2e Preliminary Design and Cost Analysis of a Distillation Column for Separation of Methanol, Water and a High Boiling Point Inorganic Acid

Nicole L. Felgenhauer, Robert R. Thompson, Mani Firouzain, and Sawinder Singh. Chemical Engineering, Cal Poly Pomona University, 3801 W. Temple ave., Pomona, CA 91768

Cal Poly Pomona Chemical Engineering Team has undertaken a combined experimental/design and economic distillation study for the recovering of methanol from an aqueous mixture of inorganic acid and methanol. This recovered methanol will be reused in other operation or will be resold to outside vendors. The design will also help the team meet strict environmental regulations. To recover the entire methanol in the feed and to meet the environmental regulations, the methanol concentration of the bottoms product cannot exceed 0.1 wt% and the concentration of water in the distillate is not to exceed 5 wt%. In this investigation, both a pilot and commercial scale distillation column are being designed. A series of batch distillations have been performed to determine the extent to which the inorganic acid will appear in the distillate. The team is using PRO/II software to determine size and energy requirements for the distillation system. The design study includes investigations of alternative phase equilibrium relationships and column efficiency models. The effects of process parameters such as reflux ratio, number of trays, feed tray location, and material flow rate is being studied in order to optimize the efficiency and cost effectiveness of the distillation column design.

2f Removal of Parathyroid Hormone by Immunospecific Adsorption during Kidney Dialysis

Shu Xia[1], **Nichole Y. Hodge**[1], Melvin Laski[2], and Theodore F. Wiesner[1]. (1) Department of Chemical Engineering, Texas Tech University, Lubbock, TX 79409, (2) Department of Internal Medicine, Health Science Center, Texas Tech University, Lubbock, TX 79409

Nearly all patients on kidney dialysis suffer from elevated levels of parathyroid hormone (PTH), a condition known as secondary hyperparathyroidism (SHPT). The symptoms of SHPT include a form of osteoporosis called renal osteodystrophy. As a consequence of renal osteodystrophy, kidney dialysis patients suffer from high rates of bone fracture, vascular calcification, and a rate of cardiovascular morbidity ten times that of the healthy population.

The restoration of normal PTH levels for kidney failure patients has to date proven to be a very challenging clinical problem. Conventional hemodialyzers are optimized to remove small molecular weight solutes such as urea and creatine (~100 Daltons). Middle molecular weight toxins such as PTH are of the order of 10,000 Daltons, and are poorly dialyzed. Meanwhile, drug therapies require weeks to months to lower serum PTH levels, allowing long periods of bone exposure to the hormone.

Affinity chromatography is a well-established technique to remove proteins from complex mixtures with high efficiency and specificity. To make use of its characteristic of high specific affinity between the antigen and antibody, a chromatographic column has been designed to reduce PTH levels to normal within one dialysis session. The removal is accomplished by a monoclonal antibody attached on the surface of the packing. The column is placed in series with a conventional hemodialyzer.

A mathematical model of the modified dialysis circuit was developed which reproduces an in vitro experiment with high accuracy. A simple, closed form, solution to the model was obtained, which was scaled up to the clinical scenario. The model indicates that the immunosorptive column will be effective in lowering PTH levels.

2g Seed-Mediated Growth of Platinum Nanoparticles Via Green Chemical Reduction of Platinum Salts

Tyler M. Gunn, Chemical & Biochemical Engineering, University of Iowa, 1011 Sheridan Ave, Iowa City, IA 52240

Monodisperse platinum particles of various diameters were grown from platinum seeds utilizing a stepwise addition of calculated amounts of platinum salt and ascorbic acid. Ascorbic acid, being a weak reducing agent, prevented new seed nucleation while fully reducing the platinum salt in solution. It was discovered that as the platinum seeds were grown into much larger particles they became more faceted; increasing their surface area-to-volume ratio. This new nanoparticle synthesis technique allows the creation of uniquely faceted platinum nanoparticles of controllable size that are ideal for catalysis.

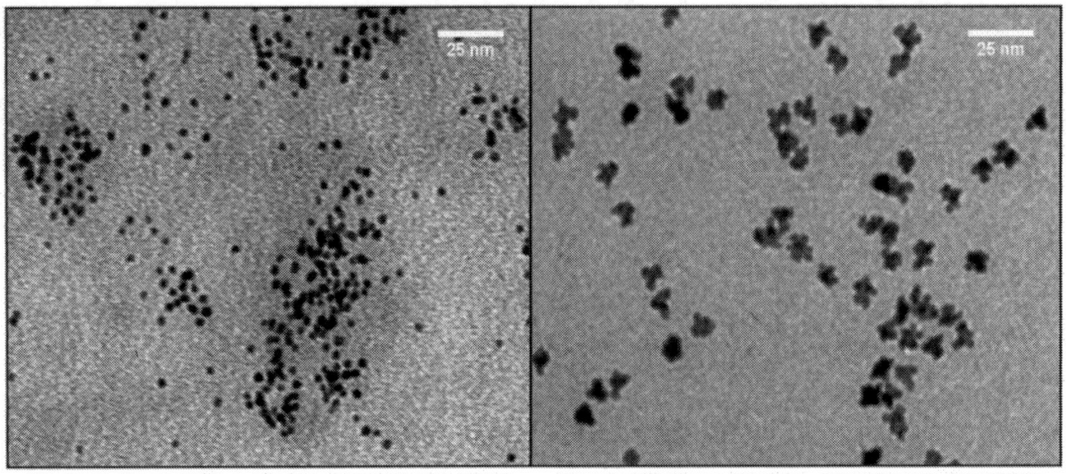

Platinum Seeds Before Mediated Growth 160,000X   Faceted Platinum Particles After Growth 160,000X

2h Synthetic Routes to Photoactive Nanostructured Hydrogels

Alexander Thomas Leonard and Travis S. Bailey. Chemical and Biological Engineering, Colorado State University, 2324 Coventry Ct., Fort Collins, CO 80526

In this work we present preliminary results aimed at the fabrication of nanostructured hydrogels based on self-assembled amphiphilic block copolymers (BCPs). While the current work is focused on the use of BCPs containing polystyrene and poly(ethylene oxide), the approach presented can be generalized to any amphiphilic system and extended to materials containing multiple block sequences. The emphasis of the work presented is concerned with the integration of photoreversible agents into the framework of the nanostructured material, in an attempt to gain external control (through wavelength dependent responses) over various properties of the hydrogel, including size, water content, elasticity, mechanical integrity, permeability, and self-adhesion. The ultimate goal is to design a system in which one can tune these properties remotely without direct contact being made with the hydrogel, itself. The results presented will focus on the synthetic strategies to form the component block copolymers, the morphological characteristics of the zerogels (unswollen materials), and preliminary studies involving the spatially controlled integration of photoreactive elements into these structured materials.

2i Vapor Deposited Polymer Coatings for Implantable Neuroprosthetic Devices

Kyle D. Stephens, Chemical Engineering, Northeastern University, Boston, MA 02120

Neuroprosthetic implants are silicon based microelectronic devices that are designed to send and receive electrical signals to and from neurons in the central nervous system. An important part of the fabrication and performance of these devices is the protection of the integrated circuitry from the corrosive effects of the in-vivo environment. In addition, the implantation of these devices is often accompanied by the formation of scar tissue around the implant, which effectively insulates the stimulating and recording electrodes of the implant, thereby interfering with performance.

With these facts as motivation, the goal of the current research is to create polymer thin films that are biocompatible, well-insulating, have good adhesion to silicon, and actively mitigate scar tissue formation. A cyclic siloxane-acrylic acid copolymer is chosen for this purpose; the siloxane providing the desired biocompatibility, insulating characteristics and silicon adhesion, and the acrylic acid providing COOH tether points for the attachment of proteins known to mitigate scar tissue formation and enhance neural proliferation. Chemical Vapor Deposition, a vapor phase polymerization technique, is chosen to avoid the surface tension limitations associated with liquid phase polymerization techniques.

5a Characterizing Water in Oil Emulsions with Application to Methane Hydrate Formation

Clint P. Aichele, Chemical and Biomolecular Engineering, Rice University, 6100 Main St. (MS 362), Houston, TX 77005-1892 and Walter G. Chapman, Chemical and Biomolecular Engineering Department, MS-362, Rice University, 6100 Main St., Houston, TX 77005.

Water in oil emulsions have significant applications during the production and processing of crude oil. The ability to predict and measure drop size distributions is important for developing effective flow assurance strategies. Specifically, water in oil emulsions are particularly important regarding methane hydrate blockages during crude oil production. The drop size distribution of the emulsion, and the relationship between drop size and methane hydrate formation are significant when mitigating methane hydrate blockages. Currently, there are speculations about the exact structure of water drops during hydrate formation, and this is propagated by the paucity of water droplet size measurements in crude oils.

This work utilizes nuclear magnetic resonance to directly measure emulsion properties of both model and crude oils. Nuclear magnetic resonance is particularly useful for these systems because it is not constrained by the optical properties of the systems, and it considers the entire emulsion. Specifically, drop size distributions are obtained for both crude oil and model oil emulsions using the pulsed field gradient with diffusion editing (PFG-DE) technique. This technique is useful for directly measuring drop size distributions because it does not assume a form of the drop size distribution. A one dimensional imaging technique is also employed that yields quantitative information about the spatial distribution of components in the emulsions. Finally, transverse relaxation measurements are used to directly measure the conversion of water to methane hydrate. This work provides quantitative information that describes the structure of water droplets during methane hydrate formation. In addition, this work provides insight about the mechanisms that govern drop breakup in crude and model oil emulsions.

5b Physics and Fabrication of Nanostructured Materials

Mustafa Akbulut, Chemical Engineering, Princeton University, Olden st., E-quad A-306, Princeton, NJ 08544

Nanoscience and nanotechnology have an enormous potential to be beneficial to various areas of research and application, and funding from governmental and businesses has been expanding. Hence, nanoscience and nanotechnology have rapidly been evolving, and nowadays numerous subfields of nanoscience and nanotechnology have emerged. One of the key research directions in these fields is the fabrication of well-defined functional nanostructured materials for a specific purpose, which requires an understanding of the fundamentals of how the production of such nanostructured materials works. This poster describes specific projects that I want to tackle in this context.

Related Publications:

•M. Akbulut, N. Belman, Y. Golan, J. Israelachvili, "Frictional properties of confined nanorods", Advance Materials, 2006, 18(19), 2589-2592

•M. Akbulut, A.R.G. Alig, Y. Min, et al., "Forces between surfaces across nanoparticle solutions: Role of size, shape, and concentration", Langmuir, 2007 23 (7): 3961-3969

•Y. Min, M. Akbulut, N. Belman, Y. Golan, J. Zasadzinski, J. Israelachvili, "Normal and lateral forces across confined straight and curved nanowires", Nanoletters. 2008, 8 (1), 246 -252

•Y. Min, M. Akbulut, K. Kristiansen, Y. Golan, J. Israelachvili, "Role of interparticle and external forces on the assembly and properties of nanoparticle materials", Nature Materials, (In press)

•A.R.G. Alig, M. Akbulut, Y. Golan, J. Israelachvili , "Forces between surfactant-coated ZnS nanoparticles in dodecane: Effect of water", Advanced Functional Materials, 2006, 16(16), 2127-2134

5c Quantitative, Multinuclear Magnetic Resonance Microscopy Methods for the Study of Multiphase Systems

Belinda S. Akpa, Department of Chemical Engineering, University of Illinois at Chicago, 810 S Clinton Street, MC-110, Chicago, IL 60607

Magnetic resonance microscopy (MRM)

MRM is a non-invasive, non-destructive technology that can be used to probe optically opaque systems. By judicious design of magnetic resonance (MR) methodologies, this mode of visualization can be made sensitive to a number of phenomena including chemical composition and both coherent (velocity, acceleration) and incoherent molecular motion (diffusion, dispersion). While it is better known in the context of medical diagnosis, MR imaging has found application in fields ranging from Physics to Engineering and Geology – with applications as varied as microfluidics, catalysis, materials characterization, multiphase flows, fluidization, oil recovery, and drug delivery.

Multinuclear MR and multiphase systems

The extent to which chemically-specific information can be resolved in a magnetic resonance experiment depends on both the physical attributes of the system being studied and the nature of the particular NMR technique employed. Frequently, in a magnetic resonance imaging experiment, the ^1H nucleus is observed, because it has the highest detection sensitivity of all the NMR-active nuclei. However, in the study of heterogeneous porous media, observation of the ^1H nucleus has some distinct disadvantages. Although NMR signal is inherently both chemical and nucleus-specific, the frequency characteristics that allow individual resonances to be identified can easily become obscured by a number of effects. Here, a multinuclear approach to chemical mapping has been adopted as a means of improving the versatility of the MRM toolkit for the study of heterogeneous systems. Advantages of the multinuclear approach will be highlighted; these include: (i) improved frequency resolution via the inherent superior chemical shift dispersion of X-nuclei, (ii) reduced vulnerability to magnetic susceptibility-induced field gradients at interfaces, (iii) longer signal lifetimes, permitting observation of transport phenomena on longer time-scales (and hence, length-scales), and (iv) inherent or highly efficient solvent suppression in aqueous systems. The studies reported here focus primarily on the detection of ^{13}C at natural abundance. However, the techniques presented are applicable to other spin-½ nuclei (e.g. ^{31}P, ^{15}N).

Expanding the MRM toolkit to access new information about complex systems

Tailored MRM methods have been implemented to achieve the following.

Heteronuclear imaging methods for the study of ^{13}C at natural abundance, more specifically: *In situ*, *quantitative mapping of conversion and selectivity in heterogeneous catalytic reactors*[1-3].

Hybrid polarization transfer and pulsed-field gradient (PFG) methods for the study of species-specific diffusion and dispersion in multicomponent systems, including: *Flow in packed beds*[4], *mass transport in biofilm systems*[5], *and characterization of emulsion structure*[6] *with superior phase resolution and extended observation time-scales.*

Characterization of anomalous diffusion as probed by MR, focusing on: *The relevance of observation time-scale*[5], *reconciliation of proposed physical models with the observed displacement dynamics, and a new approach to characterizing structure and improving image contrast by application of fractional calculus*[7].

Quantitative visualization of mixing processes in a microfluidic device, including: *Concentration mapping in optically opaque microchannels of arbitrary cross-section and mapping of evolving flow fields*[8, 9].

References:

1. B. S. Akpa, M. D. Mantle, *et al.* In situ ^{13}C DEPT-MRI as a tool to spatially resolve chemical conversion and selectivity of a heterogeneous catalytic reaction occurring in a fixed bed. Chemical Communications. 21, 2741 (2005)

2. L. F. Gladden, B. S. Akpa, *et al.* (2005). In situ reaction imaging in fixed-bed reactors using MRI. NMR Imaging in Chemical Engineering. S. Stapf and H. Song-I. Weinheim, Wiley-VCH: 590-606.

3. A. J. Sederman, M. D. Mantle, *et al.* In situ MRI study of 1-octene isomerisation and hydrogenation within a trickle-bed reactor. Catalysis Letters. 103, 1-8 (2005)

4. B. S. Akpa, D. J. Holland, *et al.* Enhanced C-13 PFG NMR for the study of hydrodynamic dispersion in porous media. Journal of Magnetic Resonance. 186, 160-165 (2007)

5. D. A. Graf von der Schulenburg, B. S. Akpa, *et al.* Non-invasive mass transfer measurements in complex biofilm-coated structures. Biotechnology and Bioengineering. Accepted Article, (2008)

6. B. S. Akpa, M. L. Johns, *et al.* (2008). Heteronuclear PFG-NMR methods for the sizing of emulsion droplets using natural abundance ^{13}C. 49th Experimental NMR Conference, Asilomar, CA.

7. B. S. Akpa, O. Abdullah, *et al.* (2008). Anomalous diffusion expressed through fractional order differential operators in the Bloch-Torrey equation. Magnetic Resonance in Porous Media, Cambridge, MA.

8. B. S. Akpa, S. M. Matthews, *et al.* Study of miscible and immiscible flow in a microfluidic device using magnetic resonance imaging. Analytical Chemistry. 79, 6128-6134 (2007)

9. S. P. Sullivan, B. S. Akpa, *et al.* Simulation of miscible diffusive mixing in microchannels. Sensors and Actuators B. 123, 1142-1152 (2007)

5d Overview of a Vapor Deposited Thin Silica Film

A. Anderson and W. Robert Ashurst. Chemical Engineering, Auburn University, 212 Ross Hall, Auburn, AL 36849

The deposition of silica layers from the vapor phase hydrolysis of silicon tetrachloride will be discussed. These silica layers, which support the formation of organic monolayers formed from standard organosilane chemistries, can be deposited onto a wide variety of surfaces, including silicon, cellulose, zinc selenide, polystyrene, and poly (methyl methacrylate), offering the potential to drastically expand the types of materials that silane surface modification chemistry can be applied to. This talk will present an overview of our studies on this silica layer, including morphological and spectral characterization, dehydration and rehydration properties, interactions with water, and interactions with bacterial cells.

5e Polymer Thin Film Nanoassemblies for Electrochemical Energy Conversion Devices

Avni A. Argun, Chemical Engineering, Massachusetts Institute of Technology, 77 Massachusetts Avenue, 66-525, Cambridge, MA 02139

The increasing global focus on alternative energy sources has led to an interest in electrochemical energy conversion devices such as proton-exchange membrane fuel cells (PEMFCs) and dye-sensitized solar cells (DSSCs). Layer-by-layer (LBL) assembly is a robust technique for the fabrication of solid state ultrathin film electrolytes for these devices. This approach presents strong advantages as it allows the incorporation of many different functional materials within a single thin film at a full range of compositions with exceptional homogeneity. Application of these functional thin films toward a number of electrochemical device systems such as PEMFCs and DSSCs are discussed. In addition, the electrical and optical properties of electrochromic devices (ECDs) based on conjugated polymers are discussed, along with the application of a number of patterning techniques to prepare electrodes for organic electronic devices. Truly all-organic ECDs are demonstrated by replacing conventional transparent metal oxide electrodes with a highly conducting polymer.

5f Water Interaction and Diffusion on Platinum Surfaces

Liney Arnadottir, Department of Chemical Engineering, University of Washington, Box 351750, Seattle, WA 98195, Eric M. Stuve, Chemical Engineering, University of Washington, Box 351750, Seattle, WA 98195, and Hannes Jónsson, Chemistry, University of Iceland, Hjarðarhaga 2-6, Reykjavik, Iceland.

Density functional theory (DFT PW91) calculations were used to study diffusion of water molecule on flat Pt(111) terrace as well as on kinks and steps edges. Diffusion on the (111) terrace is predicted to become active on the time scale of seconds at 74 K in reasonable agreement with experimental results of Kay and coworkers (Daschbach et al. J. Chem. Phys. 120 (2004) 1516). The binding at steps and kinks on the Pt(111) surface is substantially stronger than at the flat terrace, by 0.16 eV and 0.25 eV, respectively.

The diffusion barrier of a single water molecule moving on the flat terrace, up to a kinks side or along the step edge are all comparable while the barrier from the flat terrace up to a step is significantly lower then all the other barriers calculates. Consequently the barrier for diffusion from the step edge down to the terrace is substantially larger then on the flat terrace. At low coverage, clusters of water molecules would, therefore, be expected to line the upper side of step edges. This is consistent with experimental STM images taken by Morgenstern et al. (Morgenstern et al. Phys. Rev. Lett. 77 (1996) 703).

5g Conductive Polymer and Hydrogel Composites as Functional Biomaterials

Carolyn L. Bayer, Biomedical Engineering, The University of Texas at Austin, 1 University Station, CO400, Austin, TX 78712 and Nicholas A. Peppas, Chemical Engineering, The University of Texas at Austin, 1 University Station, CO400, Austin, TX 78712.

Conductive polymers have been extensively researched for the creation of polymer electronics, such as thin film transistors, photovoltaics and organic light emitting diodes. In these devices, the processability and mechanical properties of conductive polymers are improved in comparison to traditional inorganic materials. Though less explored, these same properties encourage the use of conductive polymers in biomedical applications. In particular, conductive polymers may be chemically and physically tailored to be similar to established biomaterials, with the added feature of possessing unique electrical properties. The incorporation of these electrical properties in biomaterials presents opportunities for systems integration in biomaterials.

Through this research we seek to develop a new sensing mechanism for the detection of human proteins and peptides relevant to specific disease states. Development of fabrication methods, materials characterization, functionality testing and assessment of biocompatibility have all been performed on a type of water-dispersible polyaniline (PANI). This material is highly conductive and solutions-processable due to the synthesis process, which involves templating the aniline monomer on a polyelectrolyte chain, in this case poly(2-acrylamido-2-methylpropane sulfonic acid) (PAAMPSA), to synthesize PANI-PAAMPSA. The processability enables integration into of the PANI-PAAMPSA as a component in a water-based hydrogel polymerization to form a semi-IPN with conductive properties.

This work was supported by the NSF-IGERT program grant DGE-0333080.

5h How to Make An Information Processing Device out of RNA

Chase L. Beisel, Chemical Engineering, California Institute of Technology, Pasadena, CA 91125 and Christina Smolke, Chemistry and Chemical Engineering, California Institute of Technology, 236A Spalding, Mail Code 210-41, Pasadena, CA 91125.

RNA is a rich and versatile building block for the construction of information processing devices. These devices detect the levels of specified intracellular biomolecules and control cellular behavior accordingly. With few constraints on the identity of the recognized biomolecule or the targeted gene, RNA-based information processing devices can be rapidly implemented toward various applications in medicine and biotechnology. However, current examples serve as single instances of successful design that do not immediately lend to the construction of novel devices. My thesis work has focused on delineating design principles for the construction of information processing devices out of RNA and applying these principles to the design of devices that operate through RNA interference (RNAi).

Ligand-responsive RNAi substrates RNAi is an endogenous pathway present in humans and other eukaryotes that mediates targeted gene silencing and has garnered recent interest as a revolutionary biological research tool and a powerful therapeutic strategy. While RNAi has left an indelible mark on the scientific community, exerting greater control by engineering ligand responsiveness would advance the applicability of this already impressive pathway. Toward this goal, we developed functional composition frameworks for two distinct types of RNAi substrates, small hairpin RNAs and microRNAs. Both of our frameworks enabled facile replacement of the biomolecule sensory and gene targeting domains, thus mediating rapid implementation as biosensors or autonomous control devices. Experimental and computational characterization studies provided a comprehensive understanding of device behavior, thereby facilitating forward design.

Design principles for riboswitch design Riboswitches are genetic regulatory elements that serve as a dominant class of RNA-based information processing devices. Various experimental characterization studies of riboswitches have shown that kinetic factors such as ligand binding and RNA folding have a tremendous impact on device performance, although these factors have not been comprehensively evaluated or considered when formulating design principles for device construction. We explored the contribution of kinetics factors to riboswitch performance in silico, where model predictions match experimental observations, including results from our ligand-responsive RNAi substrates. From our modeling results, we developed a universal set of design principles that guide riboswitch assembly and performance tuning. Current work is focused on applying these principles to the redesign of existing riboswitches for improved performance.

Harnessing the many faces of RNA Beyond my thesis work, I am interested in understanding how nature utilizes RNA and translating this knowledge into the improved design of biological systems. Nature has devised ingenious ways to interface RNA with natural regulatory architectures as exemplified by bacterial small RNAs, riboswitches, alternatively spliced RNAs, and microRNAs. These

RNAs are involved in many fundamental cellular processes including stochastic noise dampening, metabolism, stress response, development, and carcinogenesis, although characterization of these RNAs is still ongoing. I plan to investigate these processes from a systems design perspective and directly apply these insights to the fields of synthetic biology, metabolic engineering, and disease therapeutics.

5i Advanced Material Architectures for Tissue Engineering and Drug Delivery Applications

Danielle S.W. Benoit[1], Patrick S. Stayton[1], and Kristi Anseth[2]. (1) Bioengineering, University of Washington, 1705 NE Pacific St., Foege Hall, Seattle, WA 98105, (2) Department of Chemical and Biological Engineering, University of Colorado, Howard Hughes Medical Institute, ECCH 128, Campus Box 424, Boulder, CO 80309-0424

Biomaterials with advanced, nano-scale organizations are of great interest for both drug delivery and tissue engineering applications. Current approaches in tissue engineering emphasize the control over cell behaviors and tissue formation by nano-scale topography that closely mimics the natural extracellular matrix (ECM). With the understanding that the natural ECM is a multifunctional, dynamic architecture, this research focused on understanding how osteoprogenitor cells receive information from their external environment (e.g., scaffolds). We utilized a synthetic poly(ethylene glycol) (PEG) platform that provides a highly regulated microenvironment that is permissive (i.e., permits cell survivability), and can be systematically altered biochemically to affect cell signaling pathways. Specifically, how cell-matrix interactions, local presentation of cell-internalized signals, and temporal changes in the scaffold microenvironment regulate functions, such as differentiation, proliferation, and extracellular matrix (ECM) production were investigated.

Further, macromolecular drug delivery systems can be synthesized with controlled composition, shape, size and morphology. siRNA, similar to other macromolecular drugs, once endocytosed, is predominantly trafficked to the lysosome and degraded enzymatically. In addition, serum proteases are capable of rapid siRNA degradation. Therefore, an optimal therapeutic siRNA delivery approach would combine serum-stability, long circulation times, and tissue-specific targeting, with efficient endosomal uptake and escape prior to lysosomal trafficking. We have developed advanced polymer architectures based on reversible addition fragmentation chain transfer (RAFT) polymerizations that enable siRNA-complexation and stabilization and pH-dependent endosomal-disruption. These carriers have been further functionalized with moieties to provide tissue-specific targeting that result in increased efficacy and decreased dosing in vitro and in vivo. Clearly, designing biomaterials with controlled organizations at the nanometer scale can dramatically enhance the biological functions of encapsulated drugs and cells and this approach allows the design of appropriate materials for cell delivery in applications such as tissue engineering and targeted therapeutic delivery.

Applicable peer-reviewed publications:

D.S.W. Benoit, A.J. Convertine, C.L. Duvall, A.S. Hoffman, and P.S. Stayton. Development of a novel endosomolytic diblock copolymer for siRNA delivery. Submitted.

D.S.W. Benoit, M.P. Schwartz, A.R. Durney, and K.S. Anseth. Small molecule functional groups for controlled differentiation of human mesenchymal stem cells encapsulated in poly(ethylene glycol) hydrogels. Submitted.

A.R. Rydholm, N.L. Held, D.S.W. Benoit, C.N. Bowman, and K.S. Anseth. Modifying network chemistry in thiol-acrylate photopolymers through post-polymerization functionalization to control cell-material interactions. Journal of Biomedical Materials Research, Part A, In press.

D.S.W. Benoit and K.S. Anseth. 2007. Multifunctional hydrogels that promote osteogenic hMSC differentiation through stimulation and sequestering of BMP2. Advanced Functional Materials, 17 (13):2085-2093.

D.S.W. Benoit, M.C. Tripodi, J.O. Blanchette, S.J. Langer, L.L. Leinwand, and K.S. Anseth. 2007. Integrin-linked kinase production prevents anoikis in human mesenchymal stem cells. Journal of Biomedical Materials Research, Part A, 81A(2):259-268.

D.S.W. Benoit, A.R. Durney, K.S. Anseth. 2007. The effect of heparin-functionalized PEG hydrogels on three-dimensional human mesenchymal stem cell osteogenic differentiation. Biomaterials, 28(1):66-77.

V. Khire, D.S.W. Benoit, K.S. Anseth, and C.N. Bowman. 2006. Ultrathin gradient films using thiol-ene polymerizations. Journal of Polymer Science Part A: Polymer Chemistry, 44(24):7027-7039.

D.S.W. Benoit, C.R. Nuttelman, S.D. Collins, and K.S. Anseth. 2006. Synthesis and characterization of a fluvastatin-releasing to modulate hMSC differentiation and function for bone regeneration. Biomaterials, 27(36): 6102-6110.

D.S.W. Benoit, A.R. Durney, and K.S. Anseth. 2006. Manipulations in hydrogel degradation behavior enhance osteoblast function and mineralized tissue formation. Tissue Engineering, 12(6): 1-11.

C.R. Nuttelman, D.S.W. Benoit, M.C. Tripodi, and K.S. Anseth. 2006. The effect of ethylene glycol methacrylate phosphate in PEG hydrogels on mineralization and viability of encapsulated hMSCs. Biomaterials, 27(8): 1377-1386.

D.S.W. Benoit and K.S. Anseth. 2005. Heparin functionalized PEG gels that modulate protein adsorption for hMSC adhesion and differentiation. Acta Biomaterialia, 1(4): 461-470.

D.S.W. Benoit and K.S. Anseth. 2005. The effect on osteoblast function of colocalized RGD and PHSRN epitopes on PEG surfaces. Biomaterials, 26(25): 5209-5220.

D.S. Wentworth, D. Skonberg, D.W. Donahue, and A. Ghanem. 2004. Application of chitosan entrapped β-galactosidase in a packed bed reactor system. Journal of Applied Polymer Science, 91(2): 1294-1299.

5j Homogeneous and Heterogeneous Acid Catalysis of Lipids to Produce Green Fuels: Advancing Biofuels in a Fossil Fuel World

Tracy J. Benson, Chemical Engineering, Mississippi State University, Mississippi State, MS 39759

Development of renewable fuels is essential to global sustainability as Earth's fossil fuels are depleted, resulting in decreased supplies for transportation fuels and energy generation. Scientists have identified new renewable resources capable of displacing significant quantities of petroleum, the chief feedstock for transportation fuels and commodity chemicals. For example, lipids can be extracted from plants, animals, and microorganisms and used for fuel and specialty chemical production. Our research group has estimated that approximately 10 billion gallons of lipids could be generated and converted into fuels using current agricultural practices, wastewater treatment, and refining infrastructure.

During my master's work, I explored various bio-adsorbents for the removal of water from ethanol/water mixtures (1). Using a temperature swing adsorber, mixtures of 90, 95, and 97 wt% ethanol were volatilized and passed through a column bed packed with either hardwood sawdust, kenaf core, or bleached wood pulp. A glass adsorption column with an inside diameter of 2.54 cm was used to generate breakthrough curves to determine the effectiveness of the adsorbents and to allow comparability with starch based adsorbents that are currently used within the ethanol industry. Experimental results indicated that water was preferentially adsorbed allowing for complete dehydration of ethanol. Bleached wood pulp was a superior adsorbent over the kenaf core and hardwood sawdust in adsorbing water due to its large percentage of hydroxyl groups stemming from a high concentration of xylans. The bleached wood pulp was the only material that demonstrated similar water adsorbing capacities as compared to starch-based adsorbents such as corn grits.

My doctoral thesis has been devoted to the advancement of science in the cracking of lipid molecules into diesel-like and gasoline-like fuels. First, oleic acid, an unsaturated free fatty acid in many lipid sources, was reacted with triflic acid, a liquid superacid catalyst at 0°C (2). Instrumentation employed for sample analyzes included 1H-NMR, 13C-NMR, FTIR, and GC/MS (both EI and CI) to identify reaction products and to hypothesize a reaction pathway. The resultant products were saturated fatty acids with carbon lengths of C9-C14, C16, and C18. This work concluded that a Bronsted acid protonates the double bond causing methyl and hydride shifts that succumb to β-scissions resulting in linear and branched chain isomers of smaller carbon lengths than the original starting material. The cracking of oleic acid was in stark contrast to palmitic acid, a saturated fatty acid, which yielded no discernable products when reacted with triflic acid at temperatures up to 100°C.

The second step in this research was to evaluate the heterogeneous catalytic cracking chemistry of model lipid compounds, chiefly oleic acid, mono-, di-, and tri-olein. For this phase, a reactor/analyzer was constructed using a Varian 3600 GC which was reengineered to allow for reactions with online mass spectrometry analysis (3). This unit, which is capable of accepting a gas, liquid, or solid reactant, was equipped with a Saturn quadrupole ion trap mass spectrometer for mass spectral analysis in real-time as lipids were passed over a bed of highly acidic ZSM-5 catalyst. As this was an integrated reactor/ analyzer, the Rxi-1ms GC column was exposed in the inlet to temperatures in excess of 400°C. This temperature is beyond the degradation point of the column, and so selective ion storage RF waveform was used to remove the siloxane masses m/z (207, 281, and 355) from the spectra. This produced lower detection limits and full scan data for identification. Electron impact and selected ejection chemical ionization were used for product identification and molecular weight confirmation. Hexane was reacted over H-ZSM-5 catalyst for instrument validation. This produced a series of alkanes and alkenes with distributions consistent with that reported for the cracking of hexane.

The heterogeneous cracking of model compounds resulted in mixtures that were within the ranges of diesel and gasoline organics. Oleic acid was found to form chiefly BTEX and trimethylbenzene as initial cracking products. Decarboxylation and decarbonylation are responsible for yielding CO2 and CO, respectively (4-6).

Along with research, I have had the opportunity to teach a chemical engineering core course (CHE 3213 - Fluid Flow Operations) for two semesters. My average student evaluations (4.23/5.00) were above the average for the College of Engineering. I also had numerous opportunities to teach laboratory practices and experimental procedures to undergraduate and high school students who have worked with me to develop this research. My goal is to contribute to the education of future chemical engineers through research, teaching, and service to the community.

Refereed Publications

(1) T Benson and C George (2004), "Cellulose Based Adsorbent Materials for the Dehydration of Ethanol Using Thermal Swing Adsorption," Adsorption Journal Vol.11.

(2) T Benson, R Hernandez, T French, E Alley, and W Holmes (2007) "Reactions of Fatty Acids in Superacid Media: Identification of Equilibrium Products" Journal of Molecular Catalysis A: Chemical, V 274, 173 – 178.

(3) T Benson, R Hernandez, T French, and E Alley, W Holmes (2007) "Development of a Heterogeneous Catalytic Cracking Reactor Utilizing Online Mass Spectrometry Analysis" Journal of Chromatography A, V 1172, 204 – 208.

(4) T Benson, R Hernandez, M White, T French, E Alley, W Holmes, and B Thompson, "Heterogeneous Cracking of an Unsaturated Fatty Acid and Reaction Intermediates on H+ZSM-5 Catalyst" Clean – Soil, Air, Water (Accepted for publication, April 2008).

(5) T Benson, W Holmes, T French, E Alley, and R Hernandez, "Elucidation of Reaction Pathways for Unsaturated Acylglycerols on H+ZSM-5 Using 13C Tracer Compounds, Journal of Molecular Catalysis A: Chemical (Expected Submission: June 2008).

(6) T Benson, W Holmes, E Alley, T French, and R Hernandez, "Production Distribution for the Catalytic Cracking of Triacylglycerols on Silica-Alumina and Zeolite Y Heterogeneous Catalysts, Journal of Molecular Catalysis A: Chemical (Expected Submission: July 2008).

(7) T Benson, S Shuttleworth, W Holmes, A Brown, T French, and R Hernandez, "Transport of Metals from Microbial Growth Media to Biofuel Lipids: Utilizing Inductively Coupled Plasma/Mass Spectrometry for Identification and Quantitation, (Expected Submission: Fall 2008).

5k Protein and Surfactant Interactions: Applications to Protein Crystallization, Stability and Design

Bryan Berger, Biochemistry and Biophysics, University of Pennsylvania School of Medicine, 1009 Stellar-Chance Laboratory, 36th & Hamilton Walk, Philadelphia, PA 19104

My general interests lie in understanding the forces that control protein-protein interactions. My work has focused on developing and applying novel experimental methods to characterize these interactions in areas such as protein crystallization, protein-surfactant interactions, membrane protein folding and engineering specificity of transmembrane helix-helix interactions.

My graduate work at the University of Delaware (advisors: Abraham Lenhoff, Eric Kaler) was focused on characterizing protein and surfactant interactions as they relate to protein crystallization. I used a variety of experimental techniques, including dynamic light scattering, self-interaction chromatography and isothermal titration calorimetry, to study the phase behavior of protein-surfactant mixtures in the presence of a wide range of additives and precipitants. By relating surfactant phase behavior to interactions between protein-detergent complexes (PDCs) in terms of the osmotic second virial coefficient (B_{22}), I established patterns of crystallization for both the integral membrane protein bacteriorhodopsin (BR) and soluble protein calcium- and integrin-binding protein (CIB). My results led to a greater understanding of the various influences additives and precipitants have on the interplay between PDC interactions and surfactant phase behavior, and how these interactions can be manipulated to favor PDC crystallization.

As a post-doctoral fellow at the University of Pennsylvania School of Medicine (advisors: Joel Bennett, William DeGrado), I have been investigating factors that contribute to transmembrane (TM) helix-helix interactions and stability. Much of this work has focused on integrins, a family of heterodimeric, cell-surface receptors involved in cell adhesion and signal transduction. I have developed genetic methods to examine the specificity of the homo- and heterooligomeric interactions involving integrin TM domains in membranes, identified key residues and interactions motifs through scanning mutagenesis and characterized effects of these mutations on integrin function *in vivo*. I have also used these methods to examine the specificity of designed peptides that target TM domains such as integrins.

Representative Publications:

1. **Berger BW**, Bennett JS, WF DeGrado*. Homo- and Heterooligomerization of Integrin Transmembrane Domains. In preparation (2008)

2. Yin H, Slusky JA, **Berger BW**, Walters RS, Vilaire G, Litvinov RI, Lear JD, Caputo GA, Bennett JS, WF DeGrado*. Computational Design of Peptides That Target Transmembrane Helices. Science 315 (2007) 1817-1822

3. **BW Berger***, Gendron CM, Lenhoff AM, EW Kaler*. Effects of additives on surfactant phase behavior relevant to bacteriorhodopsin crystallization. Protein Science 15 (2006) 2682-2696 (feature article, see also accompanying commentary: MC Wiener. When worlds colloid. Protein Science 15 (2006) 2679-2681)

4. **Berger BW**, Gendron CM, Robinson CR, Kaler EW, AM Lenhoff*. The role of protein and surfactant interactions in membrane-protein crystallization. Acta Crystallographica D61 (2005) 724-730

5. **Berger BW**, Blamey CJ, Bahnson BJ, AM Lenhoff*. Role of additives and precipitants in crystallization of calcium- and integrin-binding protein. Crystal Growth and Design 5 (2005) 1499-1507

5I Formation and Pinch-off of Viscoelastic Filaments: Numerical Analysis and Ink-Jet Experiments

Pradeep P. Bhat, School of Chemical Engineering, Purdue University, 480 Stadium Mall Dr, West Lafayette, IN 47907

When a mass of liquid fragments into drops, the undergoing process essentially involves the formation, thinning, and pinch-off of filaments. Technologically, production of drops has been exploited in modern applications such as ink-jet printing and more recent printing and patterning applications such as micro-arraying of genomes or proteins, printing of biological cells, and fabrication of microelectronic components (e.g., transistors) and solar cells via certain printing techniques. Most of the liquids encountered in these applications exhibit complex, viscoelastic behavior (as opposed to simple Newtonian behavior shown by, e.g., water or glycerol) due to the presence of macromolecules such as DNA or synthetic polymers in them. Among other effects, the presence of polymers is known to cause (1) a delay in the breakup of filaments [Amarouchene et al. 2001] and (2) the formation of a topology that consists of a number of droplets interconnected by thin ligaments or the so-called beads-on-a-string structure during filament pinch-off [Goldin et al. 1969]. Thus, a clear understanding of the formation and pinch-off of viscoelastic filaments is critical in the analysis and better design of these processes. In addition, the dynamics in the region close to where pinch-off occurs is known typically to exhibit behaviors rich in physics such as universality and self-similarity [Eggers 1993].

Most studies to date have used the 1-D long-wavelength (or slender jet) approximation of the flow equations to probe the self-similar dynamics of thinning viscoelastic filaments. This approximation is clearly invalid in regions where slenderness is lost. A full 3-D time-dependent axisymmetric (2-D) numerical analysis of the problem is presented here. Fluid viscoelasticity is captured using a conformation tensor approach [Pasquali and Scriven 2002] and the governing equations are solved using a fully-coupled finite element method that has been well benchmarked against experiments [Chen, Notz, and Basaran 2001, 2002]. The dynamics of thinning viscoelastic filaments in the low capillary number regime is also studied. The results from this study are valuable in assessing experiments that are based on the breakup of filaments of low viscosity, water-like liquids (e.g., the filament stretching rheometry and the capillary breakup extensional rheometry). Experimentally, formation of drops of polymeric liquids is studied using drop-on-demand (DOD) ink jet dispensers at room and elevated temperatures and the potential of the DOD technology in applications such as deposition of drugs on different substrates is explored.

5m Molecular Logic Circuits for Information Processing in Human Cells

Leonidas Bleris, FAS Center for Systems Biology, Harvard University, 7 Divinity Avenue, Cambridge, MA 02138

Components of a living organism, from organs and tissues to single cells and subcellular compartments, exchange and process numerous molecular signals in order to coordinate their activity. When these components fail, they generate characteristic signals that often trigger self-repair processes but can also cause disease when left unchecked. In the not-so-distant future, engineered biomolecular circuits will process information in human cells monitoring in parallel multiple inputs, detecting minute changes, rapidly assessing a patient's condition, and responding in infinitesimal time. Such systems will be used for diagnosing, preventing, treating, and monitoring disease in ways that achieve optimal and highly specific individual health-care, redefining personalized medicine and opening the path to new technologies.

Today, scientists in the cross-sections of disciplines such as biology, chemistry, mathematics, and engineering strive to produce molecular circuits with novel and useful functionalities, in a strikingly similar manner to physicists and engineers that built the fundamental building blocks of computers and modern electronic devices during the 19th century. Similar to a transistor, the basic component of an electrical circuit, with the voltage indicating binary high and low output, in cells a gene can have a binary high and low state depending on the protein concentration. Towards this direction, there are several prototype "biodevices" that operate in cells, such as oscillators, toggle switches, and circuits implementing basic Boolean operations (i.e. AND, OR, NOT logic gates). These devices comprise of genetic and biochemical components such as RNA, DNA fragments, proteins, and inducer molecules.

We have constructed a de novo molecular information-processing gene network that operates in human kidney cells. This molecular circuit is based on RNA interference (RNAi), a mechanism for RNA-guided regulation of gene expression. We show that the RNAi pathway in human cells can form a molecular computing module capable of evaluating arbitrary Boolean expressions on endogenous cues. We experimentally demonstrate in human kidney cells the direct evaluation using exemplary expressions in standard forms with up to five logic variables.

Web Page: www.sysbio.harvard.edu/csb/benenson/bleris.html

5o Understanding Multiscale Single-Molecule Dynamics with New Computational Statistics Methods

Christopher P. Calderon, Computational and Applied Mathematics, Rice University, 6100 Main St, Houston, TX 77251-1892

A large amount of physically useful information is contained in the vast data sets coming from single-molecule experiments. Numerical methods which estimate stochastic differential equations (SDEs) using observational data are presented. The information in these models can be used to approximate the dynamics, indirectly infer how structure influences kinetics, and quantitatively understand how noise from different time scales influences small scale systems relevant to targeted drug design using modern statistical analysis tools. Local maximum likelihood ideas are used throughout and the computational tools can also be used to quantitatively compare data-driven models to purely theoretical models to either refine existing and/or guide the development of new models. Results and insights provided by applying these methods to experimental single-stranded DNA and titin I27 time series data are presented.

Web Page: www.stat.rice.edu/~cpc1/

5p Bio-Based Materials from Renewable Resources

Alejandrina Campanella, Department of Chemical Engineering and Center for Composite Materials, University of Delaware, 150 Academy St, Newark, DE 19716

Due to the increased cost and depletion of petroleum-based materials, combined with the desire to move towards greener products, there has recently been renewed interest to replace part, if not all, of the conventional products with those made from natural resources. The use of these products have both economic and environmental advantages, thereby making them attractive alternatives to petroleum-based materials – namely, natural oils are a renewable resource, environmentally friendly, biodegradable, and are relatively inexpensive. In this poster I will highlight my doctoral and postdoctoral research that was primarily focused on the development of new processing technologies based on the modification of vegetable oils for their conversion into higher value-added products.

During my doctoral studies I worked in the production of epoxidized vegetable oils and synthesis of ring-opening reactions, trying to improve the production by heterogeneous catalyst. Research efforts were on multiphasic systems (liquid-liquid and liquid-liquid-solid) analyzing the impact of the main process variables and studying the overall reaction rate, modeling separately all the transport and intrinsic kinetic steps, and identification in each case the constants of reaction.

My postdoctoral work focused on the design of synthetic pathways for the production of monomers from plant triglycerides, using a broad range of chemical routes to produce polymeric materials; more specifically, in the development and characterization of bio-based polyurethane foams, studying the morphology to better understand how the reactions taking place during the production of this product affect the foam structure. Also, design strategies were implemented to replace styrene in the production of triglyceride-based thermosetting polymers. All resins currently used employed in industrial production of polymers and composites contain high concentrations of reactive diluents to allow resin transfer molding. Styrene is typically employed, but it is both a hazardous pollutant and a volatile organic compound. Therefore, non-volatile reactive diluents, such as fatty acids offer an alternative over current reactive diluents. Investigation focused in the synthesis of chemically modified fatty acids and the characterization of their polymer analogs to produce materials with properties comparable to commercial products.

5q Electrochemically Induced Charge Transfer to Wide Band Gap Semiconductors and Its Implication to Nanowire Based Solar Cells

Vidhya Chakrapani, Chemical Engineering, University of Lousiville, Room 311, Ernst Hall, Univeristy of Louisville, Louisville, KY 40292

High quality diamond is an extremely good insulator with a band gap of 5.5 eV. However, for undoped, hydrogen-terminated diamond exposed to air, a p-type conductive surface layer appears that is mediated by adsorbates from the ambient atmosphere. In our recent work we showed that this observed conductivity is due to the charge transfer induced by an electrochemical reaction occurring in adsorbed water film on the surface[1]. The electron transfer depends on the band edge positions of the semiconductor with respect to the redox reactions and affects both its electronic and optical properties.

Although these effects were first observed at diamond/water interphase, they can ocuur other wide band gap semiconductors such as those used in photoelectrochemical cells (PEC). In PEC cells, the photocatalytic activity is strongly influenced by the surface electronic properties of the semiconductor. These properties in turn depend on the nature of defect states and band edge positions of the semiconductor with respect to the redox couple. Thus by controlling these properties one can tailor the performance of photoelectrodes. Nanowire based electrodes in this regard show huge promise because of their surface area. Furthermore, they offer a unique platform for designing hybrid architectures.

The proposed research would be focused on dye sensitized solar cells and photolysis of water. Several different strategies will be adopted for designing novel photoelectrodes that are stable against photocorrosion and can efficiently harvest solar light. Specifically, they are: 1. band gap reduction of oxide semiconductors, 2. protective coatings for existing visible light absorbers against photocorrosion based on the band edge engineering concepts. The studies will couple synthesis and characterization of inorganic nanowires with fundamental studies of energy absorption and transport, for high performance energy conversion device.

[1]. V. Chakrapani, J.C. Angus, A.B. Anderson, S.D. Wolter, B.R. Stoner and G. Sumanasekera, Science 318, 1424 (2007)

5r Complete Flux Elucidation Using Metabolic Flux Analysis

YoungJung Chang[1], Patrick F. Suthers[2], and Costas D. Maranas[2]. (1) Chemical Engineering, The Pennsylvania State University, 147B Fenske Laboratory, The Pennsylvania State University, University Park, PA 16802, (2) Department of Chemical Engineering, The Pennsylvania State University, University Park, PA 16802

Metabolic flux analysis (MFA) aims at quantifying the magnitude of intracellular fluxes in metabolic networks. A key question in this analysis is choosing a set of measurements (external fluxes, concentration and isotopic distribution of metabolites) that is capable of yielding a unique flux distribution (identifiability). Since the system of equations that need to be solved to infer intracellular fluxes is nonlinear, this question is very difficult to answer even under a steady-state assumption.

Recently we introduced an optimization-based framework that uses incidence structure analysis to determine the smallest (or most cost-effective) set of measurements leading to complete flux elucidation using stationary MFA (S-MFA) (Chang et al., 2008). This approach relies on an integer linear programming formulation OptMeas that allows for the measurement of external fluxes and the complete (or partial) enumeration of the isotope forms of metabolites without requiring any of these to be chosen in advance. Since OptMeas approximately satisfies the identifiability, we subsequently query and refine the measurement sets suggested by OptMeas to minimize the measurement cost while ensuring a unique flux distribution. We showed that OptMeas is capable of generating optimal measurement sets for small to medium-sized demostration examples. Moreover, OptMeas suggested a number of additional measurements that lead to maximum flux elucidation in a large-scale amorphadiene producing *E. coli* strain. OptMeas is particularly useful in identifying essential measurements that are critical for the unique flux elucidation.

Although S-MFA has been extensively used to estimate flux distributions, it could be sometimes impractical because 1) isotopic steady state is very difficult to attain (e.g., miniaturized experiments), or 2) steady-state measurements are not enough to infer all intracellular fluxes. Isotopically nonstationary MFA (IN-MFA) can overcome some of these limitations by measuring isotopic distributions (mass isotopomer distributions in general) of internal metabolites at a number of different time points. We extend the OptMeas procedure in order to account for the added concentration variables and differential isotopic balance equations. The extended OptMeas is used to identify the essential metabolites that need to be measured to ensure unique flux elucidation under isotopically nonstationary conditions. The identified essential measurements are then queried and refined for identifiability and optimality. It is important to note that this procedure calls for the repetitive solution of the inverse problem of IN-MFA, which is a dynamic optimization (DO) problem. The DO problem is converted to a nonlinear programming (NLP) problem by discretizing the time domain. In order to efficiently solve the resulting large-scale NLP problem, we combine a suitable network decomposition scheme with a Lagrangean decomposition based global optimization algorithm. The proposed approach was tested with a small network example involving eight metabolites and ten fluxes, and then applied to medium-scale demonstration examples including 1,3-propanediol producing *E. coli* strain. We found that the proposed approach is both scalable and reliable in predicting flux distribution and suggesting essential measurements.

Reference:

(1) Chang, Y., Suthers, P. F., Maranas, C. D., Identification of optimal measurement sets for complete flux elucidation in MFA experiments, *Biotech Bioeng*, accepted, 2008.

5s Multiscale Modeling for Energy Applications

Abhijit Chatterjee[1], Dion Vlachos[2], Blas P. Uberuaga[3], and Arthur F. Voter[1]. (1) Theoretical Division, Los Alamos National Laboratory, T-12 MS B268, Los Alamos, NM 87545, (2) Director of Center for Catalytic Science and Technology (CCST), University of Delaware, Newark, DE 19716, (3) Material Science and Technology Division, Los Alamos National Laboratory, Los Alamos, NM 87545

The goal of our research is to understand thermodynamics, mechanics and transport in materials for energy applications, i.e., energy generation (fuel cells) and energy storage (batteries and capacitors) devices, using multiscale modeling. This understanding can be employed for a) improving efficiency and performance, and reducing cost of current technologies and b) discovering new materials to counter challenges faced by existing technologies.

Materials used in energy generation/storage devices typically entail phenomena involving several atomic/ molecular species and thermodynamic phases and span multiple length and time scales. The device material structure-property relationships are generated using reliable multiscale models that we have developed in our group, and systems analysis methods. Four different multiscale models are employed. The temperature accelerated dynamics (TAD)-based kinetic Monte Carlo (KMC) method is used to efficiently generate an on-the-fly KMC process catalog with a given confidence. Adaptive coarse-grained Monte Carlo (CGMC) method is derived from a KMC model for performing accelerated KMC simulations. Information theory based a posteriori error estimates are derived for error control in the CGMC method. The tau-leap method is developed for accelerated KMC simulations in time. The continuum mesoscopic equations are derived from KMC models in the limit of infinite interaction lengths. The accuracy and computational requirements of these multiscale models decreases in the order mentioned.

The strengths of our multiscale framework enable predictive materials modeling for energy application. The correct atomistic physics, such as interactions, mechanisms, thermal fluctuations and correlations are incorporated. This is essential for accurately predicting material properties, such as phase behavior and dynamics. A systematic derivation of each of the multiscale models directly from an atomistic model ensures seamless exchange of information between different multiscale models. Thus, depending on the accuracy requirements different model descriptions can be employed in different regions of the material. The aforementioned multiscale models overcome the inherent material length and time scale separation. Studies on kinetics and thermodynamics in multicomponent materials at device length and operation time scales are therefore possible.

Application of the multiscale models to three broad problem areas in rational design of materials for energy generation and storage are illustrated next.

Role of material structure in ionic mobilities

An understanding of ionic transport is essential for designing better materials for energy storage applications. However, the atomic mechanisms prevalent at time scales of interest are typically not accessible to molecular dynamics, and are therefore not well understood. Diffusion and ionic conduction in oxide materials is studied using the TAD-based KMC method. The role of nanostructure up to microstructure, e.g., the grain size and interfaces, in determining ionic mobilities is elucidated.

Role of defects interactions in materials structure, phase diagram and stability

Defect formation and propagation in materials is a complex phenomenon that spans atomic length and time scales (nm and ps, respectively) to macroscopic scales at the microstructure level (mm and min, respectively). Large concentration of defects in a material crystalline structure can be created by intentional addition of substitutional/dopant atoms. Depending on the defect source and the material, different defect shape and size distributions are observed, which ultimately determines the physico-chemical, mechanical and electrical properties of the material. We present fundamental insights obtained from the multiscale framework regarding material structure, phase diagram, segregation and stability.

Control of material structure for tuning material properties

Material properties, such as ionic transport, can be controlled indirectly by tuning nanostructure features. Bottom-up fabrication strategies based on self-assembly offer exciting possibilities for inexpensively manufacturing nanostructures. However, control of self-assembly to obtain the desired nanoparticle size, shape, density, and lateral arrangement is still a major challenge. We present a multiscale modeling based optimal control strategy that we developed recently, for controlling self-assembled structures.

5t Molecular Mechanisms for the Aggregation of Proteins and Therapeutic Antibodies

Naresh Chennamsetty, Chemical Engineering, Massachusetts Institute of Technology, 77 Massachusetts Ave, E19-528, Cambridge, MA 02139

My overall research goal is to understand the mechanisms behind protein and therapeutic antibody aggregation, and design proteins with enhanced stability. Antibody aggregation leads to immunogenicity and loss of activity in the pharmaceutical formulations used in the treatment of numerous cancers, chronic inflammatory diseases, and infectious diseases. Furthermore, protein aggregation in vivo is associated with a number of neurodegenerative disorders such as Alzheimer's and Creutzfeldt-Jakob diseases, and type II diabetes. We study these aggregation mechanisms using multi-scale modeling and simulation techniques in close collaboration with experiments.

I. Mechanisms for the Aggregation of Therapeutic Antibodies

In my current post-doctoral position with Prof. Bernhardt L. Trout at the Massachusetts Institute of Technology, we study the mechanism behind the aggregation of therapeutic antibodies and design antibodies with enhanced stability. Antibodies constitute the most rapidly growing class of human therapeutics. These antibodies are, however, thermodynamically unstable under conditions required for storage and administration and degrade due to aggregation. We use molecular simulation tools in collaboration with experimental techniques to identify the regions of the antibody most prone to aggregation. Subsequently, we perform mutations in these specific aggregation prone regions to engineer antibodies with enhanced stability.

II. Development of Multi-scale Modeling Techniques for Soft Condensed Matter Systems

During my PhD work with Prof. Keith E Gubbins at North Carolina State University, we developed a multi-scale methodology connecting the atomistic and meso-scale models useful in understanding the long length and time scale phenomena involved in surfactant self-assembly, polymers, colloids and protein solutions. In addition, we have developed a novel technique to determine the rigorous effective pair potentials for a simple binary A/B mixture by integrating out the degrees of freedom of solvent B using Widom's particle insertion method. We have also explored the effect of cosolvents and cosurfactants such as alcohols on surfactant self-assembly and solubilization capacity in supercritical carbon dioxide with applications in separation processes in the food industry such as extraction of water-soluble vitamins and proteins, dry cleaning and polymerization.

Web Page: web.mit.edu/naresh/www/

5u Photoreactive Ultrafiltration Membranes for Prevention of Biofouling

Shannon Ciston, Department of Chemical and Biological Engineering, Northwestern University, 2145 Sheridan Rd., Rm A225, Evanston, IL 60208, Richard M. Lueptow, Mechanical Engineering, Northwestern University, 2145 Sheridan Rd., Evanston, IL 60208, and Kimberly Gray, Department of Civil Engineering, Northwestern University, 2145 Sheridan Road, Evanston, IL 60208-3109.

Growing stresses on water supplies worldwide necessitate the efficient use and reuse of fresh water. Membrane filtration can be used to purify even low-quality sources. However, biofouling and biofilm growth are major obstacles for membrane filtration because they cause dramatic losses in efficiency and system flux. By coating ceramic ultrafiltration membranes with TiO2, we have created a photoreactive membrane that resists fouling and biofilm growth by improving hydrophilicity, inactivating microorganisms, and degrading organics. In recent work we have identified Degussa P25 as a strong candidate for biofilm prevention based on its excellent bacterial cell attachment resistance and high organic degradation activity. In this poster, we will present work on the selection of catalyst materials, results that show biofilm prevention and reduced flux decline, as well as the incorporation of this membrane material into a rotating reactive filtration system.

5v Engineering Approaches in Neuroscience

David Colby, Institute for Neurodegenerative Diseases, UCSF, Box 0518, San Francisco, CA 94143

In science, the most exciting discoveries often occur at the interface between two disciplines. In both my doctoral and postdoctoral work, I have been applying my training in chemical engineering to problems in neurodegeneration. During my graduate studies, I worked with Dane Wittrup at MIT to engineer antibody fragments to counteract disease in cellular models of Huntington's Disease. I also developed stochastic models to describe the kinetics of huntingtin protein nucleation in cells. Since completing my doctorate, I have been working with Stanley Prusiner at UCSF, engineering "synthetic prion" strains, which are comprised of recombinant prion protein refolded into an infectious form *in vitro*. In the process of doing this work, I have developed an assay for prion detection with femtogram sensitivity. My work in the Prusiner lab is also shedding light on the role of prion protein conformation in disease pathogenesis. In the next phase of my career, I would like to establish a lab at the interface of engineering and neuroscience, applying cell and biomolecular engineering approaches to the most pressing problems in neurodegeneration and neuroscience more broadly.

5w Colloidal Suspensions: Fundamental Physics and Engineering Applications

Jacinta C. Conrad, Materials Science and Engineering, University of Illinois at Urbana-Champaign, 1304 W. Green St., Urbana, IL 61801

Colloidal suspensions are ubiquitous in industrial and technological applications, and serve as excellent model systems for a variety of complex fluids. Engineering functional materials from colloids requires a thorough knowledge of the relationship between the microscopic properties of the suspension and the bulk physical properties.

My graduate work at Harvard University focused on the phase behavior, structure, and rheology of model nonequilibrium colloidal suspensions, including both glasses and gels.

1.1. Colloidal microstructure and dynamics near the colloidal glass transition (with F. W. Starr, P. P. Dhillon, E. R. Weeks, D. R. Reichman, and D. A. Weitz): We use confocal microscopy and simulation to investigate the structure and dynamics of model colloidal suspensions near the glass transition. First, we investigate the relationship of local static structure to dynamics in suspensions of nearly-hard-sphere poly-(methyl methacrylate) colloids. We calculate the Voronoi volume for all particles and show that this quantity is only weakly correlated to local particle displacements; we obtain qualitatively similar results in a simulation of a polymer melt. These results suggest that static local structure alone does not predict dynamical behavior in experimental colloidal suspensions. Second, we investigate the behavior of the slowly-relaxing particles. These particles form spatially correlated clusters that evolve over time; furthermore, the structure of these clusters is correlated to the macroscopic properties of the suspension. In supercooled fluids, the largest cluster spans the system on short time scales but breaks up on longer time scales; in glasses, a percolating cluster exists on all accessible time scales. Using molecular dynamics simulation, we show that these clusters make the dominant contribution to the bulk elasticity of the sample.

1.2. Phase behavior, structure, and rheology of weakly attractive colloidal suspensions (with S. Manley, H. M. Wyss, P. J. Lu, K. Miyazaki, V. Trappe, L. J. Kaufman, A. B. Schofield, D. R. Reichman, and D. A. Weitz): We investigate the phase behavior, structure, dynamics, and rheology of colloidal suspensions with weak attractions. We induce a depletion weak attraction between poly-(methyl methacrylate) colloids by adding linear polystyrene. We investigate the phase behavior of these suspensions as a function of colloid volume fraction and polymer radius of gyration and concentration, and show that we can obtain interconnected gels, stable fluids of clusters, and colloidal fluids. We investigate in detail one region of the phase space, with intermediate volume fraction (0.15-0.35) and a fixed range of the interparticle attraction. In this region colloid-polymer mixtures can undergo spinodal decomposition into colloid-rich and colloid-poor regions. Gelation results when the interconnected colloid-rich regions solidify; we show that this occurs when these regions undergo a local attractive glass transition. This arrest scenario bridges the gap between colloidal gelation and the colloidal glass transition; the resulting structures display hallmarks of both glasses and gels. The resultant macroscopic rheology of these structures is strikingly different than that observed for either colloidal glasses or gels; while the relaxation time depends strongly on colloidal volume fraction, the storage modulus is relatively insensitive to changes in volume fraction.

My postdoctoral work at the University of Illinois focused on studying colloidal suspensions under flow. We investigate the structural and dynamic properties of flowing suspensions, and exploit the flow properties to assemble novel colloidal granules and to pattern colloidal films.

2.1 Flow of attractive colloidal suspensions in microchannels (with J. A. Lewis): We use confocal microscopy to investigate the behavior of suspensions of attractive colloids flowing in microchannels. We flow suspensions of attractive silica colloids through microchannels and directly image both their structure and dynamics during flow. We study two different model systems: polyelectrolyte-flocculated gels, and hydrophobic gels. We investigate the flow properties as a function of applied pressure, microchannel geometry, and colloid volume fraction.

2.2 Microfluidic assembly of homogeneous and Janus colloid-filled hydrogel granules and shells (with R. F. Shepherd, A. Cote, S. M. Menke, S. K. Rhodes, D. R. Link, M. Marquez, D. A. Weitz, and J. A. Lewis): We use microfluidic techniques to assemble colloid-hydrogel granules and shells. We suspend silica microspheres in an aqueous acrylamide suspension and use microfluidic devices to emulsify this suspension in a continuous oil phase. By varying the microchannel geometry, we can form spheres, discoids, and shells; by varying the input streams, we can form both homogeneous and chemically heterogeneous (Janus) surfaces. This approach offers a facile route for assembling granules and shells of controlled size, shape, and composition.

2.3 Patterning colloidal films via evaporative lithography (with D. J. Harris, H. Hu, and J. A. Lewis): We investigate evaporative lithography as a route for patterning aqueous colloidal films. Films are dried beneath a mask that induces periodic variations between regions of free and hindered evaporation; fluid and entrained particles migrate towards the regions of highest evaporative flux, leading to lateral segregation of particles within the film. By tuning the colloidal suspension composition, separation distance between the mask and underlying film, and mask geometry, we can control the pattern formation within the dried films.

5x Catalytic Strategies for the Molecular Engineering of Fuels

Steven P. Crossley and Daniel E. Resasco. Cbme, University of Oklahoma, 100 East Boyd, Room T-335, Norman, OK 73019

The concept of molecular engineering to fit a particular application is of vast importance in the fuels industry. This idea has allowed our group to create strategies for improving diesel, gasoline, and alternative fuels. The first step is accomplished by identifying molecules or types of molecules which have potential for improvement, such as naphthenics in LCO, aromatics in gasoline, or glycerol from biofuels. Representative model compounds are then chosen and converted via heterogeneous catalysis with all reaction products tracked. The desired reaction products are then identified and maximized by optimization of the catalyst and reaction conditions. The best catalyst is then tested with real feeds and a successful strategy is born. This has been implemented for a variety of situations in both the oil and alternative fuels energy industries involving hydrogenation, ring contraction/ring opening, isomerization, decarbox/onylation, cracking, as well as others. In each case, careful attention is paid to catalyst characterization in order to both learn about important reaction intermediates and provide direction towards catalyst optimization.

Stemming from this approach is the need to know important fuel properties of each potential reaction product in order to better determine which products should be maximized. Because of this, quantitative structure-property relationships (QSPRs) have been heavily utilized in order to predict a variety of fuel properties, including cetane number (CN), octane number (ON), threshold sooting index (TSI), and micropyrolysis index (MPI). These relationships are created by correlating molecular descriptors to fuel properties through the use of genetic algorithms, artificial neural networks, principal components, and other statistical techniques. The end result is a reliable fuel property of each potential reaction product which has been calculated based only upon its molecular structure.

Quantitative structure-activity relationships (QSAR) have also been utilized to predict primary product distributions. By measuring the primary product distributions of several different feed molecules, and creating models which predict several different ratios of products, we now have the ability to predict for several compounds what their reaction behavior will be over specific catalysts as well as the fuel properties of each of these potential reaction products. This is a great tool for determining how a particular catalyst will influence the properties of a fuel.

A new fuel property was also developed in order to determine the sooting tendency that a specific fuel will produce through pyrolysis alone. This technique, coined the micropyrolysis index (MPI), involves injecting 20μ L of a fuel across a hot bed of α-Al_2O_3 in a flow of helium. By measuring the amount of carbon that has deposited on the alumina via temperature programmed oxidation (TPO), one is measuring the kinetic tendency of that particular fuel to form soot via pyrolysis. This is an essential tool in molecular engineering as a reliable sooting measurement is very important for nearly every fuels upgrading situation. This has also led to a better understanding of how molecular structure impacts sooting tendency, and the ability to characterize the soot which was formed has led to further studies with SEM, Raman, and same spot TEM.

5y Classical Chemical Engineering Is Alive and Well

Benjamin J. Davis, UCLA Chemical Engineering Department, 5531 Boelter Hall, Los Angeles, CA 90095

"Any chemical process, on whatever scale conducted, may be resolved into a coordinate series of what may be termed 'unit operations'. . . the number of these basic unit operations is not large and relatively few of them are involved in any particular process. The complexity of chemical engineering results from the variety of conditions as to temperature, pressure, etc., under which the unit operations must be carried out in different processes, and from the limitations as to material of construction and design of apparatus imposed by the physical and chemical character of the reacting substances."

-Arthur D. Little, 1915

Unit operations are the fundamental building blocks of chemical process design; they are the conceptual tool which defines the field of chemical engineering and separate it from other disciplines. It is the unique job of the chemical engineer to choose which types of units to include, how many of each, how they will be run, and how they are connected together to produce a final flowsheet which performs some purpose. This is the science behind chemical engineering and it is still an active and productive research field.

As there are an infinite number of ways to produce a product from some set of inputs, a chemical engineer must in the course of his or her job answer the question "How good is my flowsheet?" The evolution of computer-aided process design has given this designer a multitude of different tools to answer this question. As our field and our society mature, the definition of "good" also changes and evolves; a flowsheet which may have been "good" in 1960 (or even 2005) may now be "bad" due to changes in the cost of raw materials, rising energy costs, stricter emissions standards, increased safety of operation, etc. What this means is that a process designer must be in tune with this "multi-objective" world and do his job using the latest tools available.

The goal of my research is threefold: first, to enhance these tools using the techniques of optimization so that they perform more efficiently and can solve bigger and more general problems; secondly, to identify bounds on the performance of units and flowsheets to give real process designers benchmarks to compare their work to; thirdly, to advise and educate future process engineers in the most state-of-the-art methods for designing and operating their plants in the most energy efficient, cost-effective, and sustainable way possible. My thesis work has focused on the development of one such process design method, the Infinite DimEnsionAl State-space framework (IDEAS). IDEAS, in short, is a way to represent all possible flowsheets simultaneously to eliminate infeasible flowsheets, identify properties of "good" or optimal flowsheets, and to give lower bounds on the best flowsheet for a given design problem. I have applied this method to networks of batch reactors [Davis et al. "Identification of the Attainable Region for Batch Reactor Networks" Ind. Eng. Chem. Res., in press], networks of PEM fuel cells, power generation cycle flowsheets, and cryogenic liquefaction flowsheets.

5z Fischer Tropsch Synthesis and Related Technologies

Ed Durham, Department of Chemical Engineering, Auburn University, 212 Ross Hall, Auburn, AL 36849

The goal of my research is to use novel catalysts and solvents to enhance the Fischer Tropsch reaction and related processes (cracking, isomerization, oligomerization ,etc). The increasing price and scarcity of crude oil is making the utilization of coal for fuel and chemical markets a viable alternative while concerns about global warming have made biomass utilization (for CO_2 neutrality) an important area of research.

As a teacher, it is my belief that Chemical Engineering, while very complex, is built almost entirely out of a few basic concepts. Having been both a TA and an instructor, it is my belief that my primary challenge is to prevent students from taking shortcuts and memorizing equations and instead getting them to always return to the basic concepts and work out each problem from there. I am passionate about Chemical Engineering and looking forward to the opportunity to teach it. My experiences as a student, a teacher, and as a member of Auburn's Academic Honesty Committee have built in me a commitment to help students grow into good engineers and people.

5aa Investigating Dynamic Penetrant Transport in Glassy Polymers for High-Tech Applications

Adam K. Ekenseair and Nicholas A. Peppas. Department of Chemical Engineering, The University of Texas at Austin, Austin, TX 78712

Glassy polymers are increasingly used in the construction and fabrication of high-tech devices. These devices are often used in applications, such as microfluidics, membrane separations, food packaging, artificial tissues, bones, and organs, and controlled drug delivery, where the polymers come into contact with fluids capable of penetrating the polymer network. To examine the behavior of such systems both on the macroscopic and molecular level, it is important to study the penetrant diffusion process. In such transport, the macromolecular chains rearrange toward new conformations, where the rate of relaxation depends on the penetrant concentration. The relative rates of penetrant diffusion and macromolecular chain relaxation determine the nature of the transport process and lead to a wide variety of penetrant transport phenomena, such as Fickian, anomalous (non-Fickian), and Case II sorption behavior.

With trends leading to smaller device scales and increasingly complex polymer structures, there is a need for both a qualitative and quantitative understanding of the manner in which a polymer's network structure alters not only the rate, but also the mode of penetrant transport. To this end, poly(methyl methacrylate) (PMMA) was synthesized in an iniferter-mediated radical polymerization process to produce a variety of well-characterized polymer network structures. The effects of basic network parameters, including the degree of crosslinking, polymer mesh size, and the crosslink size, on the transport process were studied. The effects of sub-Tg annealing/aging, temperature, and the presence of un-reacted monomer were also investigated.

Utilizing both gravimetric sorption data and in-situ ultra-high-resolution X-ray computed tomography studies, the transport of methanol in PMMA discs was investigated. Changing the polymer network structure is shown to not only influence the rate of sorption but also the nature of the transport process.

5ab Analytical Ultracentrifugation of Colloidal Nanomaterials for Disease Detection and Drug Delivery Applications

James B. Falabella, Process Measurements, National Institute of Standards and Technology, 100 Bureau Dr. MS 8362, Gaithersburg, MD 20899

Few methods exist to monitor important parameters such as aggregation, molecular surface coverage, and molecular surface conformation of biomedical nanodevices. Seemingly minor changes in the formulation such as reversible aggregates, surface coverage of the functional molecules, and molecule conformation can incite life threatening side effects or reduce the effectiveness of the device in the most benign case. Examining the device in its native suspending liquid simplifies sample preparation and helps ensure that reversible aggregation can be detected before the devices are used. Analytical ultracentrifugation (AUC) is an established technique that has been used extensively to study molecular conformation and reversible interactions between biomolecules such as proteins. The high sensitivity of AUC to protein shape and state of aggregation make it ideal to investigate the subtle changes that can make the nanodevices unsafe. The flexibility to use different matrix fluids makes AUC ideal for performing systematic studies on factors that initiate aggregation and alter the apparent size of the device. Sedimentation velocity AUC was performed on model nanodevices consisting of 10 nm gold particles derivatized with thiol terminated thymidine homooligomers to demonstrate how systematic changes in molecular surface coverage and conformation could be reliably detected.

5ac Enhanced CO2 Solubility in Aromatic and Alkyl Chain Substituted Imidazolium-Based Room Temperature Ionic Liquids

Alexia Finotello, Jason E. Bara, Suguna Narayan, Douglas L. Gin, and Richard D. Noble. Department of Chemical and Biological Engineering, University of Colorado, UCB 424, Boulder, CO 80309

Room-temperature ionic liquids (RTILs) are molten organic salts that possess a number of desirable properties for gas separations applications. Their high thermal and chemical stability, negligible vapor pressures, and tunable chemistry make them great candidates to replace less environmentally friendly materials such as volatile organic compounds. The versatile functionality of RTILs allows for their potential usage for gas separations as a bulk fluid, a supported liquid membrane, or as a polymerized RTIL solid membrane. Gas separations in these materials are driven by large solubility differences between gases, rather than by diffusion differences. Therefore, it is important to investigate the gas transport properties in RTILs to better understand their gas separation capabilities.

This study focuses on the solubility (S), diffusivity (D), permeability (P), and selectivity behaviors of CO2, CH4, and N2 gases in bulk fluid RTILs at constant temperature (22C). The bulk fluid RTILs tested are 1-methyl, 3-tolyl imidazolium bis(trifluoromethanesulfonyl)imide ([mtim][Tf2N]), 1-ethyl, 3-paraxylyl imidazolium bis(trifluoromethanesulfonyl)imide ([epxim][Tf2¬N]),1-butyl, 3-paraxylyl imidazolium bis(trifluoromethanesulfonyl)imide ([bpxim][Tf2¬N]), 1-hexyl, 3 paraxylyl imidazolium bis (trifluoromethanesulfonyl)imid ([hpxim][Tf2N]), 1-octa, 3 paraxylyl imidazolium bis (trifluoromethanesulfonyl)imid ([opxim][Tf2N]), and 1-deca, 3 paraxylyl imidazolium bis (trifluoromethanesulfonyl)imid ([dpxim][Tf2N]). Our initial results show that gas solubility increases, diffusivity has a small variance, and permeability slightly increases with increasing alkyl chain length. Solubility and permeability selectivity decreases while diffusion selectivity remains relatively unchanged with increasing chain length. The addition of the aromatic ring substitution shows enhanced solubility over analogous alkyl-chain only substituted RTILs. Also, our work demonstrates that regular solution theory is valid for these RTILs.

5ad Photopolymerized pH-Responsive Nanomatrices for Applications in Nanomedicine

Omar Z. Fisher[1], Timothy Kim[1], Stephen R. Dietz[1], and Nicholas A. Peppas[2]. (1) Biomedical Engineering, University of Texas at Austin, 1044 Camino La Costa, apartment 1070, Austin, TX 78752, (2) Biomedical and Chemical Engineering, University of Texas at Austin, 1044 Camino La Costa, apartment 1070, Austin, TX 78752

There has been tremendous interest in developing nanoparticulate drug delivery vehicles for the intracellular delivery of therapeutics. Nanoparticles on the order of hundreds of nanometers in diameter have been shown to be endocytosed by numerous cell types. This mechanism is dependent on particle size, surface charge and hydrophobicity. Uptake of particulate matter is quickly followed by conversion of the endosome to a lysosome via the introduction of hydrolytic enzymes and a proton pump induced drop in pH. Clearly, this is not a friendly environment for biomacromolecules such as proteins and DNA. The use of polybasic hydrogels with a lower critical swelling pH as a carrier would allow for lysosomal escape and subsequent cytosolic drug delivery. This work describes the UV initiated free radical emulsion polymerization of steric stabilizing (and toxicity decreasing) poly(ethylene glycol) grafts onto poly(2-(diethylamino)ethyl methacrylate) polycationic hydrogels (P(DEAEM-g-PEG)). Specifically, microgel and nanogel suspensions were prepared by UV-initiated emulsion polymerization of P(DEAEM-g-PEG). Tetraethylene glycol dimethacrylate and Irgacure 2959 were used as crosslinker and photoinitiator. The particle size was characterized by dynamic light scattering (DLS) and scanning electron microscope (SEM). Changes in particle zeta potential were characterized as a function of pH and PEGylation using electrophoretic light scattering. For potential transmucosal drug delivery applications, the in vitro biocompatibility of these carriers was determined using a Caco-2/HT29-MTX coculture model. Insulin loading and release were monitored using HPLC. Control over particle size was achieved by varying polymerization time (15–60 minutes), surfactant type and concentration, and crosslinking ratio. The processing methods were successful in achieving particles with diameters ranging from 300 nm to 50 ìm in diameter. DLS measurements showed a lower critical swelling pH between 7-7.3 A minimum DEAEM/PEG weight ratio of 1:1 was successful in keeping particles from flocculating through changes in pH. Increasing the PEG content achieved a minimum zeta potential of 15-20 mV. This also lowered the insulin loading efficiency profile and cytotoxicity of the carriers. Current strategies for adding cell targeting specificity to these carriers through the use of different heterobifunctional crosslinking agents and ligands will be outlined.

5ae Supercritical and Expanded Solvent Systems for the Sustainable Utilization of Biomass Feedstocks

Jackson W. Ford[1], Raghunath V. Chaudhari[1], Bala Subramaniam[1], and Charles A. Eckert[2]. (1) Center for Environmentally Beneficial Catalysis and Department of Chemical & Petroleum Engineering, University of Kansas, Lawrence, KS 66045, (2) School of Chemical & Biomolecular Engineering and Specialty Separations Center, Georgia Institute of Technology, 311 Ferst Drive, NW, Atlanta, GA 30332-0100

Effective and economical biomass utilization is one of the key challenges for ensuring a secure and sustainable supply of fuels and chemicals. The recent food crisis further underscores the importance of developing novel conversion routes for lignocellulosic biomass. Supercritical fluids and gas-expanded liquids have the potential to improve reaction kinetics and heat transfer, enhance mass transfer in multiphase systems, and reduce coking in heterogeneously catalyzed reactions involving biomass substrates. In addition, the tunability of these solvents with pressure can facilitate less energy intensive separations such as precipitation and decantation to recover products and catalysts.

During my doctoral research at the Georgia Institute of Technology with Prof. Charles Eckert and Prof. Charles Liotta, I used kinetic and spectroscopic techniques to examine solvent property changes in gas-expanded liquids. My postdoctoral work at the University of Kansas with Prof. Bala Subramaniam and Prof. R.V. Chaudhari has focused on catalytic hydrogenations of biomass substrates in supercritical media. My future research will examine the use of supercritical and expanded solvents for both raw biomass processing to usable feedstocks and the production of fuels, fine chemicals, and intermediates from biomass-derived compounds.

5af Nanobiomaterials and Therapeutic Systems

Harvinder S. Gill, Department of Microbiology and Immunology, Emory University, Atlanta, GA 30332

My research interests are broadly in the areas of nanobiomaterials and therapeutic systems. In this poster presentation I will give an overview of my proposed research plans and summarize my research experiences from my doctoral work done at Georgia Institute of Technology under the guidance of Professors Mark Prausnitz (Chemical and Biomolecular Engineering) and Mark Feinberg (Vaccine Research Center, Emory Univesity)and my postdoctoral experience under the mentorship of Professor Richard Compans (Microbiology and Immunology) at Emory University.

The focus of my research is to integrate principles and techniques of engineering, immunology and recombinant DNA technology to rationally design and engineer novel responsive nanobiomaterials and microsystems to understand and probe fundamental issues in nanoscience, biomedical science and pharmaceutical science.

One of the fundamental issues in nanoscience that I will tackle involves development of architectural frameworks and scaffold for directed assembly of nanomaterials. While significant research has been done to develop nanomaterials, their directed and controlled assembly into 2-D and 3-D nanostructures and long-range organization and assembly for integration to the mesoscale continues to be a bottleneck. I will describe one of my proposed approaches to tackle this issue. Other research directions that I will describe include development of sensitive microneedle based painless biosensors, templating nanobiomaterials such as viruses and virus-like particles for developing hybrid nanomaterials and investigating recombinant peptide polymers as smart and tunable nanobiomaterials. Recombinant peptide polymers are a relatively new class of biopolymers synthesized in bacteria using recombinant DNA technology and they have potential to be used for drug delivery, as tissue engineering scaffolds, for development of biointerfaces and as transducers in sensors. I will focus on investigating structure-function relationships in peptide polymers to enable their controlled assembly as functional and smart nanobiomaterials and to use them as scaffolds for hybrid nanomaterial assembly.

5ag Polymer-Directed Self-Assembly as a Highly Flexible Route to Multifunctional Nanomaterials

Marian E. Gindy, Chemical Engineering, Princeton Unversity, E-Quad, Olden Street, Princeton, NJ 08544 and Robert K. Prudhomme, Princeton University, Dept of Chemical Eng Olden St, Princeton, NJ 08544.

With recent advances in nanotechnology, a large variety of single-component nanomaterials (i.e. polymer nanoparticles, carbon nanotubes, gold nanoshells, and quantum dots) have been developed for therapeutic and diagnostic applications. The next phase of development aims to construct multicomponent nanoscale entities which combine several functional properties into an integrated system. Nonetheless, the efficient and reproducible preparation of strategically engineered multifunctional nanomaterials is a significant challenge. Conventional approaches typically rely on unique synthetic routes tailored for the chemical linkage of individual components, making the development process lengthy and prohibiting independent optimization of individual components. Chemical strategies may additionally complicate facile incorporation of newly discovered materials into already existing constructs.

To overcome these limitations, a modular technology for the preparation of multifunctional nanoparticles using directed self-assembly of components is developed. The technology allows for the integration of materials of various properties based solely on thermodynamic driving forces, negating the need for explicit chemical modification of components. The key to the process is the control of times scales for micromixing, self-assembly, flocculation, and nucleation and growth, enabling particles of complex composition to be formed. Using this technology, stable and uniform composite nanoparticles simultaneously encapsulating nanocrystalline imaging contrast agents and organic drug molecules could be prepared with tunable sizes between 50-500 nm, high concentrations of encapsulated components, and precise control of component compositions, enabling multifunctional applications in drug delivery, medical imaging, and diagnostics. Based on this technology, novel particulate systems for MRI contrast enhancement and photodynamic therapy were developed. Finally, controlled functionalization of particle surfaces with disease-targeting ligands was demonstrated, allowing for directed delivery via a synergistic combination of passive and active targeting strategies. The technology provides a comprehensive and highly flexible platform for the tailored design and preparation of multifunctional nanomaterials in an economical and scaleable manner.

5ah Molecularly Engineered Nanoporous Adsorbents: Synthesis, Characterization, and Application

T. Grant Glover, Department of Chemistry and Biochemistry, University of California Los Angeles, 607 Charles E. Young Drive East, Los Angeles, CA 90095-1569

In order to design effective adsorption systems for energy storage and gas purification it is necessary to develop novel adsorbent materials that are tailor-made on a nanoscale. It is critical to characterize not only the materials physical characteristics and gas phase thermodynamics, but also the mass transfer fundamentals of these novel systems. In addition, it is important to understand how both thermodynamics and mass transport impact the performance of fixed-bed systems. This research addresses these areas by examining three specific topics: the synthesis and characterization of novel nanoporous molecularly engineered adsorbent materials including templated carbons, metal organic frameworks (MOFs), and zeolite imidazolate frameworks (ZIFs); the determination of mass transport mechanisms in nanoporous materials via concentration-swing frequency response; and the sensitivity analysis and subsequent optimization of fixed bed adsorption systems.

5ai a Novel Three-Stage Treatment for Brackish Water Reverse Osmosis Concentrate: Treatment and Analysis of a Real Brackish Water

Lauren F. Greenlee[1], Elise Barbot[2], Benny D. Freeman[1], Desmond F. Lawler[3], Benoit Marrot[2], and Philippe Moulin[2]. (1) Chemical Engineering, The University of Texas at Austin, 1 University Station C0400, Austin, TX 78712, (2) Laboratoire Mécanique, Modélisation et Procédés Propres, Aix en Provence, France, (3) Civil, Environmental and Architectural Engineering, The University of Texas at Austin, 1 University Station C1786, Austin, TX 78712

Reverse osmosis (RO) membrane desalination has been increasingly used to produce drinking water from salt water. Many regions lack sufficient fresh water resources and are turning to alternate resources, such as seawater or brackish water, to sustain water needs. In particular, a growing number of inland communities have both insufficient fresh water and unused brackish water (500 – 10,000 mg/L total dissolved solids) resources. A key financial and technical limitation to inland RO desalination is disposal of the waste stream (concentrate); typically, 10 – 25% of the influent volume becomes the RO concentrate stream. This waste volume is large when compared to the waste volume produced by traditional fresh water treatment (less than 1%). To improve the feasibility of RO desalination, RO system recovery (volume of product water per volume of feed water) must be increased to decrease the concentrate volume.

Brackish water RO treatment recovery is limited by sparingly soluble salt ($CaCO_3$, $CaSO_4$, $BaSO_4$, $SrSO_4$, silica) precipitation. Specifically, calcium carbonate ($CaCO_3$) is known to be a key, omnipresent precipitate. Antiscalants are typically synthetic organic phosphonates, acrylic polymers, or polymer blends and are used, along with pH adjustment, to prevent precipitation. However, as recovery is increased, antiscalant control is overcome and precipitation occurs. An alternate approach is thus required for further recovery augmentation.

Previous research using precipitation and separation to treat concentrate has shown that significant increases in total system recovery are possible [1]. However, the presence and influence of antiscalants and natural organic matter (NOM) during RO concentrate treatment have not been investigated.

This study presents the development of a novel three-stage process to treat the concentrate from a brackish water RO system. The process achieves problematic salt removal through three treatment steps: antiscalant deactivation, precipitation, and solid/liquid separation. Antiscalant deactivation is performed using ozone (O_3) and hydrogen peroxide (H_2O_2). pH elevation is used to precipitate salts, and solid/ liquid separation is achieved through sedimentation and filtration. While technologies for solid/liquid separation are well-established, the combination of antiscalant oxidation and precipitation represents a new system; research on antiscalant oxidation has been limited [2], and the effect of ozonation on precipitation has not been investigated.

Previous results with synthetic concentrates have shown that ozonation prior to precipitation increases calcium precipitation. For a simplified concentrate containing only sodium chloride (NaCl), sodium bicarbonate ($NaHCO_3$), and calcium chloride ($CaCl_2*2H_2O$), ozonation times as small as one minute increased calcium precipitation from 94% to 99.6%, for precipitation performed at pH 10.5 for 1 hour. In comparison, precipitation of the same solution without antiscalant results in 99.7% calcium precipitation. For a more complex water containing magnesium, sulfate, and other metals, 10 minutes of ozonation increased calcium precipitation from 81% to 87% after 30 minutes precipitation at pH 10.5. These results show the three-stage process allows increased calcium removal and thus a larger possible recovery for the overall process.

After tests and process optimization with synthetic RO concentrates, experiments were performed on a real water sample. The purpose of testing a real water sample was to evaluate the effect of NOM on the concentrate treatment process. The water sample contained similar concentrations of calcium, magnesium, and carbonate, with larger concentrations of sodium, chloride, and sulfate. A laboratory-scale RO pilot was used to make concentrate; a Koch ultra-low pressure spiral wound RO membrane module was used during experiments (operating pressure = 12 bar). Several antiscalants were tested, including phosphonate and acrylic polymer blend products. Measures for dissolved metal ions, total carbonate, conductivity, total organic carbon and COD were taken. In addition, changes in the precipitate were evaluated; particle size and particle number distributions were obtained, using a laser granulometer Mastersizer S (Malvern Instruments) and a laser particle counter (Met One).

[1] Rahardianto, A.; Gao, J.; Gabelich, C.J.; Williams, M.D.; Cohen, Y., High recovery membrane desalting of low-salinity brackish water: Integration of accelerated precipitation softening with membrane RO. *Journal of Membrane Science* **2007,** 289, 123-137.

[2] Yang, Q.; Ma, Z.; Hasson, D.; Semiat, R., Destruction of Anti-Scalants in RO Concentrates by Electrochemical Oxidation. *Journal of Chemical Industry and Engineering (China)* **2004,** 55(2), 339-340.

5aj Novel Fabrication Methods for the Production of Polymer Films: Initiated Chemical Vapor Deposition and Templated Formation of Ionotropic Gels Using Patterned Paper

Malancha Gupta, Department of Chemistry and Chemical Biology, Harvard University, 12 Oxford Street, Cambridge, MA 02138

This poster will describe two methods for the production of polymer films. My graduate work at MIT focused on the use of initiated chemical vapor deposition (iCVD) to make a wide variety of polymer coatings such as poly(2-(perfluoroalkyl)ethyl methacrylate) (hydrophobic) [1], poly(glycidyl methacrylate) (adhesive) [2], and poly(furfuryl methacrylate) ("click chemistry") [3]. Vapor deposition has the environmental benefit of using no solvents and the process can be used to conformally coat substrates with complex geometries such as fabrics and wires since there are no surface tension problems. Deposition rates as high as 300 nm/min can be achieved. The proposed polymerization mechanism is the classical free radical polymerization mechanism of vinyl monomers. Monomer and initiator gases are fed into a vacuum chamber where resistively heated wires are used to thermally decompose the initiator molecules into free radicals. The free radicals then attack the vinyl bonds of the monomer molecules. Propagation occurs on the surface of a cooled substrate. We have demonstrated that the iCVD process can be used to modify the surfaces of high-aspect-ratio (~100) polymeric membranes [4] and electrospun fiber mats [5, 6].

My postdoctoral work in the Whitesides group at Harvard focuses on the use of paper templates to fabricate shaped films of ionotropic hydrogels [7,8]. Solutions of polymers such as alginic acid, carrageenan, and carboxymethyl cellulose form films with defined shapes when brought into contact with patterned templates of paper wetted with aqueous solutions of multivalent cations. This method allows the production of topographically and topologically complex 3D shapes, such as interlocking rings and Möbius strips. The shaped films can be made magnetically responsive by using paramagnetic ions like holmium as the cross-linking ions or by suspending ferrite microparticles in the hydrogels. Heterogeneous films of ionotropic hydrogels can be fabricated through the use of multiple templates. These heterogeneous structures include single films where a pattern of one hydrogel is surrounded by another hydrogel ("gel-in-gel" structures), multilayered hydrogels, and hydrogels that contain a gradient in the concentration of cross-linking agent.

Relevant Publications:

[1] **Gupta, M.**; Gleason, K. K. "Mechanistic Study of the Initiated Chemical Vapor Deposition (iCVD) of Poly(1H,1H,2H,2H-Perfluorodecyl Acrylate) (PPFDA) Thin Films", *Langmuir*, 22, 10047, 2006.

[2] **Gupta, M.**; Gleason, K. K. "Large Scale Initiated Chemical Vapor Deposition of Poly(glycidyl methacrylate) Thin Films", *Thin Solid Films*, 515, 1579, 2006.

[3] Chen, G.; **Gupta, M.**; Chan, K.; Gleason, K. K., "Initiated Chemical Vapor Deposition of Poly (Furfuryl Methacrylate)", *Macromolecular Rapid Communications*, 28, 2205, 2007.

[4] **Gupta, M.**; Kapur, V.; Pinkerton, N. M.; Gleason, K. K. "Surface Modification of High Aspect Ratio Pores and Surface Features by initiated Chemical Vapor Deposition (iCVD)", *Chemistry of Materials*, 20, 1646–1651, 2008.

[5] Ma, M.; **Gupta, M.**; Li, Z.; Zhai, L.; Gleason, K. K.; Cohen, R. E;. Rubner, M.F.; Rutledge. G. C. "Superhydrophobic fabrics produced by electrospinning and chemical vapor deposition", *Advanced Materials*, 19, 255, 2007.

[6] Ma, M.; Mao, Y.; **Gupta, M.**; Gleason, K. K.; Rutledge. G. C."Superhydrophobic fabrics produced by electrospinning and chemical vapor deposition", *Macromolecules*, 38, 9742, 2005.

[7] Bracher, P. J.; **Gupta, M.**; Whitesides, G. M. "Shaped Films of Ionotropic Hydrogels Fabricated Using Templates of Patterned Paper", *Advanced Materials*, submitted.

[8] **Gupta, M.**; Bracher, P. J.; Whitesides, G. M. "Heterogeneous Films of Ionotropic Hydrogels Fabricated Using Paper Substrates", in preparation.

5ak Colloidal Assembly in Microfluidic Devices

Ramin Haghgooie, Massachusetts General Hospital, 114 16th Street, Room 1239, Charlestown, MA 02129

In recent years, self-assembled systems have garnered much attention due to their potential for creating order on very small scales as well as their obvious analogy to biological systems. When considering these systems, one must be aware of the ever present issue of boundary effects which can strongly influence the behavior and structure of the self-assembly. In many cases, the boundary effects are caused by a confined system in which the assembly is taking place. For instance, colloidal assembly in microfluidic devices has been shown to be a promising route for manufacturing micron, or sub-micron, structured systems. Furthermore, the study of confined self-assembly has revealed much information about factors which influence structure and dynamics of the system. We have focused upon the self-assembly of magnetorheological (MR) fluids in confinement. Self-assembled MR fluids have been used as structural components in microfluidic systems offering a fast, inexpensive alternative to traditional photolithographic techniques. The characteristic length-scales in microfluidic devices have continued to shrink down to the colloidal size, making studies of self-assembly under extreme confinement a timely undertaking. It has become essential to study these systems in order to enable the design of meaningful applications using such technology.

My doctoral thesis focused upon the self-assembly of MR fluids in confined geometries. In order to study this problem, I developed a versatile Brownian dynamics (BD) code capable of simulating MR colloids of any shape interacting in any general geometry. I used the BD code to study the self assembly of spherical colloids confined in two-dimensional (2D) channels and discovered a number of interesting phenomena. In addition to the simulation work, I performed an extensive set of experiments on the self-assembly of MR colloids in 2D microfluidic channels and showed the first experimental observation of re-entrant melting as a function of confining geometry. Aside from the 2D studies, I investigated the factors controlling self-assembly in the thin-slit geometry. I elucidated the important physical phenomena affecting the self-assembly of dilute MR fluids in this common microfluidic geometry. I showed how the system transitions from 2D to 3D behavior as the confinement is relaxed from a monolayer to a channel of finite thickness.

During the course of my Ph.D. I also completed a Masters of Chemical Engineering Practice in which I gained invaluable industrial experience. Since graduating, I took a position as a Lecturer of Chemical and Biological Engineering at Tufts University and spent the 2006-2007 academic year teaching graduate level Transport Phenomena, graduate level Thermodynamics, two undergraduate laboratory courses, and a graduate level course in microfluidic technology. This experience has enabled me to sharpen my teaching skills and given me a great head-start in my development as an academic researcher and teacher. In February of 2007, I was awarded the MIT/MGH postdoctoral fellowship in translational research (2 awards out of more than 50 applicants) and I am currently a postdoctoral fellow in the BioMEMS Resource Center at Massachusetts General Hospital

In my postdoctoral work, I have developed a technique for fabricating non-spherical polymeric microparticles with tunable deformability. I have studied the properties of these particles under confinement and demonstrated that they exhibit bio-mimetic properties similar to red blood cells. I have also explored novel biomedical applications for these particles.

5al Electrodeposited Nanowires for Sensing, Energy Conversion, and Logic Devices

Carlos Hangarter, Chemical and Environmental Engineering, University of California, Riverside, Bourns Hall A242, University of California, Riverside, Riverside, CA 92521

One-dimensional (1-D) nanostructures, such as nanowires (NWs) and nanotubes (NTs), are attractive materials for electronics because of their potential for device miniaturization and unique properties which arise from size dependent quantum confinement effects. These building blocks of nanotechnology can be further complexed with axial and radial interfaces, for segmented, multilayered and core/shell structures. However, to fully exploit these higher level configurations, nanomaterials must be engineered for specific properties and interfacial characteristics that give rise to enhanced functionalities for communications, data storage, logic operations, sensing, and energy conversion. This ardent task requires a fundamental understanding of the synthesis approach on the structure-property relationship as well as size dependent effects. Yet the true values of these physical properties are not always accurately measured at this level, as many can only be assessed by interrogation of individual nanostructures. Additionally, the commercialization of nanowire-based devices has been severely stunted by stunted by controllable integration, including contact and alignment, into existing technologies.

During my time at UCR I have employed various electrochemical techniques to synthesize a barrage of nanostructures for sensing, spintronic and thermoelectric application. The bulk of my work focused on conjugated polymers and the development of hybrid organic/inorganic nanowires. To describe fundamental characteristics of these materials we have developed methodologies for isolating single nanowires and characterizing their true magneto and electron transport properties to distinguish interfacial and size effects. Rational assembly of such nanoentities is however, a continuing challenge in the field. To this end, the inherent magnetic and electric properties of these materials were utilized for aligning single nanowires and electrical contact has also been explored with annealing, lithography, electron beam lithography (e-beam), and focused ion beam (FIB).

5am Nanoengineering Systems for Targeted Drug Delivery, Cell-Based Therapy, and Microfluidic Biosensors/Chips

Hongyan He, The Ohio State University, 140 WEST 19TH AVE, Columbus, OH 43210

Many characteristics of polymers and ease of processing make them attractive for various biomedical applications. Proper combination of functional polymers and a number of manufacturing modules can offer tailored properties for sophisticated technologies used in targeted drug delivery, cell-based therapy, and microfluidic biosensors/chips.

1. Multifunctional delivery systems for oral and implant administrations

Many protein- and DNA-based drugs need to be properly protected and precisely controlled during administration. A robust manufacturing protocol has been developed to produce particulate-like polymeric devices based on functional hydrogels to achieve multi-functionalities, such as drug protection, self-regulated oscillatory release, enhanced mucoadhesion and targeted unidirectional release. The delivery devices will be of great benefit to the advancement of oral administration of existing therapeutic agents. A nanoporous miniature device made of biodegradable poly(lactide-co-glycolide acid) (PLGA) has also been designed based on the integration of a number of micro-manufacturing modules, such as phase inversion, a sacrificial template nano-imprinting, and carbon dioxide (CO_2) -assisted bonding. The biocompatibility, biodegradation, and release performance of this device have been evaluated in animal studies. Such devices are able to enhance the bioavailability of existing therapeutic and provide a continuous, pre-designed release profile within the local tumor environment, potentially without the need for surgical implantation.

2. Cell-based delivery systems by micro-/nanofabrication

Cell-based delivery devices holds great promise for applications requiring site-specific and sustainable drug delivery of cell-synthesized molecules. On the other hand, embryonic stem (ES) cell therapies have been proposed for regenerative medicine and tissue replacement after injury or disease. An integration of ES cells within immunoprotective and biodegradable devices will provide great potential for novel delivery platforms. Our recent work focus on the development of nanoporous PCL-based chambers for immunoisolation of embryonic stem cells. A series of advanced techniques such as composite membrane preparation by phase inversion, surface coating by plasma polymerization, and nitrogen (N_2) -assisted bonding, have been combined to improve the device functionalities. Studies described herein also include the interactions of PCL-based substrates with differentiated ES cells in terms of viability, proliferation, and functionality.

3. Design and evaluation of microfludic biochips for Enzyme-Linked Immunosorbent Assay (ELISA)

A polymeric microfluidic biochip has been designed and fabricated for enzyme-linked immunosorbent assays based on the integration of manufacturing modules: polyaniline-based surface modification, the conceptual design with splitters, and the oxygen plasma-PEI-TR protein A modification. The proper flow sequencing was achieved on a CD-like microfluidic chip by integrating the necessary microfluidic functions such as capillary valving treated with polyaniline-based surface modification. It has been successfully demonstrated that the splitters indeed work very well to even distribution of the fluid to branch flows. To enhance the sensitivity of miniature detection area on the PMMA biochips for ELISA immunoassay, a TR-catalyzed protein A antibody immobilization technique was successfully developed. By treating the PMMA surface with this method, not only the antibody binding efficiency, but also the specific capture capacity towards target protein of the antibody was greatly enhanced, leading to a significant improvement of the fluorescence signal ELISA in PMMA chips. The microfluidic biochips we developed here would be applicable to a variety of clinically relevant disease conditions. And the modification technologies in this study can be extensively implemented in lab-on-a-chip systems, drug/gene delivery carriers and other immunoassay biosensor applications.

5an Understanding the Catalytic Activity of Bi- and Monometallic Pd and Au Catalysts for Aqueous Reactions

Kimberly N. Heck, Chemical and Biomolecular Engineering, Rice University, 6100 S. Main St., MS 362, Houston, TX 77005

Understanding of the interactions between a catalytic surface with reactants and intermediates under reaction conditions can be used to tailor catalytic surfaces for increased activity or selectivity. Moreover, the feasibility of a catalyst for commercial use depends partly on the activity of the catalyst in the presence of possible deactivants in the reactant stream.

However, in situ spectroscopic characterization of heterogeneous catalysts in aqueous systems has long been a challenge. While methods such as electron energy loss spectroscopy have been used successfully to monitor to the adsorption of compounds to model catalyst sites, they rely on ultra high vacuum environments. While FTIR coupled with ATR techniques can offer more realistic conditions, the penetration depth of such spectroscopy is on the bulk level, and signals produced suffer from the strong absorption of the aqueous environment. Surface-enhanced Raman spectroscopy (SERS) using gold nanoshells, on the other hand, could provide the requisite sensitivity for analyzing aqueous-phase catalyzed reactions. Au nanoshells (NS's) are a class of SERS substrates, whose surface plasmon resonance (SPR) can easily be tuned by modifying core particle size and shell thickness, resulting in signals up to 10^{14} times greater than normal Raman alone.

To determine the applicability of SERS using Au NS's for aqueous systems, we examined two systems. Recent efforts have shown that glycerol, a major byproduct of biofuels production, can be converted over Au catalysts in alkaline media with great selectivity to fine chemicals. Additionally, previous studies show that Pd-on-Au (Pd/Au) nanoparticle (NP) catalysts have been shown to be 70x more effective than conventional Pd catalysts in the hydrodechlorination of chlorinated ethylenes in groundwater. While mechanisms exist for these two reactions, they there is little spectroscopic proof.

In-situ SERS monitoring of aqueous glycerol oxidation over Au NS's

In this study, we report the in-situ time-resolved identification of adsorbates and reaction intermediates in the basic aqueous-phase oxidation reaction of glycerol via surface-enhanced Raman spectroscopy (SERS) over a Au nanoshell (NS) catalyst. Coupled with batch experiments using a Au/C catalyst, we gain important insights into the effect of pH and O_2 on the surface reaction.

In-situ SERS monitoring of aqueous HDC over Pd-on-Au NS's

We also report the synthesis of Pd-decorated Au NS's to mimic the HDC catalytic behavior of Pd/Au NPs during SERS analysis. SERS spectral analysis indicates that chemisorption and HDC surface reaction events of dichloroethylene (DCE) probe molecule can be observed. It appears that 1,1-DCE adsorbs to the Pd catalyst via pi-bonding with Pd atoms on the surface, for example. The results of this study highlight the promising use of SERS spectroscopy of metal-catalyzed reactions in water in situ.

Kinetic deactivation study of Pd/Au NPs with chloride and sulfide

Finally, to test the feasibility of Pd/Au NPs for use in the field, we examined the effect of two common Pd deactivants found in groundwater, chloride and sulfide. Results show that the catalyst is completely resistant to chloride, while monometallic Pd NPs and Pd/Al_2O_3 deactivate to 30% its initial rate at the highest chloride concentration tested. Additionally, the Pd/Au catalyst show improved resistance to deactivation of sulfide. While the Pd catalysts deactivated at the theoretical limit of $S:Pd_{surf}$ of 0.5, the Pd/Au NPs showed resistances of up to $S:Pd_{surf}$ of 1.5. The resistance is found to be compositionally dependent, with higher Pd wt% catalysts deactivating more rapidly. Additionally, the shift in the deactivation profile from a linear decrease in rate with increasing $S:Pd_{surf}$ for low Pd wt% catalysts to a nonlinear profile with higher Pd wt% suggests the formation of different active sites made of larger Pd ensembles for the higher wt% catalyst.

5ao Relating Structure, Function, and Mechanics in Surfactant-Colloid Mixtures

Matthew H. Helgeson, University of Delaware, Newark, DE 19711, Norman J. Wagner, Department of Chemical Engineering, University of Delaware, 150 Academy St., Newark, DE 19716, and Eric Kaler, Stony Brook University, Stony Brook, NY 11794.

Self-assembled surfactant micelles are ubiquitous to a wide variety of products and industrial processes, ranging from consumer products to enhanced oil recovery, where phase behavior and rheology are critically linked to material performance. These applications require a particularly detailed understanding of the hierarchical nature of micellar self-assembly, whereby molecular interactions, mesoscale structure, and macroscopic properties are intimately connected. Recent interest in novel surfactant fluids involves the incorporation of colloidal particles, specifically on the nanoscale, into self-assembled surfactant solutions in order to impart additional structure and functionality to materials of interest. Despite the demand for such formulations, the affect of colloid addition on the structure and properties of micellar solutions (and vice versa) is poorly understood. The goal of this work is to understand the structure and rheology of model mixtures of nanoparticles and wormlike micelles (WLMs), in order to better engineer fluids for targeted applications. The work shown here focuses on two aspects of this research that combine methods that probe length scales ranging from the particle surface to macroscopic micellar networks to develop a detailed knowledge of novel colloid-surfactant mixtures.

Modification of viscoelastic wormlike micellar solutions via nanoparticle incorporation

This work demonstrates that the addition of nanoparticles to like-charged WLM solutions allows for unique tunability of the structure and rheology of such fluids. Specifically, nanoparticle addition enables the formation of entangled micellar solutions under normally dilute conditions, and enhances the existing viscoelasticity of semi-dilute solutions. Studies of the structure by scattering methods and energetics using isothermal titration calorimetry (ITC) at the surfactant-nanoparticle interface show that this rheological modification results from association of micelles with an adsorbed layer at the particle surface. This results in the formation of micelle-nanoparticle junctions which impart additional network formation to the micellar fluid. Additionally, these network junctions restrict the local orientational mobility of WLMs, and can be used as a method to suppress flow instabilities such as shear banding in concentrated solutions. Studying these phenomenon for a variety of particle sizes and surface chemistries leads to an engineering understanding of the structuring of WLMs by nanoparticle addition.

Interactions and phase behavior of colloids dispersed in wormlike micellar media

The unique structuring of surfactants at the particle interface leads to modified colloidal interactions between nanoparticles suspended in WLM solutions. Specifically, the presence of WLMs leads to aggregation and thermoreversible phase separation of nanoparticles at sufficient concentrations. We have successfully modeled these WLM-mediated colloidal interactions using a simple statistical mechanical description of reversibly breaking, end-adsorbing micellar chains. The resulting interaction potential contains model parameters that can be experimentally determined, and can be used to simultaneously describe the observed phase behavior as well as structural measurements using small angle neutron scattering. The results suggest that the model is a more appropriate description than previously used potentials, and describes the complex colloidal interactions observed in WLM-colloid mixtures.

5ap Reconstruction of a New Genome-Scale Model of B. Subtilis

Christopher Henry[1], Jenifer Zinner[1], Matthew Cohoon[1], Edmond Jolivet[2], Kosei Tanaka[2], Sabine Autret[2], Philippe Noirot[2], and Rick Stevens[1]. (1) Mathematics and Computer Science, Argonne National Laboratory, 9700 S. Cass Avenue, Argonne, IL 60439, (2) Microbial Genomics, INRA, 147 rue de l'Université, Paris, France

Bacillus subtilis is a gram positive bacteria often utilized in industry as a producer of high quality enzymes and proteins [1]. One of the primary challenges involved in the use of *B. subtilis* in industry is the extensive regulatory pathways in the cell, making the flux through the metabolic reactions of the cell extremely resistant to alteration by genetic manipulation [2]. It has already been demonstrated that removing a portion of the regulatory genes in *B. subtilis* results in significantly enhanced protein production by the cell [3]. Now, we are endeavoring to produce a minimal strain of the *B. subtilis* genome. This minimal strain will lack every dispensable alternative metabolic pathway and every dispensable regulatory gene, making the strain much more amenable to alteration for industrial use. *B. subtilis* was selected as the platform organism for the construction of this minimal strain because the natural competence of *B. subtilis* allows for rapid knockouts of single genes and intervals of genes [4]. Additionally, the extensive information available about *B. subtilis* will allow for a systematic and planned approach to be used during the construction of the minimal strain. To facilitate the development of our minimal organism, we have constructed a new genome-scale metabolic model of *B. subtilis* based on the annotations available in the SEED subsystems-based annotation environment [5] and supplemented by data included in two previously developed *B. subtilis* models [6, 7]. The new model includes elements of the *B. subtilis* regulation when necessary to properly predict the effect of gene knockouts. The thermodynamic properties of the model reactions were estimated to predict the reversibility of model reactions [8]. The new model also includes numerous genes and pathways not found in any previous models. The new model was validated against a variety of experimental observations including biolog phenotyping array results [7], gene essentiality data [9], 60 published gene interval knockout experiments [3], and over 300 new gene-interval knockout experiments. When model predictions did not match experimental observations, a variety of methods were developed and applied to improve the model accuracy. As a result of these efforts and the inclusion of features from currently published models, this is the most complete and accurate metabolic model of *B. subtilis* constructed to date. In this poster, we will present the methods used to assemble and correct this new genome-scale model; we will compare the accuracy and content of the new model with previously published models; and we will describe the application of the model to the design of a systemic and optimized strategy for combining gene interval knockouts to produce a minimal strain of *B. subtilis*.

1. Zweers, J.C., et al., Towards the development of *Bacillus subtilis* as a cell factory for membrane proteins and protein complexes. Microb Cell Fact, 2008. 7: p. 10.

2. Fischer, E. and U. Sauer, Large-scale in vivo flux analysis shows rigidity and suboptimal performance of *Bacillus subtilis* metabolism. Nat Genet, 2005. 37(6): p. 636-40.

3. Morimoto, T., et al., Enhanced Recombinant Protein Productivity by Genome Reduction in *Bacillus subtilis*. DNA Res, 2008. 15(2): p. 73-81.

4. Fabret, C., S.D. Ehrlich, and P. Noirot, A new mutation delivery system for genome-scale approaches in *Bacillus subtilis*. Mol Microbiol, 2002. 46(1): p. 25-36.

5. Aziz, R.K., et al., The RAST Server: rapid annotations using subsystems technology. BMC Genomics, 2008. 9: p. 75.

6. Goelzer, A., et al., Reconstruction and analysis of the genetic and metabolic regulatory networks of the central metabolism of *Bacillus subtilis*. BMC Syst Biol, 2008. 2: p. 20.

7. Oh, Y.K., et al., Genome-scale reconstruction of metabolic network in *Bacillus subtilis* based on high-throughput phenotyping and gene essentiality data. J Biol Chem, 2007. 282(39): p. 28791-9.

8. Jankowski, M.D., et al., Group contribution method for thermodynamic analysis of complex metabolic networks. Biophysical Journal, 2008. Accepted.

9. Kobayashi, K., et al., Essential *Bacillus subtilis* genes. Proc Natl Acad Sci U S A, 2003. 100(8): p. 4678-83.

5aq Biological Production of Renewable Energies from Waste Materials

Sage R. Hiibel, Chemical and Biological Engineering, Colorado State University, 100 Glover Building, 1370 Campus Delivery, Fort Collins, CO 80523-1370

In my research, I seek to apply fundamental engineering concepts and advanced biological processes to help solve meaningful societal problems. One significant challenge the world faces is the need for clean, renewable energy sources. My research will focus on the development and optimization of biological processes for the production of affordable, environmentally-friendly renewable energy. My goal is to consistently complete hypothesis-driven research based on scientifically sound experimental design and implementation while maintaining the integrity of the engineering profession.

5ar Dynamics of Complex Fluids and Complex Flows near Surfaces

Nazish Hoda, Department of Chemical Engineering and Materials Science, University of Minnesota, 151 Amundson Hall, 421 Washington Ave. SE, Minneapolis, MN 55455

Complex fluids and complex flows near surfaces are encountered in biological systems and in numerous technological applications, such as microfluidics and coatings. Understanding the dynamics of such fluids and flows is of both technological and fundamental importance. In my PhD dissertation, under the supervision of Prof. Satish Kumar, we investigated three different problems involving complex fluids and complex flows.

a) The dynamics of polyelectrolytes near complex surfaces is important for the development of polyelectrolyte multilayers, the understanding of protein complexation, and microfluidics. In the first problem, we probe the effects of external fields, such as electrostatic and hydrodynamic, on the static and dynamic properties of polyelectrolyte molecules near complex surfaces [1-7]. We found that surface heterogeneities not only affect the condition required for adsorption but also the conformation of adsorbed chains. In the presence of an imposed flow, polyelectrolyte adsorption is determined by the competition between electrostatic and hydrodynamic forces. We propose a desorption criterion and put forward numerous strategies for controlling the adsorption/desorption of polyelectrolyte molecules. The diffusion dynamics of an adsorbed polymer chain was also probed, and we found that it is sensitive to chain flexibility and solvent quality.

b) Linear dynamical systems can have solutions that grow substantially at short times, even though they decay at long times. One method to study this so-called transient growth, which can put the system into a regime where nonlinear interactions are no longer negligible, is to study the response of the linearized Navier-Stokes equations to external disturbances. In the second problem, we apply this method to study the effects of external disturbances on channel flows of Oldroyd-B fluids [8-9]. We found that in addition to inertia, even elasticity can amplify disturbances. Our analysis of the Reynolds-Orr equation demonstrates that the energy-exchange term involving the streamwise/wall-normal polymer stress component and the wall-normal gradient of the streamwise velocity is the main driving force for amplification in flows with strong viscoelastic effects. We also found that viscoelastic effects introduce additional timescales and promote development of flow patterns with smaller time constants than in Newtonian fluids.

c) In gravure printing and coating processes, the print quality as well as coating thickness is affected by the amount of liquid that gets transferred from the cells, engraved on the surface of the gravure roll, to the substrate. In the last project, we aim to better understand the cell-emptying process, and to suggest ways of increasing liquid removal from the cell [10]. We found that the liquid removal from the cell can be significantly increased either by increasing the ratio of cell width to cell height, or by increasing the vertical velocity of the substrate.

References

1. Nazish Hoda and Satish Kumar, "Brownian dynamics simulations of polyelectrolyte adsorption onto charged patterned surfaces", Langmuir, 23, 1741 (2007)

2. Nazish Hoda and Satish Kumar, "Kinetic theory of polyelectrolyte adsorption in shear flow", J. Rheol., 51, 799 (2007).

3. Nazish Hoda and Satish Kumar, "Brownian dynamic simulations of polyelectrolyte adsorption onto topographically patterned surfaces", Langmuir, 23, 11747 (2007)

4. Nazish Hoda and Satish Kumar, "Brownian dynamics simulations of polyelectrolyte adsorption in shear flow with hydrodynamic interactions", J. Chem. Phys., 127, 234902 (2007)

5. Nazish Hoda and Satish Kumar, "Theory of Polyelectrolyte adsorption onto surfaces patterned with charge and topography", J. Chem. Phys., 128, 124907 (2008)

6. Nazish Hoda and Satish Kumar, "Brownian dynamics simulations of polyelectrolyte adsorption in shear flow: Effect of hydrophobicity and charge patterning", J. Chem. Phys., 128, 164907 (2008)

7. Nazish Hoda and Satish Kumar, "Parameters influencing diffusion dynamics of an adsorbed chain", Phys. Rev. Lett. (submitted)

8. Nazish Hoda, Mihailo R. Jovanovic, and Satish Kumar, "Energy amplification in channel flows of viscoelastic fluids", J. Fluid Mech., 601, 407-424 (2008)

9. Nazish Hoda, Mihailo R. Jovanovic, and Satish Kumar, "Frequency Responses of the 2D/3C Model in Channel Flows of Oldroyd-B Fluids", J. Fluid Mech. (submitted)

10. Nazish Hoda and Satish Kumar, "Boundary integral simulations of gravure cell emptying", Phys. Fluids (submitted)

5as Fundamental Interactions at Biomolecular Interfaces: An Integrated Simulation and Experimental Approach

Jason C. Hower, University of Washington, Chemical Engineering, Seattle, WA 98195 and Shaoyi Jiang, Chemical Engineering, University of Washington, Benson Hall, Box 351750, Seattle, WA 98195.

The success of biomaterials is controlled by the interface between the foreign material and the tissue or contacting body fluid. Therefore, it is vital to control the properties of the material and the interactions occurring at this complex interface. However, the biointerface can be difficult to study as the surface of interest is buried in a solution of potentially unstable and variable chemistry that impacts the nature of the interactions possible. My research focuses on studying the physical and thermodynamic mechanisms of interactions at the biointerface using an integrated approach of molecular simulations and fundamental experimental techniques.

Motivated primarily by the need for robust and stable surface coatings that prevent non-specific protein adsorption and cell adhesion for the marine, sensing, and biomedical industries, my research has linked interfacial properties like surface hydration structure and dynamics to repulsive forces generated on probe proteins. My research has also investigated calculating the change in free energy during protein adsorption from simulations as a predictor of surface nonfouling ability. Simulations predict a strong correlation between surface hydration and nonfouling ability. Therefore, the intrinsic hydration capacity of candidate nonfouling materials was evaluated by measuring the partial molal volume at infinite dilution. When normalized by solute size, the measured hydration capacity correlates directly with protein adsorption data to surfaces presenting similar chemical groups. This strong relationship suggests that simple molal volume measurements could be used to screen candidate nonfouling materials.

My future research goals are to apply the integrated simulation and experimental approach to elucidate molecular-level mechanism of biological interfacial phenomena. Significant amounts of biological interfacial research focuses on characterization and description and offers little insight into the fundamental mechanisms. However, by pairing molecular simulations and basic experimentation techniques, the mechanisms of interactions can be discovered which would lead to rational material design and property-targeted progress.

5at Plastic Solar Cells from Nano-Patterned Fullerenes and Semi-Conducting Polymers

J. J. Huang, Plastics Engineering and NSF Nanoscale Science and Engineering Center for High-rate Nanomanufacturing, University of Massachusetts, One University Avenue, Lowell, MA 01854

Bulk heterojunction plastic solar cells, as a type of organic photovoltaics, have attracted great attention in developing inexpensive renewable energy sources due to their low cost and ease of manufacturing. Recent work in the field has focused on using systems based on fullerene with its derivatives and poly(3-hexylthiophene)(P3HT) because of relatively higher conversion efficiency. By utilizing different methodology from prevailing approaches, e.g., spin coating, we recently fabricated plastics solar cells by nano-patterning fullerence/P3HT via dip-pen nanolithography. Our goal is to obtain a structure that leads to high mobility charge carrier transport with reduced recombination. Here, we would show some of our preliminary results.

5au Phase Behavior of Polymer Tethered Nanoparticles

Christopher R. Iacovella, Chemical Engineering, University of Michigan, 2350 Hayward Street, 3440 G.G. Brown, Ann Arbor, MI 48109 and Sharon C. Glotzer, Chemical Engineering and Materials Science and Engineering, University of Michigan, 2300 Hayward Street, Ann Arbor, MI 48109.

Block copolymers and surfactants have long been known to self-assemble into a wide variety of complex microstructures where the assembly is driven by immiscibility between chemically distinct blocks in the polymers and between the head and tail groups in surfactants. These ordered microstructures are highly sought for applications at the nanoscale, ranging from photonic-bandgap materials to templates for nanoparticle assembly. Hybrid building blocks have recently been created that resemble block copolymers where the individual blocks consist of nanoparticles and polymers; these hybrid building blocks, or tethered nanoparticles, constitute a class of "shape amphiphiles" where microphase separation occurs due to the immiscibility between the nanoparticle and polymeric tether resulting in mesostructured equilibrium phases that resemble the morphologies of block copolymers and surfactants.

Specifically, we have simulated the phase behavior of nanospheres each functionalized with a single polymer tether -- analogous to a diblock copolymer or surfactant -- finding hexagonally packed cylinders, the double gyroid, perforated lamella, and lamella [1,2], as expected [3]. However, due to the geometry of the nanoparticles, these tethered systems are able to adopt complex local order not normally found in block copolymer and surfactant systems. For example, in our simulation studies of monotethered nanospheres, we find distinct icosahedral and HCP ordering between nanospheres [2]. This local ordering has additional consequences. For example, when considering the bicontinuous, double gyroid structure, we find that the icosahedral ordering between nanospheres reduces the packing frustration at the nodes, potentially leading to a greater range of stability as compared to a flexible block copolymer system [2,4]. Similarily, in our simulations of nanorods functionalized by a single tether, we find that these tethered nanorods also order into the double gyroid structure where packing frustration is also reduced at the nodes, here as a result of the nanorods adopting liquid-crystalline, splayed-hexagonal bundles [4].

The similarities between tethered nanospheres and block copolymers also extends into the area of triblock copolymer systems. Linear and star triblock copolymers are known to self assemble into an even wider array of complex morphologies than typical diblocks. We have investigated the nanoparticle analog of a triblock copolymer by replacing the middle block of an ABC triblock with a nanosphere. We investigated the phase behavior under melt conditions as a function of both nanoparticle diameter and dihedral angle, the angle separating the two tethers. We found a variety of phases that resemble those of linear and star triblock copolymers, including the alternating diamond, alternating gyroid, and alternating orthorhombic tricontinuous lattices, tetragonal and [8,8,4] columnar phases, and a unique phase of NaCl ordered spherical micelles not seen for triblocks [5]. The ability to adjust the geometry of the tethers allows for additional control over the ordering and placement of the nanospheres within the system. Additionally, we have explored the phase behavior (i) as a function of block fraction under melt conditions and show evidence of a transition from tricontinuous to lamellar structures, and (ii) as a function of volume fraction under selective solvent conditions for which we find the double gyroid, hexagonally packed cylinder, and BCC ordered micelle phases [5]. Overall, we find we have a wide array of control over the placement and organization of nanoparticles in both mono- and ditethered systems.

Building from our research results, we created The Soft Matter Wiki, an community driven online encyclopedia of soft matter and simulation terminology, with the help of the Materials Digital Library Pathway (MATDL) [6]. This has been used as an educational resource within numerous classes within the University of Michigan. In related work, we have created interactive simulation modules based on our research codes for use in both the undergraduate and graduate curriculum and distributed these through the Soft Matter Wiki.

[1] C.R. Iacovella, M.A. Horsch, Z-L Zhang, S.C. Glotzer. "Phase diagrams of self-assembled mono-tethered nanospheres from molecular simulation and comparison to surfactants" Langmuir, 21 (21), 9488-9494, (2005)

[2] C.R. Iacovella, A.S. Keys, M.A. Horsch, S.C. Glotzer "Icosahedral packing of polymer-tethered nanospheres and stabilization of the gyroid phase" Phys. Rev. E 75, 040801(R) (2007)

[3] M.A. Horsch, Z-L. Zhang, C.R. Iacovella, S.C. Glotzer. "Hydrodynamics and microphase ordering in blockcopolymers: Are hydrodynamics required for ordered phases with periodicity in more than one dimension?", Journal of Chemical Physics, 121(22): 11455-11462, (2004)

[4] C.R. Iacovella, M.A. Horsch, S.C. Glotzer, "Local Ordering Of Tethered Nanospheres And Nanorods And The Stabilization Of The Gyroid Phase", preprint.

[5] C.R. Iacovella, and S.C. Glotzer, preprint.

[6] L.M. Bartolo, C.S. Lowe, S.C. Glotzer, C.R. Iacovella •
"Development of a wiki-based, expert community-driven nanosystem vocabulary."•
Proceedings of the International Conference on Dublin Core and Metadata Applications 211-214, (2006)

5av Molecular Simulation and Computational Design of Materials for Energy and Environmental Applications

Ahmed E. Ismail, Performance Assessment and Decisions Analysis Department, Sandia National Laboratories, 4100 National Parks Hwy, Carlsbad, NM 88220

The rapid increase in computational power obtained in recent years now allows for the design and targeting of materials to specific applications. Two major application areas that will be of increasing importance in the coming decades will be energy and the environment. In particular, there will be a need to study the transport and storage of advanced fuels for nuclear and consumer applications, as well as the design of materials useful for environmental remediation of targeted contaminants. Experience with both Monte Carlo and molecular dynamics simulations of polymers and polymer-water interfaces [1-5], as well as algorithm development and implementation [6-8], provides us with the tools and experience necessary to branch into these application areas and the flexibility to extend into new fields as they arise.

References

[1] **A. E. Ismail**, G. C. Rutledge, and G. Stephanopoulos. Topological coarse-graining of polymer chains using Wavelet-Accelerated Monte Carlo. I. Freely-jointed chains. *J. Chem. Phys.*, **122**, 234901 (2005).

[2] **A. E. Ismail**, G. C. Rutledge, and G. Stephanopoulos. Topological coarse-graining of polymer chains using Wavelet-Accelerated Monte Carlo. II. Self-avoiding chains. *J. Chem. Phys.*, **122**, 234902 (2005).

[3] **A. E. Ismail**, G. S. Grest, and M. J. Stevens. Structure and dynamics of water near the interface with oligo(ethylene oxide) self-assembled monolayers. *Langmuir*, **23**, 8508 (2007).

[4] **A. E. Ismail**, M. Tsige, P.J. in 't Veld and G.S. Grest. Surface tension of normal and branched alkanes. *Mol. Phys.*, **105**, 3155 (2007).

[5] **A. E. Ismail**, G. S. Grest, and M. J. Stevens. Capillary waves at the liquid-vapor interface and the surface tension of water. *J. Chem. Phys.*, **125**, 014702 (2006).

[6] P. J. in 't Veld, **A.E. Ismail**, and G. S. Grest. Application of Ewald summations to long-range dispersion forces. *J. Chem. Phys.* **127**, 144711 (2007).

[7] **A. E. Ismail**, G. C. Rutledge, and G. Stephanopoulos. Multiresolution analysis in statistical mechanics. I. Using wavelets to calculate thermodynamic properties. J. Chem. Phys., **118**, 4414 (2003).

[8] **A. E. Ismail**, G. Stephanopoulos, and G. C. Rutledge. Multiresolution analysis in statistical mechanics. II. The wavelet transform as a basis for Monte Carlo simulations on lattices. J. Chem. Phys., **118**, 4424 (2003).

5aw Engineering Cells and Tissues for Therapeutic Applications Via Simulation and Experiment

Ehsan Jabbarzadeh, Chemical Engineering, California Institute of Technology, 1200 E. California Blvd., Pasadena, CA 91125

I will present an overview of my PhD and postdoctoral research, which were aimed at (1) engineering an artificial bone tissue and (2) gaining a fundamental understanding of the underlying mechanisms by which cells sense microenvironmental cues. Specifically, the first focus of this poster is the fabrication of polymeric matrices for bone tissue engineering applications. This part also focuses on the applications of stem cells and drug delivery approaches to control blood vessel formation in engineered matrices. The second focus of this poster is the development of a probabilistic mathematical model for the simulation of blood vessel formation in biomaterials. The third focus of this poster is the design of instructive extracellular matrix (ECM) substrates for programming cell behavior.

5ax Creation of Chemical Patterns for Directed Assembly of Block Copolymers

Shengxiang Ji, Department of Chemical and Biological Engineering, University of Wisconsin-Madison, 1415 Engineering Drive, Madison, WI 53705 and Paul Nealey, Department of Chemical Engineering, University of Wisconsin-Madison, 1415 Engineering Drive, Madison, WI 53705.

In the photolithographic process, information in the form of exposed and unexposed material is transferred to a sacrificial photoresist film in the exposure tool. Through a series of processing steps patterned features are ultimately created from information that defines components of devices. We are investigating the insertion of block copolymer lithography into the current lithographic process for improved information transfer, process control, and resolution. My researches focus on developing methods for creating neutral and chemically nanopatterned substrates for block copolymer lithography. In this presentation I will describe a new strategy for creating and replicating nanoscale chemical surface patterns, which will eliminate the necessity of lithographic patterning at or near the dimension of the block copolymer material for each chip and provide a path for improved or sub-lithographic resolution, and is suitable for high-volume manufacturing.

5ay 3-Dimensional in Vitro Model of Hepatic Tissue for Investigating Liver Physiology and Pathophysiology

Rohit Jindal, Center for Engineering in Medicine, Massachusetts General Hospital, Harvard Medical School, 51 Blossom Street, Boston, MA 02114

In vivo, cells are organized in complex tissue architecture with various extracellular matrix (ECM) and cellular signals governing cell behavior. I am interested in developing in vitro models of tissue that mimic this 3 dimensional arrangement of ECM and different cell types. These models offer the potential of capturing the complex signaling environment found in vivo and thus provide more physiologically relevant systems for investigating tissue physiology and pathophysiology. Currently, I am working on the development of coculture model of hepatic tissue that mimics the sinusoid structure found in the liver. This in vitro model is being used for evaluating the role of cell-cell signaling and cell-ECM interactions in a variety of physiologically relevant settings including maintenance of hepatocytes differentiated function, ischemia- reperfusion injury, and inflammation.

In this presentation, I will also briefly describe my doctoral work that included patterning of polymers in the microfluidic channels; and on chip electrochromatography using sol-gel immobilized stationary phase particles.

5az Controlling Nanoparticle Location in Block Copolymers Using External Fields: Simulations and Experiments

Vibha Kalra, School of Chemical and Biomolecular Engineering, Cornell University, 120 Olin Hall, Ithaca, NY 14853

Spatially patterning nanoparticles using block copolymer self assembly has been demonstrated as an effective way to achieve synergistic effects of novel properties such as optical, magnetic and mechanical. My research goals are to combine the functionalities of block copolymer nanocomposites with the advantages of nanofiber mats (e. g. high surface area, open pore structure, ability to be woven into a fabric) to greatly widen the scope of potential applications.

My Ph.D. work with Prof. Yong Joo at Cornell University focused on integration of nanomaterial fabrication, nano-characterization and molecular simulations to develop self assembled nanocomposite nanofiber-based materials. These nanofibers with nanostructures hold great technological promise for applications in separations, catalysis, sensing and photonic smart fabrics, to name a few. In this work, I designed the first ever tailored nanofiber mats with internal confined assembly used as template to disperse functional nanoparticles.

Self Assembly in Block Copolymer Nanofibers [1,2,6]

Nanofiber mats with desired microphase-separated poly(styrene-*block*-isoprene) (PS-*b*-PI) structures have been fabricated. The fibers are obtained using coaxial electrospinning with self-assembling block copolymer (BCP) as the core and silica (synthesized using sol gel technique) as the shell. Silica shell provides thermal stability to the fibers and helps to anneal the materials at temperatures higher than polymer glass transition to obtain equilibrium confined assembly without destroying the fiber morphology. It also serves as a process aid for polymers that are difficult or impossible to electropsin. The shell is later etched using HF solution to obtain continuous nanofibers (100-400 nm diameters) with periodic internal structures. A systematic study was conducted on the effect of cylindrical confinement on internal self assembly for various block volume ratios with comparisons to the corresponding bulk phase diagram (see figure 1 for selected results). All materials were characterized using scanning electron microscopy (SEM), small angle x-ray scattering (SAXS) and transmission electron microscopy (TEM). To view the internal structures, the fibers were microtomed into 60 nm thick sections both along the fiber axis and perpendicular to the axis and then analyzed under TEM. These materials serve as the first step towards controlling the location of nanoparticles in electrospun fibers for many potential applications. For example, alternate concentric layers, formed by a symmetric diblock copolymer under cylindrical confinement, can create photonic band gaps at certain wavelengths when dielectric contrast between layers is enhanced by addition of high refractive index nanoparticles in one block (figure1). Such materials can be woven into fabrics to produce high performance radiation shields and filters.

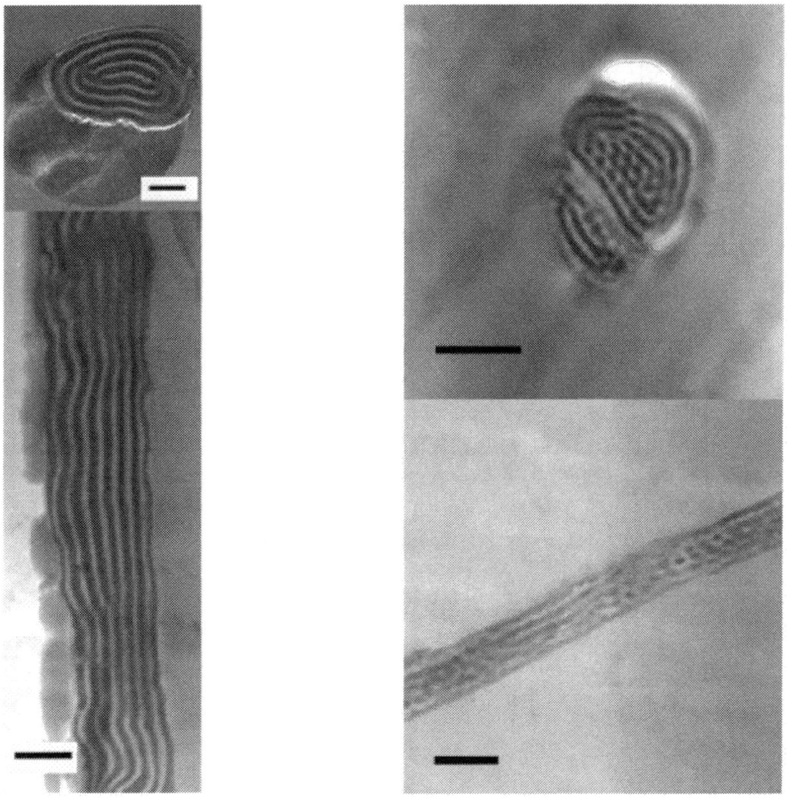

(Left) Alternate concentric layers of styrene and isoprene and (Right) mixtures of hexagonal cylinders and alternate concentric layers are formed due to the effect of cylindrical confinement.

Figure1. . TEM images of cylindrically confined self sssembly in PS-*b*-PI nanofibers with styrene:isoprene volume ratio of (left) 50:50 (right) 69:31. Dark regions are isoprene and light regions are styrene. Scale bars are 200 nm. Click image to see full view

Patterning Nanoparticles using Cylindrically Confined Self-assembly in Nanofibers [5,6]

To demonstrate an example of the technological applications of such self assembled materials and understand and utilize the effects of processing conditions during electrospinning on nanoparticle (NP) location, I have incorporated magnetite NPs into the PS-*b*-PI core material. Monodisperse magnetite NPs (4 nm) were synthesized and surface coated with oleic acid to provide slight selectivity towards isoprene domain while preserving their magnetic moment. Magnetite nanoparticles have been shown to display superparamagnetic behavior due to their small size making them ideal for magnetic field driven transport of drugs and bioseparations. However, controlling the location of NPs and prevention of aggregates has been the biggest challenge in fabrication of nanocomposites where inter-particle interactions (magnetic in this work) are dominant.

In this work, I showed that strong deformation (~10000 s^{-1}) and fast solvent evaporation (200 nl/s) during electrospinning enables selective dispersion of magnetite NPs in the isoprene domain for as high as 20wt% NPs with respect to isoprene. The effect of different nanoparticle fractions on nanocomposite self assembly was studied (fig. 2). No aggregates or loss of domain selectivity was found in spite of strong magnetic interactions between nanoparticles. Magnetic properties were measured using superconducting quantum interference device (SQUID) magnetometer and all fiber samples exhibit superparamagnetic behavior (fig. 3).

Good control over location of magnetic nanoparticles at such high loading, opens up a plethora of possible future applications including electromagnetic shielding, high density memory, magnetic recording, drug delivery and separations. With the freedom to vary nanoparticle surface coating, BCP self assembly, processing conditions (e.g. deformation rate) and nanoparticle size, among other parameters, these results demonstrate the possibility of developing an entire spectrum of multifunctional materials with precisely controlled properties. My work provides a robust methodology to fabricate high surface area multifunctional nanocomposites including cases where particle-particle interactions are dominant.

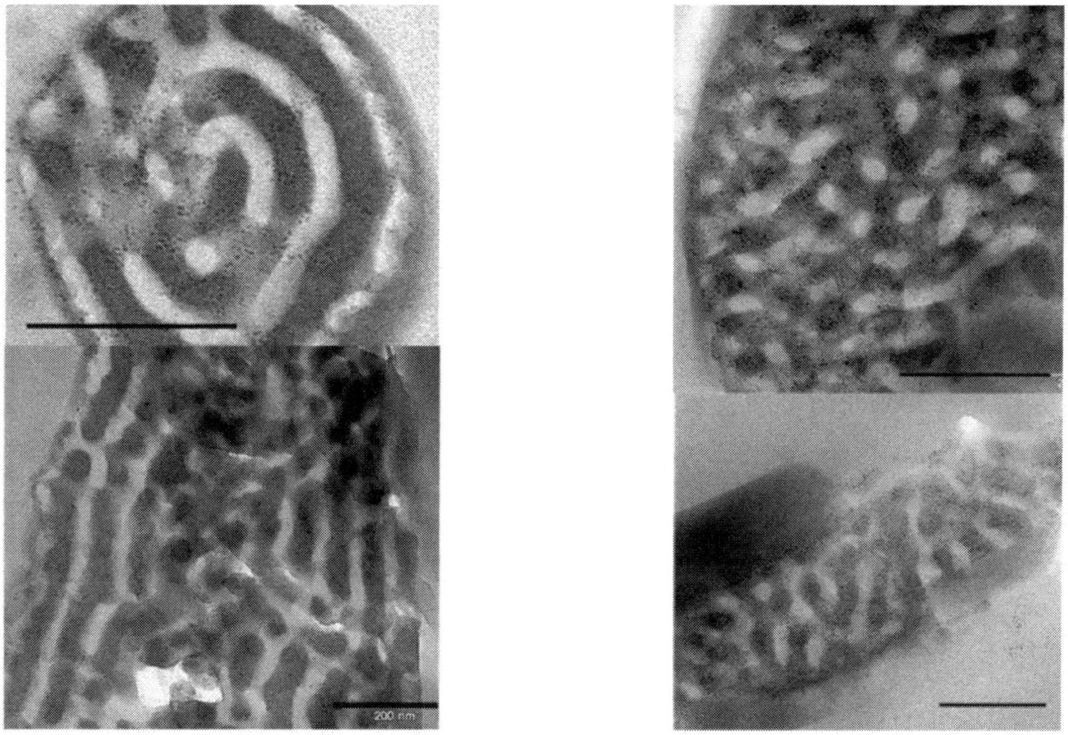

Figure 2. TEM images of sections of nanocomposite nanofibers. PS-*b*-PI with 50:50 styrene:isoprene volume ratio is used. (Left) 8 wt% magnetite NPs with respect to isoprene are used. Lamellar self assembly of block copolymer is preserved. (Right) 20 wt% magnetite NPs with respect to isoprene are used. Self assembly changes to bicontinuous phase due to swelling of isoprene domain. Scale bars are 200 nm. Click image to see full view.

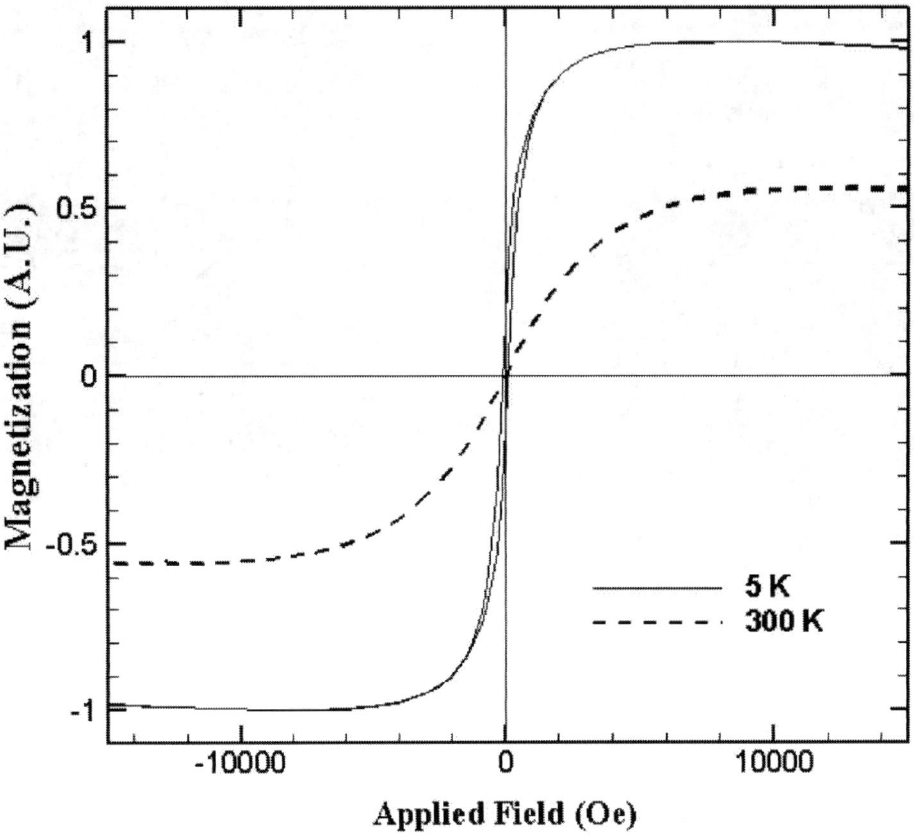

Figure 3. Magnetic properties of above samples (20 wt% magnetite) measured using SQUID showing superparamagnetic behavior (no hysterisis).

Study of Polymer Nanocomposites under deformation using Molecular Simulations [3,4,7,8]

I have integrated my experimental work with coarse grained molecular dynamics (MD) simulations to elucidate experimental findings and to serve as a guide for discovery of new nanocomposite based materials. First, I modeled and simulated the behavior of symmetric diblock copolymer/nanoparticle systems under simple shear flow. I considered three systems such that; a) NPs have selective interactions with one block, b) NPs have equal interactions with both blocks and c) NP-NP interactions are dominant. The aim of this study was to address the following questions: 1) *How does shear flow affect the spatial distribution of NPs?;* 2) *How does the presence of NPs affect shear induced BCP self-assembly?;* 3) *How is the rheology of BCPs affected by inclusion of NPs?; and 4) How does the effect of shear vary for different NP diameters, interaction potentials and polymer chain lengths?*

One of my key findings was that shear has a significant effect on prevention of NP aggregation in cases where NP-NP attraction is dominant and this effect of shear is a strong function of both nanoparticle size and polymer chain length. These results qualitatively match with my experimental results on P S-*b*-PI/magnetite NP nanofibers. In addition, this work provided the first ever systematic study on the effect of deformation on block copolymer nanocomposites and established the possibility of beneficial use of flow to further control nanoparticle location.

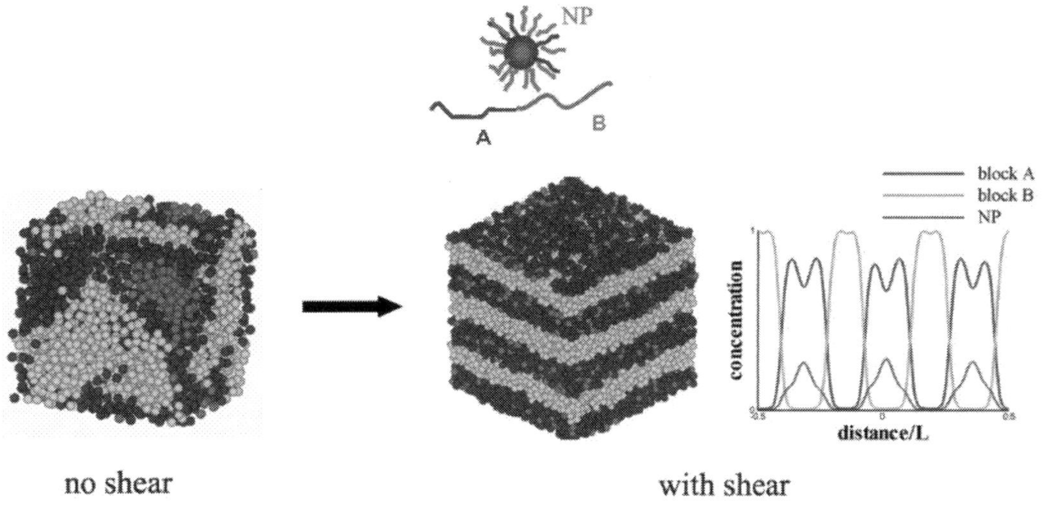

Figure 4. Effect of shear on prevention of aggregation in nanocomposite systems where NP-NP attractions are dominant. The above cartoon depicts that particles are strongly attracted to each other and slightly attracted to A domain (blue). The rightmost image shows the concentration profile of nanoparticles and A and B domains for system under shear flow. The NPs exhibit a maxima in the center of preferred domain A where most chain ends lie to minimize the entropic penalty of chain stretching.

Although shear flow does not represent dominant elongational deformation, it serves as a first step towards my study on nanocomposite behavior during electrospinning using computational modeling and simulation. Moreover, this work would enable a comparison between the effects of different flow types and discovery of more optimized methodology for fabrication of high-end nanocomposites. To draw a closer comparison with my present electrospinning experiments, I have implemented the algorithm for molecular dynamics simulations of elongational flow. Unlike the conventional MD simulations of elongational flow, this algorithm is temporally periodic and allows the simulation to continue indefinitely, which is a necessary condition to study slow relaxing systems like polymers under strong deformation. Study on the effect of elongational deformation on nanocomposite behavior is currently underway.

References

1. **V. Kalra**, P. Kakad, S. Mendez, T. Ivannikov, M. Kamperman, and Y. Joo, "Self Assembled Structures in Electrospun Poly (styrene-*block*-isoprene) Fibers", *Macromolecules,* 2006, 39, 5453.

2. **V. Kalra** , S. Mendez, J. Lee, H. Nguyen, M. Marquez, and Y. Joo, "Confined Assembly in Coaxially Electrospun Block Copolymer Fibers" *Advanced Materials*, 2006, 18, 3299.

3. **V. Kalra**, S. Mendez, F. Escobedo, and Y. Joo, "Coarse-grained Molecular Dynamics Simulation on the Placement of Nanoparticles within Symmetric Diblock Copolymers under Shear Flow" *J. Chem. Phys.,* 2008, 128, 164909.

4. **V. Kalra**, Y. Joo, "Using External Fields to Control the Location of Nanoparticles in Polymers: Simulations and Experiments" *Proceedings of XVth International Congress on Rheology, 2008, in press.*

5. **V. Kalra**, J. Lee, J. H. Lee, M. Marquez, U. Wiesner, and Y. Joo, "Controlling Nanoparticle Location via Confined Assembly in Electrospun Block Copolymer Nanofibers" *Small,* 2008, *submitted.*

6. **V. Kalra**, J. Panels, and Y. Joo, "Utilization of Silica Sol-gel Synthesis in Multi-axial Jet Electrospinning" *Advanced materials,* 2008, *to be submitted.*

7. **V. Kalra**, F. Escobedo, and Y. Joo, "Effect of shear on Nanoparticle dispersion in Polymer Melts: A Molecular Dynamics Study", *J. Chem. Phys., 2008, to be submitted.*

8. **V. Kalra**, F. Escobedo and Y. Joo, "Effect of Elongational Flow on Nanoparticle Placement in Polymer Matrices using Molecular Dynamics Simulations", *in preparation.*

5ba The Development and Reactivity of Oxide Phases on Pd(111) Utilizing Gas Phase Oxygen Atom Beams in Ultra High Vacuum

Heywood H. Kan, Jose A. Hinojosa, Jr., R. Bradley Shumbera, Robert J. Colmyer, and Jason F. Weaver. Chemical Engineering, University of Florida, Department of Chemical Engineering, Gainesville, FL 32611

Palladium oxide is an excellent catalyst for the oxidation of hydrocarbons and carbon monoxide under oxygen-rich conditions. However, the fundamental mechanisms governing oxide growth and reactivity still remain unclear due to difficulties associated with oxidizing palladium under the ultra high vacuum conditions necessary for most surface analytical techniques. Over the past two years we have made strides in elucidating the mechanisms for the growth and reactivity of palladium oxide by utilizing gas phase oxygen atoms beams to simulate high pressure oxidizing environments while operating under clean ultra high vacuum conditions. We present our recent findings concerning the growth and properties of palladium oxide on Pd(111) in addition to studying the effects of gas phase oxygen atom flux and substrate temperature on oxide growth. Notably, we have found that low temperature oxidation of Pd(111) with gas phase oxygen atoms enables the formation of a single crystalline PdO(101) thin film which has never been observed using molecular oxidants. This ability to generate PdO(101) has enabled us to be the first to study the interaction of molecular oxygen, water, hydrogen, and carbon monoxide with a single crystalline facet of PdO which we will also present.

5bb Functional Nanotechnology for Imaging and Therapy

Efstathios Karathanasis, W.H. Coulter Department of Biomedical Engineering, Georgia Institute of Technology, 313 Ferst Dr NW, Atlanta, GA 30332

Nanoscale control in fabricating nanoparticles allows for multi-functional carriers that can potentially enable personalized therapies that facilitate diagnosing, sensing, treating and monitoring the progress of treatment for each individual patient. To realize this vision, my research has focused on the development of (1) imaging nanoprobes and methods of use for early cancer detection and prediction/monitoring of cancer therapies by non-invasively probing vascular permeability and molecular markers (postdoctoral work), and (2) nanostructured glucose-sensing particles for insulin delivery (doctoral work).

Imaging nanoprobes for prediction of chemotherapy outcomes: Besides the cytotoxic effect at the molecular level, the success of nanoparticle-based cancer chemotherapy is primarily dependent on the access these agents have to tumors via the tumor leaky vasculature. Yet, the extent of vascular permeability to nanoparticles in individual tumors varies widely resulting in a correspondingly wide range of responses to the therapy. To address this need, we developed a long-circulating liposomal contrast nanoprobe for X-ray imaging. Imaging of a breast tumor model in rats using this nanoprobe and a clinical mammography system, demonstrated a wide range of vascular permeability. The same animals, when treated with a long circulating liposomal doxorubicin preparation exhibited a range of responses, which strongly correlated with the vascular permeability measured by the nanoprobe imaging. *Such prediction can therefore facilitate personalized therapy, and spare potential non-responders from the rigors of a chemotherapy regimen.* In a subsequent study, a multifunctional nanocarrier was developed co-encapsulating a contrast agent and a chemotherapeutic to predict and monitor chemotherapy. Following multiple treatments, it was observed that specific tumors that exhibited high uptake of the nanocarrier as visualized by imaging were mostly benefited from the treatment showing low tumor growth and long survival. In other efforts, nanoprobes for x-ray and MR imaging have been advanced to characterize brain tumors and cardiovascular diseases by probing vasculature and molecular markers.

Nanostructured glucose-sensing particles for insulin delivery: Recognizing the need for a system able to adapt a release rate based on the occasion, an inhalable glucose-sensing insulin nanostructured particle was developed with long residence times in the deep lungs. The nanostructured particle consists of insulin-loaded nanoparticles with glycosyl-presenting external leaflets which were agglomerated by a sugar binding moiety, con A . Upon exposure of the particle to elevated glucose, the release of insulin was triggered by binding of con A to free glucose and liberation of nanoparticles. The glucose-sensing particle was intratracheally instilled in the lungs of rats with suppressed pancreatic functions. Hyperglycemic events triggered an acceleration of the release of insulin achieving normoglycemia shortly after "sensing" the elevated systemic glucose. *This work is a demonstration of glucose-sensing insulin release mimicking the function of a healthy pancreas.*

Future research plans: My research interests lie in the intersection of nanotechnology with molecular markers and imaging to enable non-invasive characterization of angiogenesis and pathology of diseased vasculature. My laboratory will exploit the multifunctionality of nanoparticles to design nanoscale contrast agents for vascular anatomical, functional and molecular imaging to provide prognostic and diagnostic tools enabling personalized therapies. Among the numerous diseases where vasculature plays a critical role such as cancer, atherosclerosis, stroke, aneurism, and ischemia, my research will focus on solid tumors and vulnerable atherosclerotic plaques.

Selected publications

1. E. Karathanasis, L. Chan, S.R. Balusu, I. Sechopoulos, R.V. Bellamkonda, Multifunctional nanocarriers for mammographic quantification of tumor dosing and prognosis of breast cancer therapy, *Biomaterials* (accepted)

2. E. Karathanasis, S. Suryanarayanan, S.R. Balusu, K. McNeeley, I. Sechopoulos, A. Karellas, A.V. Annapragada, R.V. Bellamkonda, Imaging nanoprobe for prediction of nanoparticle chemotherapy using mammography, *Radiology* (accepted)

3. E. Karathanasis, J. Park, A. Agarwal, V. Patel, F. Zhao, A.V. Annapragada, X. Hu, R.V. Bellamkonda, MRI mediated, non-invasive tracking of intratumoral distribution of nanocarriers in rat glioma, *Nanotechnology* 19 315101 (9pp) 2008

4. E. Karathanasis, A. Agarwal, A.V. Annapragada, Triggered release of post-administered drugs, In: Handbook of Particulate Drug Delivery, R. Kumar Ed., American Scientific Publishers 2007, 243-260

5. E. Karathanasis, R. Bhavane, A.V. Annapragada, Glucose-sensing pulmonary delivery of human insulin to the systemic circulation of rats, *International Journal of Nanomedicine* 2(3) (2007) 501-513

6. R. Bhavane, E. Karathanasis, A.V. Annapragada, Triggered release of ciprofloxacin from nanostructured agglomerated vesicles, *International Journal of Nanomedicine* 2(3) (2007) 407-418

7. E. Karathanasis, R. Bhavane, A.V. Annapragada, Triggered release of inhaled insulin from the agglomerated vesicles: Pharmacodynamic studies in rats, *Journal of Controlled Release* 113(2) (2006) 117-127

8. E. Karathanasis, A. Ayyagari, R. Bhavane, R. Bellamkonda, A.V. Annapragada, Preparation of in vivo cleavable agglomerated liposomes suitable for modulated pulmonary drug delivery, *Journal of Controlled Release* 103(1) (2005) 159-175

9. R. Bhavane, E. Karathanasis, A.V. Annapragada, Agglomerated Vesicle Technology: A new class of particles for controlled and modulated pulmonary drug delivery, *Journal of Controlled Release* 93(1) (2003) 15-28

5bc Transport Processes at the Micron Scale

Aditya S. Khair, Chemical Engineering, University of California at Santa Barbara, Santa Barbara, CA 93106-5080

Transport processes and fluid dynamics at micron (or smaller) length scales play a crucial role in microfluidics, self assembly, biological transport, and energy conversion strategies, to name but a few areas. In this poster, I present selected vignettes from my research on transport processes at the micro- and nano-meter scale. Specifically, the following topics will be discussed.

1. **Surprising consequences of ion conservation in electro-osmosis** [1]: A variety of microfluidic technologies utilize electrokinetic transport over surfaces with rapid variations in charge. Here, as a paradigmatic model system for such scenarios, we consider electro-osmosis over a flat wall possessing a sudden jump in surface charge. The requirement of ion conservation at the jump gives rise to gradients in the bulk electric field and fluid flow (i.e. outside of the nanometer sized screening cloud adjacent to the wall). Remarkably, we show that these bulk variations may persist over length scales much greater than the screening-cloud thickness, with the implication that they may be readily observed in microfluidic channels.

2. **Active and nonlinear microrheology** [2]: Microrheology aims to infer the rheological properties of complex, microstructured materials from the motion of embedded colloidal probes. In passive microrheology, the probe motion is driven by thermal (Brownian) fluctuations, and the linear viscoelastic response (moduli) of a sample is measured via a generalized Stokes-Einstein relation. In contrast, in active or nonlinear microrheology, one actively forces a probe through a material (via e.g. optical tweezers) with the aim of inferring nonlinear rheological properties (e.g. shear thinning, normal stress differences). As a theoretical model for nonlinear microrheology, we investigate the actively driven motion of a probe particle immersed in a colloidal dispersion. A major goal is to establish under what conditions the computed microrheological properties are in agreement with those from macroscopic rheometry.

3. **Hydrodynamics of Janus particles** [3]: The last several years have witnessed significant progress in the design of nanometer sized "Janus" particles, possessing patterned surface properties. Here, we propose and investigate a hydrodynamically Janus particle: the "slip-stick" sphere, whose surface is partitioned into two regions. On one region fluid sticks to the particle, whereas on the other fluid slips past it. The Janus nature of a slip-stick sphere leads to a number of interesting findings. In particular, when placed in a shear flow, a slip-stick sphere undergoes either periodic translational motion or a net displacement along the flow direction. Moreover, the rheology of a dilute suspension of Janus spheres is non-Newtonian to even first order in particle volume fraction.

4. **Electrophoretic mobility of a colloidal particle redux** : The calculation of the electrophoretic mobility of a spherical particle is a fundamental problem in colloid science, with practical relevance in

separation of macro- and bio-molecules. We report on ongoing work aimed at extending the standard model for computing the mobility (O'Brien and White, J. Chem. Soc. Faraday Trans. II **74**.) to capture the effects of fluid slip and finite ion size.

For further information, please visit: www.engineering.ucsb.edu/~akhair/

References:

[1] A. S. Khair and T. M. Squires, "Surprising consequences of ion conservation in electro-osmosis over a surface charge discontinuity," J. Fluid Mech. (In press).

[2] A. S. Khair and J. F. Brady, "Single particle motion in colloidal dispersions: a simple model for active and nonlinear microrheology," J. Fluid Mech. 557, 73 (2006).

[3] A. Ramachandran and A. S. Khair, "The dynamics and rheology of a dilute suspension of hydrodynamically Janus spheres in a linear flow," J. Fluid Mech. (Submitted)

5bd Engineering Novel Surfaces for Tissue Engineering and Drug Delivery Applications

Srivatsan Kidambi, Center for Engineering in Medicine, Massachusetts General Hospital, Shriners Hospital, Harvard Medical School, 51 Blossom Street, Boston, MA 02114

Over the past decades, the development of new methods for fabricating thin films that provide precise control of the three-dimensional topography and cell adhesion has generated lots of interest. These films could lead to significant advances in the fields of tissue engineering, drug delivery and biosensors which have become increasingly germane areas of research in the field of chemical engineering. The ionic layer-by-layer (LbL) assembly technique called "Polyelectrolyte Multilayers (PEMs)", introduced by Decher in 1991, has emerged as a versatile and inexpensive method of constructing polymeric thin films, with nanometer-scale control of ionized species. PEMs have long been utilized in such applications as sensors, eletrochromics, and nanomechanical thin films but recently they have also been shown to be excellent candidates for biomaterial applications. During my doctoral research, I engineered these highly customizable PEM thin films to engineer in vitro cellular microenvironments to control cell adhesion and for drug delivery applications.

PEM films were engineered to control the adhesion of primary hepatocytes and primary neurons without the aid of adhesive proteins/ligands. We capitalized upon the differential cell attachment and spreading of primary hepatocytes and neurons on sulfonated polystyrene (SPS) and poly diallyldimethylammoniumchloride) (PDAC)surfaces to make patterned co-cultures of primary hepatocytes/ fibroblasts[1,2] and primary neurons /astrocytes[3] on the PEM surfaces. We further demonstrated that PEM-coated PDMS surfaces with different topographies affect the attachment, spreading and even proliferation of three types of mammalian cell, transformed 3T3 fibroblasts (3T3s), HeLa (transformed epithelial) cells and primary hepatocytes.[4] In addition, we developed self-assembled monolayer (SAM) patterns of m-d-poly(ethylene glycol) (m-dPEG) acid molecules onto PEMs.[5] The created m-dPEG acid monolayer patterns on PEMs acted as resistive templates, and thus prevented further deposits of consecutive poly(anion)/poly(cation) pairs of charged particles and resulted in the formation of three-dimensional (3-D) patterned PEM films or selective particle depositions atop the original multilayer thin films.[6] These new patterned and structured surfaces have potential applications in microelectronic devices and electro-optical and biochemical sensors. The PEG patterns developed are tunable at certain salt conditions and be removed from the PEM surface without affecting the PEM layers underneath the patterns. These removable surfaces provide an alternative method to form patterns of multiple particles, proteins and cells. This new approach provides an environmentally friendly and biocompatible route to designing versatile salt tunable surfaces. Finally, we illustrate the use of PEM films to engineer aptamer and siRNA based drug delivery systems.[7]

References:

1."Controlling Primary Hepatocyte Adhesion and Spreading on Protein Free Polyelectrolyte Multilayer

Films" Srivatsan Kidambi, Ilsoon Lee, Christina Chan. **J. Am. Chem. Soc.**, 126 (50), 16286 -16287, 2004.

2."Selective Adhesion of Primary Hepatocytes on Polyelectrolyte Multilayers: Template for Patterned Cell Co-culture" Srivatsan Kidambi, Lufang Sheng, Mehmet Toner, Martin Yarmush, Ilsoon Lee, Christina Chan. **Macromol Biosci**, 7(3), 344-53, 2007.

3."Patterned Co-culture of Neurons and Astrocytes on Polyelectrolyte Multilayer Films for Studying Astrocyte Mediated Oxidative Stress in Neurons" Srivatsan Kidambi, Ilsoon Lee, Christina Chan. **Advanced Functional Materials**, 18, 294-301, 2008.

4."Cell Adhesion on Polyelectrolyte Multilayer coated PDMS Surfaces with Varying Topographies" Srivatsan Kidambi, Natasha Udpa, Stacey Schroeder, Ilsoon Lee, Christina Chan. **Tissue Engineering**, 13 (8), 2105-2117, 2007.

5. "Selective Depositions on Polyelectrolyte Multilayers: Self-Assembled Monolayers of m-dPEG Acid as Molecular Templates" Srivatsan Kidambi, Christina Chan, Ilsoon Lee. **J. Am. Chem. Soc.**, 126, 4697-4703, 2004.

6."pH and Salt Responsive m-dPEG Acid Self Assembled Monolayers on Polyelectrolyte Multilayers" Srivatsan Kidambi, Christina Chan, Ilsoon Lee., **Langmuir**,24(1),224-30, 2008.

7. "Aptamer Incorporated Polyelectrolyte Multilayer Films Targeting Influenza Virus's Hemagglutinin Bnding Region" Srivatsan Kidambi, Ilsoon Lee, Christina Chan., submitted **ChemBioChem**.

8. "Effect of propofol on neuronal gene expression" Srivatsan Kidambi, Joel Yarmush, Wayne Fong, Sangeetha Kamath, Joseph SchianodiCola, Yaakov Nahmias, submitted **Neuroscience Letters**

9. "Effect of FMRP KO on the pluripotent genes in embryonic stem cells" Monica Casali, Srivatsan Kidambi, John Levine, Martin Yarmush, submitted **Stem Cells**

5be High-Throughput Functional Screening of Human cDNA Libraries for Novel Regulators of Innate Immune Responses

Hwijin Kim and Brian Seed. Center for Computational and Integrative Biology & Department of Genetics, Massachusetts General Hospital, Harvard Medical School, 185 Cambridge Street, Boston, MA 02114

Virus infection rapidly initiates an intracellular signalling cascade leading to the production of type-I interferon (IFN-α/β) and proinflammatory cytokines, which further activates a subset of genes that confer an antiviral state on the host cell, thereby constituting the first line of defense against invading viruses. Despite major advances over the last decade in our understanding of the innate immune signaling pathways, the human gene components involved in the pathways have not yet been fully elucidated. Here, we have begun to systematically identify additional pathway components by high-throughput cell-based screening of human cDNA libraries that represent about 50% of the human coding transcriptome. Using luciferase-based transcriptional reporter assays that measure the activation of IFN-β and/or NF-κB promoters, combined with data mining, we have been able to identify a number of novel positive and negative regulators in the pathways. Hit validities were confirmed by various biochemical, genetic and biophysical methods, including shRNA knockdown and/or knockout mouse-based methods, RT-PCR, immunoblotting and co-immunoprecipitation. We expect our findings will offer novel targets for therapeutic manipulation of the innate immune signaling pathways, leading to the development of more effective agents to treat virus- and inflammation-related diseases.

5bf Surface Chemistry of Oxide Clusters as Model Catalyst

Yu Kwon Kim, Department of Chemistry, University of Texas at Austin, Austin, WA 78712

A systematic study on $(WO_3)_3$ clusters as a model catalyst was performed by the experimental techniques temperature-programmed desorption (TPD) and reflection-absorption infrared spectroscopy (RAIRS), as well as by density functional theory (DFT) calculations. The model catalyst was prepared by deposition of $(WO_3)_3$ vapor onto the well-characterized oxide surfaces $TiO_2(110)$ and FeO(111)/Pt (111) in ultra-high vacuum. A series of aliphatic alcohols from methanol to t-butanol were used to identify various partial oxidation channels such as dehydration, dehydrogenation, and alcohol condensation. Our results indicate that the active sites are Lewis acidic tungstyl (W=O) groups where RO-H heterolytically dissociates into alkoxy (RO-) and hydroxyl (HO-). The alkoxy bound to W^{6+} undergoes further reactions into alkene, aldehyde, and ether. Dehydration is found to be the dominant channel for secondary and tertiary alcohols, while the other channels compete with dehydration in the oxidation of primary alcohols. The reaction mechanism for each reaction channel is discussed based on DFT calculations.

The research described in this presentation was performed in the Environmental Molecular Sciences Laboratory, a national scientific user facility sponsored by the Department of Energy's Office of Biological and Environmental Research and located at Pacific Northwest National Laboratory.

5bg Studies of Polymer Dynamics, Interfaces, and the Impact of Small Molecule Functionality on Electronic Properties

Robert J. Klein, Organic Materials, Sandia National Laboratories, P.O. Box 5800, MS-1411, Albuquerque, NM 87122

My research interests lie in the area of polymer interfaces, with particular emphasis on polymer dynamics, molecular transport, and specific interactions. Specifically, I intend to focus on two research topics, which will build on my postdoctoral and graduate experiences, and address exciting problems currently faced in the polymer science and engineering field:

(1) The application of broadband dielectric spectroscopy, focused on a nm-sized interface using special techniques, to characterize the impact of an interface on bulk dynamics and the interface dynamics itself. There are a variety of applications (e.g., interfacial coupling agents in polymer nanocomposites, membrane transport for water purification, columnar design to increase fracture toughness, relaxor ferroelectric domains in electroactive polymers and composites, among many others) where such a technique would improve both the fundamental understanding and lead to specific technological advances.

(2) The use of properly-placed functional groups to enhance interactions between polymers and small molecules, with relevance to chemical sensors, impregnation of polymers with pi-conjugated small molecules for enhanced electronic properties, and the diffusion and partitioning of molecules across interfaces.

My postdoctoral studies at Sandia National Laboratories with Joe Lenhart have encompassed a variety of organic materials projects, including: self-assembled monolayers for anti-stiction in microelectromechanical machines, characterized by near-edge x-ray absorption spectroscopy (NEXAFS); rheological and mechanical properties of nano- and micron-filled composites; the incorporation of pi-conjugated small molecules into dielectric films for improved electronic properties, from both processing and materials characterization perspectives; processing of nanoporous epoxies using removable small molecules as templates; and incorporation of oxygen groups into polystyrene by UV-ozone, characterized by NEXAFS. These projects have utilized collaborations with Daniel Fischer of NIST (at Brookhaven National Laboratory) and Mike Dugger at Sandia, among other collaborations at Sandia.

My doctoral studies at Penn State University in Materials Science and Engineering with Prof. Jim Runt focused on the dielectric properties of ion-containing polymers, using broadband dielectric spectroscopy to examine ion and polymer dynamics in low-Tg transport membranes and high-Tg structural polymers. During my master's work, also at Penn State, I studied electroactive polymers with Qiming Zhang, specifically PVDF terpolymers: this work focused on electromechanical, dielectric, and structural characterization of the polymers and the impact of processing on electromechanical efficiency.

My undergraduate degree was in Chemical Engineering at the University of California at Santa Barbara. There I spent some time working with Prof. Fred Lange to produce superhydrophobic ceramic surfaces via nano-silica coatings. I also spent a summer at Howard University, preparing (by molecular beam epitaxy) InGaAsN solar cells and performing electrical characterization.

Acknowledgement: Sandia is a multiprogram laboratory operated by Sandia Corporation, a Lockheed Martin Company, for the United States Department of Energy's National Nuclear Security Administration under contract DE-AC04-94AL85000.

5bh Nano/Microdevices for Biomedical Applications

Chee Guan Koh, Department of Chemical and Biomolecular Engineering, The Ohio State University, 140 West 19th Avenue, Columbus, OH 43210

Nano/Microfluidics is a versatile technology for proteomics, diagnostics, and therapeutics applications. An integrated plastic microfluidic system for pathogen detection and identification and a nanoparticles assembly microfluidic system for gene/drug delivery will be presented.

5bi Skin Tissue Engineering: Development of Treatment Strategies for Skin Diseases & Wound Repair and Regeneration

Piyush Koria, Center for Engineering in Medicine, Massachusetts General Hospital, Shriners Burns Hospital & Harvard Medical School, 51 Blossom Street, Boston, MA 02114

I plan to establish a research group that aims in establishing realistic tissue engineered models of skin that could be used to understand molecular mechanisms involved in skin diseases and wound repair and regeneration. I will then use this knowledge of molecular mechanisms for devising therapeutic strategies for treating these skin diseases and for accelerated wound healing. Lastly, these therapeutic strategies could then be tested on the tissue engineered models that I have developed above before taking them to clinic.

Overall the three interlinked aims in my laboratory will be:

1. Tissue Engineering & Dynamic Tissue Systems: To develop realistic tissue engineered model systems that mimic skin diseases and normal skin function.

2. Molecular and Cellular Mechanisms: To elucidate the important molecular players and molecular mechanisms involved in skin development and skin diseases.

3. Therapeutic Engineering: To develop strategies and platforms for the intervention and perturbation of the identified molecular mechanisms in 2 for treatment of skin diseases.

5bj Factors Influencing Product Selectivity in Heterogeneous Catalysis: Examination of Monodisperse Platinum and Rhodium Nanoparticles

John N. Kuhn, Chemistry, University of California, Berkeley, D54 Hildebrand Hall, Berkeley, CA 94720

Identifying and understanding the key catalyst features that influence selectivity is a common goal in heterogeneous catalysis. The ability to control fundamental properties in order to design heterogeneous catalysts with high selectivity would greatly reduce energy consumption. For monometallic catalysts, features include surface structure, nature of the metal-oxide interface, and particle size. In the present work, monodisperse Pt and Rh nanoparticles were synthesized using colloid chemistry and influences of the catalysts' properties upon catalytic activity and selectivity were examined.

Ultrasmall (as small as ~ 1 nm) monometallic and bimetallic Pt and Rh nanoparticles were synthesized by using fourth generation PAMAM dendrimers as the template and capping agent. Nanoparticles between 2 and 10 nm were synthesized using polymer (PVP) capping. Shape controlled nanoparticles (5 nm and above) were also synthesized by various colloid chemistry routes. Nanoparticles were loaded onto mesoporous silica, multiwalled carbon nanotubes (MWCNTs), and commercial zirconia and alumina supports by sonication. Characterization data, by techniques such as TEM, DRIFTS, elemental analyses (ICP-MS), XPS, and chemisorption, are also discussed.

Using well studied reactions such as ethylene hydrogenation and CO oxidation, catalytic activities were screened and treatments prior to reaction studies were optimized. Then, influences of the catalysts' properties were examined for pyrrole hydrogenation, a model reaction for the study of hydrodenitrogenation (HDN). Hydrogenation of the ring was observed at lower temperatures, regardless of size, over Rh catalysts than Pt ones. However, the scission of C-N bonds occurred more readily over Pt catalysts. Moreover, C-N bond breaking was more rapid when the Pt particles were larger than 2 nm, and consequently n-butylamine and butane were more likely to form over larger nanoparticles. These results are believed to be caused by Rh and smaller Pt nanoparticles being poisoned by strongly bound amine groups. Additionally, in the study of (Pt,Rh) bimetallic catalysts, Pt-rich nanoparticles were able to break C-N bonds whereas Rh-rich and even mixtures were unable to. Further multipath reactions, such as CO hydrogenation and hydroisomerization, are being investigated.

5bk Soft Materials Mechanics towards Development of High Performance Structural Materials, Bio-Implants, and Energy Storage Materials

Santanu Kundu, Polymer Science and Engineering Department, University of Massachusetts-Amherst, 120 Governors Drive, Amherst, MA 01002

My research interest is achieving a fundamental understanding of various aspects of soft materials such as flow, interfacial, mechanical properties and their interrelationship with microstructure. Soft materials are found in many places, ranging from all life forms such as blood, cells, tissues, mucus to commonly used paints, cosmetic products, hydrogels and food materials. Properties of these materials depend on their microstructure, which can be altered using flow field or stress. Also, materials with novel microstructures (in various length scales) are fabricated using top-down and bottom-up approaches [1, 2] and a structure-property-processing relationship of these materials needs to be developed for efficient materials design.

Presently, in my postdoctoral work I am investigating surface, near surface and mechanical characterization of hydrogel materials. The mechanical characterization of hydrogels is extremely difficult due to their inherent compliance and fragility. To overcome these problems a new technique, cavitation rheology, is being developed for the measurement of mechanical properties on small length scales, e.g. 1-100 μm, at any arbitrary location within a soft material. The technique involves growing a cavity at the tip of a syringe needle and monitoring the pressure of the cavity at the onset of instability. This critical pressure is directly related to the local modulus of the material. The technique is presently being investigated for polyacrylamide, a chemically crosslinked gel[3, 4]. Successful development of this technique will lead to measurement of modulus of biological tissues in vivo. Furthermore, the simplicity of this technique not only allows easy implementation into variety of materials systems, but also for high throughput testing.

For some applications it is important to know the adhesion and frictional behavior of hydrogels and one such example is contact lenses. In an industry sponsored project, I have developed an experimental set-up to measure the adhesion behavior of contact lenses and other hydrogels using a contact mechanics approach. Our set-up is also equipped with a liquid cell and in an ongoing study we are measuring low adhesion forces typically experienced in a liquid environment. We are also investigating the effect of pattern surfaces (on both hydrogels and elastomers) on the adhesion behavior at various environmental conditions [5, 6]. The patterned surfaces are created using novel fabrication techniques such as, wrinkling, micro-molding, lithography, etc.

In my Ph.D. research I have studied the flow and three dimensional microstructure of a liquid crystalline carbonaceous material (mesophase pitch) for different flow conditions, such as steady shear flow, dynamic flow, and processing flow using polarizing optical microscopy and X-ray diffraction[7-11]. The evolution of microstructure in different flow conditions was uniquely studied in three orthogonal planes. The systematic understanding of flow and its effect on microstructure helped us to predict/model the complex flow behavior and microstructural evolution of this material, which in turn will help to design carbon materials, such as carbon fibers, carbon-carbon composites, and fuel cell separators in more efficient ways. My dissertation, 'Investigation of flow and microstructure in rheometric and processing flow conditions for liquid crystalline pitch', has been awarded the Best Dissertation in Carbon Science, 2007 for *'outstanding scientific achievement'* by the *Elsevier-Carbon Journal*.

The research program that I plan to develop will largely build on my postdoctoral and doctoral research and specifically, I would like to develop a research program in the following areas:

• Flow and microstructure of anisotropic particle systems,

• Adhesion behavior of hydrogels at various environmental conditions,

• Large scale development of graphitic thin films from carbonaceous precursors and their transfer on elastic materials for flexible electronics applications,

• Development of porous and graphitic carbon materials with hierarchical structure for energy storage applications, such as, supercapacitors and hydrogen storage.

In addition to research, I am passionate about teaching and am inspired to educate and to mentor future scientists and engineers to meet the challenges of today's ever-changing technologies. My teaching philosophy has been developed through my experience as a lab and teaching assistant during my graduate school days, and as a supervisor during my industrial tenure. I can teach traditional undergraduate and graduate level chemical engineering courses. In addition, I intend to offer advanced graduate level courses in *polymer adhesion, fabrication and characterization of nanostructured materials, viscoelasticity of polymers and biopolymers: theory and simulation.*

References:

1. Ladet S, Laurent David L and Alain Domard A. Multi-membrane hydrogels. Nature 452, 76-79 (2008).

2. Dzenis Y. Structural Nanocomposites. Science 319, 419-420 (2008).

3. **Kundu S**, Zimberlin JA, Crosby AJ. Cavitation rheology and fracture mechanics of polyacrylamide hydrogels. American Physical Society -March meeting, New Orleans, USA, 2008.

4. **Kundu S**, Zimberlin JA, Crosby AJ. Cavitation rheology and fracture behavior of polyacrylamide gels, in preparation.

5. **Kundu S**, Crosby AJ, Sharma R. Adhesion behavior of a model hydrogel in various environmental conditions, in preparation

6. **Kundu S**, Crosby AJ. Adhesion of patterned surfaces under water, in preparation.

7. **Kundu S**, Grecov D, Rey AD, Ogale AA. Shear flow induced microstructure of a synthetic mesophase pitch, submitted.

8. **Kundu S**, Ogale AA. Rheostructural studies on a synthetic mesophase pitch during transient shear flow, Carbon 44(11): 2224-2235 (2006).

9. **Kundu S**, Ogale AA. Microstructural effects on the dynamic rheology of a discotic mesophase pitch, Rheologica Acta 46(9):1211-1222 (2007).

10. **Kundu S**, Naskar AK, Ogale AA, Anderson D, Arnold JR. Observations on a low-angle x-ray diffraction peak for AR-HP mesophase pitch, In press, Carbon, doi:10.1016/j.carbon.2008.03.014.

11. **Kundu S**, Ogale AA. Microstructure development of a synthetic mesophase pitch during processing flow, in preparation.

5bl Potential Applications of Laser-Capture Microdissection-Based Proteomics in the Study of Disease Progression

Carla M. R. Lacerda, Department of Clinical Sciences, Colorado State University, 1678 Campus Delivery, Fort Collins, CO 80523-1678

My research interests focus on applications of proteomics to the analysis of small tissue samples. In the biomedical field, a variety of tissues and their respective proteomes are of limited availability. In these cases it is convenient to separate the phenotypes of interest from the main tissues by using laser-capture microdissection (LCM), as it reduces samples variability when compared to whole tissue analysis. LCM-based -omics can be applied to diverse pathological cases as well as cancer research, and technologies such as LCM-transcriptomics are relatively well-developed due to the ease of RNA amplification. LCM-proteomics is a field in its infancy and there is a need for optimization of protein extraction and purification techniques for proteomic analyses. Insights into the applications of LCM-proteomics in the biomedical field are presented here.

5bm Single Molecule Chemically Amplified Photoresists for Nanolithography

Richard Lawson[1], Clifford L. Henderson[1], and Laren M. Tolbert[2]. (1) School of Chemical & Biomolecular Engineering, Georgia Institute of Technology, 311 Ferst Drive NW, Atlanta, GA 30332-0100, (2) School of Chemistry and Biochemistry, Georgia Institute of Technology, Atlanta, GA 30332

The shrinking of feature sizes in microelectronics is the primary driving force for smaller, faster, more powerful, more energy efficient computers and electronics. The critical enabling factor for the reduction in feature size is advances in photolithography, the process wherein well-defined areas are patterned onto a semiconductor substrate. Current state-of-the-art semiconductor manufacturing prints features as small as 45 nm. To continue shrinking feature sizes toward the molecular scale, multiple improvements must be made in photoresists.

Our design approach for next generation chemically amplified photoresists is termed single molecule resists. The name does not imply that the entire resist film is made of a single molecule, but rather that the film is composed of multiple copies of a single molecule that contains all the functionality desired in a chemically amplified resist. This differs radically from conventional photoresists which are composed of blends of polydisperse co-polymers, small molecule photoacid generators (PAGs), and small molecule base quenchers. Single molecule resists combine the advantages of multiple types of resists and have additional unique advantages. Since they are small molecules instead of polymers, they are purely monodisperse, can be made with full synthetic control of stereochemistry and regiochemistry, and have much smaller molecular size relative to the features being printed. Additionally, since there are no physically blended additives, this approach allows for a molecularly homogeneous resist.

We have taken a combined experimental/modeling approach in the design and synthesis of single molecule resists. Mesoscale modeling of the lithographic process suggests that using very high PAG loading, such as in our system, provides the best combination of resolution, sensitivity, and pattern fidelity. The modeling also shows the trade-off between photoacid diffusion length and number of protecting groups which directs synthetic design for performance. Experimentally, multiple different families of single molecule resists have been designed, synthesized, and exercised lithographically. Results have shown that resolution can be maintained and even improved by this approach, while providing significant enhancement in pattern fidelity. Two completely different single molecule designs have shown a line-edge-roughness of 3.9 nm, a significant improvement over the limit of 5 nm found in conventional designs. The single molecule resist approach shows great promise to provide a path forward in photoresist design, leading to ever smaller and faster integrated circuits.

5bn Engineering Nanomaterials for Bioimaging and Therapeutic Applications

Dongwon Lee, Department of Biomedical Engineering, Georgia Institute of Technology, 315 Ferst Drive, Atlanta, GA 30332

Hydrogen peroxide is an essential oxygen metabolite in living organisms, and is emerging as a newly recognized messenger in cellular signal pathways. However, the overproduction of hydrogen peroxide leads to oxidative stress and functional decline of organ systems, and has been observed in several life-threatening diseases such as atherosclerosis, cancer, and Alzheimer's and related diseases. Despite the significance of hydrogen peroxide in human health, at present, there are no contrast agents that can image hydrogen peroxide in vivo with sensitivity and selectivity. Therefore, there is great need in developing contrast agents that can image hydrogen peroxide in vivo.

I have developed peroxalate nanoparticles which detect hydrogen peroxide based on peroxalate chemiluminescence. Peroxalate nanoparticles were composed of peroxalate ester containing polymers and fluorescent dyes, and were designed to image hydrogen peroxide by performing a three component chemiluminescence reaction; first, hydrogen peroxide diffuses into the nanoprobes and reacts with peroxalate esters to generate a high energy intermediate dioxetane, second, dioxetane chemically excites fluorescent dyes, leading to light emission and imaging of hydrogen peroxide. Peroxalate nanoparticles with a mean size of 550 nm were capable of detecting hydrogen peroxalate with high sensitivity and specificity and were also capable of imaging hydrogen peroxide in the peritoneal cavity of mice, after a direct intraperitoneal injection. I have also developed peroxalate micelles in order to enhance physicochemical properties needed for intravenous applications. Peroxalate micelles had peroxalate esters and fluorescent dyes in their hydrophobic core and a stealth PEG corona. Peroxalate micelles had a mean diameter of 33 nm and were capable of detecting hydrogen peroxide at concentrations as low as 50 nM. It is anticipated that peroxalate-based nanoprobes have great potential for imaging of hydrogen peroxide in vivo, given their physicochemical properties and nanomolar sensitivity.

5bo Simulation and Experimental Studies of the Interactions of Membranes with Peptides, Synthetic Polymers and Nanoparticles

Hwankyu Lee, Lab. of Computational biology, National Heard, Lung and Blood Institute, National Institutes of Health, Rockville, MD 20892-9314

All-atom and coarse-grained (CG) molecular dynamics (MD) simulations were performed to investigate (1) peptide-peptide, peptide-monolayer, and peptide-micelle interactions; (2) effects of the dendrimer surface, size, and concentration on pore formation in lipid bilayers; and (3) hydrodynamics and shape anisotropy of polymers as they pass through the pores of membrane proteins. Circular dichroism spectroscopy and single-channel current recording experiments were also performed to complement simulation results. Future research directions are presented.

(1) Peptide interactions with lipid monolayers and micelles.

SP-$B_{1\text{-}25}$ is a fragment of 79-amino acid lung surfactant protein B, an important component for preventing respiratory distress syndrome. Multiple copies of the peptide in palmitic acid monolayer were simulated with all atoms, showing that the final peptide conformation favorably compared with experiments. The peptides are anchored by hydrogen bond interactions between the cationic residues Arg-12 and Arg-17 and the hydrogen bond acceptors of the ionized monolayer headgroup, and the peptide tilt angle is modulated by the interactions of Tyr-7 and Gln-19 with the monolayer headgroup. These results can be applied to the rational design of synthetic lung surfactant peptides. [1]

Coiled-coil peptides [2,3] and the BAR peptide-SDS micelle [4] were simulated to investigate models of membrane curvature by BAR domains. Both simulation and experiment indicate that the N-terminal region (helix-0) is disordered, and that the peptide curves to adopt the micelle shape. This is consistent with notion that helix increases the peptide-membrane binding affinity.

(2) Nanoparticle induced-pore formation in the lipid bilayer: Effects of the nanoparticle size, surface charge, and concentration.

Polyamidoamine (PAMAM) dendrimers, which consist of regularly branched monomeric building blocks and surface terminal groups, are good candidate nanoparticles for use as anti-tumor therapeutics and drug delivery. To investigate the effect on pore formation of the dendrimer properties (size, surface charge, and concentration) and solution conditions (temperature and salt concentration), we simulated PAMAM dendrimers in DMPC bilayers with explicit water using the CG model. When initially clustered together near the bilayer, neutral acetylated dendrimers aggregate, whereas cationic un-acetylated dendrimers disperse, in agreement with experiments. Bilayers interacting with un-acetylated dendrimers of higher concentration are significantly deformed and show pore formation on the positively curved portions, while acetylated dendrimers are unable to form pores. Larger un-acetylated dendrimers bring more water molecules into the pores than do smaller ones. At higher salt concentration (~500mM NaCl) or lower temperature (277 K), un-acetylated dendrimers do not insert into the bilayer. These results highlight the applicability of CG simulations, and help explain why charged dendrimers deform membranes and form pores at high concentration. [5,6,7]

(3) Hydrodynamic radius and shape anisotropy of polymers as they pass through the pores of membrane proteins

Polyethylene oxide (PEO) and polyethylene glycol (PEG) are used as probes of pore sizes of membrane channels, although many molecular details are unclear. PEO and PEG were simulated to examine anisotropy of the shape and translational diffusion tensors of these polymers as a step toward understanding their interaction with membrane pores. The calculated persistence length, the exponent u relating the radius of gyration and molecular weight ($R_g \mu M_w^{\nu}$), and hydrodynamic radii are in quantitative agreement with experiment values. The dimension of the middle length for each of the polymers nearly equals the hydrodynamic radius obtained from diffusion measurements in solution. This implies that a polymer diffuses with its long axis parallel to the membrane channel without substantial distortion. This quantitative study also helps development of coarse-grained models of PEG and PEO for ongoing investigation of the effects of the molecular flexibility (soft linear vs. hard sphere) in membrane channels at the microsecond time scale. Single-channel current recording experiments of PEG and dendrimers are also being performed to complement simulation results. [8]

References

[1] **Lee H**, Kandasamy SK, and Larson RG, Molecular dynamics simulations of the anchoring and tilting of the lung-surfactant peptide SP-$B_{1\text{-}25}$ in palmitic acid monolayers. ***Biophysical J.*, 2005**, 89:3807-3821

[2] Sayer JA, Otto EA, O'toole JF, Nurnberg G, Kennedy MA, Becker C, Hennies HC, Helou J, Attanasio M, Fausett BV, Utsch B, Khanna H, Liu Y, **Lee H**, Larson RG, and Hildebrandt F et al., The centrosomal protein nephrocystin-6 is mutated in Joubert syndrome and activates transcription factor ATF4. ***Nature Genetics*, 2006**, 38:674-681

[3] **Lee H**, and Larson RG, Prediction of the stability of coiled coils using molecular dynamics simulations. ***Molecular Simulation*, 2007**, 33:463-473

[4] Low C, Weininger U, **Lee H**, Schweimer K, Pastor RW, Balbach J. Structure and dynamics of Helix-0 of the N-BAR domain in lipid micelles and bilayers. ***Biophysical J.***, submitted.

[5] **Lee H**, Baker JR, and Larson RG, Molecular dynamics studies of the size, shape, and internal structure of 0% and 90%-acetylated G5 PAMAM dendrimers in water and methanol. ***J. Physical Chemistry B.*, 2006**, 110:4014-4019

[6] **Lee H**, and Larson RG, Molecular dynamics simulations of PAMAM dendrimer-induced pore formation in DPPC bilayers using a coarse grained model. ***J. Physical Chemistry B.*, 2006**, 110:18204-18211

[7] **Lee H**, and Larson RG. Coarse-grained molecular dynamics studies of the concentration and size dependence of fifth- and seventh-generation PAMAM dendrimers on pore formation in DMPC bilayer. ***J. Physical Chemistry B.* 2008,** In press.

[8] **Lee H**, Venable RM, MacKerell AD, Pastor RW. Molecular dynamics studies of polyethylene oxide and polyethylene glycol: Hydrodynamic radius and shape anisotropy. ***Biophysical J.* 2008,** In press.

5bp New Biotherapy and Drug Delivery for Inflammatory Diseases

Sungmun Lee, Biomedical Engineering, Georgia Institute of Technology, 315 Ferst Drive, Atlanta, GA 30332

Many inflammatory diseases are preventable and curable, provided that the pathogenic mechanism is well characterized and the proper drugs can be delivered efficiently. This presentation will focus on (1) investigation of pathogenic mechanism and a new therapeutic approach for Alzheimer's disease, and (2) a novel drug delivery system for chronic or acute inflammatory diseases.

Aβ, a main protein in Alzheimer's disease, is known to be toxic when aggregated; however, the relationship between neurotoxicity and Aβ• aggregates is still under investigation. In this presentation, I will discuss the effect of shear stress, one of key environmental factors, on Aβ aggregation and toxicity. As a new biotherapy, small heat shock proteins will be introduced to prevent Aβ aggregation and toxicity. Small heat shock proteins, a family of chaperone, are overproduced under stressed condition and prevent incorrect protein folding and aggregation. In order to enhance the efficacy of anti-inflammatory drugs, polyketals, a new family of acid degradable polymers, are used as a drug delivery vehicle. Key features of polyketals are that degradation products are non-immunogenic and hydrolysis rate is controllable depending on monomers. Two polyketal polymers having different hydrolysis rate will be presented for treating chronic and acute diseases.

5bq Systems Biology: From Identifying Toxicity Pathways to Therapeutic Engineering

Zheng Li[1], Christina Chan[2], and James J. Collins[1]. (1) Biomedical Engineering, Boston University, 44 Cumminton St, Boston, MA 02215, (2) Chemical Engineering and Material Science, Michigan State University, 2527 Engineering Building, East Lansing, MI 48824

High throughput profiling of biomolecules, including genes, proteins, miRNAs, and molecular interactions, including protein-protein and protein-DNA interactions, has made it possible to study the complex cellular regulatory programs at the systems level. Here I present how systems biology approaches can be developed and applied to study problems from identifying toxicity pathways to discovering therapeutic targets for complex multi-factor diseases such as cancers and neurodegenerative diseases.

A systems approach to identifying toxicity pathways

In this study, a systems approach was developed to integrate gene expression and metabolic profiles to identify pathways related to a metabolic function. The approach first extracts features from the metabolite profiles to discriminate different phenotypes; and then selects genes relevant to these features with genetic algorithm coupled partial least squares analysis and independent component analysis; and finally reconstructs the pathways from the selected subset of genes using Bayesian network analysis.

We applied the approach to investigate the cytotoxcity related pathways in a human hepatoblastoma cell line (HepG2/C3A). In this cellular system, saturated fatty acid and acute TNFa exposure were found to be cytotoxic, although chronic TNFa exposure was found to be cytoprotective. Genes encoding biomolecules in pathways involved in regulating cytotoxicity were selected and the pathways were reconstructed. Several pathways were experimentally validated by perturbing the pathways and measuring their protein levels. Finally, the BN network analysis was applied to predict toxicity level and the effects of perturbing genes, e.g. stearoyl-CoA desaturase and Bcl-2 on toxicity level were simulated using the model. The results illustrate that the integration of gene expression and metabolite data by our framework provides informative and testable hypothesis on the mechanism(s) of palmitate-induced cytotoxicity.

A hierarchical framework was further developed to incorporate a priori knowledge in a framework to identify the toxicity pathways. The hierarchical framework consisted of three stages. 1) Different phenotypes induced by TNF-a and FFA were characterized, for example, a cytotoxic phenotype was characterized by the level of lactate dehydrogenase (LDH) release. Important metabolites that contribute to separating the phenotype were identified using a discriminant analysis of the twenty seven metabolites. Relevant metabolic pathways were suggested by the identified metabolites. 2) cDNA microarray was used to measure the global gene expression changes induced by FFAs and TNF-a. Gene set enrichment analysis (GSEA) of the microarray data identified the relevant pathways in addition to those identified by the metabolite analysis. The pathways identified to be enriched included mitochondria-related pathways, oxidative-stress-related pathways and fatty acid metabolism. 3) Finally, the gene expression and metabolite profiles were integrated with a regression model to identify the genes most relevant to the metabolites identified in the first step to be highly correlated with the cytotoxicity. The cytotoxic and cytoprotective pathways identified, as well as the experimentally validated pathways will be discussed.

Finally we studied the dynamic properties of the toxicity pathways. Cells continuously reprogram gene regulatory networks when they sense the changes in the environmental conditions. Reconstruct the dynamic network is essential to understand the regulatory mechanism of progression of diseases. Gene expression profiles of HepG2/C3A cells were measured after 24, 48, and 72 hours of exposure to FFA and TNF-a. To identify the underlying pathways, module map analysis was used to identify the important genes. 522 biologically meaningful gene sets were first defined based upon their functional category or pathway, as defined by the MsigDB database. The functional gene modules were extracted from the 522 gene sets based upon the expression profile with Genomica. The module maps at different time points were compared to identify the dynamics of the modules that are important to the cytotoxic phenotype. The profiles of the genes in the modules identified to be involved in activating/inducing cytotoxicity along with the profile of the phenotype were used for network reconstruction to identify potential targets that can be perturbed to modulate the cytotoxicity. Some of the targets identified to be important to palmitate induced cytotoxcity were experimentally validated.

Therapeutic engineering

Combination therapy for cancer

Described herein is a network biology approach useful for the identification of multiple therapeutic targets, which can be targeted simultaneously using an agent (or a plurality of agents) to modulate cellular phenotypes, or in combination with pharmaceutical compounds to improve the drug sensitivity and/or reduce the drug doses for minimal side effects. The approach described herein relies on first identifying the mediators of a condition of interest, and second, selecting gene combinations that are in competing/parallel pathways.

Systems biology analysis of neurodegenerative diseases

Amyotrophic lateral sclerosis (ALS) is one of the most common neurodegenerative diseases, characterized by degeneration of motor neurons in brain and spinal cord, leading to muscle atrophy and eventually death. Extensive studies have been undertaken to understand the pathogenesis of ALS. However, the underlying mechanisms of selective degeneration of motor neurons in ALS are still incompletely understood so far. It is the objective of this study to apply systems biology approaches to integrate interactome data with mircroarray gene expression profiles to identify the causal factors of ALS. Interactome and transcriptome profile of both ALS patients and mice model were integrated to identify active pathways that were differentiated in ALS disease condition. Therapeutic targets were predicted and validated experimentally.

Applicable peer reviewed publications

1. Li, Z., Srivastava, S., Findlan, R., Chan C., "Using dynamic gene module map analysis to identify targets that modulate free fatty acid and tumore necrosis factor (TNF)-a induced cytotoxicity" (2008) Biotechnology Progress, 24, 29-37.

2. Srivastava, S.#, Li, Z.#, Yang, X., Yedwabnick M., Shaw S., Chan C. "Identification of genes that regulate multiple celluar processes/response in the context of lipotoxicity to hepatoma cells" (2007) BMC Genomics, 8:364. # co-first author.

3. Li, Z. Srivastava, S., Mittal, S., Yang, X., Sheng, L., Chan, C., " Identifying active pathways by integrating gene expression and phenotypic profiles", (2007) BMC Bioinformatics, 8:101.

4.Li, Z.#, Srivastava, S.#, Yang, X., Mittal, S., Norton, P., Resau, J., Brian, H., Chan, C., "A Hierarchical Approach to Identify Pathways that Confer Cytotoxicity in HepG2 Cells from Metabolic and Gene Expression Profiles", (2007) BMC Systems Biology, 1:21.# co-first author.

5.Li, Z., Shaw, S. M., Yedwabnick, M. J., Chan, C. "Using a state-space model with hidden variables to infer transcription factor activities", (2006) Bioinformatics 22(6): 747-754

6.Walton, S.P., Li, Z., and Chan, C., "Biological network analyses: computational genomics and systems approaches" (2006) Molecular Simulation, 32 (3-4): 203-209.

7.Patil, S., Li, Z., and Chan, C., "Approaches to Optimizing Cellular Function of Engineered Tissue: Cellular to Tissue Informatics", in Advances in Biochemical Engineering /

Biotechnology, eds. K. Lee and D. Kaplan, Springer, 2006, volume 102, p. 139-159.

8.Li, Z., and Chan, C., "Inferring pathways and networks with a Bayesian framework", (2004) FASEB J 18(6): 746-8 (published on-line Feb 6, 2004, doi:10.1096/fj.03-0475fje).

9.Li, Z., and Chan, C., "Integrating gene expression and metabolic profiles", (2004) Journal Biological Chemistry 279: 27124-27137.

10.Li, Z., Yarmush, M. L., and Chan, C., "Insulin concentration during pre-conditioning mediates the regulation of urea synthesis in primary rat hepatocytes during exposure to amino acid supplemented plasma", (2004) Tissue Engineering 10(11):1737-1746.

11.Li, Z., and Chan, C., "Extracting novel information from gene expression data", (2004) Trends in Biotechnology, 22: 381-383.

5br Implementation of Green Chemistry and Engineering Principles into Nanomaterial Synthesis and Processing

Juncheng Liu, Department of Chemical Engineering, Auburn University, 230 Ross Hall, Department of Chemical Engineering, Auburn, AL 36849 and Christopher B. Roberts, Chemical Engineering Department, Auburn University, 212 Ross Hall, Auburn, AL 36849.

Green chemistry and engineering aims toward the design of environmentally-benign products and processes using ecologically friendly materials and solvents, with limited waste generation. During the past six years of my research career, I have been working on the implementation of green chemistry principles into advanced nanomaterials synthesis and processes.I have developed the use of environmentally friendly capping agents, reducing agents, and solvents as well as green engineering processes to synthesize advanced metal nanoparticles[1-8] and subsequently assemble these monodisperse metal nanoparticle building blocks into supermolecular structures such as order arrays and thin films.[3,4] In addition, I have developed methodologies that use non-toxic and biodegradable sugars (as capping agent)and organic acid (as reducing agent) to precisely control the synthesis of nanoparticle building blocks for desired supermolecular assembly.[1] Moreover, these metal nanoparticles were used for novel ground water treatment[2] and as catalysts for alternative energy synthesis.[9]

1.Liu, J.; Roberts, C. B. "Green" and Precisely Size-controlled Seed Growth of Pd Nanoparticles, Journal of American Chemistry Society, submitted for publication, 2008

2.Liu, J. C.; He, F.; Durham, E.; Zhao, D.; Roberts, C. B. Langmuir. 2008, 24, 328.

3.Liu, J.; Sutton, J.; Roberts, C. B. J. Phys. Chem. C. 2007, 111, 11566.

4.Liu, J. C.; Anand, M.; Roberts, C. B. Langmuir, 2006, 22, 3964.

5.Liu. J. C.; Raveendran, P; Ikushima Y, Chemistry – A European Journal, 11, 2005, 1854 -1860.

6.Liu. J. C.; Qin, G. W.; Ikushima, Y. Chemistry – A European Journal, 2005, 12, 2131.

7.Liu. J. C.; Qin, G. W.; Ikushima, Y. Chemical Communication, 2005, 2972-2974.

8.Liu. J. C.; Raveendran, P; Ikushima Y, Chemical Communication, 2004, 2582-2583.

9.Research Proposal, "Green and Size-controlled Synthesis and Processing Cobalt Nanocrystals Deposited on Carbon Nanofiber/Nanotube Using CO2 Expanded Liquids as Novel Fischer-Tropsch Catalysts for Synthesis of Long-Chain Hydrocarbons and Clean Fuels" Christopher B. Roberts (PI), Juncheng Liu (Co-PI), Funding agency: Green Chemistry Institute (GCI)-ACS Petroleum Research Fund

5bs Computational Chemical Engineering

Eric L. Maase, Chemical Engineering Department, Lafayette College, Acopian Engineering Center Rm 268, 740 High St., Easton, PA 18042

Developing, applying, and implementing computational tools for research and education in Chemical Engineering of the future.

5bt Nano-Scale Materials for Energy and Environmental Applications

Mainak Majumder, Chemical and Biomolecular Engineering and Richard E. Smalley Instiute for Nano-scale Science and Technology, Rice University, 6100 Main Street, MS-362, Houston, TX 77005

In my PhD research with Professor Bruce J. Hinds (Chemical and Materials Engineering, University of Kentucky), I utilized a macroscopic membrane platform consisting of billions of vertically oriented carbon nanotubes (CNTs) encapsulated in a polymer matrix for understanding molecular transport through the inner-core of CNT. In this work, I showed that liquids such as water are transported extremely fast (almost four orders of magnitude faster than conventional predictions of flow through pipes) through the hydrophobic cores of the nanotubes, with ~7 nm diameter, and explained by a slip-at-the-wall mechanism. On the environmental front, I have served as staff scientist at a National Laboratory in India with a focus on developing a membrane based process for decontamination of arsenic from ground water

Currently, I am expanding my expertise on carbon nanotubes, through my post-doctoral research with Professor Matteo Pasquali (Chemical and Biomolecular Engineering, Rice University) on synthesis of CNT dispersions and their application in making transparent conductive electrodes and nanowires .

My research interests are at the interface of materials science and chemical engineering. Of immediate interest is application of nano-scale materials, particularly CNTs, in water purification and desalination, in enhancing oil-recovery from under-ground reservoirs, in interfacing micro-fluidics with nano-fluidic channels, and in functional films and coatings - all these applications will be backed by chemical engineering fundamentals. I am certain that my research program will be potentially attractive to funding agencies, yet diverse enough to suit the needs and interests of students - the key components to establish a highly productive academic research program.

I also had the good fortune in training several under-graduate students on a one-to-one basis; their work often resulting in peer-reviewed articles. Considering this rewarding experience, I plan to include under-graduate students in my research program. My teaching interests are in materials science, mass transport and fluid mechanics and I propose to develop a graduate level course on 'Nano-materials for Chemical Engineering Applications'. To obtain further teaching experience, I am planning to deliver several lectures for an under-graduate fluid mechanics course with Professor Pasquali in Fall 2008.

5bu Biocompatible Thin Films for Use in Tissue Engineering

Holly J. Martin, Dave C. Swalm School of Chemical Engineering, Mississippi State University, 330 Swalm - President's Circle, Mississippi State, MS 39762 and Kirk H. Schulz, Vice President for Research and Economic Development, Mississippi State University, 617 Allen Hall, Mississippi State, MS 39762.

Performing novel research and getting solid, reproducible results are the main goals of any graduate student and post-doctorate researcher, including myself. Besides the novel research that I performed as a graduate student and as a post-doctorate researcher examining ways to improve coating adhesion on implant metals, I have also had many opportunities to collaborate with other researchers, mentor graduate students, teach two different classes, write proposals based on my research, and perform research on a topic different from my graduate work. As an assistant professor, I plan to develop my research program on ways to improve implant coating adhesion without sacrificing biocompatibility.

Modifying the surface of implant quality titanium to improve bone cell attachment and growth was the focus of my research as a graduate student and as a post-doctorate researcher. We were able to greatly improve the bond between chitosan and commercially pure grade 4 titanium, using two novel reaction schemes that I developed. X-Ray Photoelectron Spectroscopy (XPS) was used to examine the titanium surface following the metal treatments, passivation or piranha, and following the individual reaction steps, including the deposition of aminopropyltriethoxysilane or triethoxsilylbutyraldehyde. XPS was also used to ensure that no changes occurred to the chitosan film based on the four treatment combinations. Nano-indentation was used to further ensure that the mechanical properties of the chitosan film were unchanged, while tensile testing confirmed that the chitosan films were bound to the titanium surface more strongly than previously published research. Biological testing, including cell attachment and growth, were performed to ensure that the toluene solvent and/or the silane molecules did not change the biological properties of the chitosan.

As both a graduate student and a post-doctorate researcher, I have been the laboratory manager for the XPS equipment at Mississippi State University. This position has allowed me opportunities to collaborate with many different professors, departments, schools, and companies and to mentor graduate students involved in surface science. In addition to running and analyzing samples for professors within Chemical Engineering, I have had the opportunity to work with people from the departments of Plant and Soil Sciences and Electrical and Computer Engineering. I have also had the opportunity to collaborate with people from the University of Texas Health Science Center and Georgia Tech. As a laboratory manager, I am also responsible for teaching graduate students the proper way to prepare their samples, load and run samples, and analyze the data gathered. I am also currently teaching two people how to maintain the equipment so that a smooth transition can be made upon the completion of my post-doctoral work.

I envision developing a broad, interdisciplinary research program incorporating surface science, mechanical testing, and biological studies as an assistant professor. Surface science equipment, such as XPS, is a necessary part of my research, but will not be solely dedicated to my research nor will it be the only equipment needed. I believe a strong collaboration could be developed between myself and other professors, including those using XPS, nano-indentation, other mechanical testing equipment, and biological studies. I believe these collaborations should take place both within Chemical Engineering and between other departments, such as Mechanical Engineering, Biology, and Biological Engineering.

As part of this collaboration, I believe that I need to be able to assist in other research areas besides modifying implant surfaces to improve coating adhesion. One area in which I believe I can be a solid contributor is in the area of Wastewater Treatment. As a post-doctorate researcher, I have had the opportunity to work on a project completely different from my dissertation. Since I obtained a minor in Microbiology as an undergraduate student, this project allowed me to use my microbiology background and combined it with new laboratory methods with which I was previously unfamiliar. The project involved testing a new chemical, peracetic acid, to disinfect wastewater and was sponsored by the Environmental Protection Agency (EPA) and FMC Corporation. The hope of this research was to be able to treat wastewater without having to store the dangerous chemical associated with chlorination and also to reduce or eliminate the production of hazardous disinfection by-products. Using peracetic acid at two concentrations, we were able to determine that the permitting levels required by the EPA for wastewater treatment plants could be met. Following the successful completion of this project, FMC requested that proposal be created to allow my group to perform more funded research. The proposal is currently in development and is expected to be submitted in the middle of May.

In addition to the research I performed as a graduate student, I also had the opportunity to teach two classes. The first class I taught was an entirely new laboratory for an established class, focusing on process design simulation. As part of my responsibilities, I created all of the materials needed, including the syllabus, Powerpoint presentations covering individual unit operations, homework, and a final project combining the individual unit operations into a working flowsheet. I also was responsible for grading both the homeworks and the final projects. I was responsible for teaching this class for the Fall semester and the Spring semester and was evaluated by the students, receiving 4.7/5.0 for both semesters. I also taught Unit Operations Laboratory I and II in the Spring semester. I was responsible for teaching the class, including assigning the groups to the experiments each week, for grading the laboratory reports, and for assigning final grades. As with the class I developed, I was evaluated by the students, receiving a 4.5/5.0 average.

As an assistant professor, I plan to continue collaborating with professors in both my research area, biomedical implants and biomaterials, and in other research areas and departments. I truly enjoy teaching and look forward to teaching both established undergraduate and graduate classes and classes I create based on my research interests.

5bv Development of Reactive Nanostructured Systems for Energy, Environmental and Biomedical Applications

Karen S. Martirosyan, Chemical Engineering, University of Houston, 4800 Calhoun, Bldg. 1, S222, Houston, TX 77054

The main topic of this presentation will focus on the fundamental aspects of strongly reactive/ exothermic nano-particulate systems and its potential for fabrication of advanced nanomaterials and devices for applications of power systems, energy storage, environmental, national security and health care. I will discuss novel reactive systems that have the potential to enable a more concentrated energy release and potentially can be used for various military applications. In addition, a novel cost-effective and energy efficient fabrication method of nanostructured complex oxides that we referred to as Carbon Combustion Synthesis of Oxides (CCSO) will be presented. In this process, the reactive oxidation of carbon/graphite nanoparticles generates a steep thermal wave (temperature gradient of up to 500 C/cm) that propagates through the solid reactant mixture (oxides, carbonates or nitrates) converting it to the desired product. The high rate of CO2 release enables synthesis of highly porous, crystalline single phase complex oxides having a particle size in the range of 50-800 nm. The experimental results of fabrication of various devices such as hard and soft magnetic systems, bulk ceramic resistors, capacitors, catalysts, MRI contrast agents will be presented. Key factors that affected to the phase composition, particle size, and morphology of compositions as well as coercivity, magnetization, conductivity and other device characteristics will be discussed.

5bw Rational Design of Catalysts for Energy through Fundamental Research

Carl A. Menning and Jingguang G. Chen. Department of Chemical Engineering, University of Delaware, 150 Academy St., Newark, DE 19716

The increasing demand for prompt energy engineering solutions requires the use of systematic approaches for efficiently designing solutions for tomorrow's energy needs. These systematic approaches of research are founded on the fundamental knowledge of the chemical and electronic behavior of physical matter. As an example of the systematic, rational design of catalysts, the study of the stability of bimetallic electrocatalysts for proton exchange membrane fuel cells is discussed. Considering the number of ways that the first 30 transition metals can be combined, 435 possible combinations, rational design of catalysts is a crucial approach in order to determine the few promising catalysts efficiently. This study focuses on using fundamental surface science methods in conjunction with density functional theory predictions to quantify the thermodynamics and kinetics of preferential metal segregation, an electrocatalyst degradation mechanism. Through this work, a general correlation has been developed for the screening of the stability of bimetallic catalysts.

5bx Systems Biology Approaches to Studying Intracellular Signaling Interactions Between Virus and Host

Kathryn Miller-Jensen[1], Douglas A. Lauffenburger[2], and David V. Schaffer[1]. (1) Chemical Engineering, University of California at Berkeley, Berkeley, CA 94720-3220, (2) Biological Engineering, Massachusetts Institute of Technology, 77 Massachusetts Ave, 56-341, Cambridge, MA 02139

Human host cell responses to viral pathogens are complex processes of immense importance to society. These processes also play a crucial role in viral vector-based gene therapy. My research interest is to use quantitative, systems-oriented approaches to understand how viruses alter the host cell signaling networks that control cell function. Here I describe investigations of two key problems in viral pathogenesis: induction of cell death by adenoviral vectors and the establishment of latency in HIV infection.

Recombinant adenoviruses are used extensively as delivery vectors in clinical gene therapy and in molecular biology, but little is known about how the viral carrier itself contributes to cellular responses. I previously demonstrated that adenoviral vectors (Adv), by "rewiring" intracellular signaling networks, alter human epithelial cell death responses to the inflammatory cytokine tumor necrosis factor alpha (TNF-α). Based on multivariate experimental measurements of kinase signaling pathways, we constructed a partial least squares regression model that describes how Adv alters TNF-induced apoptosis and cytokine secretion as a function of kinase signaling activity data. Importantly, the same functional model predicts Adv-TNF-induced phenotypic behavior in multiple epithelial cell types, given each cell types' unique set of signaling inputs. We demonstrate that cell-specific signal activation is a mechanism for achieving different phenotypes from the same initial stimulus. This property, which we call common effector processing, offers an important general biological principle for understanding and predicting cell-specific responses to non-viral treatments, such as rational drug therapies.

In addition to perturbing the upstream signaling pathways that regulate apoptosis, viruses hijack host cell transcriptional machinery to control viral replication and–in some cases–establish latent infections. In the case of the lentivirus HIV, viral replication causes host cell death, but latent infections can persist for decades and represent the most significant obstacle to eliminating HIV from a patient. I am currently studying how the availability of transcription factors within the host cell contributes to the replication-versus-latency decision. We have previously shown that, for some integration positions, HIV promoter activity can result in a bifurcating expression pattern in an infected clonal population, where some cells have high promoter activity while others have low activity. We are now exploring how the host transcription factor NF-κB and the viral transactivator of transcription Tat affect expression from the HIV promoter at fixed integration positions. Using both computational modeling and experimental measurements, we show that clonal populations will transition between specific expression profiles if the concentration of either of these factors is systematically varied. By quantitatively understanding how transcription factors affect the HIV fate decision, we hope to improve therapeutic strategies to counteract viral latency.

5by Chemical Energy Conversion: Molecular Approach towards the Discovery of Efficient and Environmentally Friendly Heterogeneous Catalysts and Electro-Catalysts

Eranda Nikolla[1], Johannes W. Schwank[2], and Suljo Linic[1]. (1) Department of Chemical Engineering, University of Michigan, Ann Arbor, MI 48109, (2) Chemical Engineering, University of Michigan, 2300 Hayward, 3014 H.H. Dow Building, Ann Arbor, MI 48109-2136

My objective is to utilize cutting edge tools, including various spectroscopies, microscopies and quantum calculations to study processes at the gas/solid and liquid/solid interfaces. The aim of these studies is the development of a molecular knowledge-base that can be utilized to develop efficient and environmentally friendly materials for chemical energy conversion. To illustrate the overall approach I will present some on my recent efforts.

Steam reforming is widely utilized to convert hydrocarbon fuels into hydrogen and oxygenated carbon species. This process is important not only for the catalytic hydrogen production but also for direct electrochemical reforming over high temperature fuel cells such as Solid Oxide Fuel Cells (SOFCs). One of the main problems associated with the process is that the commonly used reforming catalysts, such as Ni supported on oxides, deactivate due to carbon poisoning.

We have utilized Density Functional Theory (DFT) calculations to study carbon chemistry on Ni and Ni-containing alloys. [1,2] The main objective of these studies was to develop molecular insights regarding carbon poisoning on Ni and to utilize this molecular information to identify possible carbon-tolerant alternatives to Ni. The DFT calculations indicated that the Ni surface alloys should a have lower tendency toward carbon poisoning than monometallic Ni. These findings were corroborated by in-house reactor studies and various spectroscopic and microscopic studies. [2,3] The potential utility of these alloy materials as carbon-tolerant SOFC anodes was experimentally demonstrated.

Furthermore, we have also utilized electron energy-loss near-edge spectroscopy (ELNES), in-situ x-ray photoelectron spectroscopy (XPS) and augler electron spectroscopy (AES) to measure the electronic structure of supported monometallic Ni and Ni alloy catalysts. We have demonstrated that the measured electronic fingerprints can be uniquely related to the chemical activity of these materials. Our observations are in accord with the d-band theory of Norskov and Hammer.

References

[1] E. Nikolla, A. Holewinski, J. Schwank, S. Linic, J. Amer. Chem. Soc, 128 (2006) 11354.

[2] E. Nikolla, J. Schwank, S. Linic, J. Catal. 250 (2007) 85 .

[3] E Nikolla, J. Schwank, S. Linic, Catal. Today, (2008) in press.

5bz Development of Fundamental Models of Single-Site Olefin Polymerization Kinetics

Krista A. Novstrup[1], W. Nicholas Delgass[1], Mahdi M. Abu-Omar[2], and James M. Caruthers[1]. (1) School of Chemical Engineering, Purdue University, Forney Hall of Chemical Engineering, 480 Stadium Mall Drive, West Lafayette, IN 47907-2100, (2) Department of Chemistry, Purdue University, 560 Oval Drive, West Lafayette, IN 47907-2084

Single-site catalysts are revolutionizing polyolefin production, the largest by volume polymer produced in the chemical industry. Their impact could be enhanced ever further, however, by a rational method for design of new catalysts to achieve a specific molecular architecture of the polymer. In our view, such a rational design procedure must begin with fundamental knowledge of the micro kinetic mechanism for a given catalyst system and the associated rate constants; information that is difficult to obtain because the number of different mechanistic possibilities for individual steps in the polymerization (i.e. activation, initiation, propagation, chain transfer/termination) can result in potentially thousands of different polymerization models. Our strategy for solving this problem is to take full advantage of the number of experimental techniques (e.g. ^{1}H-NMR, ^{13}C-NMR, UV-VIS, and GPC) used in analyzing both the polymerization and the resulting polymer, to rapidly and robustly use that information in determining which candidate polymerization models can describe the experimental data, and to then design new experiments to discriminate between candidate models.

We will show that batch polymerization data is ideal for modeling single-site polymerization. Since a single experiment samples the rate of polymerization over a range of monomer concentrations, fewer experiments are required for model building and discrimination, although model building is more complex. Kinetic rate constants are primarily determined by fitting rich multi-response data sets consisting of (i) monomer concentration versus time profiles and (ii) the time evolution of the molecular weight distribution for batch polymerizations with (iii) constraints imposed by other measurements like end group analysis by NMR. This integrated analysis approach is unique to our group. Special computational tools have been developed for the population balance models of the polymerization that can include up to 100,000 ODEs, where a new kinetic model can be formulated, computer code automatically generated and the model parameters optimized with a set of experimental data – all within a few hours. An example of this method of kinetic analysis for the polymerization of 1-hexene with [rac-$(C_2H_4(1\text{-Ind})_2)ZrMe][MeB(C_6F_5)_3]$ will be shown. During the course of this analysis it was discovered that new mechanisms beyond those considered in the literature for this catalyst are required to fit the molecular weight distribution. Study of this catalyst system exemplifies the power of this quantitative modeling strategy in conjunction with batch polymerization data by significantly aiding the discovery of new mechanisms that would otherwise not have been considered. Finally, I will discuss how this approach can be extended to develop quantitative descriptions for other polymerization processes for producing novel and technologically interesting polymers.

5ca Optically Integrated Microfluidic Diagnostic Tools

John Oakey, Center For Engineering In Medicine, Massachusetts General Hospital, 114 16th St, Room 1239, Charlestown, MA 02129

Microfluidic tools for cellular research and diagnostics have traditionally been hampered by cumbersome optical requirements that complicate hardware design and operation. We have developed ultra-fast optical direct writing techniques with which we have fabricated and successfully integrated optical components directly into microfluidic platforms. This scheme elminates scaling issues associated with free space optics and enables on-chip optical signal processing at the same scale and density as microfluidic channels, themselves. We have applied this technology to the development of biomedical diagnostic and cell sorting platforms that utilize compact, low-cost devices. I shall present a comprehensive overview of technique and device development as well as demonstration applications for optically integrated microfluidic devices. My focus for applications of this technology has largely been upon on-chip hematology diagnostics and flow cytometry, which represent one area of emphasis in my projected research agenda.

One may read more about the work summarized in this poster in the following references:

R.W. Applegate, Jr., J. Squier, T. Vestad, J. Oakey, D.W.M. Marr, "Optical Trapping, Manipulation and Sorting of Cells and Colloids in Microfluidic Systems With Diode Laser Bars", Optics Express, 12, 4390-4398, 2004.

R.W. Applegate, Jr., J. Squier, T. Vestad, J. Oakey, D.W.M. Marr, P. Bado, M.A. Dugan, A.A. Said, "Microfluidic Sorting System Based On Optical Waveguide Integration and Diode Laser Bar Trapping", Lab on a Chip, 6, 422-426, 2006.

Applegate, R.A., Jr., Schafer, D.N., Amir, W., Squier, J., Oakey, J., Vestad, T., Marr, D.W.M., "Optically Integrated Microfluidic Systems for Cellular Characterization and Manipulation", J. Opt A: Pure Appl. Opt, 9, S122, 2007.

Applegate, R.A., Jr., Squier, J., Vestad, T., Oakey, J., Marr, D.W.M., "Fiber Focused diode bar trapping for microfluidic flow manipulation", Applied Physics Letters, 92, 013904, 2008.

5cb Heterologous Expression of Membrane Proteins in *Saccharomyces Cerevisiae* Enables Biophysical Characterization of Pharmaceutical Targets

Michelle O'Malley and Anne Skaja Robinson. Chemical Engineering, University of Delaware, 150 Academy Street, Newark, DE 19716

Membrane proteins have become an important focus of recent biochemical research, since they play a vital role in cellular communication as well as most biological functions. The G-Protein Coupled Receptors (GPCRs) represent the largest class of membrane proteins, which mediate cellular responses in the presence of extracellular stimuli. GPCRs serve as the primary detection mechanism for sight and smell, and also play direct roles in cancer, diabetes, and HIV infection among other debilitating diseases. In fact, about 50% of all pharmaceuticals on the market target GPCRs, earning annual revenues of over $30 billion. Although there are over 1000 suspected GPCRs in humans, they have proven difficult to study as they are expressed in extremely low levels in native tissues. Biophysical characterization would help elucidate GPCR structure and function, but usually requires micrograms to milligrams of functional protein, which is extremely difficult to achieve due to low natural abundance. Even though GPCRs have been the focus of therapeutic targets for some time, relatively little is known about their expression, folding, interactions, and detailed structure.

The ultimate goal of this work is to elucidate protein folding and explore conformational changes for individual GPCRs by using a non-native, yeast expression system to over-express them. Specifically, the work presented here focuses on (1) understanding heterologous expression of mammalian G-protein coupled receptors in the yeast *S. cerevisiae* and (2) using biophysical techniques to characterize functional GPCRs that have been expressed, purified, and reconstituted from this system.

Engineering *Saccharomyces cerevisiae* for the Expression of Mammalian GPCRs

Given its overall low expense, ease of genetic manipulation, and eukaryotic secretory pathway, yeast are an attractive host organism for the over-expression of complex membrane proteins. In an effort to understand potential limitations to the use of a yeast system for the expression of mammalian GPCRs, recombinant DNA techniques were used to sub-clone 12 GPCRs into the genome of *S. cerevisiae*. Trafficking of these GPCRs through the secretory pathway was monitored using confocal microscopy to follow the expression of green fluorescent protein (GFP) tagged receptors. Overall, our results indicate that GPCRs which fail to localize to the plasma membrane activate the stress response pathway within the endoplasmic reticulum (ER) and preferentially associate with an ER-resident chaperone BiP, as determined through affinity precipitation studies. Additionally, many mis-localized GPCRs trigger the heat shock response pathway within the cytosol. However, by optimizing cell culture conditions, we have managed to express milligram amounts of the human adenosine A_2a receptor (hA_2aR) in this system, which is membrane localized and properly folded. Currently, purified yields obtained for the A_2a receptor are the highest achieved from any heterologous or native expression system. Collectively, these studies have shown that the main bottleneck in yeast-based GPCR expression is located within the ER of the cell. Current efforts are underway to manipulate folding in this organelle to promote proper folding and trafficking of mammalian GPCRs in *S. cerevisiae.*

Characterization of GPCRs using Biophysical Techniques

Expression and purification of milligram amounts of the human adenosine A_2a receptor from *S. cerevisiae* has allowed for extensive biophysical characterization of this G-protein coupled receptor through biophysical studies. GPCRs, as with all membrane proteins, require surfactants for their stabilization outside of their native plasma membrane environment. Therefore, numerous surfactants and lipid additives were screened for their ability to stabilize the active conformation of hA_2aR in a protein detergent complex (PDC), as assessed through ligand binding and neutron scattering techniques. Circular dichroism and intrinsic fluorescence spectroscopy were used to study purified protein actively reconstituted in a protein-detergent complex in order to elucidate general protein stability and explore conformational changes under a variety of conditions. Binding of agonist caused a small blue shift in the emission peak of hA_2aR's intrinsic fluorescence spectra, suggesting a rearrangement of hA_2aR tryptophan residues upon agonist binding. In contrast, no detectable changes in CD spectra of hA_2aR are seen upon binding agonist or antagonist ligands, implying that hA_2aR's alpha helices may rearrange but otherwise remain unaffected during ligand binding. Thermal denaturation of purified receptors shows that hA_2aR is a highly thermostable protein, yet undergoes irreversible aggregation at higher temperatures. The secondary and tertiary structures of hA_2aR are also somewhat more resistant to thermal unfolding in a ligand-bound state. Unfolding with chemical denaturants and hydrostatic pressure leads to an apparent red-shift of the fluorescence emission maxima, which suggests that hA_2aR's tryptophan residues are solvent exposed in an unfolded state. We have also applied these methods towards the characterization of a disulfide bond deficient mutant of hA_2aR, which has suggested that disulfide bonding is critical to maintenance of overall protein stability, yet its absence preserves functionality of the receptor. These results have provided the first experimental insights into conformational changes and protein folding for the hA_2a receptor, and may further direct future folding and re-folding studies for other GPCRs.

5cc Synthesis and Applications of Organic Functionalized Nanoporous Silica

Gifty Osei-Prempeh, Chemical and Materials Engineering, University of Kentucky Extended Campus at Paducah, 4810 Alben Barkley Drive, Paducah, KY 42002

Abstract:

Direct synthesis of organic functionalized nanoporous silica using surfactant templates possess the advantage of uniform distribution of the functional group within the nanoporous silica and tailoring of the degree of functional group loading over post synthesis functionalization.

Incorporation of long chain hydrocarbon and fluorocarbon functional groups, and reactive amine functional group in nanoporous silica using fluorinated and hydrocarbon surfactants as templates was investigated in this research. Surfactant/functional group combinations and the degree of functional group loading had significant effect on the properties of the synthesized silica such as surface area, pore size, structure and order of the pores and surface hydrophobicity.

The wide application areas of these functionalized silicas, such as adsorbent for separation processes, CO2 capture, sensing and catalysis have generated a plethora of research activities. My future research will focus on understanding properties of composite materials formed from organic-functionalized nanoporous metal oxides and polymers; and their application as membranes for fuel cells, sensing, contaminant removal from waste water, CO2 capture and also as surface coatings.

In addition to research I am interested in instructing both graduate and undergraduate students in all chemical engineering curriculum. I have developed instruction material and taught Chemical Reactor Design, Separation Processes, Chemical Engineering Seminar, Engineering Thermodynamics and Process Control at University of Kentucky Extended Campus at Paducah.

5cd Engineering Colloidal Quantum Dot-Based Electronic Devices for Solar Energy Conversion and Color Display Applications

Matthew J. Panzer, Research Laboratory of Electronics, Massachusetts Institute of Technology, Room 13-3082, 77 Massachusetts Avenue, Cambridge, MA 02139

Colloidal semiconductor quantum dots (QDs) are robust, nanosized lumophores featuring narrowband absorption and emission spectra that can be tuned (via their size and/or composition) across a wide range of visible and infrared wavelengths. Since QDs can be synthesized and easily processed in solution, these unique materials naturally suggest their utility in a variety of large-area, potentially inexpensive electronic applications. By pairing microcontact-printed thin films of QDs with organic semiconductor charge transport layers, the demonstration of bilayer photovoltaic cells with large open circuit voltages will be presented. In addition, two potential color display technologies featuring QDs will also be highlighted: a transparent light-emitting diode architecture including air-stable sputtered metal oxide transport layers, as well as an AC-driven structure based on optical downconversion of a commercial blue phosphor.

My general research interests lie in the development and understanding of novel electronic/chemical processes for efficient energy harvesting, storage, and conversion using a wide variety of solution-processable materials (organic molecules, polymers, QDs, and inorganic nanostructures).

Mentors:

Vladimir Bulovi•, Electrical Engineering & Computer Science, Massachusetts Institute of Technology (Postdoc)

C. Daniel Frisbie, Chemical Engineering & Materials Science, University of Minnesota (PhD)

5ce Multiscale Biophysics: Theory and Application to the Dynamics of Actin and Actin-Related Structures

Jim Pfaendtner, The University of Utah and ETH Zürich, Salt Lake City, UT 84112

Molecular modeling and simulation have come to play an indispensable role in both interpreting experimental results as well as advancing the discovery and understanding of biological phenomena. Historically, the main challenges in applying the principles of molecular modeling to systems of biological relevance have been the poor scaling of these techniques and lack of computational power. The multiscale challenge, moreover, is a problem of dimensionality in both length and time scales. For example, the most advanced all-atom molecular simulations are on the order of 2-3 million atoms with a simulation length of up to 100 ns – far short of time and length scales investigated in most biological experiments. As many biological phenomena take place on disparate orders of magnitude, it is of the utmost importance to adopt a multiscale approach. This poster is a presentation of recent research efforts to address both these challenges with application to the study of the actin cytoskeleton.

The structure, properties and function of the cytoskeleton are determined in large part by the matrix of actin filaments, fibers and bundles that constitute the makeup of cytoskeletal architecture. Recently it has been determined that the directed assembly and growth of the network of actin filaments are responsible for cell motility. A key component of this network is the Arp2/3 complex, whose major function is to serve both as a nucleation site for new actin filaments and as an anchor/branch point between existing filaments. We have performed molecular dynamics (MD) simulations of the isolated Arp2/3 complex and probed the interaction between Arp2/3 and full actin filaments. In turn, results obtained from these simulations have been translated into coarse-grained (CG) models that are able to propagate atomic-level details to meso- and larger length scales.

The second part of this poster presents an investigation of the folding of an important alpha-helical domain within the G-actin protein. Such computational investigations present unique challenges given that many protein folding events occur on far longer timescales than those accessible by traditional molecular simulations. Transformation between the folded and unfolded states requires the crossing of a free-energy barrier that is much larger than kBT, and therefore extremely unlikely to be observed during a classical MD simulation. To address these challenges, we have used the method of metadynamics, which is able to bias a simulation along a set of collective variables that offer a reduced description of the process of interest. Using metadynamics allows one to faithfully reconstruct the free-energy surface for the folding event, and to obtain, among other things, estimates of the free-energy differences between the folded and unfolded state as well as an estimate of the free-energy barrier for folding.

5cf Micro/Nanofluidic Devices for Sensing and Reaction Engineering

Adarsh D. Radadia, Chemical and Biomolecular Engineering, University of Illinois, 600 S. Mathews St., Urbana, IL 61801

My expertise lies in fabricating and surface functionalizing micro/nano structured microfluidic components. During my PhD, I have studied mass transfer and performance of serpentine microchannels for micro gas chromatography applications. The kind of questions that I answered were, 1) how to maximize radial mass transfer and minimize axial dispersion in gases, 2) how to functionalize walls uniformly with polymers, and 3) how to avoid artifacts due to wall interactions. Chemicals of interest for separation included organophosphate and organosulfur compounds. My PhD work has greatly contributed to the microGC research by providing design rules on how to build microGC columns.

For my future research directions, I will apply the techniques developed during my PhD study to bio-applications. My research will focus on combining top-down fabrication approaches with surface biofunctionalization techniques to create near-ideal mass transfer conditions for biomicrofluidics.

The questions I will address in my future research include,

1)How to increase mass transfer, minimize dispersion in biomicrofluidics?

2)How to avoid wall interactions, i.e. reduce non-specific binding of biomolecules?

3)How to functionalize walls uniformly keeping biomolecular activity?

4)How to coat channel walls with biomolecules rather than polymeric films?

References:

1) Radadia A.D., Cao L., Jeong H.K., Shannon M.A., Masel R.I., A 3D Micromixer Fabricated with Dry Film Resist, IEEE MEMS, 361-364, 2007.

2) Radadia A.D., Shannon M.A., Masel R.I., New Column Designs for MicroGC, TRANSDUCERS, pg. 2011-2014, 2007.

3) Bae B.H., Yeom J., Radadia A.D., Masel R.I., Shannon M.A., A Fully-Integrated MEMS Preconcentrator for Rapid Gas Sampling, TRANSDUCERS, pg. 1497-1500, 2007.

4) Radadia A.D., Microfluidics for Bio- and Chemical Reactions, Encyclopedia of Micro and Nanofluidics (In Print).

5) Yeom J., Oh I., Field C., Radadia A., Ni Z., Bae B., Han J., Masel R.I., Shannon M.A., Enhanced toxic gas detection using a MEMS preconcentrator coated with the metal organic framework absorber, IEEE MEMS 2007 (In Print).

6) Lee C.Y., Sharma R., Radadia A.D., Masel R.I., Strano S.S., An On-Chip Microfluidic Gas Chromatograph Enabled by a Non-covalently Functionalized Single Walled Carbon Nanotube Sensor Array, Angewandte Chemie Intl. (Accepted).

7) Radadia A.D., Jerrell J., Shannon M.A., Masel R.I., Microcolumns for Organo Phosphonate and Sulfur Compound Separation, Analytical Chemistry (Accepted).

5cg High Recovery Desalination of Brackish Water

Anditya Rahardianto, Chemical & Biomolecular Engineering, University of California, Los Angeles, 5531 Boelter Hall, 420 Westwood Plaza, Los Angeles, CA 90095-1592

With increasing salinity levels of traditional water sources and global water shortages, desalination is emerging as a necessity to maintain sustainability, as well as a solution for exploiting underutilized brackish water sources. However, desalination at inland locations is often not feasible due to the high costs associated with the disposal of high volume of concentrate residuals. Consequently, inland brackish water desalination requires operation at high product water recovery (>90%) to minimize concentrate volume and thus to improve the options for concentrate disposal. The main operational bottleneck to high recovery is membrane mineral scaling, a direct consequence of mineral salts (e.g., $CaSO_4$, $BaSO_4$, $CaCO_3$) being concentrated above their solubility limits, resulting in membrane surface blockage, reduced membrane productivity, and eventual membrane damage. Moreover, the wide variety of brackish water composition suggests that mineral scaling, the type of mineral scalants and the mineral scaling process, are highly site specific. A new approach is thus required to overcome this operational bottleneck to high recovery desalting, utilizing fundamental and systematic methodologies that can robustly transform site-specific challenges into site specific solutions.

This poster will describe the development of brackish water desalting at high recovery using a multi-prong approach, considering both membrane-desalting operations and the integration of accelerated chemical demineralization in accordance to site-specific conditions. The development of membrane-desalting operations from initial assessment of water recovery limits to systematic selection of operating conditions will be presented, underlining the importance of both theoretical analysis and experimental evaluations. Novel membrane scaling evaluation techniques were developed, allowing determination of scale-free operating conditions. These techniques also allowed, for the first time, real-time observation of membrane scale formation, and therefore direct quantification of surface crystal growth kinetics during membrane-desalting operations, as well as antagonistic/synergistic mineral scaling phenomena that are not predicted by theoretical solubility analysis, the current standard in conventional practice

The poster will also present a new approach of integrating chemical demineralization with membrane-desalting operations, which were developed and demonstrated through successive bench-, pilot-, and production-scale operations. Fundamental understanding of the impact of chemical additives on both scale suppression and on mineral precipitation is an integral aspect in the development of the present high recovery approach. An example will highlight how such fundamental understanding can lead to a novel, less chemical-intensive process that may be suitable for high recovery desalting of certain source waters. The present approach enabled water recovery enhancement up to very high levels (at or above 95% water recovery). The work clearly demonstrates that successive switching of strategies (from scale-suppresion during membrane desalting operation to accelerated chemical precipitation during chemical demineralization) is both technically and economically feasible, and can be readily implemented in large-scale desalting.

Relevant publications include:

1. Shih, W.-Y., A. Rahardianto, R.-W. Lee, and Y. Cohen, Morphometric characterization of calcium sulfate dihydrate (gypsum) scale on reverse osmosis membranes, J. Membrane Science 252 (2005) 253

2. Rahardianto, A., W.-Y. Shih, R.-W. Lee, and Y. Cohen, Diagnostic characterization of gypsum scale formation and control in RO desalination of brackish water, J. Membrane Science 279 (2006) 655

3. Shih, W.-Y., J. Gao, A. Rahardianto, J. Glater, Y. Cohen, and C. J. Gabelich, Ranking of antiscalant performance for gypsum scale suppression in the presence of residual aluminum, Desalination 196 (2006) 280

4. Rahardianto, A., J. Gao, C. J. Gabelich, M. D. Williams, and Y. Cohen, High recovery membrane desalting to low-salinity brackish water: integration of accelerated precipitation softening with membrane RO, J. Membrane Science 289 (2007) 123

5. Uchymiak, M., A. Rahardianto, E. Lyster, J. Glater and Y.Cohen, A novel RO ex situ scale observation detector (EXSOD) for mineral scale characterization and early detection, J. Membrane Science 291 (2007) 86

6. Gabelich, C.J., M.D. Williams, A. Rahardianto, J.C. Franklin, and Y. Cohen, High-recovery reverse osmosis using intermediate chemical demineralization, Journal of Membrane Science 301 (2007) 131

7. Rahardianto, A., B. McCool, Y. Cohen, RO desalting of inland brackish water of high gypsum scaling propensity: kinetics and mitigation of membrane mineral scaling, Environmental Science and Technology, 10.1021/es702463a (Web release on February 6, 2008)

8. Zach-Maor, A., R. Semiat, A. Rahardianto, Y. Cohen, S. Wilson, S.R. Gray, Diagnostic analysis of the technical feasibility of RO desalting of treated wastewater, Desalination 230 (2008) 239

Web Page: www.polysep.ucla.edu/andi.htm

5ch The Impact of Microstructure on the Gravitationally-Induced Phase Separation in Concentrated Vesicle Dispersions

Arun Ramachandran, Chemical Engineering, University of California at Santa Barbara, 3235 Engineering II, Santa Barbara, CA 93106

It is well known that most processes for synthesizing vesicles such as lipid film hydration, ethanolic lipid injection in aqueous buffers and reverse phase evaporation result in vesicles that are in kinetically trapped states. As such a concentrated vesicle suspension attempts to shift towards some equilibrium, the size and lamellarity distributions of the vesicles evolve with time. The changes in these distributions have a special implication for non-neutrally buoyant vesicle systems in that a shift towards larger and denser vesicles leads to a greater propensity for them to separate from the suspending fluid on account of gravity. Since the shelf-life expected of vesicle based formulations can be as long as a couple of years, the knowledge of the time scale of coarsening of vesicle dispersions is extremely valuable.

The time required to observe gravitationally-induced separation in vesicle dispersions depends broadly on the following two factors:

(a) The exact procedure used to synthesize the vesicle dispersion is crucial in determining the long term stability of vesicle dispersions, since this establishes the initial size and lamellarity distributions of the vesicles. Poorly designed synthesis techniques which produce large and multilamellar vesicles at the outset lead to high rise velocities of the vesicles, and consequently, shorter separation times. A popular method of preparing a concentrated vesicle suspension is by mixing a concentrated solution of lipids in ethanol with an aqueous buffer. This technique is particularly attractive in the industry because of its capability of being scaled up into a continuous process. However, there is no information in the literature on the size and lamellarity distributions that result from this mixing process for concentrated systems. There are several important process variables that determine these distributions: the properties of the constituent surfactants, the concentration of the ethanolic lipid solution, the flow rates of the aqueous and ethanolic lipid solutions, the design of the mixing unit and the operating temperature. The role of each of these factors will be discussed in the poster.

(b) Any mechanism that leads to bigger and/or more multilamellar vesicles can play an important role in the phase separation process. Even though the vesicles may be small right after synthesis, such mechanisms may quickly produce larger and denser vesicles which gravitationally separate more easily. The usual suspects for vesicle growth are aggregation, such as produced by the presence of a suitable non-adsorbing polymer, and fusion, and these have been addressed in some detail in the literature. However, in our studies of the evolution of these vesicle suspensions, there is evidence of mechanisms such as encapsulation and deflation/invagination leading to larger and denser vesicles, and these new mechanisms will be presented.

The results and questions posed in the poster provide plenty of scope for future experimental and theoretical work in this area. These will be briefly outlined in the poster.

5ci Control of Heterogeneous Polymer Properties through the Use of Nanoparticle and Antiplasticizer Additives

Robert A. Riggleman and Juan J. de Pablo. Department of Chemical and Biological Engineering, University of Wisconsin, Madison, 1415 Engineering Drive, Madison, WI 53706

As polymeric materials such as photoresists and nanocomposites are designed with features approaching the order of tens of nanometers, it becomes crucial for one to control the transport and mechanical properties of polymeric solids on such length scales. It is well known that amorphous materials exhibit dynamic heterogeneity, and several recent studies have shown that amorphous solids also exhibit mechanical heterogeneity on the nanoscale. In my work, I have employed molecular simulations to study how nanoparticles and molecular additives can be used to control the heterogeneous properties of polymers on the nanoscale.

When one adds a low molecular weight additive to a polymeric system, one expects to observe a class of behaviors known as plasticization, whereby the glass transition temperature (Tg), density, and elastic constants are all decreased. However, there is a second class of materials, known as antiplasticizers, which will decrease (Tg) while simultaneously increasing both the density and the stiffness of the polymer. Using molecular dynamics simulations, I have explored the fundamental origins of antiplasticization and its consequences on the nanoscale properties of glass-forming polymers. We have found that the addition of antiplasticizers improves the packing efficiency in bulk polymeric materials, and this leads to a more dynamically homogeneous material. When a dynamically heterogeneous, pure polymer is confined to a free-standing thin film, the dynamics of the polymer near the free surfaces lead to substantial changes in the properties of the entire thin film. However, when the homogeneous, antiplasticized polymer is confined to a free-standing thin film, the effects of the free surface are greatly reduced, leading to improved properties in nanoscale thin polymer films.

More recently, my research has focused on understanding how aging and deformation changes the properties of polymer nanocomposite materials, which is essential to understand as nanocomposite materials are finding increased use in critical applications where they will be subject to deformation. Recent experiments and simulations have shown that the dynamics in glassy solids can be enhanced by several orders of magnitude. Using molecular dynamics simulations on a model nanocomposite material, I have studied how the presence of nanoparticles changes the dynamic enhancement. I have shown that the changes due to the presence of the particles arise from the nanoparticles' ability to stiffen the polymer. Additionally, by examining the entanglement network in a polymer glass and in the nanocomposite, we have found that nanoparticles have the ability to increase the entanglement density in the polymer, leading to increases in the plateau modulus, a key design parameter in many applications.

5cj Mechanistic Investigations of Crystallization: From Self-Assembled Nanomaterials to Pathological Biomineralization

Jeffrey D. Rimer[1], Michael D. Ward[1], Raul F. Lobo[2], and Dion Vlachos[3]. (1) Molecular Design Institute, Department of Chemistry, New York University, 100 Washington Square East, New York, NY 10003, (2) Center for Catalytic Science and Technology, Department of Chemical Engineering, Univesity of Delaware, 150 Academy St., Newark, DE 19716, (3) Director of Center for Catalytic Science and Technology (CCST), University of Delaware, Newark, DE 19716

Developing new scientific initiatives centered on the design and fabrication of nanocrystalline materials, in areas ranging from photovoltaic technologies to therapeutic drug delivery, relies upon expanding fundamental, molecular-level understandings governing their self-assembly, application, and/or physiological function. Design is often challenged by the necessity for predictable and tailored control of material properties through a priori knowledge of growth kinetics and thermodynamics, intra-and inter-particle interactions and interfacial phenomena that regulate particle size, morphology, and polymorphism. Along these lines, research in my doctoral (Department of Chemical Engineering, University of Delaware) and postdoctoral (Molecular Design Institute, New York University) studies merge along a centralized theme of mechanistic investigations of crystallization. Collective results from these studies will be presented in this poster, highlighting (I) my doctoral research on developing predictive models for microporous silica growth from precursor solutions of self-assembled silica nanoparticles, and (II) postdoctoral research on identifying interactions and pathways governing kidney stone formation.

I. Mechanistic Investigations of Microporous Silicate Nucleation and Growth (Ph.D. Research at the University of Delaware)

Research efforts were focused on developing a molecular-level understanding of microporous silicate (i. e., zeolite) crystallization through combined experiments and modeling, concentrating largely on silica nanoparticles (2-6 nm) that are precursors, and possible building units [1] in the synthesis of the all-silica zeolite, silicalite-1. Comprehensive analyses of nanoparticle self-assembly, nucleation, and growth were performed to gain mechanistic insight into their potential role in silicalite-1 crystallization. Protocols were established to measure silica phase behavior [2], identifying the underlying physics governing nanoparticle formation, while simultaneously using a modeling strategy [3] that incorporates chemical equilibria and electrostatic interactions to quantify silica speciation, predict nanoparticle self-assembly, and provide a viable explanation for their colloidal stability. Small-angle scattering and calorimetry experiments were coupled to (i) examine the influence of structure-directing agents on nanoparticle self-assembly [4], (ii) measure kinetic rates and enthalpies of growth and dissolution, and (iii) probe compositional [5] and structural [6] changes that occur during the initial stages of nucleation.

A predictive silicalite-1 growth model was derived from experimental rate constants and kinetic rate expressions that incorporate chemical speciation, linking the mechanism to solution-mediated processes and quantitatively capturing growth rates over a wide range of reaction conditions. This poster will address these findings within the context of mechanisms proposed in the literature, highlighting advancements made toward characterizing processes integral to silicalite-1 formation. These research efforts also provided a methodology that we applied to studies of mesoporous silicate [7] and germanium-oxide nanoparticle 8 syntheses, ultimately serving as a framework toward rational syntheses of tailored nanomaterials and a template for expanding these techniques to areas such as biomimetics and biomineralization to better understand nature's ability to generate a host of exquisitely-structured materials.

II. Pathological Biomineralization of Kidney Stones (Postdoctoral Research at New York University)

Research in my postdoctoral work explores molecular-scale interactions at the surface of calcium oxalate monohydrate (COM) and cystine crystals (two common constituents of kidney stones) using atomic force microscopy (AFM) in an effort to ascertain their underlying mechanisms of formation. Kidney stones are polycrystalline aggregates that assemble through a series of events that include: nucleation, growth, aggregation, and crystal and/or aggregate attachment to epithelial cells. Integral and peripheral membrane proteins can mediate crystal-crystal contact, while a multitude of urinary proteins, polysaccharides, lipids, and cellular debris adsorb to crystal surfaces, likely acting as adhesive glue among aggregates.

Investigations of kidney stone pathogenesis involved (i) in vitro analyses of crystal growth to examine face-specific inhibitory effects in the presence of protein and macromolecular additives [9], and (ii) adhesion force measurements using AFM probes modified with biologically relevant functional groups to study surface-adsorbate interactions involved in crystal adhesion and aggregation [10]. The latter were performed with functional groups that mimic L-arginine and L-glutamic acid residues, and in the presence of additives deemed to influence crystal growth and aggregation, focusing on two predominant urinary proteins: Tamm-Horsfall protein (THp) and osteopontin (OPN). Systematic studies of THp were performed to identify the affect of sialylated and glycoslated side chains on adhesion and growth, while dynamic protein unfolding measurements were conducted on OPN – a protein known to exhibit uncharacteristically strong adhesion to stone surfaces. Using established protocols in the literature for binding proteins to AFM tips, we have expanded the use of single amino-acid functionalized tips to directly measure protein-crystal and protein-protein interactions on stone surfaces. These combined results reveal new insights on physiological processes that regulate kidney stone formation, serving as a basis for future development of targeted drugs capable of inhibiting single crystal nucleation, dramatically reducing crystal growth rates or retention times, and/or preventing interparticle aggregation; thereby offering a generalized approach applicable to a variety of stone-forming crystals as well as analogous physiological diseases, such as gallstones.

References:

1. Kragten, D.D., Fedeyko, J.M., Sawant, K.R., **Rimer, J.D.**, Vlachos, D.G., Lobo, R.F., J. Phys. Chem.

B 2003, 107, 10006

2. Fedeyko, J.M, **Rimer, J.D.**, Lobo, R.F., Vlachos, D.G., J. Phys. Chem. B 2004, 108, 12271

3. **Rimer, J.D.**, Lobo, R.F., Vlachos, D.G., Langmuir 2005, 21, 8960

4. **Rimer, J.D.**, Trofymluk, O., Navrotsky, A., Lobo, R.F., Vlachos, D.G., J. Phys. Chem. C, 2008, submitted

5. **Rimer, J.D.**, Vlachos, D.G., Lobo, R.F., J. Phys. Chem. B 2005, 109, 12762

6. **Rimer, J.D.**, Trofymluk, O., Navrotsky, A., Lobo, R.F., Vlachos, D.G., Chem. Mater. 2007, 19, 4189

7. **Rimer, J.D.**, Fedeyko, J.M., Vlachos, D.G., Lobo, R.F., Chem. Eur. J. 2006, 12, 2926

8. **Rimer, J.D.**, Roth, D.D., Lobo, R.F., Vlachos, D.G., Langmuir 2007, 23, 2784

9. **Rimer, J.D.**, Lee, M.H., Wesson, J.A., Ward, M.D., In Preparation 2008

10. **Rimer, J.D.**, Viswanathan, P., Wesson, J.A., Ward, M.D., In Preparation 2008

5ck Physical Aging and Gas Permeability in Ultra-Thin Glassy Polymer Films

Brandon W. Rowe, Benny D. Freeman, and Donald R. Paul. Chemical Engineering, The University of Texas at Austin, 1 University Station CO400, Austin, TX 78712

Most gas separation membranes are formed from glassy polymers because of their exceptional permeability-selectivity properties. However, glassy polymers are non-equilibrium materials that will spontaneously, but usually slowly, change over time towards an equilibrium state by a process known as physical aging. The physical aging rate becomes orders of magnitude more rapid when film thickness is decreased below one micron, indicating a strong influence of the free surface on aging behavior.1 This phenomenon is an intrinsically fascinating scientific issue, and understanding physical aging has broad impacts in several technologies including gas separation materials, sensors, and microelectronics.

Previous studies have investigated the aging behavior of free-standing films with thicknesses down to approximately 400 nm. These studies showed accelerated aging rates, as compared to bulk-like films, for samples with thicknesses below one micron, and the aging rates increased with decreasing film thickness. 1 Creating defect-free films below this limit for gas permeability measurement is extremely difficult. A coating technique has been developed in this work to enable the study of gas transport properties in glassy films with thicknesses down to 20 nm. This thickness is similar to, if not below, the thickness of the selective skin in state-of-the-art hollow fiber membranes. A thin coating of a highly permeable, rubbery polymer, poly(dimethylsiloxane) (PDMS), effectively blocks the convective flow through the selectivity-destroying defects in the glassy film. The resistance to mass transfer of the PDMS layer is constant, due to its equilibrium nature at the testing conditions, and accounting for this resistance enables calculation of the glassy layer properties. This is the first report, to our knowledge, on the aging behavior of ultra-thin (<100 nm) glassy polymer films monitored by gas permeability.

The gas transport properties of films with thicknesses down to 20 nm were monitored during physical aging at 35 °C. All films exhibited decreasing permeability and increasing pure gas selectivity with time, as expected from the densification caused by physical aging. As film thickness was decreased, the initial permeability decreased while the pure gas selectivity increased. These trends are consistent with the idea of a rapidly aging region near the surface, whereby the thinner films would exhibit a greater aging response before the first measurement could be completed, ~ 1 hour. The aging rate increases with decreasing film thickness, down to ~75 nm. Below 75 nm, the aging rate depends less on thickness and even slightly slows as film thickness decreases. It is suggested that this behavior is caused by the very rapid initial aging in the near-surface region of the film, i.e. the apparent aging rate of the ultra-thin films is reduced because the film is initially closer to the equilibrium state than the thicker films. These results are consistent with the findings of a recent variable energy positron annihilation spectroscopy study where the size of free volume elements was found to be smaller near the surface of these glassy films.

(1) Huang, Y.; Paul, D. R. Polymer 2004, 45, 8377-8393.

5cl Neural Nanotechnology: Nanomaterials for Molecular Imaging and Therapy

Gang Ruan, Chemical and Biomolecular Engineering, the Ohio State University, 140 W 19th Ave, Columbus, OH 43210

The potential of nanotechnology in biology and medicine has been sufficiently demonstrated on a general level by a number of proof-of-principle studies. My proposed research program seeks to apply nanotechnology in tackling significant problems in neural systems. With this overall goal in mind, I will pursue the following specific thematic areas: (1) Probing neuron cell biology by nanotechnology. The optical properties of quantum dots will be used to probe the transport phenomena in neurons; the electric properties of quantum dots will be used to interface with neurons; the mechanical properties of iron oxide nanocrystals will be used to apply force on neurons. (2) Treating neural diseases by nanotechnology. Insights into the workings of neural systems obtained from fundamental research will be applied to develop newer treatment strategies. For example, quantum dots will be used to examine how nanoparticles improve drug delivery across the blood brain barrier. Further, multifunctional nanoparticles will be used for combined imaging and therapy. (3) Platform nanotechnology. Efforts will be made to optimize the surface coating of nanoparticles for decreased non-specific binding and improved biocompatibility, to reduce the size of nanoparticles for minimized interference with the biological events, and to decrease the fluorescence intermittency of quantum dots for non-stopped tracking of biomolecular transport. Combined with newer delivery, targeting, imaging and magnetic manipulation techniques, these nanoparticles are expected to ultimately achieve single molecule-level probing at regular basis in live cells.

My previous research forms the foundation for the above-proposed work, for example: (1) Engineering nanomaterials. I have created drug delivery devices based on combinations of polymer-based and lipid-based systems. I have assembled composites of semiconductor quantum dots and iron oxide nanoparticles. (2) Molecular imaging. I have used quantum dots to image Tat peptides in live cells, especially the downstream events of cellular uptake of Tat-quantum dots. I have been the first to obtain the image of single QDs in live cell cytoplasm. I have successfully used QDs to track motor proteins in live cells. (3) Transport phenomena in biological systems. Experimentally, I have made discoveries in the intracellular trafficking and cellular secretion of Tat-quantum dots, as well as the dynamics of motor proteins. Theoretically, I have investigated the transport of nanoparticles in cytoplasm and the transport of drugs in drug delivery devices. The predictions of the theoretical studies have been confirmed by experimental results.

5cm Challenges in the Modeling of the Gas Plasma-Biomaterial Interaction

Yukinori Sakiyama and David B. Graves. Chemical Engineering, University of California at Berkeley, Berkeley, CA 94720

The state of the art gas plasma technology is now ready to be used in the biomedical field. Several research groups have already demonstrated the capability of atmospheric pressure gas plasmas to inactivate bacteria, transport living cells, and even coagulate blood. The research field, however, is still in the early stages. Fundamental research is indispensable for a deeper understanding of the nature of the phenomena and for further development of the technology in the field of biomedicine.

The ultimate goal of this project is to find an answer to the question: 'how does gas plasma inactivate bacteria/cells or coagulate blood?' It is well known that gas plasma mainly consists of charged particles, neutral species, photons, and an electric field. No general agreement, however, has been reached in addressing the question due to the complex nature of the plasma. A reliable numerical model therefore would play an important role in exploring the mechanisms of the plasma-biomaterial interaction. From the viewpoint of numerical modeling, the plasma-biomaterial interaction is a multi-physics and multi-scale problem. The interaction involves various research fields such as medicine, biology, thermo-fluid dynamics, and plasma physics. In addition, the interaction occurs over a wide range of time and spatial scales. For instance, electron collision frequency is on the order of picoseconds, whereas the time scale of neutral gas flow is on the order of milliseconds. The plasma treatment takes from minutes to days to take effect, depending on the biomaterials being treated. Thus, the plasma-biomaterial interaction is quite a challenging problem. As of yet, no group has succeeded in developing such a numerical model.

Our numerical model using finite element method demonstrates that the plasma needle operates in two discharge mode: the corona-mode under low power condition and the glow-mode under high power condition. The mode transition observed in our simulation is consistent with experimental results. Also, the present study reveals that neutral gas flow plays a very important role in the experimentally observed killing pattern of bacteria by gas plasma at atmospheric pressure.

5cn Green Nanotechnology, Engineering and Synthesis for Energy, Biomedical and Environmental Applications

Carl D. Saquing, Department of Chemical and Biomolecular Engineering, North Carolina State University, 911 Partners Way, Raleigh, NC 27606

My research interests lie in the exploitation of green nanotechnology, engineering and synthesis for the generation of innovative nano- and biomaterials for alternative energy, biomedical, and environmental applications. My masters project obtained fundamental insight on the molecular interactions of model drug isomers in SC CO_2 fluid media [1]. In my doctoral studies, I have developed novel supercritical (SC) deposition methods to synthesize effectively metal (Pt, Ru) nanoparticles (1-20 nm) supported on various nanoporous substrates such as carbon and silica aerogels with potential application as fuel cell electrocatalysts. The resulting nanocomposites were characterized by TEM, SEM, XRD, TGA, Physisorption and Chemisorption Analyses, FTIR, GC-MS and membrane electrode assembly (MEA) testing.[2-5] In a collaborative project, techniques were developed to prepare stabilized drug-loaded biodegradable polymeric nanoparticles for drug delivery and foams for medical devices using supercritical fluids. The morphology and composition of drug-loaded polymeric nanoparticles by TEM, SEM and HPLC, and the physical properties of the foam by sorption, SEM and OM techniques were evaluated.[6-7] In this project, I also successfully performed ring opening polymerizations to synthesize biodegradable poly(lactic acid) and poly(lactic-co-glycolic acid) in SC CO_2 and $CHClF_2$ at different temperatures, pressures and catalyst concentrations and characterized the polymerization products by NMR and GPC techniques. In my postdoctoral work, one set of projects focused on the (1) extraction of lignin from woody biomass [8], (2) use of novel supercritical gas and gas-expanded liquids to perform oxidations on this biomass to produce high-value chemicals and potential biofuels [9] and (3) use of SC CO_2 in the solubilization of amino acids and polypeptides via reverse micelle formation that has an attractive implication on protein processing such as separation from and impregnation into solid matrices.[10] Another set of project in my postdoctoral research involves the use of the electrospinning method to fabricate porous [11-12] functional and hybrid nanofibers[13] for various applications such as wound dressing[14] , tissue scaffolds[15] , drug delivery [16], filtration for capture and recovery of toxic metals[17] and catalysis of nitrophenolic compounds[18] . A few highlights of my contribution through these projects are (1) the development of a one-step green process in the generation of electrospun nanoparticle-nanofiber composites with the use of the electrospinning polymer as the reductant and stabilizer for the nanoparticles at ambient conditions, that is, without the use of usually toxic reducing agent or the application of external energy (2) the utilization of surfactants to aid in the electrospinnability of difficult-to-electrospin polysaccharides and the use of sacrificial species for the formation of porous nanofibers and (3) the coupling of atomic layer deposition and electrostatic self-assembly with electrospinning to generate hybrid nanofibers.

References

1."Steric effects and preferential interactions in supercritical CO_2" C.D. Saquing, F.P. Lucien and N.R. Foster. Ind Eng Chem Res, 1998, 37, 4190-4197.

2."Preparation of platinum/carbon aerogel nanocomposites using a supercritical deposition method" C. D. Saquing, T.T. Cheng, M. Aindow and C. Erkey. J Phys Chem B, 2004, 108, 7716-7722.

3. "Investigation of the supercritical deposition of platinum nanoparticles into carbon aerogels" C.D. Saquing, D. Kang, M. Aindow and C. Erkey. Micropor Mesopor Mater, 2005, 80, 11-23.

4. "Supported platinum nanoparticles by supercritical deposition" Y. Zhang, D. Kang, C.D. Saquing, M. Aindow and C. Erkey. Ind Eng Chem Res, 2005, 44, 4161-4164.

5. "High-resolution TEM characterization of carbon aerogels as catalyst supports", D. Kang, Y. Zhang, C.D. Saquing, C. Erkey and M. Aindow. Mater Res Soc Symp Proc, 2003, 800, 367-372.

6. "Preparation of drug delivery biodegradable PLGA nanocomposites and foams by supercritical CO_2 expanded ROP and by rapid expansion from $CHClF_2$ supercritical solutions" A.D. Asandei, C. Erkey, D. J. Burgess, C.D. Saquing, G. Saha and B.S. Zolnik, Mater Res Soc Symp Proc, 2005, 845 (Nanoscale Materials Science in Biology and Medicine), 243-248.

7. "Preparation of Drug-Biodegradable Polymer Nanocomposites for Drug Delivery via Rapid Expansion from Supercritical Solutions", with A.D. Asandei, C. Erkey, D.J. Burgess, G. Saha and B.S. Zolnik, ACS Polymer Preprints, 2004, 45(2), 770.

8. "Toward a better understanding of the lignin isolation process from wood" A. Guerra, I. Filpponen, L. A. Lucia, C.D. Saquing, S. Baumberger, D.S. Argyropoulos. J. Agric. Food Chem. 54, 5939 -5947, 2006.

9. "Oxidative chemistry of lignin in supercritical carbon dioxide & expanded liquids" C.D. Saquing, D. S. Argyropoulos, A.R. Gaspar, N.U.Soriano, L.A. Lucia, O.J. Rojas, Materials,Chemicals and Energy from Forest Biomass, Argyropoulos, D. S., Ed., ACS Symposium Ser., Washington, DC. Book Chapter, 2006.

10. "Solubilizing amino acids and polypeptides in supercritical carbon dioxide via reverse micelle formation" N.U. Soriano, R. Venditti, C.D. Saquing, D. Bushey, D.S. Argyropoulos. Colloids and Surfaces A: Physicochem. Eng. Aspects, 2008, 315, 110–116.

11. "Preparation and characterization of silica nanoparticulate–polyacrylonitrile composite and porous nanofibers" L. Ji, C.D. Saquing, S.A. Khan and X. Zhang, Nanotechnology, 19 (085605), 2008.

12. “Porous nylon 6 fibers via a novel salt-induced electrospinning method” A. Gupta.; M. Afshari C.D. Saquing, R. Kotek, S.A. Khan, X. Zhang, A. Tonelli. in preparation for submission to Macromolecules (draft available)

13. “Metal nanoparticles-loaded Al_2O_3 microtubes by atomic layer deposition on electrospun nanofiber templates” C.D. Saquing, Q. Peng, J. Manasco, G. Parsons and S.A. Khan. in preparation. [A portion of this paper was presented at the 235th ACS Spring National Mtg, New Orleans, LA, Apr 6-10, 2008.]

14. “Electrospun nanoparticle-nanofiber composites with the electrospinning polymer as reducing, protecting and templating agent” C.D. Saquing, J. Manasco, and S.A. Khan, in preparation for submission to Nano Letters (draft available). [This paper was presented at the AIChE 2007 Annual Meeting, Salt Lake City, UT, Nov 3-9, 2007.]

15. “Surfactant-assisted electrospun alginate-based polymer blend nanofibers for biomedical applications” C.D. Saquing, J. Manasco, J. M. Saquing, C. Bonino, F. delos Reyes and S.A. Khan, in preparation for submission to Biomacromolecules. [This paper was presented at the 235th ACS Spring National Mtg, New Orleans, LA, Apr 6-10, 2008.]

16. High throughput nanofiber production via electrospinning of polycaprolactone melts” J. Manasco, C. D. Saquing, S. Talwar, J. Hinestroza, and S.A. Khan, in preparation for submission to Macromolecules (draft available) [This paper was presented at the 235th ACS Spring National Mtg., New Orleans, LA, Apr 6-10, 2008.]

17. "Polysaccharide-based polymer blend nanofibers for environmental applications” C.D. Saquing, J. M. Saquing, F. delos Reyes and S.A. Khan, in preparation.

18. “pH activated nanoparticle self-assembly in polymer nanofiber surfaces as catalysts for nitrophenolic compounds” C.D. Saquing, J. Genzer and S.A. Khan, in preparation for submission to Adv Materials.

5co Incorporation of Non-Linear Methods into Genome-Scale Models: Expanding the Model for the Butanol Producer *Clostridium Acetobutylicum*

Ryan S. Senger, Department of Chemical Engineering, University of Delaware, Delaware Biotechnology Institute, 15 Innovation Way, Newark, DE 19711 and Eleftherios T. Papoutsakis, Dept. of Chemical Engineering, Delaware Biotechnology Institute, 15 Innovation Way, Newark, DE 19711.

Traditional flux models of metabolic processes, including flux balance analysis and metabolic flux analysis, have been an important computational element of metabolic engineering since their development, nearly 25 years ago. Central to model development is the pseudo steady-state approximation argument, which allows all differential equations describing a metabolic process to be written as linear equations (forming a stoichiometric matrix and a vector of reaction fluxes). The resulting system of equations is then traditionally solved using linear programming. We will present developments in which non-linear metabolic processes are incorporated into a genome-scale model. For example, the selectivity of weak acid (acetate and butyrate) re-uptake by *Clostridium acetobutylicum* for the production of solvents (acetone and butanol) is known to be a non-linear function dependent on extracellular medium conditions. Real-coded genetic algorithms were applied to this problem to replace linear programming in solving the system of equations. More generally, we will show how genetic algorithms can be used with traditional stoichiometric matrices to locate singularities, which may then be identified as targets for further experimental study. Many singularities exist in the recently developed genome-scale model for *C. acetobutylicum*. Here, we also focus on a selected set of singularities of the stoichiometric matrix and the roles of thermodynamics calculations, available transcriptional data, and design of further experiments for developing resolutions.

5cp Designer Emulsions Using Microfluidics and Their Applications

Rhutesh K. Shah, School of Engineering and Applied Sciences (SEAS), Harvard University, 40 Oxford Street, Room 213, Cambridge, MA 02138

I currently investigate fundamental and applied problems related to soft materials, more specifically, colloids, polymers, and emulsions. The research work, which is mainly experimental, can be broadly divided into two main projects (i) Designer emulsions using droplet-based microfluidics and their applications, and (ii) Intermolecular and interparticular interactions of crude oil asphaltenes. This poster will describe the first project on emulsions using microfluidics.

An emulsion is a mixture of two immiscible liquids, where one liquid is dispersed in the form of small drops in another liquid. Most conventional methods for making emulsions involve drop breakup using shear stresses. However, such stresses are not uniform across the system, nor are they finely controlled; emulsions formed in these ways thus consist of drops that are highly polydisperse in size. Microfluidic devices offer an alternate and versatile route to produce emulsions. In contrast to bulk emulsification methods, an emulsion in a microfluidic device is made by precisely fabricating one drop at a time: this process results in a highly monodisperse emulsion. One of the most attractive features of microfluidic techniques is that they enable the fabrication of double, triple, and even higher order emulsions, where the size and number of the encapsulated droplets can be manipulated with unprecedented accuracy. [1-2]

Such monodisperse emulsions can serve as templates for the fabrication of monodisperse particles, vesicles, and "supraparticles". We demonstrate the fabrication of particles with different internal structures [3-4], and second order particles (supraparticles) by guided self assembly of sub-micron size colloidal polymeric solids in droplets [5-6]. The diameters of these particles can be precisely controlled over a range of 10-1000 microns by controlling the droplet size which is a function of the fluid flow rates and device geometry. Such designer particles can be used in numerous applications as rheological probes, particle surfactants, micro-valves, and bio-vehicles for encapsulating and delivering macromolecules in pharmaceutical and cosmetics formulations.

Relevant Publications

1. Shah R.K., Shum H.C., Rowat A., Lee D., Agresti J.J., Utada A.S., Chu L.Y., Kim J.W., Fernandez-Nieves A., Martinez C.J., and Weitz D.A. Designer emulsions using microfluidics; MATERIALS TODAY 2008; 11(4): 18-27.

2. Chu L.Y., Utada A.S., Shah R.K., Kim J.W., and Weitz D.A. Controllable monodisperse multiple emulsions; ANGEWANDTE CHEMIE INT. ED. 2007; 46(47): 8970-8974.

3. Shah R.K., Kim J.W., Agresti J.J., Weitz D.A., and Chu L.Y. Fabrication of monodisperse microgels and gel microcapsules in microfluidic devices; SOFT MATTER 2008; submitted.

4. Chu L.Y., Kim J.W., Shah R.K., and Weitz D.A. Monodisperse thermo-responsive microgels with tunable volume-phase transition kinetics; ADVANCED FUNCTIONAL MATERIALS 2007; 17(17): 3499-3504.

5. Shah R.K., Kim J.W., and Weitz D.A. Fabrication of monodisperse biphasic Janus supraparticles using microfluidics – in preparation;

6. Shah R.K., Lee D., Kim J.W., and Weitz D.A. Monodisperse thermo-responsive microgel vesicles by self assembly of microgels in droplets – in preparation;

5cq Sustainable Energy and Environmental Resource Management: a Systems Theory Approach

Yogendra Shastri, Agricultural and Biological Engineering, Energy Biosciences Institute, University of Illinois at Urbana Champaign, 1304 W. Pennsylvania avenue, Urbana, IL 61801

The intensity and scale of human activities on planet Earth has led to severe depletion and deterioration of natural resources. Since this has strong implications on the continued existence of the human race, understanding and maintaining the ecosystem services is of paramount importance. Sustainability has emerged as a new paradigm of analysis and management. Two aspects that are critical for human and ecosystem sustenance are energy and environment. While energy is the most important driver behind our economic development of the past century, environment plays a critical role in ensuring the well being of various species, including humans. The research presented in this poster focuses on developing sustainable management policies for these two important resources. A systems theory approach based on information theory, efficient modeling and problem solution techniques (such as optimization and control) forms the basis of the work.

In the area of energy management, the work focuses on developing engineering solutions of biomass feedstock production system since biomass feedstock production (BFP) is a critical subsystem within the overall bio-based energy production and utilization system. It is an important precursor for the success of bioenergy initiative and includes the operations of agronomic production of energy crops and physical handling/delivery of biomass, as well as other enabling logistics. The research work will focus on the engineering solutions for its subsystem of biomass feed stock production while keeping in mind its "external" interactions and influencing factors, such as social/economic considerations, environmental impact, and policy/regulatory issues.

In the area of environmental resource management, this work analyzes an integrated ecological-economic-social model. The model is an integration of a food web model and a macro-economic model and is an abstract representation of the world (a 'mini world'). The work conducts scenario studies on the model, where scenarios are possible future developments, to identify potential catastrophic events and the ensuing system-wide consequences. The manifestation of inter-compartmental dependencies is a valuable asset of this analysis. The work then uses Fisher information based objective functions (based in information theory) to derive time dependent management strategies to ensure system stability. The results from the dynamic optimization problem give qualitative guidelines for policy making.

The overall theme of the research work thus is to implement systems theory based approaches to develop sustainable ecosystem.

5cr Engineering Protein Assemblies for Physicochemical Studies and Biomedical Applications

Hadley D. Sikes, Chemical Engineering, California Institute of Technology, Mail Code 210-41, Pasadena, CA 91125

Research interests: electron transport on the nanoscale, enzyme engineering, self-assembled systems, biomaterials

After over a decade of cross-disciplinary training in chemical and biomolecular engineering and physical chemistry, I am prepared to lead a group in several innovative lines of inquiry.

(1) Redox polymerization for low cost, high throughput detection of biomarkers, epigenetic states of genes, and rare cell types

(2) Exploring coupling of domains in redox enzymes using directed evolution combined with spectroscopic and electrochemical techniques

(3) Redox enzymes for tunable, controlled delivery of reactive oxygen species to undesirable cell types

(4) Multi-enzyme assemblies via genetically encoded, self-assembled scaffolds

Publications:

Polymerization-based signal amplification for molecular diagnostics:

H.D. Sikes, R.R. Hansen, L.M. Johnson, R. Jenison, J.W. Birks, K.L. Rowlen, and C.N. Bowman. "Using polymeric materials to generate an amplified response to molecular recognition events" Nature Materials 2008 7, 52. (Highlighted in Science 2007 318:5852, 888, in C&EN 2007 85:46, 45)

R.R. Hansen, H.D. Sikes, C.N. Bowman. "Sub-attomole detection of labeled oligonucleotides using visible light photopolymerization for signal amplification on glass microarray surfaces" Biomacromolecules 2008 9 (1), 355.

Electron transfer:

H.D. Sikes, J.F. Smalley, S.P. Dudek, A.R. Cook, M.D. Newton, C.E.D. Chidsey, S.W. Feldberg. "Rapid electron tunneling through oligophenylenevinylene bridges" Science 2001 291,1519.

S.P. Dudek, H.D. Sikes, C.E.D. Chidsey. "Synthesis of ferrocenethiols containing oligo (phenylenevinylene) bridges and their characterization on gold electrodes" J. Am. Chem. Soc. 2001 123, 8033.

H.D. Sikes, Y. Sun, S.P. Dudek, C.E.D. Chidsey, and P. Pianetta. "Photoelectron spectroscopy to probe the mechanism of electron transfer through oligophenylenevinylene bridges" J. Phys. Chem. B 2003 107:5, 1170.

J. F. Smalley, S. B.Sachs, C.E.D. Chidsey, S. P. Dudek, H. D. Sikes, S. E. Creager, C. J. Yu, S. W. Feldberg, M.D. Newton. "Interfacial electron-transfer kinetics of ferrocene through oligophenyleneethynylene bridges attached to gold electrodes as constituents of self-assembled monolayers: Observation of a nonmonotonic distance dependence" J. Am. Chem. Soc. 2004 126:5, 14620.

Phase transitions in self-assembled systems:

H.D. Sikes, J.T. Woodward, D.K. Schwartz. "Pattern formation in a substrate-induced phase transition during Langmuir-Blodgett transfer" J. Phys. Chem. 1996 100:21, 9093.

H.D. Sikes, D.K. Schwartz. "Two-dimensional melting of an anisotropic crystal observed at the molecular level" Science 1997 278, 1604.

H.D. Sikes, D.K. Schwartz. "A temperature-dependent two-dimensional condensation transition during Langmuir-Blodgett deposition" Langmuir 1997 13, 4704.

J.T. Woodward, I. Doudevski, H.D. Sikes, D.K. Schwartz. "Kinetics of self-assembled monolayer growth explored via submonolayer coverage of incomplete films" J. Phys. Chem. B 1997 101:38, 7535.

5ct Generation of Renewable Fuels and Chemicals from Lipids Via Supercritical Fluid Processing

Darrell L. Sparks, Chemical Engineering, Mississippi State University, Box 9595, Mississippi State University, MS 39762

Identifying renewable sources for fuels and chemicals is currently a paramount global research initiative. Lipids derived from renewable feedstocks such as vegetable oils, animal fats, and microorganisms are a promising alternative to petroleum. For example, lipids derived from soybean oil are used to produce biodiesel, which is currently the dominant fuel alternative to petroleum diesel. In order to generate fuels and chemicals from lipids, typically the lipids are first extracted from lipid-containing biomass. Traditionally, this unit operation is accomplished by extraction with an organic solvent such as hexane. However, hexane is a volatile, flammable material that can leave behind a toxic residue. Hence, if the extracted lipids are to be further processed into food-grade products, extra processing costs can be incurred to ensure that any remaining hexane is at a safe level. Extraction of lipids with supercritical carbon dioxide is an appealing alternative because carbon dioxide is inexpensive, readily available, non-toxic, and non-flammable. Also, supercritical carbon dioxide is considered a tunable extraction medium because changes in temperature or density of the solvent can greatly alter its solvating effectiveness.

Supercritical carbon dioxide can also be an effective reaction medium. Many lipids are composed of unsaturated fatty acids, and once lipids have been extracted from a biomass matrix, the unsaturated fatty acids can be oxidized in a controlled manner (as opposed to autoxidation) to produce a variety of valuable chemical intermediates. The oxidation products are useful in a spectrum of products such as pharmaceuticals and polymers. By conducting the reaction in a supercritical fluid, several advantages can be realized. For example, the reaction rate and conversion can be improved by increasing pressure. Also, if the solubilities of the reaction products in the supercritical fluid are different, then a supercritical fluid separation can be used to fractionate the products. For oxidation reactions, the use of carbon dioxide as the supercritical fluid is very appealing because it is already completely oxidized; therefore it will not break down during the reaction.

This poster will emphasize the author's graduate work under the guidance of Dr. Rafael Hernandez at Mississippi State University. In the first phase of the author's research, supercritical carbon dioxide was evaluated as an alternative lipid extraction media to hexane. The first lipid source considered was rice bran, a component of rice kernels removed during whitening and polishing. Extraction of rice bran lipids with supercritical carbon dioxide provided similar oil yields as hexane on a gravimetric basis; however, use of supercritical carbon dioxide resulted in a much higher solvent efficiency [1]. The other lipid source was municipal wastewater sewage sludge [2]. Once again, supercritical carbon dioxide was comparable to hexane in terms of oil yield. However, addition of a small amount of methanol to the carbon dioxide increased extraction yield significantly [3]. In the second phase, the oxidation of oleic acid (a common lipid component) was performed in supercritical carbon dioxide. Ozonolysis of oleic acid is the most common oxidation reaction performed commercially to generate azelaic acid and pelargonic acid. However, it was found that oleic acid could be oxidized to azelaic and pelargonic acids via potassium permanganate in the presence of supercritical carbon dioxide. In the absence of the supercritical fluid, this reaction occurred at a much lower rate and showed lower selectivity for the products of interest [4]. Therefore, incorporation of supercritical carbon dioxide provided definite advantages. Although solubility data for oleic acid are readily available, the solubility of azelaic acid and pelargonic acid had to be determined. Pelargonic acid had a much higher solubility in supercritical carbon dioxide than azelaic acid [5, 6]. In order to fully understand the solubility results, several models were utilized, including the Peng-Robinson equation of state and Chrastil's equation. The effectiveness of these models was evaluated in predicting the solubility behavior of a range of compounds [7].

Presently, the author is working as a Postdoctoral Research Associate in the Renewable Fuels and Chemicals Laboratory at Mississippi State University. The author's current projects include evaluating separation technologies for extraction of lipids from microorganisms in aqueous media and determination of small-chain fatty acid solubility in supercritical carbon dioxide. The author also serves as a Method Development Specialist for the Bio-Analytical Mass Spectrometry Laboratory (BAMSL) of the Mississippi State Chemical Laboratory. Current BAMSL projects include development of an online analytical technique for determining lipid oxidation products and formulation of standard operating procedures for lipid analysis.

The author's primary research interests are in the areas of renewable fuels and chemicals, supercritical fluids, and lipids. As part of completing the Ph.D. qualifying exam, the author wrote and successfully defended a NSF-style proposal. As a result, the author has several proposal ideas in the aforementioned areas. Potential funding sources for these areas are the Department of Energy, the Department of Agriculture, and the National Science Foundation.

Though confident in the ability to teach any chemical engineering course at both the undergraduate and graduate levels, the author's teaching interests include mass transfer operations, reaction kinetics, thermodynamics, and elective courses in renewable fuels and analytical techniques. The author served as a tutor for Mass and Energy Balances for several years and also taught a junior-level undergraduate course entitled Mass Transfer Operations. Hence, the author has experience interacting with students and is excitedly anticipating the opportunity to teach students on a permanent basis.

[1] Sparks, D., R. Hernandez, M. Zappi, D. Blackwell, and T. Fleming, 2006, Extraction of Rice Bran Oil Using Supercritical Carbon Dioxide and Propane, JAOCS 83: 885-891.

[2] Zappi, M., French, T., Hernandez R., Dufreche S., and Sparks, D., "Production of Biodiesel and Other Valuable Chemicals from Wastewater Treatment Plant Sludges," United States Patent Application 20050112735.

[3] Dufreche, S., R. Hernandez, T. French, D. Sparks, M. Zappi, and E. Alley, 2007, Extraction of Lipids from Municipal Wastewater Plant Microorganisms for Production of Biodiesel, JAOCS 84: 181-187.

[4] Sparks, D., L.A. Estévez, and R. Hernandez, 2008, Oxidation of Oleic Acid with Potassium Permanganate in a Supercritical-Fluid Medium, Lipids (In Preparation).

[5] Sparks, D., L.A. Estévez, N. Meyer, and R. Hernandez, 2007, Solubility of Azelaic Acid in Supercritical Carbon Dioxide, J. Chem. Eng. Data 52: 1246-1249.

[6] Sparks, D., L.A. Estévez, K. Barlow, and R. Hernandez, 2008, Solubility of Pelargonic Acid in Supercritical Carbon Dioxide, J. Chem. Eng. Data 53: 407-410.

[7] Sparks, D., L.A. Estévez, and R. Hernandez, 2007, Evaluation of Density-Based Models for the Solubility of Solids in Supercritical Carbon Dioxide and Formulation of a New Model, Chemical Engineering Science (Under Review).

5cu Genome-Scale Model of a Minimal Organism: *Mycoplasma genitalium*

Patrick F. Suthers, Department of Chemical Engineering, The Pennsylvania State University, University Park, PA 16802

With a genome size of ~580 kb and approximately 480 protein coding regions, *Mycoplasma genitalium* is the smallest known organism that can be grown in pure culture and has extremely fastidious nutrient requirements. The reduced genomic content of *M. genitalium* has led researchers to suggest that the molecular assembly represented by this organism may be a close approximation to the minimal set of genes required for bacterial growth. Furthermore, with the recent announcement of the *de novo* synthesis and assembly of a complete *M. genitalium* genome and advances made in transplanting the genome from one mycoplasma species to another, the construction of a completely synthetic genome seems an achievable target.

Here, I present a systematic approach for the construction and curation of a genome-scale *in silico* metabolic model for *M. genitalium*. Key challenges included estimation of biomass composition, handling of enzymes with broad specificities and the lack of a defined medium. Computational tools were subsequently employed to identify and resolve connectivity gaps in the model as well as growth prediction inconsistencies with gene essentiality experimental data. The curated model, *M. genitalium* *i*PS189 (189 genes, 264 reactions, 276 metabolites) was 87% accurate in recapitulating *in vivo* gene essentiality results for *M. genitalium*. Approaches and tools described herein provide a roadmap for the automated construction of *in silico* metabolic models of other organisms.

Furthermore, this curated metabolic reconstruction allows for the rational design of a defined culture medium and an *in silico* exploration of the minimal number (and alternatives) of metabolic genes required for growth.

5cv Preparation of Heterogeneous Nanostructures for Catalytic and Magnetic Applications

Xiaowei Teng, Center for Functional Nanomaterials, Brookhaven National Laboratory, Upton, NY 11973

The unique catalytic, electronic, and magnetic properties of nanostructured noble metals and their alloys have assured their widespread usage. The performance of these nanomaterials is strongly related to their size, shape, chemical composition, interaction with support, and, to a great extent, to their structural diversity. Although significant advancements were made in controlling the size and size-distribution of metal and alloy nanomaterials, it is no trivial to prepare these materials with well-controlled morphology, chemical composition, and heterogeneous structures. The design, characterization, and applications of these materials will continue to be the focus of vigorous, multi-disciplinary investigations in the future.

My research focuses were on the morphology control of noble-metal-containing nanomaterials such as Pt, FePt, AuPt, Pd, AuPd. The novel catalytic and magnetic performance of these nanostructured materials may afford a solution to present and future energy-related problems. The three distinct research objectives described below spearhead this effort.

„« Synthesis of nanoalloys with controlled distribution of atoms via a wet chemistry approach.

Detailed studies of various two-metal structures, such as intermetalic alloy, random alloy, non-random alloy, or phase segregation, are essential for their applications. Recently, non-random alloys have shown promise in designing catalysts. Our research demonstrated the successful synthesis of a Pd/Au non-random alloy with an Au-rich core and Pd-rich shell by galvanic replacement between Pd nanostructures and Au cations in solution at room temperature, offering great advantages over conventional methods of synthesis that employ the complex UHV system.

„« Designing nanostructures with improved catalytic properties in Direct Methanol Fuel Cell (DMFC) reactions, and the water gas shift (WGS) reaction.

Nanostructures with controlled morphology and chemical composition demonstrated improved catalytic properties. Beyond the conventional synthesis of monoshapes with some basic morphology defined by the crystal's configuration, our established methodology provided a new growth model based on single crystal seeds or multiple twinned crystal seeds that was applicable to the design of, but not limited to, various noble metals nanostructures such as Pt multipods, and Pt and PtRu porous dendrites. In particular, these as-made Pt and PtRu dendrites had a high surface area (14-20 m2/g), and showed improved catalytic activities towards DMFC reactions.

„« Studying ferromagnetic properties of one-dimensional 4d and 5d metals and alloys nanowires.

Magnetic material at the nanoscale presents surprising behavior that does not appear in its bulk form. In confined nanoscale systems, such as a few atomic layers, ultrathin nanowires, or nanoclusters, the reduced coordination and bonding favor more localized electronic states, as well as narrower bands and larger densities of states, bringing very exotic features to the magnetism. Two-dimensional atomic layers and zero-dimensional nanoclusters in several 4d and 5d metals display ferromagnetism, such as palladium, rhodium, ruthenium, and gold, in contrast to diamagnetism or paramagnetism in their bulk counterparts.

Our data from Pd and Pt nanowires revealed the appearance of strong "localized" ferromagnetism in these conventional paramagnetic materials. The unique features of one-dimensionality and localization of electronic states also are highly anticipated in other 4d and 5d metals; such materials likely will carry great potential for designing nano-spintronics devices.

5cw Synthesis of Reactive Copolymer Thin Films by Initiated Chemical Vapor Deposition and Applications

Wyatt E. Tenhaeff and Karen K. Gleason. Department of Chemical Engineering, Massachusetts Institute of Technology, 77 Massachusetts Avenue, Cambridge, MA 02139

Initiated chemical vapor deposition (iCVD) enables the fabrication of chemically well-defined polymeric nanofilms on both planar and three-dimensional substrates. In iCVD, a variant of hot filament chemical vapor deposition, monomer and initiator vapors are introduced into a vacuum chamber, and polymerization occurs on the substrate surface. By including divinyl monomers in the reactor feed, films with controlled crosslinker densities can be deposited. Reactive copolymeric thin films containing maleic anhydride (MA) have been deposited, and many applications have been developed that utilize the chemical versatility of the anhydride functional group.

Copolymerization was necessary for MA since the monomer does not hompolymerize readily. By flowing styrene and MA simultaneously into the iCVD reactor along with tert-butyl peroxide initiator, the deposition rate was 75 nm/min. When MA and styrene were introduced separately, the deposition rates were 5.5 and 0 nm/min, respectively. X-ray photoelectron spectroscopy and ^{13}C nuclear magnetic resonance conclusively show that the films were composed of alternating copolymers, as expected from solution-phase copolymerizations.

By substituting N,N-dimethylacrylamide (DMAA), a monomer that does not form alternating copolymers with MA, for styrene, highly swellable hydrogel nanofilms were created. Before the iCVD process, substrates were primed with 3-aminopropyldimethylethoxysilane, imparting the surfaces with free amine functional groups. Three monomers vapors were introduced into the reactor along with tert-butyl peroxide initiator: 0.6 sccm of MA, 6 sccm of DMAA, and 0.4 sccm of di(ethylene glycol) divinyl ether crosslinker (DEVE). The composition of the polymer was 76% DMAA, 14% MA, and 10% DEVE, as determined by X-ray photoelectron spectroscopy (XPS). The DEVE crosslinker created a network by linking all of the polymer chains together, while MA groups formed covalent bonds with amine groups at the interface between the substrate and hydrogel film. Hydrolysis of the anhydride functionality imparted anionic character to the film. In pH = 8 buffer solution, the film's swelling ratio was 14, while in pH = 1 buffer, the ratio was 7.

Reactive copolymeric nanofilms have also been integrated into ultrathin microelectromechanical switches for chemical sensing. 75 nm of crosslinked copolymer films were applied to 100 nm-thick silicon nitride microcantilevers. Two crosslinking densities were compared; it was shown that higher crosslinking densities yielded greater cantilever deflection upon the reaction of the polymer with analytes. Upon exposure to a hexylamine vapor-phase concentration of 0.9 mol%, the resistance dropped by over six orders of magnitude in less than 90 seconds. The sensor did not respond to nitrogen vapor saturated with heptane or 2-propanol, showing that chemical reaction is necessary for signal transduction. Multiple polymer chemistries have been investigated, and potential defense applications for the sensor will be discussed. Furthermore, alternative device designs will be described.

5cx Studies at the Interface of Materials Science and Bioengineering: New Tools to Assess Conformational Changes in Protein Solution Structure

Efrosyni Themistou, Paschalis Alexandridis, and Sriram Neelamegham. Department of Chemical and Biological Engineering, University at Buffalo - The State University of New York (SUNY), 303 Furnas Hall, Buffalo, NY 14260-4200

The integration of materials science and nanotechnology with biological sciences and medicine is expected to result in major advances in basic science research and it is expected to yield novel therapeutics. With extensive training in biomaterial synthesis, isolation and characterization (polymers, proteins and blood components), I intend to develop a research program at the interface of materials science and bioengineering. The overall goal is to apply engineering principles, chemical synthesis tools, and physicochemical measurements along with biochemical assays to gain new insight in the field of biomedical sciences. Potential application areas will lie in the fields of cardiovascular research and drug delivery. To this end, the investigation of the detailed structure of biomaterials and its stability or possible conformational changes under different physiological conditions is mandatory. As an example, in this poster, I describe the development of new tools that can be used to examine conformational changes in protein solution structure. The focus is on a single polymeric blood protein called von Willebrand Factor (VWF).

VWF is a large soluble protein in blood. This protein plays an important role in arterial thrombosis, a major cause of death worldwide. The protein plays a key role in this disease processes by aiding platelet deposition at sites of vascular injury. Structural changes in VWF regulate the physiological and pathological function of this protein. We applied two complementary strategies to study the effect of fluid forces on the solution structure of VWF: small angle neutron scattering (SANS) and fluorescence methods. Here, we present a fluorescence study performed in VWF solutions by using the fluorescence probe 4,4'-dianilino-1,1'-binaphthyl-5,5'-disulfonic acid dipotassium salt (bis-ANS) that partitions into protein hydrophobic domains. We demonstrate that shear-dependent VWF conformational changes in solution are accompanied by exposure of hydrophobic domains within the protein. A marked increase was obtained in bis-ANS binding at shear rates greater than 2300-6000/s and shear times greater than 1min. VWF protein relaxation was observed in those studies over the course of minutes following shear stoppage. In our higher resolution SANS studies structural changes in VWF were detected at shear rates below 3000/s and at length scales less than 10nm.

Overall, the data collected by these studies suggest that local rearrangements at the domain level under physiological shear rates likely precede unfolding of the protein and changes at larger length scales in higher shear rates that accompany exposure of protein hydrophobic pockets. This study introduces new tools to assess structural changes and stability of soluble biomaterials, a major regulator of the function of these molecules.

5cy Effects of Nanoparticle Addition on the Surface and Bulk Properties of Polymers

Anish Tuteja[1], Wonjae Choi[2], Joseph M. Mabry[3], Gareth H. McKinley[2], and Robert E. Cohen[4]. (1) Department of Chemical Engineering, Massachusetts Institute of Technology, 77 Massachusetts Avenue, Bldg. NE-47, Room 583, Cambridge, MA 02139, (2) Mechanical Engineering, Massachusetts Institute of Technology, Building 3-252, 77 Massachusetts Avenue, Cambridge, MA 02139, (3) Edwards Air Force Base, RZSM, U.S. Air Force, Air Force Research Laboratory, 10 East Saturn Blvd, Edwards AFB, CA 93524, (4) Chemical Engineering, Massachusetts Institute of Technology, 77 Massachusetts Ave., 66-554, Cambridge, MA 02139

Polymer–nanoparticle composite materials offer the unique opportunity to synergistically combine the properties of the polymer and filler on a nanoscale. My research during my Ph.D. and postdoctoral work has been aimed to discover and understand some of the unusual and unexpected enhancements in both the bulk and surface properties caused by the addition of nanoparticles to polymers.

My postdoctoral work with Prof. Gareth H. McKinley and Prof. Robert E. Cohen at Massachusetts Institute of Technology has been aimed at using nanoparticles to modify the surface properties of various polymers in order to develop surfaces with extreme liquid repellency. The various projects that I have been involved with as a postdoc are listed below:

Designing super-oleophobic surfaces.[1, 2] The combination of surface chemistry and roughness' on the micron and nanoscale imbues enhanced repellency to the lotus leaf surface when in contact with a high surface tension liquid such as water (surface tension $\gamma = 72.1$ mN/m). This understanding has led to the creation of a number of artificial superhydrophobic surfaces (water contact angles greater than 150°, low hysteresis). However, researchers so far have been unsuccessful in producing super-oleophobic surfaces for liquids with much lower surface tensions; for example alkanes such as decane ($\gamma = 23.8$ mN/m) or octane ($\gamma = 21.6$ mN/m). Theoretical calculations suggest that a super-oleophobic surface would need to have a surface energy < 5 mN/m, whereas the lowest solid surface energies reported to date are in the range of ~6 mN/m. In this work, we explain how a third factor, surface curvature (apart from surface chemistry and roughness), can be used to significantly enhance liquid repellency, by studying electrospun polymer fibers containing very low surface energy perfluorinated nanoparticles (FluoroPOSS). Increasing the POSS concentration in the elecrospun fibers allows us to systematically transcend from super-hydrophilic to super-hydrophobic and finally to the first ever super-oleophobic surfaces (exhibiting low hysteresis and contact angles with decane and octane greater than 150°).

Design parameters for superhydrophobicity and superoleophobicity.[3] Recent experiments have revealed that the wax on the lotus leaf surface by itself is weakly hydrophilic, and conventional understanding would suggest that such a surface should not be able to support a composite interface, leading instead to a fully wetted interface and low contact angles. Here, we show that this unexpected superhydrophobicity is related to the presence of 're-entrant texture' on the surface of the lotus leaf. We extend this understanding to enable the development of superoleophobic surfaces (i.e. surfaces that repel extremely low surface tension liquids, such as various alkanes), where in most cases no naturally-oleophobic materials exist. We also develop general design parameters that allow us to evaluate the robustness of the composite interface on a particular surface, thereby allowing us to rank various superhydrophobic or superoleophobic substrates in the literature, with particular emphasis on surfaces developed from inherently hydrophilic or oleophilic materials.

My Ph.D. work was done in the lab of Prof. Michael E. Mackay at Michigan State University. Some of the projects that I was involved with during my Ph.D. are described below:

Novel strategies for nanoparticle dispersion.[4] We demonstrated that the relative size of the nanoparticle with respect to the linear chain is the key factor governing miscibility of nanoparticles in polymers. Phase stability is possible in chemically dissimilar systems like polystyrene nanoparticles [5] in polymethylmethacralate or polyethylene nanoparticles in polystyrene (both of these are classical phase separating systems) as long as the nanoparticle radius is less than the radius of gyration of the linear chain; on the other hand, if the nanoparticles become bigger than the linear chain, even polystyrene nanoparticles phase separate from linear polystyrene.

Effects of nanoparticle addition on the viscosity of polymer melts and solutions.[6, 7] Traditionally the addition of particles to suspensions and melts causes an increase in their viscosity (this was first predicted by Einstein about 100 years ago). We demonstrated for the first time non-Einstein-like behavior in polymer melts by establishing that the addition of nanoparticles, even at low volume fractions (less than 5%), to polymers can reduce the melt viscosity by as much as 90%, contradicting Einstein's prediction for an increase in viscosity of solutions on the addition of particles.

Diffusion of nanoparticles in polymer melts.[8] One of our key predictions after the study on the viscosity decrease caused by the addition of nanoparticles to polymer melts was that the nanoparticles must diffuse faster than predicted by the Stokes-Einstein relation, to cause the observed viscosity decrease. Recent experiments using X-ray photon correlation spectroscopy (XPCS) have confirmed this prediction as we obtained order of magnitude faster diffusion coefficients than those predicted by the Stokes-Einstein relation.

Enhanced thermal stability in polymer-nanoparticle composites.[9] We demonstrated for the first time that the addition of nanoparticles greatly improves (up to 10 fold) the thermal stability of polymers. This observation is particularly surprising in the case of polymeric nanoparticles (like polystyrene nanoparticles) where the nanoparticles and the linear polymer have a much faster degradation rate individually.

Chain expansion on the addition of nanoparticles.[10] Recently there has been a lot of controversy in the literature debating the effects of nanoparticle addition on the radius of gyration of the linear polymer. Our extensive experiments on the ideal system of polystyrene nanoparticles in linear polystyrene using small angle neutron scattering have shown conclusively that the nanoparticle addition causes an expansion in the radius of linear chains (swelling).

Rheological characterization of intramolecularly crosslinked nanoparticles.[5] In this work, novel polystyrene nanoparticles were synthesized by the controlled intramolecular crosslinking of linear polymer chains to produce well defined single-molecule nanoparticles of varying molecular mass, corresponding directly to the original linear precursor chain. These molecules were ideal to study the rheological/relaxation behavior of high molecular mass polymers in the absence of entanglements, as the high crosslink density present in these molecules potentially inhibits entanglements. It was found that the reptation model (or the presence of a tube for relaxation) does not explain many of the important rheological features seen for our systems. Instead, a coupling or cooperative motion between the crosslinked loops of the molecules was found to play a significant role in the relaxation process.

Multifunctionality in nanocomposites.[9] Many of the surprising property enhancements caused by the addition of nanoparticles to polymers, as described above, can be engineered to happen for the same volume fraction of the added nanoparticles. Our studies have shown that the addition of certain nanoparticles can simultaneously lead to better processability (reduction in viscosity of up to 80%, contrary to Einstein's prediction), higher electrical conductivity (greater than Maxwell's prediction) and improved mechanical damping and enhanced thermal stability, leading to truly multifunctional nanocomposites.

References:

1. A. Tuteja, W. Choi, M. Ma, J. M. Mabry, S.A. Mazzella, G.C. Rutledge, G.H. McKinley, R.E. Cohen, 'Designing Superoleophobic Surfaces', **Science**, 2007, 318, 1618-1622.

2. A. Tuteja, W. Choi, J.M. Mabry, G.H. McKinley, R.E. Cohen, 'Creating Super-oleophobic Surfaces', **European Polymer Journal – Special Topics**, In Press.

3. A. Tuteja, W. Choi, G. H. McKinley, R. E. Cohen, M. F. Rubner, 'Design parameters for superhydrophobicity and superoleophobicity', **MRS Bulletin**, in press (to appear August, 2008).

4. M. E. Mackay, A. Tuteja et al. 'General Strategies for Nanoparticle Dispersion', **Science**, 311, 1740-1743 (2006).

5. A. Tuteja, M. E. Mackay, C. J. Hawker, B. Van Horn, & D. L Ho, 'Molecular architecture and rheological characterization of novel intramolecularly crosslinked polystyrene nanoparticles.', **Journal Of Polymer Science Part B-Polymer Physics**, 44, 1930-1947 (2006).

6. M. E Mackay, T.T. Dao A. Tuteja et al., 'Nanoscale effects leading to non-Einstein-like decrease in viscosity', **Nature Materials**, 2, 762-766 (2003).

7. A. Tuteja, M. E. Mackay, C. J. Hawker & B. Van Horn, 'Effect of Ideal, Organic Nanoparticles on the

Flow Properties of Linear Polymers: Non-Einstein-like Behavior', **Macromolecules**, 38, 8000-8011 (2005).

8. A. Tuteja, M. E. Mackay, S. Narayanan, S. Asokan & M. S. Wong, 'Breakdown of the continuum stokes-einstein relation for nanoparticle diffusion', **Nano Letters**, 7, 1276-81 (2007).

9. A. Tuteja, P. M. Duxbury, M.E. Mackay, 'Multifunctional nanocomposites with reduced viscosity', **Macromolecules**, 2007, 40, 9427-9434.

10. A. Tuteja, P. M. Duxbury, M. E. Mackay, 'Polymer chain swelling induced by dispersed nanoparticles', **Physical Review Letters**, 2008, 100, 077801.

Web Page: web.mit.edu/atuteja/www/home.html

5cz From Fundamentals to Functionality in Fluidic Systems

O. Berk Usta, Chemical and Petroleum Engineering, University of Pittsburgh, 1249 Benedum Hall, 3700 Ohara St., Pittsburgh, PA 15261

The interactions of rigid and deformable solids and a flowing fluid medium determine behavior and functionality in most biological and industrial systems. These interactions between different phases give rise to a variety of very complex phenomena and hence vital functions such as hearing, smelling, reproduction and threat sensing. Understanding these complex systems is not only important in solving problems with the imminent system but also to create artificial devices that can mimic similar functionality. In my research, I combine simple principles of fluid mechanics and solid mechanics to gain insight into the diverse mechanisms as a result of the fluid-solid interactions in physical and biological systems ranging from hard spheres and polymers to microcapsules and hair cells. My ultimate aim is to solve problems that are relevant to both the scientist in the lab and the broader community using fundamentally sound, state of the art yet simple approaches.

In this session, I will primarily focus on mesoscopic computational methods that help solve the afore-mentioned complex problems in a highly competitive manner and will describe results on some recent exciting problems that have garnered significant attention. These include the migration in dilute polymer solutions; sorting of microcapsules; chemical signaling and self propelled motion; and deformation of viscous and viscoelastic drops.

5da Organ Engineering

Korkut Uygun, Center for Engineering in Medicine, Harvard Medical School/Massachusetts General Hospital, 51 Blossom Street, Boston, MA 02114

There are currently ~100,000 patients on the organ transplant waiting list in the US, a number that far exceeds the supply of available organs, and that continues to grow ~5% each year. The most promising strategies that are being explored as means for addressing this critical shortage are: 1) bioartificial tissue and organ construction (which aims to manufacture tissue and organ analogues in vitro) and 2) donor organ re-engineering methodologies (the goal of which is to recondition marginally damaged organs for transplantation). In either case, success requires the understanding of the interaction of multi-scale components (such as cell types, matrix and vascularization) of the actual tissue. Development of such a systems-level understanding of tissues and organs is limited by two main factors: i) the cell culture systems are very limited in mimicking the tissue microstructure, extracellular matrix interactions, and incorporating multiple cell types; ii) at this level of complexity, use of systems engineering approaches becomes a must in order to model and predict; metabolic engineering toolbox, such as flux analysis is ideal for this purpose, , but have not been well utilized since they have been largely focused on microorganisms - in large part due to the nonexistence of appropriate model organ systems. In this study, we describe an organ culture system development for the dual purposes of i) introducing an accessible experimental platform for development of metabolic engineering models and algorithms for tissue/biomedical engineering applications, and ii) increasing the donor organ pool by reconditioning ischemic organs through reconditioning in a model-controlled system.

In this study, we focus on the reconditioning or marginal livers for transplantation. Liver transplantation is currently the only established treatment for end-stage liver disease, but is limited by a severe shortage of viable donor organs; roughly 3,000 patients per year perish due to lack of transplantable donor organs. Livers obtained from donors after cardiac death (DCD) are currently an untapped donor source for liver transplantation and could increase the supply of donor livers by an estimated 6,000 per year. However, preservation of DCD livers by conventional methods, namely simple cold storage (SCS) in University of Wisconsin (UW) solution, is associated with a higher risk of primary non-function and delayed graft failure. Here we demonstrate a novel Normothermic Extracorporeal Liver Perfusion (NELP) system where rat livers subjected to one hour of ex-vivo warm (34°C) ischemia and preserved for 5 hours using NELP show stable metabolic parameters. Furthermore, following normothermic preservation these ischemically damaged livers could be orthotopically transplanted into syngeneic recipients with close to 100% survival (N=10) after 4 weeks, which was comparable to control animals that received healthy livers preserved by conventional SCS for 6 hours (N=6). On the other hand, animals that received ischemically damaged livers that were stored for 5 hours in UW solution all died within 12 hours post-operatively (N=6). Additionally, animals that received ischemic livers that were directly transplanted without having undergone preservation all died within 36 hours after transplantation (N=6). These results suggest that NELP has the potential to reclaim warm ischemic livers that would not be transplantable otherwise.

Further, the future impact of the organ culture platform is discussed: i) as a practical model for studying fibrosis, hepatitis as well as any other liver disease that can be induced in animal models, iii) for development and early screening of therapeutic methods for such diseases, where the delivery and dosage optimization of the therapeutic agents can be assessed prior to the involvement of the whole body, iii) as a functional liver storage technology where the donor organ can be preserved under direct model-based metabolic control, and iv) as a tool for development of multi-scale models for organs.

5db Electronic Device Fabrication from Engineered Inorganic Nanomaterial-Conducting Polymer Composites

Sreeram Vaddiraju and Karen K. Gleason. Department of Chemical Engineering, Massachusetts Institute of Technology, 25 Ames Street, Bldg. 66, Rm. 419, Cambridge, MA 02139

Strategies for engineering the integration of inorganic nanomaterials on conducting polymer films for the fabrication of stable, cheap and robust (i) chemical sensors and (ii) Light Emitting Diodes (LEDs) are presented. These strategies involve the use of oxidative Chemical Vapor Deposited (oCVD) conducting polymer films for integrating either inorganic quantum dots or metal nanoparticles for device fabrication. These strategies do not disturb the active quantum dot or metal nanoparticle area and thus require no post-processing for obtaining functional devices.

In the first strategy, the –COOH functional groups on top of the oCVD conducting polymer films of poly (3,4-ethylenedioxythiophene-*co*-thiophene-3-acetic acid), P(EDOT-*co*-TAA), were used for tethering metal nanoparticles. Good electronic interaction between the inorganic and organic layers was ensured by using conjugated linker materials, such as 4-aminothiophenol. For the fabrication of a volatile organic compound (VOC) sensor, the property of a metal of exhibiting a change in work function on exposure to gases was exploited. These work function changes could be translated into conductivity changes upon integration of metal nanoparticles on top of the functionalized conducting polymer films. Selectivity in sensing chemicals was achieved by varying the material of the metal integrated on top of these conducting polymer films. Nanoparticles of palladium and toluene integrated on top of P(EDOT-*co*-TAA) were observed to selectively detect acetone and toluene, respectively. Concentrations of acetone and toluene as low as 171 ppm and 51 ppm, respectively, were detected using this technique.

In the second case, cadmium selenide/Zinc sulfide (CdSe/ZnS) quantum dots were assembled on top of the oCVD PEDOT films for the fabrication of red LEDs. Electrostatic self-assembly, using poly(diallyl dimethyl ammonium chloride) (PDDA), was employed for obtaining a uniform layer of quantum dots on top of the conducting polymer film. A sputtered thin film of gold served as cathode. The fabrication process was entirely performed under ambient conditions and resulted in an LED stack composed of ITO/PEDOT/CdSe/ZnS/Au. These red LEDs had a threshold voltage of 3.5 V, and exhibited a luminance of 136.6 Cd/m^2 at an applied voltage of 6 V.

5dc Novel Liquid Crystal Networks

Rafael Verduzco, Center for Nanophase Materials Science, Oak Ridge National Laboratory, 1 Bethel Valley Road, PO Box 2008, MS 6494, Oak Ridge, TN 37831

The combination of liquid crystals and polymers results in fascinating materials in which the elasticity of polymers is coupled to liquid crystal (LC) order. In this work, we explore a variety of novel LC gels and elastomers and explore their electro-optical response, mechanical actuation, and flexoelectric behavior. Physical gels made by polymer self-assembly in an LC solovent are thermoreversibly crosslinked. These gels can be made sufficiently dilute that they show a fast electro-optical response in the gelled state. Due to the coupling of polymer network elasticity and LC order, the gel exhibits fascinating texture transitions with temperature. Furthermore, dynamic light scattering studies reveal multiple director relaxation modes, in contrast to covalent gels which show a single relaxation mode. Dilute chemically crosslinked gels can be prepared via "Click" chemistry of a telechelic SGLCP with azide end-groups and a tri-functional alkyne crosslinker. In contrast to the physically crosslinked networks, the electro-optical behavior is fully reversible. Networks swollen in LC solvents have a low-threshold electro-optic response and show changes in size with temperature due to swelling by the LC solvent. Finally, liquid crystal elastomers based on bent-core liquid crystals have the potential for enhanced flexoelectricity, an electro-mechanical phenomenon whereby a macroscopic polarization develops in response to deformations of the liquid crystal alignment. In order to incorporate bent-core liquid crystals into a polymer network, we prepared two different types of reactive bent core LCs with either an alkene or acrylate functionality. Monofunctional bent-core LCs with a reactive alkene group can be used to make aligned nematic elastomers, and these elastomers can either be studied in the bulk or swollen with bent core LC to investigate their flexoelectic behavior. Following a method demonstrated for calamitic LCs, monofunctional bent-core monomers with an acrylate functionality can also be dissolved in a nematic solvent and polymerized in the presence of a crosslinker to make bent-core LC networks and gels. These various studies of different LC networks and gels provide insight into the connection between physical properties and network structure and also reveal the diversity of behavior accessible with different synthetic approaches.

Web Page: www.ornl.gov/~vgh/

5dd Thermal and Electrochemical Activation in Hydrogen Production for Fuel Cell Applications

Saurabh A. Vilekar, Chemial Engineering Department, Worcester Polytechnic Institute, 100 Institute Road, Worcester, MA 01609

Low temperature fuel cells such as the proton-exchange-membrane (PEM) fuel cells operate well only with rather very pure hydrogen. This is a very big problem, since hydrogen, although very abundant, naturally occurs on earth only in a combined form. Industrially, hydrogen is produced in large quantities from the steam reforming of hydrocarbons. The hydrogen can, of course, eventually be produced from renewable fuels, nuclear energy, wind, or directly from solar energy. In other words, the environmental impact of hydrogen economy is expected to be far less than that of the current fossil fuel based economy, and is the main driving force. The eventual goal is of course to eliminate the use of fossil fuels all together. With hydrogen economy envisioned for automobiles, it is pivotal that technical challenges for widespread production and distribution of hydrogen be thoroughly probed into. Thus, processes for clean hydrogen production need to be rigorously investigated, both mechanistically and kinetically.

With the advent of user-friendly quantum mechanics software, theoretical study of molecular catalytic reaction mechanisms has become rather commonplace as a very insightful tool. The density functional method (DFT) is an inimitable tool for determining the reaction intermediates and building a detailed molecular mechanism for catalytic reactions. However, this is only the starting point for developing a complete understanding of a catalytic system, including the most favorable reaction pathway and the key reaction steps. What is needed is an easy to use addendum to the quantum mechanical software that can utilize its first principles predictions to construct a comprehensive picture of the catalytic system, including elucidation of all parallel pathways and the dominant reaction routes and steps. While software tools such as CHEMKIN are available for this purpose, they are based on rather brute-force numerical methodologies. A more insightful graphic approach is called for that might be readily utilized by the catalytic scientist, without the necessity of being fully conversant with its theoretical underpinnings. Such a consistent graph-theoretic approach will be presented with the focus on H2 production, e.g. steam reforming of methane, methanol decomposition, NH3 decomposition, water-gas-shift reaction (WGSR).

The overall approach involves the following steps: 1) determination of surface intermediates on a given catalyst; 2) generation of a set of plausible surface elementary reaction steps involved in the overall reaction (OR); 3) graph-theoretic generation of a reaction network that depicts the overall mechanism as well as the multitude of reaction routes (RRs); 4) prediction of pre-exponential factors and activation energies, from which energy diagrams are drawn; 5) conversion of the reaction network into an equivalent electrical network that, not only pictorially represents the "circuitry" of reaction pathways, but also allows rigorous flux and kinetic analysis by making use of the vast array of tools available for the analysis of electrical circuits, namely Kirchhoff's current and potential law. Furthermore, the reaction circuitry allows a reduction of the network by a direct and transparent comparison of the reaction step resistances. Moreover, the rate-limiting steps are identified without any ad hoc assumptions.

A major technical barrier for the application of fuel cells for automobiles is the lack of H2 distribution and storage infrastructure. On board reforming of H2 seems to be viable option, with steam reforming constituting a major source. However, CO present in the reformate poisons platinum electro-catalyst and adversely affects the fuel cell performance, which calls for efficient cleanup processes that are easily integrable with the fuel cells. An entirely novel approach involving electrochemical preferential oxidation (ECPrOx) is being developed in our laboratory, where CO electro-oxidation is achieved at the anode by rendering the process electrochemical. In addition, supplemental power is produced with no hydrogen wasted. The electrochemical enhancement of catalytic WGSR is achieved since the rate of an electrochemical reaction is not only determined by temperature, pressure, and composition, but also by the electrode potential. Moreover, the equilibrium constant of an electrocatalytic reaction is a strong function of potential. Hence both, the thermodynamics and kinetics of WGSR can be improved, if carried out electrochemically. The electrochemical device has direct application in on-board reforming process for fuel cell applications and the potential to replace both WGSR and PrOx. The details of the process will be discussed in the poster presentation.

Research Interests: Fuel cells, Fuel reforming and processing, Heterogeneous catalysis and kinetics, Graph theory

My teaching styles and philosophy are mostly based on the experiences that I have had as a student in a variety of different educational settings. Having engaged classes as a Teaching Assistant at Worcester Polytechnic Institute, I have realized that involving students with the lectures through the use of examples, inquiries, demonstrations and group activities provides a more amiable atmosphere in which the students and instructors gain immediate feedback on student comprehension when compared to strictly lecture-style classes.

My strong interest in both research and teaching are further fortified by winning campus and national awards in research (CRE Division's Best Poster Award, AIChE 2007 and Graduate Research award, Worcester Polytechnic Institute, 2007-2008) and teaching (Teaching Assistant of the year award, Worcester Polytechnic Institute, 2008).

5de Optimization of 3D Porous Scaffold Structure for Favorable Fluid Shear and Nutrient Transport Inside a Perfusion Bioreactor

Roman S. Voronov, School of Chemical, Biological, and Materials Engineering, University of Oklahoma, 100 E. Boyd, SEC Room 311, Norman, OK 73019, Samuel VanGordon, Bioengineering, University of Oklahoma, 100 E. Boyd St., SEC T-335, Norman, OK 73019, Dimitrios Papavassiliou, Chemical Engineering, University of Oklahoma, Sarkeys Energy Center, 100 East Boyd Street, Norman, OK 73019, and Vassilios I. Sikavitsas, Chemical, Biological, and Materials Engineering, University of Oklahoma, 100 E. Boyd St., SEC Room T335, Norman, OK 73019.

Due to their degradation characteristics and mechanical properties biodegradable synthetic polymer scaffolds, such as Poly-L-lactic acid (PLLA) which is used in this study, seeded with osteoblastic cells have emerged as potential replacement therapies for damaged or lost bone tissue. The goal of this work is to optimize PLLA biomimetic 3D scaffold structure for enhanced tissue regeneration with three targets in mind: maximized solid-fluid contact area for increased cell adhesion, efficient nutrient transport to osteoblastic cells inside the scaffold for promotion of cell survival and favorable internal stresses inside the scaffold pores for stimulated proliferation. 3D microarchitecture of the scaffolds is characterized using microtomographic (μCT) analysis with a 10μm resolution. Laminar flows of osteogenic media through the cell-seeded cylindrical scaffolds in a flow perfusion bio-reactor are modeled via Lattice Boltzmann Method (LBM) fluid dynamics simulations. High performance computing in conjunction with a house hybrid MPI/Open MP parallelized scheme is employed in order to take advantage inherent LBM parallelizability. Macroscopic mass transfer in the macro-scale is modeled using the Lagrangian Scalar Tracking (LST) method: the motion of scalar markers is decomposed into a convection part (velocity field obtained from LBM simulations) and a diffusion part (Brownian motion obtained from a mesoscopic Monte-Carlo approach). Velocity field and internal stress distribution fields, and parametric nutrient transport study results are presented as a part of this work. Additionally, effects on the bone tissue regenaration of varying the internal scaffold geometry as well as that of the pressure force driving the flow through the scaffold are explored in detail. The study as performed at different time points throughout a two week culturing point. The theoretical framework developed allows for improved tissue generation via: maximum solid-fluid contact area for uniform cell adhesion, efficient nutrient and waste transport inside the scaffolds for promotion of cell survival and favorable mechanical stimulation of cell proliferation.

5df Dependence of Fluid Slip Over Superhydrophobic Surfaces on Equilibrium

Roman S. Voronov, School of Chemical, Biological, and Materials Engineering, University of Oklahoma, 100 E. Boyd, SEC Room 311, Norman, OK 73019, Lloyd L. Lee, Department of Chemical and Materials Engineering, The California State University Pomona, 3801 West Temple Avenue, Building 17-2114, Pomona, CA 91768, and Dimitrios Papavassiliou, Chemical Engineering, University of Oklahoma, Sarkeys Energy Center, 100 East Boyd Street, Norman, OK 73019.

The relation between contact angle and slip length is explored using molecular dynamics for Couette flow of a Lennard–Jones (LJ) fluid between graphite-like walls. Obtained results vary from the common notion that higher contact angles lead to greater slip and to more effective friction reduction. Effects of varying LJ parameters on contact angle and slip length are investigated parametrically. Results are shown in a 3D map, where the third dimension is the square root of the solid–fluid atom size ratio. Epitaxial layer data help to explain unexpected slip length behavior in relation to the contact angle, and reported inconsistencies between slip length experiments and simulations. Therefore, it seems that solids that can produce favorable epitaxial layering of the fluid will cause larger slip. Dimensional analysis is used to elucidate the contact angle-slip length relationship. The present data can be applied to developing artificial supersolvophobic surfaces that are fundamental to friction drag reduction on interfaces.

5dg Liposomal Nanomedicine: Multi-Compartment Structure and Remotely Triggered Release

Guohui Wu, Department of Chemical Engineering, University of California, Santa Barbara, Santa Barbara, CA 93106

My research has been focusing on exploring the relations between structure, composition and function in biomembranes at the molecular scale. Combining nanotechnologies and employing biomaterials, my ultimate goal is to develop "nanomedicine", aiming at early and accurate diagnoses as well as effective targeted therapy.

For example, sub-micron liposome-based nanoparticles encapsulating therapeutic agents represent one of the most advanced classes of drug delivery systems. However, currently marketed drug delivery systems lack the abilities of both good drug retention and the active release of the carried drug. To address these challenges, we learn from nature and have constructed a multi-compartment structure consisting of drug-loaded liposomes encapsulated within another bilayer, similar to the multicompartment structure in eukaryotic cells. These liposome-based complex carriers have excellent retention of the encapsulated drug, protect the drug during circulation, prevent it from reaching healthy tissues, and permit its accumulation at the disease sites. To achieve controlled drug release, it is not feasible to rely on passive diffusion or slow non-specific degradation of the liposomal carrier. We report a novel photo-activated approach where near-complete contents release from liposomes can be initiated within seconds by irradiating hollow gold nanoshells (HGN) with a near infrared (NIR) pulsed laser. The HGNs were encapsulated within, tethered to, or suspended freely with liposomes. NIR light can penetrate up to 10 cm into the tissue, which would allow this approach to apply non-invasively within a significant fraction of the human body. The mechanisms of this triggered release is that absorbing NIR causes the HGNs to rapidly increase in temperature, leading to the formation of microbubbles whose collapse causes transient membrane rupture and content release. Therefore, this remotely triggered release is analogous to using optically triggered nano-"sonicators" deep inside the body to initiate drug release.

The research on liposomal nanomedicine has the potential to advance drug delivery for conventional anticancer treatments as well as gene-delivery, immunostimulatory, and gene-silencing applications using new genetic drugs (plasmid DNA–containing therapeutic genes, antisense oligonucleotides, and small, interfering RNA [siRNA]).

5dh Quantitative Analysis of the Effects of Hydrodynamic Shear on the Textural and Areal Parameters of *S. Aureus* Biofilms

Patrick Ymele-Leki and Julia M. Ross. Chemical and Biochemical Engineering, University of Maryland, Baltimore County, 1000 Hilltop Circle, Baltimore, MD 21250

S. aureus infection is a leading public health threat in community and healthcare settings and has been shown to develop in natural habitats within aggregated biofilms. During infection, initial cell attachment, biofilm growth and maturation, and cell detachment are decisive steps that affect the virulence and the pathogenicity of that organism. The biological, chemical and physical nature of *S. aureus* biofilms grown under hydrodynamic conditions is a critical determinant of persistence, recurrence, and metastatic spread demonstrated by this pathogen. The purpose of this study was to characterize the biophysical properties of early *S. aureus* biofilms grown under physiologically relevant shear rates ranging from 50 to 500 s^{-1}. Immunofluorescent confocal microscopy provided a basis for the quantitative analysis of the architectural attributes of *S. aureus* biofilms *in situ*. Volumetric and areal parameters such as biovolume and cluster perimeter leveled off with increased shear while textural parameters and others including fractal dimension and porosity were shear-dependent over the fluid shear range investigated. Taken together, our findings highlight the effects of fluid shear stress on the biophysical properties of staphylococcal biofilms.

Web Page: www.umbc.edu/cbe/ross/

5di High Density, Vertically-Aligned Carbon Nanotube Membranes with High Fluxes

Miao Yu, Chamical and Biological Engineering, University of Colorado at Boulder, UCB 424, Boulder, CO 80309, John L. Falconer, University fo Colorado, 1111 Engineering Drive, Boulder, CO 80309-0424, and Rich Noble, University of Colorado, UCB 424 Department of Chemical and Biological Engineering, Boulder, CO 80309.

Several studies have reported carbon nanotube (CNT) membranes that consist of aligned nanotubes sealed in a polymer or inorganic matrix1-3. These membranes had single gas selectivities that were approximately Knudsen, and they had high permeation fluxes for liquid and gas feeds in nanotubes. Because the aligned CNTs grew with a low density (~ 1011 CNTs per cm2 of surface area), only a few percent (0.08 ~ 2.7) of the membrane consisted of CNTs; most of it was the sealing material. Thus, although the fluxes per cm2 of CNT area were orders of magnitude higher than other types of membranes, the fluxes per actual membrane area (CNTs plus polymer or inorganic sealant) were much lower.

We have prepared vertically-aligned CNT membranes with a CNT density of 2.9 x 1012 CNTs/cm2, which is approximately 20 times higher than these previous studies by eliminating the need for a polymer or inorganic filler. Aligned CNTs were grown on a silicon wafer with catalyst thin films 1nm Fe/10 nm Al2O3 from e-beam evaporation, the nanotubes were removed from the silicon surface by in-situ water etching, and the nanotubes were collapsed by to about 5% of their original area by solvent evaporation. The tops of the CNT membrane are expected to be open due to water etching4, 5, and the bottoms are also open because the silicon wafers can be reused for several times for CNT growth after detaching the CNT arrays. This preparation is much simpler than that used for composite membranes, and the membranes have much higher fluxes because of the much higher CNT density and additional interstitial transport pathway between nanotubes. These membranes exhibit light gas selectivities that are equal to or greater than Knudsen selectivities, but their permeances are not independent of pressure. Instead, for most gases the permeances decrease with increasing pressure. The permeance at 1 bar pressure drop for N2 through a membrane that was 750 μm thick was 1.2 x 10-4 mol/m2•s•Pa. This corresponds to a permeability of 27 cc (STP)/ m2•s •atm. Thus, these permeabilities are 222, 391 1 and 330003 times higher than those reported for composite membranes.

Because these membranes do not contain a filler, the spaces between the nanotubes must also be available for transport, but apparently the CNTs are close enough together that the flux through these spaces has similar behavior to the flux through nanotubes. The flux of liquid n-hexane through these membranes was approximately 1,500 kg/m2•h at 1 bar pressure drop, which is 3 to 4 orders of magnitude higher than the flux of n-hexane through MFI zeolite membranes, even though these membranes are thicker (750 μm). n-Hexane flux increased linearly with pressure drop. The nanotubes were approximately 3 nm in diameter, as determined by TEM and calculated from N2 desorption isotherms at 77 K. The average space between nanotubes is appropriately 3 nm with a distribution from

1.4 to 7 nm, as calculated by BJH method from N2 desorption isotherms at 77 K.

1. Hinds, B. J.; Chopra, N.; Rantell, T.; Andrews, R.; Gavalas, V.; Bachas, L. G. Science 2004, 303 (5654), 62-65.

2. Holt, J. K.; Park, H. G.; Wang, Y. M.; Stadermann, M.; Artyukhin, A. B.; Grigoropoulos, C. P.; Noy, A.; Bakajin, O. Science 2006, 312(5776), 1034-1037.

3. Kim, S.; Jinschek, J. R.; Chen, H.; Sholl, D. S.; Marand, E. Nano Lett 2007, 7(9), 2806-2811.

4. Ci, L. J.; Manikoth, S. M.; Li, X. S.; Vajtai, R.; Ajayan, P. M. Adv Mater 2007, 19(20), 3300-+.

5. Zhu, L. B.; Xiu, Y. H.; Hess, D. W.; Wong, C. P. Nano Lett 2005, 5(12), 2641-2645.

5dj Micromagnetic - Microfluidic Biomedical Device for Sepsis Therapy

Chong Wing Yung[1], Jason Fiering[2], Mark Puder, and Donald E. Ingber[3]. (1) Surgery - Vascular Biology Program, Harvard Medical School & Children's Hospital Boston, Karp Family Research Lab, 300 Longwood Ave., 11.005D, Boston, MA 02115, (2) Microsystems, Charles Stark Draper Laboratory, 555 Technology Square, Cambridge, MA 02139-3563, (3) Surgery & Pathology, Harvard Medical School & Children's Hospital Boston, Karp Family Research Lab, 300 Longwood Ave., 11.127, Boston, MA 02115

Sepsis is a major killer of man and infected soldiers in the field because the microbial load in blood overcomes even the most powerful existing antibiotic therapies and eventually leads to multi-systems failure. This is a particularly significant problem in premature infants and immuno-compromised patients with fungal infections because use of antifungal therapies is limited by their high toxicity. Development of an extracorporeal blood cleansing device that can be inserted into peripheral blood vessels and rapidly clear the blood of pathogens without removing normal blood cells, proteins, fluids or electrolytes could help remedy this problem, especially when used with low-dosage antifungal agents. We have developed a prototype microfluidic-micromagnetic blood cleansing device that rapidly clears relevant fungal pathogens (*e.g. Candida albicans*) from flowing human whole blood with high efficiency and throughput. The cell separation device is composed of a multiplexed array of linked microfluidic channels microfabricated from poly-dimethylsiloxane (PDMS). Magnetic opsonins were made using 1-μ m superparamagnetic microbeads coated with pathogen-specific antibodies. Using a prototype of the microfluidic blood cleansing device we have selectively remove nearly 80% of living fungal pathogens from a continuously flowing stream of heparinized human whole blood (20 ml/hr/device) in a single pass. We have since further enhanced the volumetric throughput of our devices through multiplexing and optimization of our microfluidic channel designs. Using our high-efficiency and high-throughput microfluidic cell separation devices, we are able to selectively remove fungi and reduce the time necessary to decrease the pathogen load of a continuously flowing stream of contaminated human whole blood.

5dk Surface Interactions in Polymer and Biological Systems at the Micro and Nano Scales

Hongbo Zeng[1], Jacob N. Israelachvili[1], and Matthew Tirrell[2]. (1) Chemical Engineering, University of California at Santa Barbara, 3357 Engineering II, Santa Barbara, CA 93106-5080, (2) Chemical Engineering, University of California, Santa Barbara, 3357 Engineering II, Santa Barbara, CA 93106-5080

Characterization of the adhesion, tribological properties and dynamics of polymer and biological surfaces has been of great interest both experimentally and theoretically for many years. Improving our fundamental understanding of the interactions of polymer surfaces and biological systems at the micro and nano scales is needed to develop further techniques in chemical engineering and materials science. This poster will show our recent progress in:

(1) Adhesion and detachment mechanisms of polymer thin films, especially the transient surface deformations, flows and dynamics during both detachment and attachment (coalescence or spreading) of two polymer films. We find that there is a continuous transition between the failure modes of viscous liquids and glassy solids. A new type of self-organized periodic transient surface fingering pattern and instability was also found to generally exist during the adhesive contact and coalescence of two polymer films.

(2) Understanding the gecko adhesive and related biomimetic Systems. We present a recent theoretical analysis that models the quick attachment and detachment of gecko feet based on a tape mode, and demonstrate the role of the geometry in the enhancement of the adhesion and coupled friction; we also show an analysis of the adhesion and coupled friction of bio-inspired patterned surfaces. Our results, when considered with recent experiments, suggest criteria for simultaneously optimizing the adhesion and friction of patterned surfaces.

5dl Engineered Biosynthesis of Natural Products

Wenjun Zhang, Chemical and Biomolecular Engineering, University of California at Los Angeles, 420 Westwood Plaza, 5667 Boelter Hall, Los angeles, CA 90095 and Yi Tang, Chemical and Biomolecular Engineering, University of California, Los Angeles, 420 Westwood Plaza, Los Angeles, CA 90095.

Natural products are a large group of complex and structurally diverse compounds with important pharmaceutical applications, such as treating cancer, infectious disease, cardiovascular disease etc. Tetracyclines are a family of natural products with broad-spectrum antibiotic activities. Engineered biosynthesis of tetracycline analogs is an attractive option to accelerate the development of the next generation of tetracycline compounds exhibiting novel antibiotic and anticancer properties, as well as to overcome the current modes of antibiotic resistance. Tetracyclines are aromatic polyketides that are biosynthesized by bacterial type II polyketide synthases (PKS). We sequenced the entire gene cluster of oxytetracycline (oxy) PKS from Streptomyces rimosus, and identified the minimal oxy PKS, the initiation module, the immediate tailoring enzymes and the further downstream tailoring enzymes responsible for anhydrotetracycline (ATC) biosynthesis in a heterologous host. Many interesting biochemical features in tetracycline biosynthesis, including tetracyclic ring formation, amination, oxygenation and methylation, have been studied extensively by combining in vivo and in vitro analyses. In addition, tetracycline intermediates, shunt products and analogs have been generated and characterized, demonstrating our heterologous host/vector pair can be a useful platform towards the engineered biosynthesis of new tetracycline analogs. Besides tetracycline family, I have also worked on the engineered biosynthesis of various complex natural products of fungal origin.

One of the most fascinating aspects about natural product biosynthesis is the biosynthesis of the rich and diverse chemical structures, which are difficult to synthesize de novo using synthetic approaches. Therefore, biosynthetic engineering represents a potent solution towards structural diversification and analoging. One of my research interests is to understand and engineer the molecular machinery nature employs during natural products synthesis. The recent sequencing of bacterial and fungal genome provides an unprecedented opportunity for a biosynthetic engineer like me to pursue engineered biosynthesis of natural products. In addition, I am also interested in pathway engineering of other bioactive molecules, and turning a microbe into a dedicated biosynthetic machinery of bioactive molecules or even biofuels will be one of my ultimate research goals.

To train myself in the arena of enzymatic catalysis and biochemistry, I will spend the next two years at Professor Christopher T. Walsh's lab (Harvard Medical School) as a postdoctoral scholar. Combined with my doctoral training in bacterial genetics and metabolic engineering, I will position myself to address some of the most interesting scientific and engineering challenges in biomolecule biosynthesis.

5dm Polymer Films and Ribbons from Electrospinning of Polymer Solutions

Pratyush Dayal, Department of Chemical and Petroleum Engineering, University of Pittsburgh, 1249 Benedum Hall, 3700 Ohara St., Pittsburgh, PA 15261

While the concept of electro-spinning has been known for quite some time, the in-situ experimental monitoring of the development of internal microstructure is not feasible because of the extremely fast electro-jet spinning process that is completed within a few milli-seconds. The theoretical modeling/ simulation is therefore of paramount importance for mimicking the rapid development of microstructures in order to guide the actual electro-spinning process. Our preliminary simulation reveals the trajectory of the morphology evolved along the spinline of the electro-spun fiber in which intricate structural ordering processes are registered. The extension of the present theoretical treatment to the liquid crystalline or rigid rod solutions is of immense interest because of the rich variety of mesophase structures displayed by these systems and also electro-jet extrusion to produce two-dimensional micro-structured films. Given the two-dimensional structure, the electro-jet extruded films have wider utilizations relative to their counterpart – electrospun fibers, because the polymer film industry is orders of magnitude larger than the fiber industry. Moreover, optimizing the process variables such as solvent evaporation rate, flow rate, applied voltage, and viscosity (spinnability) to obtain the desired morphology of the end product not only provides an added value, but also eliminates the post processing step. The knowledge thus acquired will be applied to manufacture of micro-structured materials which would have significant impact to industries producing specialty polymers.

5dn Targetted Cell Encapsulation

Pratyush Dayal, Department of Chemical and Petroleum Engineering, University of Pittsburgh, 1249 Benedum Hall, 3700 Ohara St., Pittsburgh, PA 15261

Phagocytosis is a process used by cellular organisms to engulf food. Once the organisms like Amoeba identify food they use phagocytosis to extract its nutrients and convert it into energy. The process operates in three principal steps: 1) Encapsulation of food by cell membrane, 2) Assimilation of food and 3) Expulsion of waste.

The mechanism of this process can be used to isolate bad cells in animals thereby enabling us to target drugs to specific cells.

57a Providing An Industry-Relevant Perspective to the Process Control Curriculum with Simulation Tools

Douglas Cooper and Rachelle Howard. Chemical Engineering Program; CMBE Department, University of Connecticut, 191 Auditorium Rd., Unit 3222, Storrs, CT 06269-3222

Hands-on challenges that demonstrate and reinforce important concepts benefit the learning process. This is especially true for the often abstract subject of process dynamics and control. Hands-on challenges can be motivating, promote critical thinking, facilitate understanding in the use and limitations of the theory, and help prepare students for the challenges of the professional world.

Too often, the application of textbook theory is limited to solving questions listed at the end of the chapter. One typical question is to have the student expand or extend a mathematical development presented in the book. Another is to provide bits of data and challenge the student to select and employ a combination of formulas to obtain a desired result.

Unfortunately, even when cleverly crafted, these textbook problems fall short of providing students the depth or breadth of practice required for comprehension and mastery. Thus, like most academic institutions, the Chemical Engineering Program at the University of Connecticut supplements textbook learning with unit operations laboratory exercises. Hands-on laboratory exercises are extremely important to learning because they help students make the intellectual transition from theory to practice. The abstractions presented in textbooks are literally brought to life through the tactile nature of a lab experience.

The reality of a unit operations laboratory, unfortunately, is that each study can take many hours and even days to perform. Also, equipment failures and other problems teach the important lesson that the real world can be uncertain. However, this lesson is not usually intended to be the objective of a particular assignment. Thus, students rarely explore more than a very few central concepts in the lab.

A training simulator offers an alluring method for providing students with the significant hands-on practice critical to learning process control. The proper tool can provide virtual experience much the way airplane and power plant simulators do in those fields. It can give students a broad range of focused engineering applications of theory in an efficient, safe and economical fashion. And it can work as an instructional companion as it provides interactive challenges that track along with classroom lectures.

Process control is a subject area well suited to exploit the benefits of a training simulator. Modern control installations are computer based, so a video display is the natural window through which the subject is practiced. With color graphic animation and interactive challenges, a training simulator can offer experiences that literally rival those of the real world. These experiences can be obtained risk free and at minimal cost, enabling students to feel comfortable exploring nonstandard solutions at their desk. If properly designed as a pedagogical tool with case studies organized to present incremental challenges, learning can be enormously enhanced for process control with such a training simulator.

Instructional tool for process dynamics and control should be visually appealing, easy-to use and accepted by students and practitioners. Loop-Pro is software designed to meet these goals so students will:

- learn how to collect and analyze process data to determine the essential dynamic behavior of a process,

- experience what "good" or "best" control performance means for a particular process,

- understand the computational methods behind the different control algorithms and learn when and how to use each one to achieve best performance,

- practice how adjusting the different tuning parameters required for control algorithm implementation impact control performance and explore how to determine values for these parameters,

- become aware of the limitations and pitfalls of each control algorithm and learn how to turn this knowledge to their advantage.

Loop-Pro is composed of three modules: Case Studies, Custom Process and Design Tools. The Case Studies module provides real-world experience in modern methods and practices of process control through a collection of realistic processes to practice upon. The Custom Process module is a block oriented environment that lets students construct a process and controller architecture to their own specifications for a wide range of custom control analyses. The Design Tools module is used to fit approximating dynamic models to process data and to compute PID controller tuning values. The models from Design Tools can also be used to construct advanced control strategies that use process models internal to the controller architecture such as feed forward and model predictive control.

This paper explores the use of the Loop-Pro training simulator for process control education. In particular, the discussion focuses on how this simulator bridges the gap between textbook theory and laboratory learning. Methods and benefits of using Loop-Pro in the undergraduate curriculum are presented. Loop-Pro is currently used by more than 150 universities and colleges worldwide.

Web Page: www.engr.uconn.edu/control/bio.html

57b The Polymath-Excel-Matlab (PEM) Problem Solving System

Michael B. Cutlip, University of Connecticut, 191 Auditorium Rd., Rm. 204, Storrs, CT 06269-3222, Mordechai Shacham, Ben Gurion University of the Negev, Chemical Engineering Department, Beer-Sheva, 84105, Israel, and Michael Elly, Intel Corp., Qiryat Gat, Israel.

The PEM Problem Solving System combines the use of POLYMATH, Excel, and MATLAB in an efficient problem solving environment that greatly increases the ease and efficiency of general problem solving in Chemical and Biochemical Engineering.

The PEM approach is built upon POLYMATH which is a proven computational system that has provided easy-to-use problem solving on personal computers for over 25 years. Capabilities include Linear Equations, Nonlinear Equations, Differential Equations, and Data Analysis including Regressions. POLYMATH enables users to easily achieve effective solutions to numerical problems during interactive problem solving. Graphical and tabular results are provided for easy presentation of results and for incorporation into papers and reports. POLYMATH has a minimal learning curve and provides extensive error checking during problem entry that leads to great efficiency in problem solution.

The Excel options for problem solving within the PEM System is enabled partly by the UNIQUE capability of POLYMATH to automatically export any problem and create an Excel spreadsheet with a single key press. Thus problems can be solved completely in Polymath or exported to Excel for solution. A POLYMATH ODE_Solver Add-In is included and allows for the solution of ordinary differential equations in Excel. Automatic problem export from Polymath to Excel includes the conversion of all intrinsic functions and logical variables.

The MATLAB options within PEM enable advanced problem solving that facilitated by POLYMATH's ability to output problem solution code specifically for MATLAB. This code can be directly incorporated into "m" files. Polymath automatically translates the problem into ordered equations in MATLAB code including logical variables and intrinsic functions. Templates are available that allow easy problem solution in MATLAB, and this coding can provide the basis for more advanced MATLAB use. This approach can be very helpful for novice or experienced MATLAB users.

More details and assorted chemical and biochemical engineering problems that are solved with POLYMATH, Excel, and MATLAB are available from:

http://www.polymath-software.com/

Web Page: www.polymath-software.com/

57c Using Process Simulators in the Study, Design and Control of Distillation Columns for Undergraduate Chemical Engineering Courses

Iván Dario Gil Chaves, Tomás Castrellón, and Diana Carolina Botia Gil. Department of Chemical Engineering, Universidad de los Andes, Calle 19A 1-37 Este office ML-729, Bogotá, Colombia

Today the high school centers must teach to their students the software that will be used by them at work. The present paper shows the importance to promote a closer relation between the engineering computers programs offered by companies likes Aspen Technolgy Inc., and the chemical engineering students. The methodology consists in creating tutorials capable of showing in an interesting way, their subjects to the students through engineering software packages. The main them treated in this tutorials is the simulation and design of distillation columns including a brief look in to the control of distillation columns with dynamic simulations. Three case studies are developed, where the subjects of design and control for a binary simple system and the development of a simulation for a complex ternary system are studied.

57d Real Labs Operated Remotely 24/7

Jim Henry, Chemical Engineering, University of Tennessee at Chattanooga, 615 McCallie Avenue, Chattanooga, TN 37403

Laboratory bottlenecks with too many students are relieved by having teams conduct experiments on REAL laoratory equipment remotely. Labs at UTC and other places are available 24/7. YOur students can use them.

Also, YOU can use them IN CLASS to demonstrate or illustrate concepts in your courses.

The presentation will show the variety of experiments available including

--Distillation

--Several feedback controls systems

The presentation will show the valuable observation of transient performance as well as steady-state in chemical engineering laboratory education.

Web Page: chem.engr.utc.edu/Labs/

57e Superpro Designer: An Interactive Software Tool for Designing and Evaluating Integrated Chemical, Biochemical, and Environmental Processes

Nirupam Pal[1], **Charles Siletti**[2], and Demetri Petrides[2]. (1) California Polytechnic State University, Dept. of Civil and Envr Engr, San Luis Obispo, CA 93407, (2) INTELLIGEN, INC., 700 Walton Ave., Mount Laurel, NJ 08054

The successful design and evaluation of integrated chemical manufacturing and end-of-pipe treatment processes is a challenging task that can be facilitated by the use of advanced computer-aided process design and simulation tools. Such tools also play an important role in teaching environmental engineering and other design-oriented courses. They are particularly useful in emphasizing the importance of systems approach in the design of integrated processes. In our department we have successfully used SuperPro Designer, a comprehensive simulator from Intelligen, Inc. SuperPro carries out material and energy balances, calculates environmentally significant stream properties, estimates the size and cost of equipment, and performs economic evaluation of integrated processes. Using SuperPro, the student and tomorrow's engineer can readily analyze the effect of changes in operating conditions on the environmental impact and the economics of a process. The functionalities of SuperPro and our experience in using this tool in several courses will be presented.

Web Page: www.intelligen.com/literature

57f Simulations for Inquiry-Based Experimentation In Thermodynamics and Heat Transfer

Margot A.-S. Vigeant, Michael J. Prince, and Gavin MacInnes. Chemical Engineering, Bucknell University, Dana Engineering Building, Lewisburg, PA 17837

Inquiry-based activities, where students engage in guided discovery of concepts, have been demonstrated to be highly effective in repairing student misconceptions. For these activities to have the proper effect, students must make predictions, observe how the "real world" confirms or refutes those predictions, and then have the opportunity to restate their ideas based upon their observations, thereby correcting any misconceptions. For some key concepts in thermodynamics and heat transfer, it is difficult to create a physically realizable activity for undergraduates. Therefore we have turned to Flash-based simulations to bring the activities to life. We will demonstrate several web-accessible activities meant to clarify concepts of entropy, reversibility, temperature, and the differences between factors impacting the rate and amount of heat transfer. Students perform significantly better on concept inventory tests after using these simulations.

Web Page: www.facstaff.bucknell.edu/mvigeant/thermo_demos/Carnot_page.html

58a Nanomaterials Curricula: An Interdisciplinary Module for Chemical Engineering

Holly A. Stretz, Department of Chemical Engineering, Tennessee Technological University, Cookeville, TN 38501 and Chris Wilson, Department of Mechanical Engineering, Tennessee Technological University, Brown Hall 311, Cookeville, TN 38505.

Teaching engineering modeling to composites engineers for the performance of carbon nanotube (CNT)/ polymer nanocomposite materials precipitates at least three challenges. First, nanoparticles can alter the properties of the composite on a variety of scales (multiscale modeling). BS chemical engineers may not prepared to model molecular scale phenomema, e.g. with either quantum mechanics or surface chemistry. If the class is interdisciplinary, certainly mechanical or civil engineers will not have the background coursework to handle molecular modeling. The second challenge for engineering education is ensuring that students have appropriate laboratory tools to characterize on the nanoscale. Finally, the third challenge we note is that the field of nanomaterials is advancing so rapidly that any text will be out of date before it is published. Given these difficulties, engineering students would benefit from inquiry-based or case-study based curricula in nanomaterials as the initial offering, and the case of carbon-nanotube-based composites is certainly topical. At Tennessee Technological University an interdisciplinary team of professors is developing such nanocomposite curricula. The initial phase of curriculum development has involved planning and development of written resources; a handbook on CNT composites originally aimed at mechanical engineers. However the first modules have been beta-tested in an interdisciplinary Polymer Engineering course offered through the Department of Chemical Engineering. The modules are constructed as a challenge: by writing practice questions keyed to the literature, students are asked to evaluate the need for multiscale models based on current experimental data. In the process they are exposed to models for composite performance which are simple limits of behavior. We have isolated several material properties of interest, including modulus, conduction of heat, coefficient of thermal expansion, and permeability. In addition we introduce a module on process safety. This resource provides an introduction to composite theory as it may be applied to CNT/polymer nanocomposites. While the curriculum is still in early planning stages, we believe this beta-curriculum begins to address the questions of how to teach nanomaterials in the undergraduate program in an interdisciplinary and flexible fashion. Assessment will also be discussed.

58b Integrating Laboratory Modules Based on Scanning Probe Methods into Undergraduate Engineering Education

Daniel B. Knorr Jr.[1], **Jason P. Killgore**[1], Tomoko Gray[1], Joseph Wei[2], Yeechi Chen[2], Christopher So[3], Hanson Fong[3], David S. Ginger Jr.[2], Mehmet Sarikaya[3], and Rene M. Overney[1]. (1) Chemical Engineering, University of Washington, Box 351750, Seattle, WA 98195, (2) Department of Chemistry, University of Washington, Box 351700, Seattle, WA 98195-1700, (3) Materials Science and Engineering, University of Washington, Box 352120, Seattle, WA 98195-2120

Nanoscience and nanotechnology are clearly thought to be key in the development of next-generation technologies and are expected to contribute substantially to the US economy in the future. While research in nanoscience and nanotechnology is progressing at a frenetic pace, education of scientists and engineers to fill future nanotechnology positions is not. To a large degree this is due to the lack of established, formal education available for those interested in nanotechnology. As such, a substantial amount of educational work has occurred recently with the objective of bringing nanotechnology to universities, and particularly undergraduates.

A particular problem is introducing nanotechnology related laboratories to undergraduate curricula that emphasize the tools that precipitated nanoscience and nanotechnology, namely, scanning probe methods (SPM) such as the scanning tunneling microscope (STM) and the atomic force microscope (AFM). Two key factors that inhibit the use of SPM based modules in a laboratory setting are (1) limitations in resources like appropriate instrumentation and (2) the lack of time for instructors to develop appropriate experimental modules. To address these issues, a partnership between researchers and educators at the University of Washington, North Seattle Community College and two companies, Nanosurf, AG and nanoScience Instruments has been forged. In this collaboration we responded to the above needs by (1) developing a new paradigm for operating and maintaining SPM instrumentation to make resources more readily available and (2) designing a series of nanotechnology laboratory modules that can easily be integrated into undergraduate laboratories.

A variety of SPM techniques and topics are involved in these modules including: STM with atomic scale resolution, non-contact AFM, electrostatic force microscopy (EFM), dip-pen nanolithography, force modulation microscopy, and force spectroscopy. To date, we have successfully tested our modules in an undergraduate laboratory workshop, where students were able to have extensive experience in each of the SPM modes of operation. We have found that these modules provide a truly hands-on experience in a classroom laboratory setting with a small student to instrument ratio. Further information and laboratory modules themselves are available at: http://depts.washington.edu/nanolab/NUE UNIQUE/ NUE UNIQUE.htm.

*This material is based upon work supported by the National Science Foundation under Grant No. 0634088, by the Center for Nanotechnology (CNT) at the University of Washington through a UIF graduate fellowship, and by UW GEMSEC (a NSF MRSEC Center).

Web Page: depts.washington.edu/nanolab/NUE UNIQUE/NUE UNIQUE.htm

Outcomes of a Chemical Engineering Option within a General Engineering Program at a Liberal Arts College

Michael J. Misovich, Hope College, Holland, MI, Jericho L. Moll, University of Illinois at Urbana-Champaign, Urbana, IL, Kurtis F. Blohm, The Ohio State University, Columbus, OH and Emily J. Walsh, SmartSignal Corporation, Lisle, IL

Introduction

Hope College is a small liberal arts college of approximately 3,000 students with nationally recognized programs in the sciences. Over two decades, the physics department progressed from offering a few engineering courses to the development of an ABET-accredited general engineering program.

The renamed physics and engineering department began offering coursework in chemical engineering in 2002-03. Engineering became a separate department during the 2006-07 academic year and it currently offers emphasis options consisting of a set of elective courses in chemical, biochemical, and environmental engineering, as well as emphasis options in mechanical, civil, electrical, and computer engineering. This presentation details the development and structure of the chemical/biochemical/environmental engineering options with a focus on their unique features and the outcomes of program graduates.

The number of chemical engineering graduates has grown from one or two students per year in the first three graduating classes from 2005 to 2007 to five in 2008. Eleven class of 2011 sophomores are currently following the chemical engineering course sequence. Two of the first four graduates are currently enrolled in Ph.D. programs and another is in an M.S. program. Over 80 percent of students in the option participated in a significant undergraduate research experience before graduation.

Another major strength of the program is its natural implementation of multidisciplinary engineering experiences in projects and coursework for students. Some of these ideas may be adaptable to traditional chemical engineering departments in their efforts to promote multidisciplinary experiences. Challenges include limited faculty and laboratory resources, lack of visibility for a small new program, and the constant quest to fit the “square peg” of chemical engineering into the “round hole” of general engineering.

Background on Hope College and its Engineering Program

Chartered in 1866, Hope College is a distinguished and distinctive four-year, liberal arts, undergraduate college, affiliated with the Reformed Church in America. The curriculum offers a variety of courses in 90 majors leading to a Bachelor of Arts, Bachelor of Music, Bachelor of Science, or Bachelor of Science in Nursing degree. During the 2007-08 school year, Hope had 3,226 students from 45 states and territories and 31 foreign countries. [1]

Hope’s science programs have long been recognized for their excellence. As many as six departments in the sciences – biology, chemistry, computer science, geological and environmental sciences, mathematics, and physics and engineering – have held grants through the NSF-REU program simultaneously. Hope has frequently held more of the grants than any other liberal arts college in the country, and more than all but a handful of institutions of any type — including research universities — nationwide. [1]

The college's program in the sciences and mathematics was recognized as a "Program That Works" by Project Kaleidoscope of Washington, D.C., and identified as a model for other institutions to consider. The Research Corporation ranked Hope eighth out of 136 institutions in research publications per faculty member from 1991 to 2000. [1]

According to a study of 518 baccalaureate institutions released by Franklin and Marshall College, Hope ranked in the top six percent in the nation in producing future Ph.D. holders between 1920 and 1995. The department of chemistry was in the top one percent. Data collected by the National Center for Education Statistics of the Department of Education and reported in Chemical & Engineering News showed that Hope produced more chemistry graduates than all but one other undergraduate liberal arts college nationwide between July of 1999 and June of 2000. [1]

A report from the National Science Foundation placed Hope in the top 25 nationally among baccalaureate colleges as a source of future Ph.D. recipients in the natural, physical and social sciences, and engineering, after examining the undergraduate origins of doctorate recipients from 1991 to 1995. Hope ranked third nationally in chemistry, and 14th in psychology. [1]

Hope tied for fourth nationally in the "Undergraduate research/Creative projects" category in the America's Best Colleges 2003 guide published by U.S. News and World Report for its success in teaching through active learning. [1]

Engineering courses have been offered at Hope College since 1979. Initial offerings were instituted by the Department of Physics in response to academic interests of students who were majoring in physics but whose career goals were in engineering. At that time, two faculty members, with interests and training in engineering, began offering a limited number of courses in basic mechanical and electrical engineering topics. During the decade of the 1980s, these courses included Solid Mechanics, Electronics, Thermodynamics, Fluid Mechanics, Material Science, and Vibrations. This curriculum was designed and intended to prepare students for graduate study in engineering. Another option for engineering students was the Hope College Engineering 3-2 Program in which students combined three years of study at Hope College with two years at a traditional engineering school. Upon successful completion of this program, students received a Bachelor of Science degree from Hope College and a Bachelor of Engineering degree from the engineering school. [2]

A Bachelor of Science degree with a major in Engineering Physics was established in 1989. The objective of this degree program was to improve the preparation of Physics students for engineering graduate school. In order to meet the requirements of this new major, the curriculum was modified to offer engineering courses on an alternate year basis. This arrangement allowed efficient use of the existing engineering faculty to provide students with a course pattern which more closely resembled that of a traditional four-year engineering school. As a result of these improvements to the engineering curriculum, the popularity of the 3-2

Program diminished as a majority of engineering students decided to remain at Hope College for four years to pursue a major in Engineering Physics. Most of these students continued their studies in engineering graduate school, although a fair number of students began pursuing employment in industry directly from Hope College. [2]

In 1994, the engineering faculty increased to four members through the addition of two new hires. This growth was partially supported by a grant from the Fund for the Improvement of Post-secondary Education (FIPSE, administered by the Department of Education), which was granted to the college to develop a model for engineering programs at liberal arts colleges. The educational objectives of this expansion were to implement a capstone engineering design experience, provide core engineering classes on an every-year basis, and to increase the number of engineering topics courses offered. These objectives were successfully achieved with the implementation of several changes. These included the development of a two-course capstone sequence in engineering design, the switch of Thermodynamics and Fluid Mechanics from alternate year to every year basis, the development of a freshmen engineering course, and the offering of topics courses in engineering such as Finite Element Analysis, Multi-body Dynamics, and Advanced CAD/CAE. [2]

From 1994 to 1997, as part of the FIPSE-sponsored study of the Engineering Program, a number of external reviewers from both small and large engineering colleges served as external advisors to the Engineering Program. Based partly on their largely positive reviews of the Engineering Program, the Department requested permission from the administration of Hope College to pursue an ABET-accredited engineering degree. Approval was granted in 1997, and the Department established a new degree designation: the Bachelor of Science with a Major in Engineering. This new engineering major was designed and intended to fulfill the degree requirements as specified by the ABET 2000 criteria. In 1998, the Hope College Curriculum Committee officially approved the new engineering major, and the Department of Physics changed its name to The Department of Physics and Engineering. [2]

The engineering program completed and submitted a self-study to ABET and underwent a visit and review during the Fall 1999 semester. In 2000, The Bachelor of Science with a Major in Engineering was accredited by the Engineering Accreditation Commission of ABET, 111 Market Place, Suite 1050, Baltimore, MD 21202-4012, telephone: (410) 347-7700. [2]

In 1997, a fifth engineering faculty member was hired, and in 2001, 2002, and 2005, the department filled vacancies by hiring faculty members with expertise in electrical engineering, chemical engineering, and civil/structural engineering. Additional elective engineering topics courses were added to allow students to select among emphasis options in mechanical, electrical, chemical, computer, biochemical, civil, or environmental engineering. [2]

Design and Structure of the Chemical Engineering Emphasis Option

Prior to the development of the elective emphasis options, the Hope College engineering program focused on mechanical engineering topics and to a lesser extent, electrical engineering topics. Few students indicated an interest in chemical engineering. Those who did typically completed coursework for the engineering major and a chemistry major or minor. Because not all the courses overlapped, these students needed 20 or more additional semester credits beyond the 126 required for graduation at Hope. Furthermore, those students who did complete dual degrees in chemistry and engineering found themselves unprepared for chemical engineering graduate school or industrial positions. They lacked background in mass and energy balances applied to chemical processes, advanced thermodynamics, separation processes, reaction engineering, and other chemical engineering topics.

A chemical engineering emphasis option was designed beginning in 2002 and first appeared in the 2003-04 college catalog under the Bachelor of Science Major in Engineering. Like the existing mechanical and electrical emphasis options, the chemical emphasis was designed to meet ABET requirements for coursework in engineering topics and science topics. The development of the chemical emphasis faced two unique challenges. First, it needed to integrate with the existing engineering courses, taking advantage of overlapping topics where possible. Second, it needed to integrate with the extensive general education requirements of a liberal arts college.

The degree to which these challenges shaped the chemical emphasis option at Hope is illustrated in Table 1. The data for other institutions in Table 1 were compiled as of early 2003 and do not reflect curriculum changes at those institutions since that time; however, the data for Hope have been updated to reflect current requirements.

In this table, the semester credit hours in various topics were tabulated for thirteen chemical engineering programs chosen from three groups. (One school used a quarter system and its credits were converted to semester credits for the table.) Comparisons were made to the Hope engineering program for non-chemical emphasis students and the Hope engineering program for chemical emphasis students.

Topics were categorized as general education (GenEd) if they were general education or liberal arts course credits required of all students for graduation, exclusive of natural science and mathematics topics outside of engineering (Math/Sci). Required engineering topics credits (Engr) were a separate category. The final two categories were technical elective credits (TechElec) which were allowed to be selected from either math, natural science, or engineering topics, and free elective credits.

The first two comparison groups were based upon the 2003 U.S. News & World Report rankings of engineering programs at schools without a Ph.D. program in engineering. The first four schools – Rose-Hulman, Cooper Union, Bucknell, and Manhattan – were listed among the top five chemical engineering programs. The second four schools – Villanova, Minnesota-Duluth, Calvin, and Widener – were additional schools with a chemical engineering program that were listed among the top 50 engineering programs. The final five schools – Michigan, Michigan State, Wayne State, Western Michigan, and Michigan Tech – were all other schools in Michigan that offered chemical engineering programs.

Of the thirteen benchmark schools, only Calvin had an emphasis option within a general engineering program similar to the one being designed. All the others had separate chemical engineering programs, in most cases as their own departments.

Table 1. Benchmarking the Distribution of Topics in Chemical Engineering Programs.

School	Semester hours in					
	GenEd	Math/Sci	Engr	TechElec	FreeElec	Total
Hope non-CHE	42	32	48	0	4	126
Hope CHE	42	47	48	0	0	137
Rose-Hulman	26.0	42.0	52.0	5.3	5.3	130.7
Cooper Union	24	54	48	9	0	135
Bucknell	24	40	64	4	4	136
Manhattan	27	45	59	0	0	131
Villanova	21	44	61	12	3	141
Minnesota-Duluth	19	52	55	3	0	129
Calvin	29	48	54	0	0	131
Widener	28	50	53	6	0	137
Michigan	16	46	56	0	10	128
Michigan State	22	47	49	0	10	128
Wayne State	29	49	58	0	0	136
Western Michigan	18	49	52	17	0	136
Michigan Tech	28	44	53	3	3	131
Averages						
Top 4 CHE	25.25	45.25	55.75	4.58	2.33	133.2
Other top 50 Engr	24.25	48.50	55.75	5.25	0.75	134.5
Other CHE in Mich	22.60	47.00	53.60	4.00	4.60	131.8

The 42 general education credits required at Hope are substantially higher than any of the benchmark schools, which fall in the range 16 to 29. Unlike many colleges and universities, Hope maintains requirements in diverse subject areas including foreign languages, fine arts, and cultural heritage. In math and natural science topics, the 48 credits required for the Hope chemical engineering emphasis fell near the middle of the comparison group. Note that the non-chemical emphasis students at Hope took only 32 credits in these topics since their chemistry requirement was completed by four credits of general chemistry. The 48 credits required at Hope in engineering topics is tied for the lowest among the comparison group schools. This met ABET distribution requirements.

Because of the high general elective credit hour requirements, the chemical emphasis at Hope did not include any elective courses, neither technical electives nor free electives.

Ten of the thirteen comparison schools included either or both types of elective among the credits in their chemical engineering programs. The total number of semester hours required for the program, 137, is near the top end of the comparison group but not extreme. About half the comparison schools required between 135 and 141 credits.

The challenge of fitting the chemical emphasis within the existing engineering program was accomplished by requiring three distinct sets of coursework as shown in Table 2.

Table 2. Coursework in Engineering and Chemistry Topics for the Chemical Emphasis Option.

Required Fundamental Engineering courses: (26 credits)
ENGS 170: Computer Aided Design (1)
ENGS 221: Introduction to Solid Mechanics (4)
ENGS 222: Principles of Engineering Materials (3)
ENGS 224: Mechanics of Materials Lab (1)
ENGS 241: Electronics I (4)
ENGS 331: Dynamic Systems and Controls I (3)
ENGS 333: Dynamic Systems and Controls Lab (1)
ENGS 345: Thermodynamics (3)
ENGS 451: Introduction to Engineering Design (3)
ENGS 452: Engineering Design (3)

Elective Chemical Emphasis courses: (22 credits)
ENGS 251: Conservation Laws & Process Calculations (4)
ENGS 346: Fluid Mechanics (3)
ENGS 348: Heat Transfer (3)
ENGS 371: Chemical Reaction Engineering (3)
ENGS 375: Advanced Thermodynamics and Separations I (4)
ENGS 376: Phase Equilibrium and Separations II (3)
The remaining two elective credits may be selected
from other Engineering courses, or CSCI 160 or 225.

Additional Chemistry courses: (17 credits)
CHEM 114: General and Analytical Chem Lab II (1)
CHEM 121: General Chemistry II (3)
CHEM 221: Organic Chemistry I (3)
CHEM 255: Organic Chemistry Lab I (2)
CHEM 322: Inorganic Chemistry (3)
CHEM 343: Physical Chemistry I (3)
The remaining two Chemistry credits may be selected
from lecture or laboratory courses at the 200 level or above.

26 of the 48 credits in engineering topics are common among all the emphasis options within the engineering program. All students in the program take these courses in common; there are not separate sections. In particular, chemical emphasis students do not have specialized process control, introductory thermodynamics, or design courses, but they take generalized controls, thermodynamics, and design courses with other engineering students.

The chemical engineering emphasis students receive their specialized coursework in the remaining 22 credits of engineering topics and 17 credits of additional chemistry courses. Before the chemical emphasis was offered at Hope, the only engineering topics courses offered from this list were the Fluid Mechanics and Heat Transfer courses. These were a required course and an elective course, respectively, for the mechanical emphasis option. Currently they are taken by both chemical and mechanical emphasis students. The four new engineering topics courses reflect typical chemical engineering educational practice with one exception. Advanced thermodynamics, mass transfer, and separations topics have been integrated into two courses. This is in contrast to standard practice where thermodynamics and mass transfer are taught as distinct courses.

The seventeen credits of additional chemistry courses, when added to the four credits of general chemistry required of all engineering students, also reflect the requirements for a chemistry minor at Hope College.

In 2003, in conjunction with the computer science department, the physics and engineering department designed and added to the catalog a computer emphasis option. It was based upon the existing electrical emphasis option with substitution of some computer science coursework. Following this approach, in 2004 a biochemical emphasis option was designed and added to the catalog. It consists of the same 48 credits of engineering topics courses with a distinct set of additional coursework in chemistry and biology, listed in Table 3, substituted for the courses listed in Table 2 for the chemical emphasis option. In 2008, a chemical engineering based environmental emphasis option was designed, using the coursework in Table 4. This option is expected to be added to the college catalog in 2009.

Table 3. Coursework in Chemistry and Biology Topics for the Biochemical Emphasis Option.

Additional Chemistry/Biology courses: (22 credits)
CHEM 114: General and Analytical Chem Lab II (1)
CHEM 121: General Chemistry II (3)
CHEM 221: Organic Chemistry I (3)
CHEM 231: Organic Chemistry II (3)
CHEM 255: Organic Chemistry Lab I (2)
CHEM 311: Biochemistry I (3)
CHEM 343: Physical Chemistry I (3)
BIOL 240: Cells and Genetics (4)

Table 4. Coursework in Chemistry and Geology/Environmental Science Topics for the Environmental Emphasis Option.

Additional Chemistry/Geology/Environmental Science courses: (19 credits)
CHEM 114: General and Analytical Chem Lab II (1)
CHEM 121: General Chemistry II (3)
CHEM 221: Organic Chemistry I (3)
CHEM 255: Organic Chemistry Lab I (2)
CHEM 343: Physical Chemistry I (3)
GES 211: Earth Environmental Systems I (3)
GES 430: Environmental Geochemistry (4) or GES 450: Hydrogeology (4)

Implementation of the Chemical Emphasis Option and Outcomes Before Graduation

Because of limited department resources, typically only one of the new chemical emphasis engineering topics courses is taught each semester. The three upper-level courses – Advanced Thermodynamics and Separations I, Phase Equilibrium and Separations II, and Chemical Reaction Engineering – are taught once every two years. These are taken by juniors and seniors concurrently. Conservation Principles and Process Calculations is taught each year in the spring, taken by freshmen and sophomore students in the chemical emphasis option and occasionally by upper level students in other emphasis options as an elective course. Table 4 shows a typical course sequence for chemical emphasis option coursework. To increase flexibility for student scheduling, ENGS 331, Dynamic Systems and Controls, has been offered during Hope College's May term in 2007 and 2008. About three-quarters of students in the chemical engineering emphasis option have chosen to take ENGS 331 at that time during the academic year.

Table 4. Typical Chemical Emphasis Option Sequence in Engineering and Chemistry Topics.

	Fall Semester	Spring Semester
Freshmen	Computer Aided Design General Chemistry I General Chemistry Lab I	Conservation Laws & Process Calcs
Sophomore	Introduction to Solid Mechanics Thermodynamics	Fluid Mechanics Heat Transfer General Chemistry II General and Analytical Chem Lab II
Junior	Electronics I *Advanced Thermodynamics and Sep I Organic Chemistry I Organic Chemistry Lab I	Principles of Engineering Materials Mechanics of Materials Lab *Phase Equilibrium and Separations II Inorganic Chemistry
Senior	Dynamic Systems and Controls I Dynamic Systems and Controls Lab Introduction to Engineering Design *Chemical Reaction Engineering Physical Chemistry	Engineering Design

*Taken in alternate year by students graduating in odd years.

Interest in the chemical emphasis option has grown steadily in the six years it has been offered. Table 5 gives data regarding the number of students selecting the option or taking Conservation Principles and Process Calculations as an engineering elective for a different emphasis option. Each entry in the table lists the number of chemical emphasis students at a certain level followed by the total number of engineering students at that level. It is difficult to provide precise counts of engineering students especially at the freshmen and sophomore level, because Hope students are not required to declare a major by a certain deadline in their careers as students. On the other hand, it is easy to count students interested in the chemical emphasis by enrollment in the Conservation Principles course. The data in Table 5 show that retention of chemical emphasis option students as they advance through the program has been excellent and significantly higher than overall retention of engineering students especially at the freshmen and sophomore levels.

Table 5. Student Enrollment in the Chemical Emphasis Option. First number is chemical emphasis students; second number is total engineering students.

Year	Freshmen	Sophomores	Juniors	Seniors	Total
2002-03	1/40	1/4	0/10	0/10	2/64
2003-04	3/42	2/20	1/4	0/10	6/76
2004-05	6/38	2/20	2/15	1/4	11/77
2005-06	2/40	5/32	2/15	2/15	11/102
2006-07	2/38	4/28	5/21	1/14	12/101
2007-08	9/43	0/26	4/23	5/21	18/113
2008-09		12/40	0/21	4/23	

Specifically, a total of 23 freshman students were enrolled in Conservation Principles over the six years from 2003 to 2008, plus one sophomore who took the course the first time it was offered in 2003. The number of students graduating in the option from 2005 to 2008, plus those continuing to make progress in the option in 2009, is 25. This indicates that the number of students who took Conservation Principles as freshmen but dropped out of the option were more than counterbalanced by students joining the option in their sophomore year, an apparent retention rate of greater than 100 percent. These numbers, not explicit in Table 5, total eight students who took Conservation Principles as freshmen but did not or are not completing the option, and nine students who joined the option during their sophomore year. All students who joined the option during the sophomore year have graduated with the option or are currently in progress. Table 6 summarizes these outcomes for students who showed interest in the chemical engineering emphasis option by taking Conservation Principles as freshmen or sophomores.

Table 6. Outcomes of Students Completing Conservation Principles by End of Sophomore Year.

	Enrolled in Course in 1st Year	Enrolled in Course in 2nd Year	Total
Graduated with Chem E emphasis option	5	4	9
2008-09 Senior in Chem E emphasis option	1	3	4
2008-09 Junior in Chem E emphasis option	0	0	0
2008-09 Sophomore in Chem E emphasis option	9	3	12
Graduated with other emphasis option in Engineering	2	0	2
Changed major out of Engineering	4	0	4
Left Hope College	2	0	2
Total	23	10	33

Students in the chemical emphasis option have the opportunity to participate in multidisciplinary experiences in their coursework, especially in the senior design classes. One recent design team consisted of three students who designed a mechanical feeder system with electronic controls for a chemical reactor at Pfizer, Inc. in Holland, Michigan. [3] Other design teams including chemical engineers have worked on a sustainable water purification technology for a rural village in Cameroon [4], a test device for detecting gas leakage from defective automotive lamps, and the 2008 AIChE student design contest problem for conversion of coal to methanol. Other chemical emphasis students have designed a materials testing apparatus now used in the materials science laboratory, a shell and tube heat exchanger experiment and a pump and pipe rack system for fluid flow experimentation for the process laboratory, a bioreactor for water purification, a packed column hydraulics demonstration apparatus, and a small-scale alcohol fermentation demonstration.

One of the distinctive characteristics of Hope College is the opportunity for undergraduate research. Eight of the fourteen students who have reached junior status in the chemical engineering emphasis option so far did participate in a meaningful summer research experience (such as an NSF-REU program) no later than the summer following their sophomore year. Three others did so during the summer after their junior year. Three of the twelve current sophomores participated in such an experience during the summer following their freshman year.

Outcomes of Program Graduates

The first engineering student with a chemical emphasis option graduated in May, 2005. She also received a chemistry degree and participated in polymer chemistry research while at Hope. She is currently a Ph.D. candidate in materials engineering at the University of Illinois-Urbana-Champaign.

Two students graduated in 2006. Both participated in chemical property modeling research at Hope, both worked an engineering interns at local manufacturers, and they were coauthors of a student poster presentation given at the 2005 AIChE annual meeting. [5] One also coauthored an article which appeared in Chemical Engineering Progress [6] and he is currently completing an M.S. in chemical engineering at the University of Michigan. The second student also completed a DOE summer research internship studying membranes at Argonne National Laboratory in 2005. He is currently completing his masters degree at Ohio State University and recently took a job as a chemical engineer for an engineering firm in the Washington D.C. area.

One student graduated in 2007 and became the first program graduate to hire directly into an industry position. She is a chemical engineer with SmartSignal Corporation in the Chicago area. While at Hope, she participated in chemical property modeling research, worked as an intern for a local nuclear power electrical generating facility, and was a coauthor of a student poster presentation given at the 2005 AIChE annual meeting. [5]

Four students graduated in May, 2008 and one expects to graduate in December, 2008. All but one had received a job offer two months or more before his or her graduation. They currently work for ITT Space Science Division, the U.S. Department of the Interior, Huron Technologies, and Perrigo Corporation, with one graduate turning down several job offers to volunteer with the Peace Corps in the Dominican Republic. Four of these five students participated in undergraduate research at Hope College and two worked as engineering interns. Three of these students were coauthors of various posters and presentations given at the 2006 and 2007 Michigan Space Grant Consortium (MSGC) Annual Conferences [7,8,9,10] and one received an undergraduate research fellowship from MSGC in 2006.

Conclusions

A chemical emphasis option was designed and implemented within the ABET accredited engineering program at Hope College. The emphasis option was successfully integrated into the existing engineering program coursework and also within the framework of a liberal arts college. Enrollment in the program and/or its courses has grown from 3 students in 2003 to 18 students in 2008 with high retention. Both chemical emphasis students and other engineering students have benefited from opportunities to participate in multidisciplinary experiences in coursework especially in design classes, and to participate in undergraduate research. Graduates of the program have had success in admission to high quality graduate degree programs and in finding chemical engineering positions in industry and government.

References

[1] "Hope College – About Hope College – Overview," http://www.hope.edu/about/#overview

[2] "Hope College – Academic Departments – Engineering – History of the Engineering Program," http://www.hope.edu/academic/engineering/proginfo/history.htm

[3] "Working together to solve a feeding problem," Powder and Bulk Engineering. 2004.

[4] Best, A.A. M. DeJongh. A.J. Barton. J.R. Brown. C.C. Barney. "Models of Interdisciplinary Research and Service Learning at Hope College." CUR Quarterly XXVIII(2). 18-23. 2007.

[5] Misovich, M.J. K.F. Blohm. A.J. DeDoes. M.D. Goetz. E.J. Walsh. K. Wadelton. "Explicit Phase Equilibrium Calculations from Cubic Equations of State." AIChE Ann. Mtg. CD-ROM Proc. New York: AIChE. 2005.

[6] DeDoes, A.J. M.D. Goetz. M.J. Misovich, "A Simple Approach to Generalized Vapor Pressure Prediction." Chem. Eng. Prog. 103(1). 39-44. 2007.

[7] Estochen, M.N. A.R. Runge. J.L. Lanser. A. Miller. M.J. Misovich. "Explicit Computation Methods for Fluid Phase Equilibrium." 12th Ann. Michigan Space Grant Consortium Conf. Ann Arbor, Michigan: University of Michigan. 2007.

[8] Lanser, J.L. M.J. Misovich. "A Quadratic Limit of a Cubic Equation of State." 12th Ann. Michigan Space Grant Consortium Conf. Ann Arbor, Michigan: University of Michigan. 2007.

[9] Runge, A.R. M.N. Estochen. M.J. Misovich. "Universal Vapor Pressure Estimation." 12th Ann. Michigan Space Grant Consortium Conf. Ann Arbor, Michigan: University of Michigan. 2007.

[10] Runge, A.R. M.J. Misovich. "Equilibrium Property Predictions from Generalized Cubic Equations of State." 11th Ann. Michigan Space Grant Consortium Conf. Ann Arbor, Michigan: University of Michigan. 2006.

58d Lessons In the Operation and Control of Steam Boilers for Power Production and Heating

Douglas Cooper, Chemical Engineering Program; CMBE Department, University of Connecticut, 191 Auditorium Rd., Unit 3222, Storrs, CT 06269-3222, Robert Rice, Control Station, Inc., One Technology Drive, Tolland, CT 06084-3902, and Allen D. Houtz, Business & Industry Division, Kenai Peninsula College, 34820 College Drive, Soldotna, AK 99669.

Thermal energy for electric power generation is produced from a growing number of sources, including solar heating; waste, biomass and biofuels combustion; combustion of traditional fossil fuel sources; and heat from nuclear reactors. Thermal energy is converted into useful electric power by creating high pressure steam in a boiler and then dropping the pressure across a steam turbine to drive an electric power generator. Even when fossil fuels are used in conventional combustion turbines, the waste heat is increasingly used in a combined cycle heat recovery process to create steam to drive steam turbine generators. In northern climates, boiler plants are often used entirely for heating.

When creating steam for a turbine driven electric generator, the boiler plant must produce steam at a narrowly specified pressure and temperature. While there are numerous commercial designs, most boiler plants include a superheater, steam drum and economizer. Feed water is preheated in the economizer, which is essentially one or more heat exchangers located downstream in the exhaust gas to maximize energy recovery. The heated water then enters a boiler where the relevant thermal energy source further heats it to steam. In boiler plants serving steam turbine generators, the steam then flows into superheater coils nearest the radiant energy source to create the high temperature superheated steam employed in steam turbines.

The University of Connecticut (UConn) has a combined cycle cogeneration plant (Cogen) that generates all the electrical power for the campus under normal operating conditions. The UConn Cogen plant has a capacity of 25 MW of electricity and 200,000 lbs/hr of steam. The electric demand of the campus is met with water-cooled electric generators on three 7.5 MW gas turbines and one 5 MW steam turbine. Three heat recovery steam generators (HRSGs) located after the gas turbine generators each contain 600 psi steam systems and 125 psi steam systems. The 600 psi steam is used to run the steam turbine generator, while the 125 psi steam is used for campus heat in the winter and running chillers for campus cooling in the summer.

To provide students a motivating experience in plant operation and control, we have created dynamic simulations that mimic some of the critical process units from the UConn Cogen plant and that are typical of steam boiler plants. The simulations provide a visual presentation to the student that is similar to what the plant operators see in the UConn Cogen control room.

These simulations enable students to obtain hands-on experience in plant operation and control without risking the safety of people or equipment, wasting expensive fuel, and without distracting operations personnel from their daily jobs. Visits to the nearby Cogen facility help students make the connection between the computer simulation challenges and real world plant operations.

One unit the students study is liquid level control in a steam drum. This is a classical operating challenge in steam boiler plants because of the need to balance the interaction of control drum inventory, steam production rate and feed water makeup. Control strategies include conventional feedback loops from the PID family, as well as the potential to explore the benefits of cascade and feed forward control strategies on steam drum operation.

A second unit the students study is the operation of a heat recovery steam generator (HRSG) unit. The challenge with this operation is to run water through an economizer, steam drum and then superheater to maximize steam production at a specified temperature and pressure. Control challenges include controlling water flow to the desuperheater, as well as controlling temperature and pressure of steam exiting the HRSG.

Modern control installations are computer based, so the interactive color-graphic animations offer students a challenge that rivals that of a real plant. The simulations offer a virtual-world experience that contains a mix of theory and application as they provide an instructional companion useful in the classroom and for homework.

The courseware materials are currently employed in the process dynamics and control course at UConn and Kenai Peninsula College in Alaska. Supporting written materials and slides under development will permit others to exploit the simulations within their existing curriculum. A longer term goal it to expand the effort into the design class and across to mechanical engineering curriculum.

Web Page: www.controlguru.com/

58e Teaching and Learning Gibbs Free Energy and the Phase Equilibrium Criteria through 3D Phase Diagrams for Pure Components

L. G. Ríos-Casas[1], A. Alarcón-García[1], D. R. Téllez-Muradás[1], J. R. Flores-Tapia[1], and Nelly Ramírez-Corona[2]. (1) Chemical and Food Engineering Department, Universidad de las Americas, Puebla, Mexico, Ex-hda Sta. Catarina Martir, Sn. Andres Cholula, 72820, Mexico, (2) Depto. de Ingeniería Química y Alimentos, Universidad de las Américas, Puebla, Ex-Hda de Santa Catarina Mártir s/n, Sn. Andrés Cholula, Puebla, 72820, Mexico

This work is related to a previous one regarding the representation of Thermodynamic Properties in 3D diagrams. This time it is focused on the analysis of Free Gibbs Energy for pure components to achieve a better understanding of the equilibrium criteria of fluid phases.

The fundamental property relations are obtained from the substitution of the first and second Law of Thermodynamics to define the thermodynamic equilibrium state of a closed system, thus the independent variables are not chosen arbitrarily.

The thermodynamic properties data for water are plotted in 3D to represent the resulting equations for G (Gibbs free energy). The database was taken from Haar et. al. (1984) and introduced in Matlab®. Different colors were used to distinguish the various regions: homogeneous liquid phase, homogeneous gas phase, vapor-liquid region and supercritical fluid. The mesh is constructed with values at constant temperature and pressure and the curves for constant vapor fraction are included inside the vapor-liquid region.

The 3D diagrams obtained can be used to generate projections in 2 dimensions: P-G, G-P, T-G, G-T for fluid homogeneous phases and vapor-liquid equilibrium. The behavior of free Gibbs energy may be associated to other concepts like fugacity and fugacity coefficient to be used as additional equilibrium criteria.

It is evident the complexity that engineering students have to handle when the professor starts the study of thermodynamic properties, but a deeper understanding could be attained by adding a visual approach. In the last few years, this methodology has been applied in the Thermodynamics course for sophomore Chemical Engineering students. The professor has perceived that both, students and teachers, show good acceptance of this educational material and the evaluations have revealed a better understanding of these concepts.

The success of this approach relies on the proven fact that the learning experience is more effective when more senses are involved.

58f The Soccer Ball Model. A Powerful Pok to Approach up-Scaling In the Classroom

Jennifer Anne Pascal[1], **Cynthia Torres**[1], and Pedro E. Arce[2]. (1) Chemical Engineering, Tennessee Tech University, PH-214, Cookeville, TN 38505, (2) Chemical Engineering, Tennessee Technological University, Prescott Hall, Room 214 , 1020 Stadium Drive, Box 5013, Cookeville, TN 37083

Soccer is the world's most popular sport. It is played on the beaches of Brazil, on the grassless surfaces of Argentina, Uruguay, Chile, Africa as well as on the nap-inviting fields of Europe and North America. Therefore, soccer balls are objects, i.e. geometrical objects that are popular among college students, and can be used to connect the discrete and continuum scales through the "Soccer Ball Model.(SBM)" When studying the physics of transport, it is necessary to develop conservation equations for variables such as total mass, energy, and momentum. In order to accomplish this, the concept of a continuum scale must be introduced. Since most students have only been introduced to the physical and chemical concepts related to total mass, energy, and momentum, from a discrete scale point of view, this concept can be very challenging for the student (learner). This means that the students already have knowledge related to calculating the total mass, velocity, and momentum of a single particle (discrete scale domain). In addition to having an introduction to several scientific concepts from the discrete point of view, students have an adequate background in many complementary subjects, including calculus, integral concepts, and algebra. It appears that the instructor should use this strong background to identify an approach that catalyzes the transformations. So, how does the instructor use the students' experience with the discrete scale domain and relate it to the continuum scale domain?

Most textbooks do not address this issue, assuming that the mathematical steps and concepts are familiar to the learner [1] , when in fact, they are not. This can be frustrating to the student and does not enable them to fully understand or capture the importance of this important idea of a continuum scale. An alternative approach is to design a student-centered activity (Arce and Schreiber, 2004) to engage the student in every step from one scale to the other. Thus, we propose an approach that uses soccer balls, to help with this transformation process, in conjunction with geometrical domains, mathematical principles, and physical properties. In this contribution, the authors will discuss the soccer ball model from a student's point of view along with its advantages and compare it to more traditional, familiar methods for introducing the concept of a continuum scale. The important role of the SBM as a POK (Arce, 1994; 2000) will be highlighted.

References: Arce, P. and L. Schreiber, "High Performance Learning Environments, Hi-PeLE", Chemical Engineering Education, 286-291, Fall 2004.; Arce, P. "The Colloquial Approach: An Active Learning Technique," Journal of Science Education and Technology, 3 (3), 145, (1994); Arce, P.; Oyanader, M.; and Whitaker, S. "The Catalytic Pellet: A Rich Learning Environment for Up-Scaling," Chemical Engineering Education, 41(3), 187-194 (2007); Arce, P., "Principal Objects of Knowledge (POK's) in Colloquial Approach Environments," Annual Conference Proceedings, American Society for Engineering Education, ASEE (2000, CD-ROM).

1. In fact, textbooks suppressed the student opportunity to learn the concept by directly telling the answer. (See Arce et al, 2007).

58g Facilitating Interdisciplinary Learning through Incorporation of Biotech Applications

Claire Komives, Chemical and Materials Engineering, San Jose State University, 1 Washington Sq., San Jose, CA 95192-0082

To address the need to incorporate interdisciplinary course content in the undergraduate chemical engineering curriculum, a website of solved material and energy balance problems dealing with biological applications has been created. In addition, to facilitate faculty who have not had formal training in biotechnology, a workshop for faculty was delivered in Summer '07. Problems from the website were used at beta test sites in Fall '07 and again in Fall '08. Plans are underway for the expansion of the website to include all of the core undergraduate curriculum. The presentation will describe the website utility and design for expansion.

Web Page: www.bioemb.net/

58h Demonstration Experiments on Physical Surface Science for Middle Schools

Ranga Narayanan, Chemical Engineering Department, University of Florida, Gainesville, FL 32611

In this talk I shall report our experience on the demonstration of surface science experiments at area middle schools. These experiments demonstrate the concepts of surface tension, surface flows, surface reactions and even miscible interfaces. Experiments that demonstrate capillary effects, pressure variations on account of diameter changes, emulsions and foams will be discussed. Applications for K-12 students involving household agents such as food, garden chemicals and detergents will also be given. All of these experiments are part of a kit that we have assembled for free distribution to area rural middle schools.

84a Forging Links Between Chemical/bioengineering Programs and Middle School and High School Math/science Programs

Richard Zollars, School of Chemical Engineering and Bioengineering, Washington State University, PO Box 642710, Pullman, WA 99164-2710 and Donald C. Orlich, Science, Mathematics, Engineering Education Center, Washington State University, PO Box 642710, Pullman, WA 99164-2710.

Over the past five years we have been conducting a Research Experiences for Teachers (RET) site program within the School of Chemical Engineering and Bioengineering. This program has many goals including; enhancing the instruction of science and mathematics at the middle school and high school level, promoting engineering as a major, and forming links between K-12 teachers and university faculty. With more than 50 teachers having participated in the program we now have data on what works and what does not in this type of venture. Acting on these observations we have significantly altered our approach for this year's program. The details of these changes, and their effects, will be highlighted.

Web Page: www.eerc.wsu.edu/SWEET

84b Texts for the 21st Century

Carl R. F. Lund, Chemical and Biological Engineering, University at Buffalo, SUNY, 505 Furnas Hall, Buffalo, NY 14260

Statistics indicate that the vast majority of engineering courses are taught using a traditional lecture format. Research suggests that other formats, particularly active learning, can be much more effective. The National Research Council and others have stressed the need to create a community of education scholars who can assist other faculty seeking to implement alternative teaching approaches. However, many faculty who desire to be as effective as possible in the classroom may not seek such help if the required investment of time is substantially greater than that for a traditional lecture format. Instead, they will continue to use traditional lectures. One way the required time investment could be reduced would be through the development of a comprehensive, integrated TExT (Toolkit for Exceptional Teaching). A prototype of such a TExT is being developed in the area of kinetics and reaction engineering. The TExT includes written materials (learning objectives, information and examples) equivalent to those what would be found in a traditional textbook. It additionally includes video files that present the information and examples in the format of a traditional lecture. For students, the TExT also incorporates additional solved and unsolved examples (all including grading rubrics), computational tools, and other learning resources. The traditional textbook stops at this point (actually short of it since textbooks don't usually include a full set of lectures). The TExT also includes a number of components for the teacher. These include review slides that can be used to quickly review the information presented in the video lectures, a large set of in-class learning activities, assessment tools for student learning (quizzes, homework, exam problems), all with grading rubrics, and tools for assessing the effectiveness of the learning activities. The TExT is designed to use active learning in the classroom. Students would be assigned lecture videos and readings to be completed prior to class. In class, the teacher would review the information from the pre-class assignment and answer any questions. This normally takes only a small fraction of class time. Hence, the lecture has been moved out of the classroom, making room for active learning. The majority of the class time is then devoted to learning activities. These are varied in type and approach (individual, collaborative, inquiry, case study, problem based, object lesson, etc.). The key advantage is that the TExT provides many more learning activities than are needed, and for each one it provides a detailed lesson plan and resource list. Thus, the instructor can choose activities that suit their situation, but more importantly, the lesson plan tells them exactly how to conduct the activity in class, what resources they will need in order to do so, and possible pitfalls they may encounter (with suggestions how to proceed). This helps ensure an effective implementation of the learning activity, but it also substantially lessens the time investment required of the instructor. Because the learning activities are part of an integrated package (the TExT), the nomenclature, format, level of theory, and prerequisite knowledge are completely consistent with everything else going on in the course (the lectures, examples, homework, exams, quizzes, etc.). This presentation will describe the components of the prototype TExT, provide examples of the learning activities and discuss some preliminary qualitative assessment based on student course evaluations.

Preliminary Testing of a .NET Based Educational Heat Transfer Simulator

Peyton C. Richmond and Sidney Lin

Department of Chemical Engineering, Lamar University, Beaumont, Texas, USA

Introduction

This paper reports preliminary results of using a .NET based educational heat transfer simulation to improve learning for two adjacent undergraduate chemical engineering heat transfer classes. This simulation was developed to help improve students understanding of one-dimensional conductive heat transfer concepts by allowing them to engage in interactive exploration. Two different approaches were compared in this work. In the first semester we encouraged individual simulation exploration by asking our students to complete a homework assignment consisting of prediction-type conceptual questions using the simulation (see Appendix A). During the second semester the simulation was incorporated into an in-class active learning activity where students were grouped into 3 person teams and asked to answer those same questions. After they answered each question they were asked to check their conceptual understanding by using the simulation. Pre- and post-survey instruments were used to gauge student understanding of these concepts before and after the simulation exercise to determine the effectiveness of these two approaches.

Interactive computer simulations are widely available for inclusion in today's curriculums and can be found in many textbooks which regularly include DVDs or URLs to websites containing simulations. They can provide a correct visual mental model of scientific concepts which can advance discussions and analysis in the classroom. However, although the goal of simulations is to provide a correct model and perhaps to eliminate existing student misperceptions, some investigators have observed the creation of new misperceptions through the use of simulations [Prince, 2006]. Others have noted that that the quality of student learning facilitated by simulations also depends on the specific learning environment as a whole [Davies, 2002].

Studies have found that simulations that incorporate interactivity, animation and context can promote learning, but that the student's ability and interest in using the simulation depends greatly on the design [Adams et. al., 2007]. Some of their findings are as follows: The best start-up settings were found to include the least amount of animation and complexity possible. Real world connections are important. Color is an important visual queue. Objects should be represented consistently from simulation to simulation. Layout tips, such as limiting the number of tools/controls and arranging them in small groups, limiting text and don't use abbreviations are suggested. Also, the play area should be distinct from the control panel. Text is a distraction in the play area. Students notice large bright tabs. Students like to use play/pause buttons for no reason if they have been shown to them. Verbose help can be a deterrent to exploration.

Student users have little difficulty immediately interacting with new simulations when the tools used are consistent with interfaces that the students have learned previously. Since a majority of students use Windows computers and Microsoft software we decided to use the .NET development environment in order to develop a simulation with a consistent look and with typical Microsoft interfaces. The results of the two different approaches to exploration with the developed simulator were measured by surveys to determine the students' understanding of various conduction heat transfer concepts that were explored using the simulation.

Problem Description

The objective of this simulator was to address heat conduction. The simulator was created to represent a one-dimensional conduction heat transfer problem with Dirichlet boundary conditions. A user is allowed to change the properties of the materials using the control panel to the right side of the application (see Figure 1). The initial temperature of the semi-infinite block and the thickness of the semi-infinite plate can be specified as well as the boundary temperatures. Sliders are also provided on each side of the semi-infinite plate to allow students to change the boundary temperatures while the simulation is in progress. Two different heat transfer profiles (the temperature profile and heat flux profile) are generated using the two independent grid settings.

One of the major advantages of using Microsoft Visual Studio 2005 (VS2005) as a development platform is that ANSI C++ can exist in the same program as the C++/CLI .NET code [Microsoft, 2006]. This allows us to develop one consistent program that can handle both the technical computing requirements to solve the equations and create the visually pleasing .NET interface for the Windows operating system. The standard VS2005 toolbox also makes it easy to incorporate windows common controls like sliders which students intuitively understand how to operate. A VS2005 setup project was also created to install the application on the department's laboratory computers so that the necessary version of .NET is also installed if required.

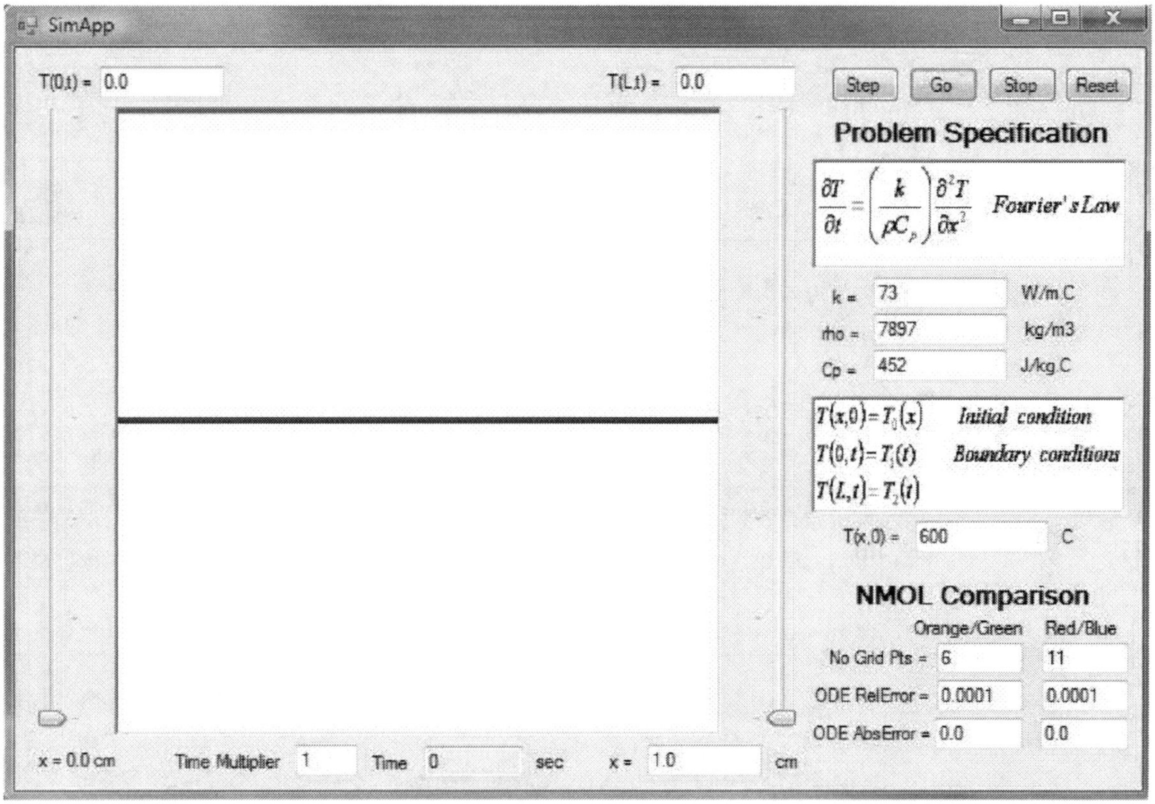

Figure 1. The .NET heat transfer simulation interface.

The transient one-dimensional heat transfer conduction problem is described by Fourier's second law for heat conduction in Cartesian coordinates as

$$\frac{\partial T}{\partial t} = \left(\frac{k}{\rho C_p}\right)\frac{\partial^2 T}{\partial x^2} \tag{1}$$

in terms of the temperature T and its spatial position x and time t. The group $(k/\rho C_p)$ represents the solid's thermal diffusivity in consistent units. One initial condition is required to prescribe the initial temperature distribution as a function of x at t=0 (typically a constant)

$$T(x,0) = T_0(x) \tag{2}$$

and two boundary conditions are required and in this case we chose Dirichlet boundary conditions specifying the dependent variable T at x=0 and x=L.

$$\begin{aligned} T(0,t) &= T_1(t) \\ T(L,t) &= T_2(t) \end{aligned} \tag{3}$$

The solution was coded in ANSI C++ using the method of lines [Schiesser, 1991]. The method of lines replaces the partial derivatives with respect to x in Equation 1 with finite difference approximations, but the time is still treated as a continuous variable. This leads to a system of ordinary differential equations (ODE) because we now have only one independent variable t and they can be solved by a standard ODE solver such as RKF45 [Forsythe, 1977].

The program was also designed to demonstrate the effect of grid size on a numerical heat transfer solution. Therefore, the graphical user interface calculates and displays two different solutions with different numbers of grid points simultaneously. Figure 2 is a screen shot of the program after the “Step” button has been pressed once. The transient temperature and flux profiles for 6 and 11 grid points at 0.2 seconds are plotted in the display window. The red line is the temperature profile for the 11 grid point solution and the orange line is the temperature profile for the 6 grid point solution. The blue and green lines are the respective flux profiles.

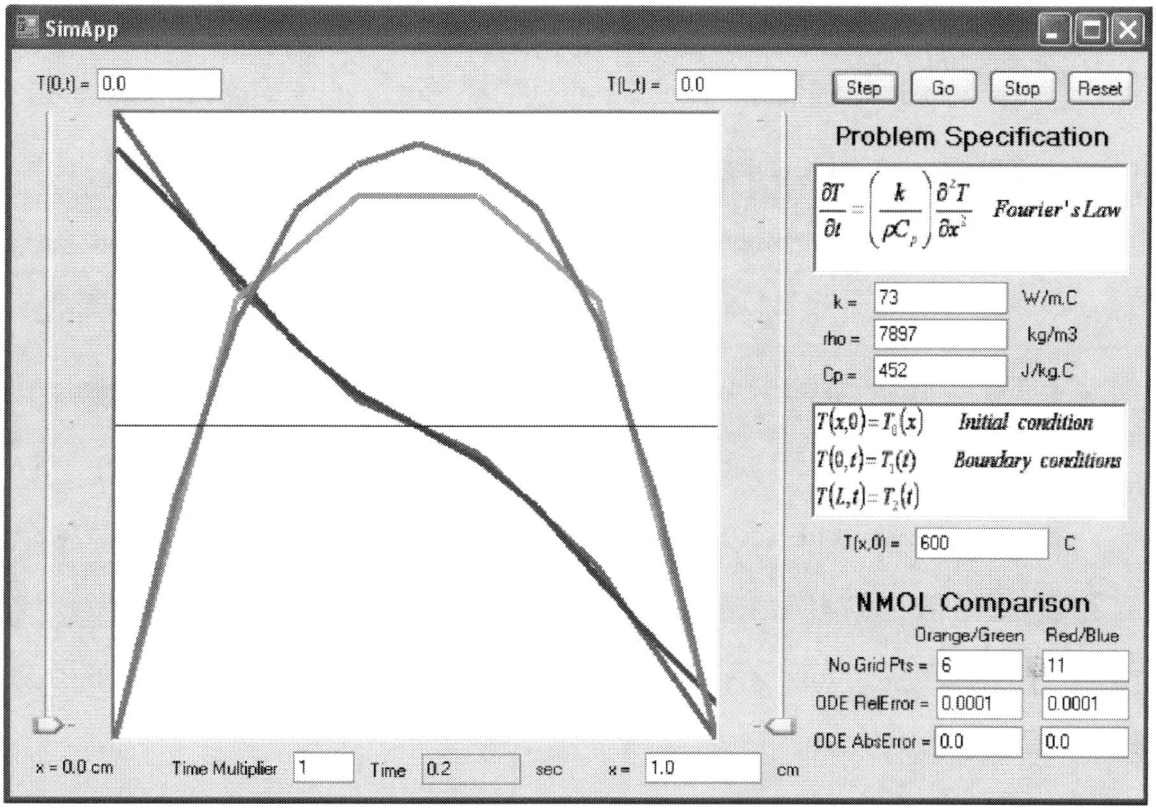

Figure 2. The .NET heat transfer simulation after pressing "Step" once.

Because this is a typical windows application, many copies can be executed simultaneously on the same machine. This allows a student to start another instance of the application, make modifications to the parameter settings and then compare the results side by side. For example, one of the exercise is to change the thickness x from 1.0 cm to 10.0 cm and then run the simulation until the computed profile matches that shown in Figure 2. To accomplish this task the student is instructed to leave the first instance of this application on his or her screen, invoke a second instance of the application, change the thickness to 10.0 cm and then run the simulation until the graphical display matches that of the first instance.

First Implementation

Seventeen students were enrolled in the Spring 2008 semester heat transfer class. For this class we encouraged individual simulation exploration guided by a prediction-type homework assignment. Students were given an overview of the simulation program during class and then provided with the homework assignments to be completed individually using the simulator (see Appendix A). Although only seven students actually returned the assignment, most indicated they had exercised the simulation. Students were given pre- and post- concept inventory surveys (Appendix B) to measure the effectiveness of the simulation as a pedagogical tool. The post-survey was provided to the students several days after the assignments. Seventeen (17) out of a total of 19 students completed and returned the surveys.

Second Implementation

In the Summer II session 2008 heat transfer class we incorporated an active learning session using the simulation. The twelve students taking this class were grouped into 3 person teams and asked to predict the answers to questions in the assignment. All groups learned and practiced how to use the simulator in class. Students were picked randomly to answer the questions in the assignment. After each group made their prediction they used the simulation to generate the answer and then discussed the results. Again, students were given pre- and post- concept inventory surveys. The post-survey was provided to the students several days following the in-class exercise and 10 of the 12 students completed and returned the post-survey.

Results and Discussion

Survey questions 1-5 are mathematical conceptual questions; questions 6-10 are visual conceptual questions; and questions 11-15 are affective questions. The answers to the mathematical conceptual questions can be deduced directly from the heat transfer formula. Where as the answers to the visual conceptual questions require knowledge of the visual shape of the temperature solution to the transient one-dimensional heat transfer problem and represent a correct mental model of the phenomena. Questions 11-15, the affective questions, were only included in the post-survey. The survey results from the Spring 2008 and Summer 2008 classes are listed in Table 1.

Table 1. Survey results before and after the simulation activity.

	Spring 2008			Summer 2008		
Question	Pre-Survey	Post-Survey	Change (%)	Pre-Survey	Post-Survey	Change (%)
Mathematical Conceptual Questions (maximun possible score = 5.0)						
1	4.8	4.8	0.0	4.6	4.3	-6.5
2	3.1	4.0	30.7	2.9	3.3	13.8
3	3.5	3.9	11.8	2.8	3.7	32.1
4	2.4	2.5	2.5	3.0	2.8	-6.7
5	2.4	2.3	-2.6	2.3	2.2	-4.3
Average	3.2	3.5	8.5	3.1	3.3	5.7
Visual Conceptual Questions (maximun possible score = 1.0)						
6	0.5	0.8	74.5	0.4	0.3	-25.0
7	0.7	0.7	0.0	0.3	0.6	100.0
8	0.5	0.4	-34.0	0.5	0.8	60.0
9	0.4	0.2	-41.5	0.3	0.4	33.3
10	0.6	0.6	0.0	0.5	0.9	80.0
Average	0.5	0.5	-0.2	0.4	0.6	49.7
Affective Questions (maximun possible score = 5.0)						
11		4.6			4.2	
12		3.8			4.2	
13		4.0			3.9	
14		3.7			3.8	
15		3.6			4.3	
Average		4.0			4.1	

From the responses of the affective questions 11-15, the students in both classes are in favor of using the simulator (average scores ≥ 4.0). It is strongly agreed that the simulator is easy to use (question 11). From question 12, it is clear that the students feel that active learning sessions help them understand heat transfer better than individual exploration. The students liked to see the usage of simulation in other classes (question 13). The active learning makes students feel more comfortable to predict the effect of their actions on the simulation (question 14) and helps students discover how heat transfer works (question 15).

Considering the responses to the mathematic conceptual questions 1-5, the Spring class using individual exploration performed equal or better in each question than the Summer class using an active learning exercise. The Spring class also improved slightly more for those questions as shown by the change between pre- and post- surveys. Given that only eight students actually returned the homework assignment we don't feel that we can state that individual exploration was more effective at improving our student's performance on answering mathematical conceptual questions than active learning. However, it is clear that the active learning activity employed for the Summer class did not improve our student's ability to answer the mathematical conceptual questions correctly.

Considering the responses of questions 6-10, the Summer class using active learning had a higher average score in the post-survey than the Spring class even though their pre-survey scores were lower. In addition, their score improved in four out of the five questions between the pre- and post- surveys while students using the individual exploration method averaged the same for the pre- and post-survey questions 6-10 with some scores higher and others lower. This indicates that although individual exploration with a simulation may correct some misperceptions other misperceptions may actually be created during that process. We feel that using an active learning exercise in conjunction with a simulation can help students understand visual conceptual problems better with less potential for creating new misperceptions.

Conclusions

A .NET based educational heat transfer simulator was developed and used to help students understand one dimensional conduction heat transfer concepts. The simulator was used both as an individual exploration tool and as part of an active learning exercise to explore these heat transfer concepts. The heat transfer simulator when coupled with an active learning approach was effective in helped students correct existing misperceptions without creating new misperceptions. Although the simulation provides students with a cursory visual mental model of the one-dimensional heat transfer phenomenon, it did not help (at least not much) students understand the theory or mathematics used in heat transfer calculation. However, this cursory understanding could serve as a basis for developing a deeper understanding of the more complex mathematical concepts.

Acknowledgements

The authors wish to thank the College of Engineering and the Office for Research and Sponsored Programs at Lamar University for their support of this work.

References

1. Prince, M. and M. Vigeant, "Using Inquiry-Based Activities to Promote Understanding of Critical Engineering Concepts," ASEE Annual Conference Proceedings, Chicago, IL, 2006.

2. Davies, C.H.J., "Student Engagement with Simulations: A Case Study," Computers & Education, 39, 271-282, 2002.
3. Adams, W.K., Reid, S., LeMaster, R., McKagan, S.B., Perkins, K.K. and C.E. Wieman, "A Study of Educational Simulations Part II – Interface Design," Journal of Interactive Learning Research…
4. URL: http://www.microsoft.com/; Microsoft Corporation, visited in 2007
5. Schiesser, W.E., The Numerical Method of Lines: Integration of Partial Differential Equations, Academic Press, San Diego, 1991.
6. Forsythe, G.E., Malcolm, M.A. and C.B. Moler, *Computer Methods for Mathematical Computations*, Prentice Hall, Englewood Cliffs, NJ, 1977.

Appendix A

Heat Transfer Interactive Computer Simulation Assignment

Name: _______________
ID#: _________________
Date: ________________

SimApp simulates the internal temperature of a semi-infinite plate with a thickness L. The plate's initial temperature is specified as T_0 at time = 0 and at time > 0 its surface temperatures T_1 and T_2 are specified on its left (x=0) and right (x=L) boundaries. The exercise that follows involves using the SimApp program to investigate how this system responds to various changes.

1. Start SimApp by using the Start-All Programs-SimApp menu command.
2. Left-click the "Step" button and identify the four lines in the graph area. The orange line is the semi-infinite plate temperature calculated using 6 grid points while the red line is the plate temperature calculated using 11 grid points. The green and blue lines represent the respective temperature derivatives. What best describes the shape of the temperature curve?
3. Consider the equation listed below the problem specification.
4. Observe the time (0.2 seconds). Then, start another copy of the program and increase the plate size from 1.0 cm to 10.0 cm thick. Left-click the "Go" button and when the shape matches the original program click the "Stop" button. How much time elapses before the second temperature profile looks the same as in step 2?
5. Start another copy of the program and decrease the thermal conductivity k value from 73 W/m.C to 7.3 W/m.C. Now how much time must elapse before the temperature profile looks like the result of step 2?
6. Start another copy of the program and increase the heat capacity Cp to 4520 J/kg.C. Again, how much time must elapse for the temperature to match the profiles above?
7. Explain the results of steps 4 through 6 in terms of Fourier's Law displayed below the problem specification.
8. Run the original program for about 15 seconds to reach steady-state and then left-click the "Stop" button. The temperature is constant (0 C) throughout the plate and the temperature derivative is zero.
9. Change the left surface temperature to 600 C while leaving the right surface temperature at 0 C and run the program for another 15 seconds to again reach steady-state. What is the shape of the final temperature curve? What is the shape of the final temperature derivative?
10. The magnitude of the temperature derivative is proportional to the amount of heat that is being transferred. Reset the program that the plate size has been increased to 10 cm. Step the solution and note where the largest derivatives occur. Left-click the start button. Do the temperature derivatives (heat flux) increase or decrease as the simulation proceeds?

11. Start another copy of the program and change the number of grid points from 6 and 11 to 11 and 21. Is the difference between the 11 and 21 grid solutions larger or smaller than the difference between the 6 and 11 grid solutions?
12. Start another copy of the program and change the number of grid points from 6 and 11 to 21 and 41. What the difference between the 21 and 41 grid solutions. Which solution is closer to the true answer?
13. Experiment and determine the different ways the program can be made to fail. Create a step-by-step procedure to re-create each method that causes the program to fail.

Appendix B

Survey
for
Improving Heat Transfer Learning by Interactive Computer Simulation

Name: ______________
ID#: ______________
Date: ______________

Consider a semi infinite plate with a thickness L has a uniform initial temperature T_0. The temperatures on both surfaces are suddenly changed to T_1 and T_2 on its left (x=0) and right (x=L) surfaces. How will the temperature inside the plate change?

1. When the surface temperatures on both sides of the semi-infinite plate are decreased, the temperature in the center of the plate will also decrease (a) strongly agree, (b) agree, (c) not sure, (d) disagree, or (e) strongly disagree

2. The time it takes for the temperature in the center of the plate to reach its final steady-state value is proportional to the square of the plate thickness (a) strongly agree, (b) agree, (c) not sure, (d) disagree, or (e) strongly disagree

3. The heat flux remains constant during the unsteady-state period (a) strongly agree, (b) agree, (c) not sure, (d) disagree, or (e) strongly disagree

4. The time it takes for the temperature in the center of the plate to reach its final steady-state value is inversely proportional to the plate heat capacity decrease (a) strongly agree, (b) agree, (c) not sure, (d) disagree, or (e) strongly disagree

5. The time it takes for the temperature in the center of the plate to reach its final steady-state value is proportional to the initial temperature decrease (a) strongly agree, (b) agree, (c) not sure, (d) disagree, or (e) strongly disagree

6. What best describes the characteristic shape of the semi-infinite plate temperature during the unsteady-state period? (a) constant, (b) linear, (c) quadratic, (d) S-shaped
7. When the surface temperatures on both sides of the semi-infinite plate are decreased symmetrically what is the heat flux at the center of the bar? (a) negative, (b) zero, (c) positive

8. If the surface temperatures on both sides of the semi-infinite plate are different, then the steady-state temperature profile is best described as (a) linear, (b) parabolic, or (c) constant

9. If the surface temperatures on both sides of the semi-infinite plate are different, then the steady-state temperature derivative is best described as (a) linear, (b) parabolic, or (c) constant

10. The location of the maximum heat flux during the unsteady-state period is at (a) one or both ends, (b) the center of the plate, (c) somewhere between the center and the end of the plate

Mark your level of agreement with the following statements:

1. Using the simulation was easy. (a) strongly agree, (b) agree, (c) not sure, (d) disagree, (e) strongly disagree

2. The simulation helped me understand heat transfer. (a) strongly agree, (b) agree, (c) not sure, (d) disagree, (e) strongly disagree

3. I would like to see other subjects use simulations. (a) strongly agree, (b) agree, (c) not sure, (d) disagree, (e) strongly disagree

4. I could predict the effect of my actions on the simulation. (a) strongly agree, (b) agree, (c) not sure, (d) disagree, (e) strongly disagree

5. The simulation helped me discover how heat transfer works in practice. (a) strongly agree, (b) agree, (c) not sure, (d) disagree, (e) strongly disagree

Note: This survey will NOT be counted towards your grade.

Concepts of Nanoscience for Non-Scientists: A Distance Education Course Coordinated Among Three Universities

Tamara Floyd Smith[1], David Baah[1], James Bradley[2], Michelle Sidler[2], Rosine Hall[3] and Christine Curtis[4]

[1]Chemical Engineering, Tuskegee University, Tuskegee, AL
[2]Auburn University, Auburn, AL
[3]Auburn University at Montgomery, Montgomery Alabama
[4]University of South Carolina, Columbia, SC

Abstract

A nanotechnology and ethics course, joint among Auburn University, Tuskegee University and Auburn University at Montgomery, was developed and offered for two semesters. The course was structured as a distance education course, and broadcasting equipment allowed students at all three campuses to view the lectures in real time. The course material was formatted for a freshman level science elective, and lecturers spanned all three campuses and various disciplines including but not limited to: chemical engineering (4 instructors), biology (2 instructors), philosophy (2 instructors), English (1 instructor) and history (1 instructor). The course overview and outcomes will be discussed.

History, Goals and Framework

The course, "Concepts of Nanoscience" began as a proposal, "Ethics of the Nanoscale" to the National Science Foundation. The proposal included several education components, but this discussion will be limited to a distance education course joint among Auburn University (AU), Tuskegee University (TU) and Auburn University at Montgomery (AUM).

The goal of the course was to offer an introduction to the science and ethics of nanotechnology to students who would not otherwise be exposed to the topics. The course was structured for freshman non-science majors. One challenge was to find a framework for the course. At AU, freshman, non-science majors can enroll in a survey course, "Concepts of Science." "Concepts of NanoScience" was offered as one section of the multi-section "Concepts of Science" course. At TU, the course was administered through the Provost's office due to its interdisciplinary nature and non-scientist target audience. At AUM, offering the course as a lower division science course without a lab did not meet the requirements of the curriculum committee. Thus, the course was offered through the biology department due to its large nano-biotechnology and environmental components.

Content

Lecturers and content spanned various disciplines including but not limited to: chemical engineering (4 instructors), biology (2 instructors), philosophy (2 instructors), English (1 instructor) and history (1 instructor). Nanotechnology: A Gentle Introduction to the Next

Big Idea [1] was chosen as the primary text for the course. Table 1 provides a list of the course topics. The course begins with an introduction to nanotechnology and proceeds to outline basic chemistry and physics principles with respect to nanotechnology. Approximately three lectures are devoted to a discussion of various nanostructures with particular emphasis placed on carbon nanotubes. Next, two lectures are designated for the discussion of instrumentation required for research in nanotechnology. For example, students are introduced to the transmission electron microscope (TEM) for characterizing nanoparticles. Next, various applications of nanotechnology are introduced that include, but are not limited to, nanocomposites, nanoelectronics, nano-electro-mechanical systems (NEMS) and nanofluidics.

The next phase of the course is dedicated to an introduction to ethical theories and the ethical implications of nanotechnology. The allocation of one-fourth of the course to this discussion reflects the interdisciplinary course that was originally proposed to the National Science Foundation. Finally, the course ends with a discussion of nanobiotechnology and environmental implications of nanotechnology. Upon the introduction of each topic, a detailed overview is required because the students are freshmen with only high school science and math preparation. For example, prior to an in-depth discussion of nano-biotechnology, students are given foundational elements such as an overview of the principles of cell biology and environmental biology.

Table 1: Concepts of Nanoscience Course Topics

Topic	Lectures	Broadcast Origin
Introduction to Nanotechnology	4	AU
Principles of Chemistry and Physics on the Nanoscale	2	AU
Nanostructures	3	AU
Instrumentation	2	AU
Applications of Nanotechnology	3	TU
Introduction to Ethical Theories	8	AU
Ethical Implications of Nanotechnology	3	AU
BioNanotechnology	7	AU
Environmental Implications of Nanotechnology	7	AUM

Structure and Execution

Once the framework for the course was identified at each institution and the content was determined, several logistical issues needed to be addressed. First, each university is on a different class schedule. Both AU and TU offer Monday/Wednesday/Friday (MWF) and Tuesday/Thursday (TTh) courses, but AU starts on the hour and ends at ten minutes until the hour whereas TU starts at ten minutes after the hour and ends on the hour. AUM does not have class on Friday. The compromise was that the course would be offered MWF with forty minutes of core content. AU handled issues like homework and announcements for ten minutes before class, and TU handled those issues for ten minutes after class. All sessions were recorded, and AUM viewed the Friday session off-line.

Another issue was the scheduling of institutional breaks. Each institution had different spring breaks, semester start/end dates, holidays, etc. Long breaks such as spring break were coordinated by viewing recorded lectures during those periods. The semester start/end dates in some cases were close enough for all three institutions to coordinate and in other cases were handled by viewing recorded lectures.

Distance education equipment was purchased (as necessary) and configured for all three universities. All three institutions had access to views of the other two institutions during lectures, but, typically, the lecturing institution was viewed unless another institution was asking a question. Because the course was viewed in real-time, it could be and was very interactive. This opportunity for an improved extended-classroom dynamic couldn't be realized for a distance education course that is recorded and viewed off-line in its entirety.

Offerings and Enrollment

The course was offered during the Spring 2007 and Fall 2007 semesters. Course enrollment data is provided in Table 2. Enrollment (pre test participation) was significantly higher at Auburn University because the course was one section of an established course. At Tuskegee University, the course was acceptable for "science elective" credit, but, despite heavy advertising, students and advisors were accustomed to more traditional courses and chose those. Enrollment at AUM was affected by the lack of a laboratory offering, since even all majors must have two laboratory science classes to meet basic curriculum requirements. At AU, the lower division Concepts of Science course, which is targeted at non-science majors, has a recitation hour instead. The curriculum committee at AUM would not allow a recitation to be substituted for a laboratory.

Outcomes

Student learning for the purpose of assigning a grade was assessed using four in-class exams and a comprehensive final exam. However, the impact of the course was assessed by administering pre-course/post-course tests to the students. The results of the pre-test and post-tests are outlined in Table 2. The pre-test was administered to establish the baseline for student knowledge of the subject matter. Typically, the post-test was administered after the final lecture but prior to the final exam. The pre/post test consisted of 32 questions (24 True/False-type and 8 short answer). Table 2 shows the number of students participating from AU, TU and AUM and their corresponding pre/post test average scores. For AU, all students who completed the pre-test did not complete the post-test, and the pre/post assessments were not matched in the end. Consequently, it was possible that the students who scored the lowest on the pre-test did not take the post-test and thus inflated the score difference. To remove this error, the pre test results reflects both the average of all the students tested and the average of the students scoring highest on the pre-test corresponding to the same number of students who took the post test at AU. The second number reported in the score difference column gives the most conservative estimate of student learning because it is calculated from the arbitrarily higher pre-test scores. Another issue is that 24/32 questions were True/False-type implying a baseline of zero knowledge at a score of 12/32⇒37.5% for random guessing. Despite the aforementioned challenges with the assessment exercise, it is clear that the students' knowledge of the subject matter improved significantly ranging from 7.8 to 29.2%.

Conclusions

A distance education course joint among Auburn University, Tuskegee University and Auburn University at Montgomery was offered to introduce non-science majors to the concepts of nanoscience. The course was offered for two semesters during 2007. The majority of the lectures were conducted in real time so that students from all three campuses could interact with the various lecturers and students at other campuses. Although several logistical issues were encountered, the course ran smoothly for two semesters. Analyzing the results of assessment tests given to students revealed that their knowledge of the concepts of nanoscience improved by 7.8-29.2% as a result of completing the course. Moreover, it was demonstrated that three universities can offer a coordinated, real-time, distance education course to expose non-science students to the science and ethics of nanotechnology.

Table 2: Concepts of Nanoscience Enrollment and Assessment Data

School	Term	Pre Test		Post Test		% Diff
		# Students	Avg. score (%)	# Students	Avg. score (%)	
AU	Spr 07	31	68.1/75.7	16	91.2	23.1/15.5
AU	Fall 07	18	67.8/72.8	11	89.1	21.3/16.3
TU	Spr 07	4	62.5	4	70.3	7.8
TU	Fall 07	2	60.9	2	90.1	29.2
AUM	Spr 07	2	72.4	2	92.6	20.2
AUM	Fall 07	1	-	1	-	-

References

Ratner, M. and Ratner, D., Nanotechnology: A Gentle Introduction to the Next Big Idea, New Jersey: Prentice Hall, 2002.

84e Moral Theory, Ethics and Economics: Decision-Making Tools

Eric L. Maase, Chemical Engineering Department, Lafayette College, Acopian Engineering Center Rm 268, 740 High St., Easton, PA 18042

Engaging students in engineering ethics remains an important and challenging subject. While there are many opportunities for integrating ethics into classroom instruction the students do need an introduction to some 'common vocabulary' that includes moral theory, engineering codes' of ethics, and the understanding that there is a connection, both explicitly and implicitly, that such concepts are often a basis for decision making. In this presentation we provide the background vocabulary necessary for the students, and then will outline approaches that have engaged students' interest in ethics, and conclude with some suggestions for integrating moral and ethical issues into chemical engineering curricula.

84f Strategies for Improving the Recruitment and Retention Rates of Women and Minority Students in Chemical Engineering

Felecia Nave, Chemical Engineering, Prairie View A&M University, PO Box 519, MS 2505, Prairie View, TX 77446

America's vitality is linked to its ability to produce a workforce capable of producing scientific and technical innovations in engineering and technology. As the US workforce becomes more diverse, increasing the participation of US citizens that are typically underrepresented in STEM disciplines is essential to the workforce of the future. With the increasing need to produce more well-trained scientist and engineers coupled with the changing demographics of the US labor pool argues for policies, programs, and resources that support greater participation by underrepresented groups in STEM education and careers (May and Chubin, 2003). If the US is to maintain its dominance, it is critical to increase the quantity and quality of students interested in pursuing careers and retaining those in the pipeline in engineering and science disciplines. This issue is particularly acute for engineering and technology disciplines. The number of minorities entering and graduating engineering and technology workforce is at best limited and at worst insignificant. Over the past five years, the Department of Chemical Engineering at Prairie View A&M University has implemented several strategic strategies aimed at improving the recruitment and retention rates of women and minority students. These efforts include 1) hiring of a female faculty, 2) updating curriculum to offer options that are of interest to women and minorities, and 3) incorporation of a freshman engineering course. The purpose of this presentation is to discuss the vital impact of each of the strategies individually and collectively. In addition, a discussion of lessons learned and future assessment will be presented.

Chemical Engineering Undergraduate Curriculum Reform, Development and Assessment: A "Strings" Approach

Lâle Yurttaş and Larissa Pchenitchnaia

In a time of rapid change, academic programs must experiment and evolve in order to keep pace with advances in knowledge, changes in professional practice, and shifting conditions in society. The need for responsive academic programs is particularly a concern in scientific and technological fields where the growth of knowledge is exponential. Three chemical engineering departments at Texas A&M University System are continuing their efforts to restructure their four-year undergraduate curricula to achieve four objectives. Students will be able to a) apply fundamental ideas in chemical engineering over a greatly expanded range of time and length scales; b) apply ChE fundamental ideas to emerging application areas; c) construct solutions for more complex, more open-ended synthesis tasks with greater facility; and d) transfer fundamentals and knowledge to novel challenges. Three major strategies for project implementation include (1) curriculum content reform and development; (2) integrated student assessment, and (3) faculty and student development initiatives. The two key strategies for curriculum content reform and development are (i) the process of reformulation of part of the curriculum using four course strings and (ii) construction of interlinked curriculum components. This paper will describe the process and implementation of the "strings" approach to curriculum reform and development and in particular to assessment.

The "strings" approach involves organizing undergraduate ChE courses into four course strings: (1) thermodynamics and kinetics; (2) emerging fundamentals and applications; (2) transport phenomena; and (4) systems design. Course string faculty committees formed address the following key issues: what must undergraduate engineers learn/accomplish in the course string to be successful throughout their academic career and in the next generation professional settings; what obstacles exist to providing the necessary educational experiences; and how can we effect change and what changes (integration) need to be made to an existing curriculum. Course string faculty committees hold regular meetings to address these questions. Strategies for implementation of course portfolios and integrated assessment of course objectives and outcomes in preparation for ABET review are also part of discussions by faculty committees. In the past academic year, the committee members redefined expectations of each course as well as educational goals of each course and measurable outcomes. The alignment of each course's educational objectives and outcomes and expectations of courses from students was evaluated. Syllabi analysis provided invaluable information to enhance the alignment of the courses. The end-of-semester faculty and student evaluations provided direct and indirect feedback for assessment. The process, experiences, and findings will be presented.

Course String Processes

The renewed chemical engineering curriculum is intended to provide a unified, coherent, effective, and efficient learning experience for students. Curricular reform often flounders because there is no consensus about shortcomings or strengths in the current curriculum due to lack of empirical data. Even if everyone wants a change, the absence of commonly shared evaluation data makes it difficult to come to consensus about the direction the changes should take. The goal of the course strings is to provide evaluation and assessment data for the current curriculum so that faculty decision making is empirically based. To achieve the desired unity and coherence, the chemical engineering department initiated a process of reformulation as part of the curriculum using four course strings. Four explicit course strings are:
String 1 - Thermodynamics and Kinetics – This string is the heart of chemical science and fundamentals and includes material and energy balances course, the two thermodynamics courses and the kinetics course.
String 2 - Emerging Fundamentals and Applications – This string provides fundamentals and applications in emerging technologies.
String 3 – Transport – The transport string builds on the current classical fluid mechanics, heat transfer, and mass transfer unit operations approach but is reinforced by new and emerging applications.
String 4 - Process Systems Engineering – This string provides an integrated approach to process synthesis, integration, and multi objective optimization.

Course string faculty committees include faculty members whose research and teaching interests reflect one of the four course strings topics. Course string committees (CSC) address the following key issues:

- What do students need to know before coming to the courses in a course string? What are the input skills of students?
- What do you want students to get out of the courses in a course string? What is important for them to learn and retain, 2-3 years after the course (s) is/are over? What kind of thinking or application abilities do you want them to develop? What the students should be able to do? What are the output skills of students?
- How are we going to get students to desired outcomes?
- How do courses in a course string fit together?
- How are courses in a course string aligned?

Course string faculty committees function like curriculum and assessment subcommittees distributing the activities related to continuous improvement of the curriculum and assessment to the appropriate faculty. Course string committees meet several times throughout a semester to discuss curriculum design and enhancement activities. First of all, faculty members discuss overall learning objectives for a course string and then proceed with discussing input skills and learning outcomes for chemical engineering courses that are part of this or that string. Figure 1 shows a schematic picture of the process for String 1.

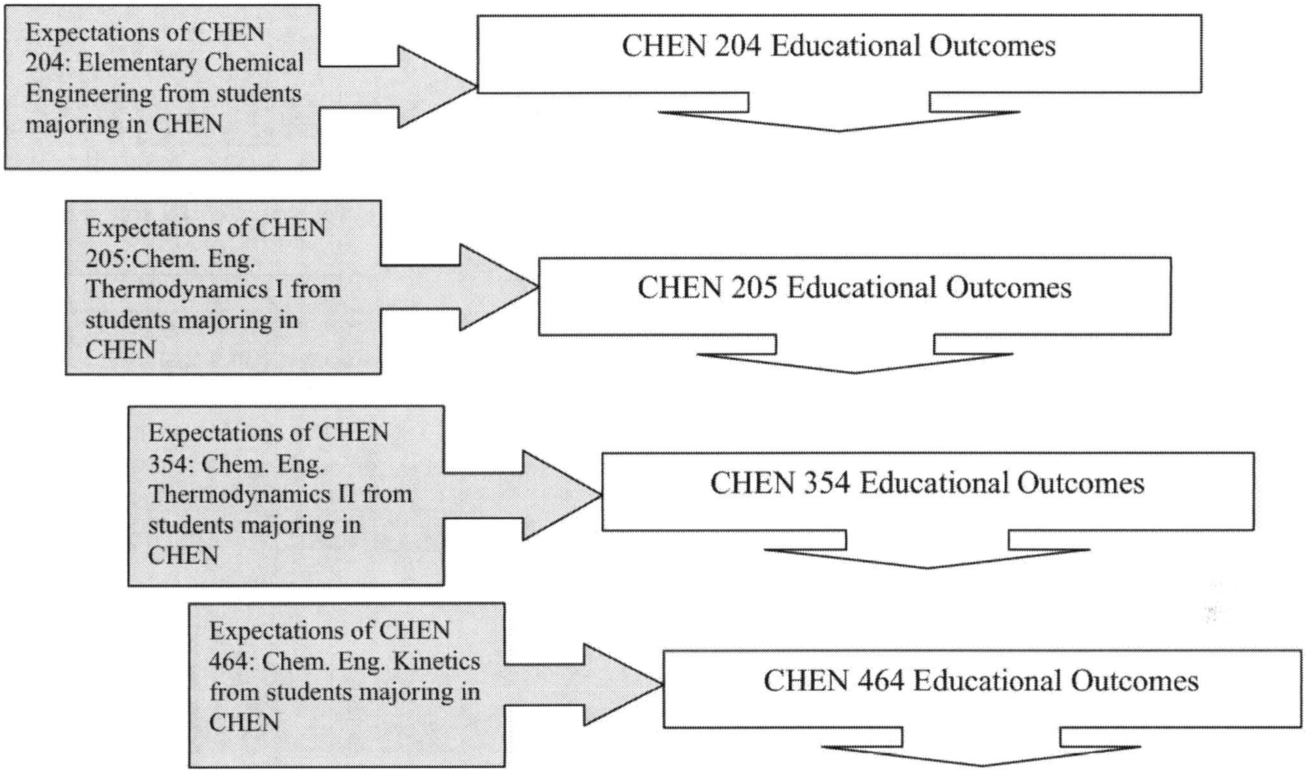

Figure 1. Thermodynamics and Kinetics Course String Input Skills and Educational Outcomes Process Flow Chart

Course string committee members define/redefine educational outcomes (intellectual content and students' skills/attributes) of each course. Faculty members discuss measurable outcomes and performance tasks following Bloom's taxonomy of educational objectives. The next step is to evaluate the alignment of each course's educational outcomes: curriculum mapping--linking each course outcomes to curriculum. CSCs identify where within the current curriculum our educational objectives and outcomes are addressed. The goals are (1) to connect what we are currently doing, whether all organizing principles are operative and move from simple to complex and (2) add new skills that are important.

Implementation

During Summer - Fall 2007 semesters CSCs met frequently to align each course content in a string, eliminate repetitious material, and add new material when it was necessary. CSCs established course string learning objectives and also defined input skills and educational outcomes for each course in a string. Most difficult step proved to be the removal of material from the content of a course to add new material. As originally proposed, thermodynamics and kinetics string was to be reduced to three courses from four current courses. Lengthy deliberations among the faculty showed the difficulty of actually removing the content material from the courses and the idea was not pursued eventually. New material was added in a few places but was decided to be done mostly by incorporation of Interlinked Curriculum Components (ICCs) (2). ICCs are Web-based resources for teaching and learning and contain relatively smaller material than a typical semester course. The scope of ICCs may range from molecular modeling to fundamental concepts such as conservation of mass. Currently, several ICCs are being constructed and tested for usability (1).

At the end of the Spring 2008 semester, the direct and indirect course assessment data were collected. The Department's annual assessment plan and the assessment instruments used may be viewed at http://www.che.tamu.edu/assessment/assessment-tools. Each course string committee evaluated the data and made a series of recommendations to the Departmental ABET Committee. Departmental committees and mode of operation are shown in Figure 2.

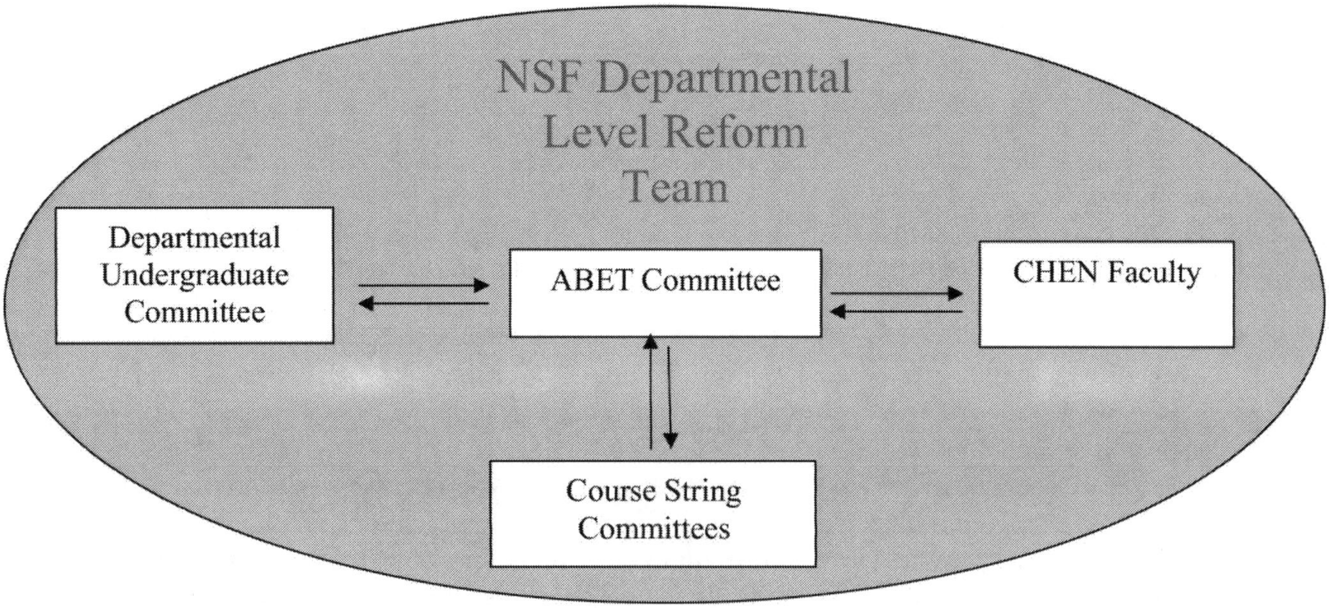

Figure 2. Departmental Committees and Mode of Interaction

Findings and Considerations

To achieve the goals of the "course string" process the faculty members first discussed overall learning objectives for a course string (e.g., Thermodynamics and Kinetics). Table 1 presents a brief example of a table that faculty members discussed and filled out together using the following assessment matrix key: "Introduced"- an objective is introduced in a class; "Emphasized"- an objective is emphasized in a class; "Utilized" - an objective is utilized in a class; and "Assessed" - an objective is comprehensively assessed in a class. The next step was to discuss how students at the department are doing against each expectation.

Table 1. Curriculum Assessment Matrix.

Course String 1 Learning Objectives	**CHEN 204**	**CHEN 205**	**…**	**Assessment**
1. Relate thermodynamics properties to observables.	Introduced	Introduced Emphasized Assessed		Meet Expectations
…	…	…	…	…

The following table provided a basis for discussing recommendations for courses and curriculum improvements. Some important recommended changes included:

1. Reinforce an important learning outcome related to demonstrating the ability to design experiments. As a result of this recommendation a Course Coordinator for a Numerical Methods course introduced this component to a course. Additionally, as part of assessing the students' ability to design experiments, Course Coordinators for this course and a Unit Ops Lab selected one of the experiments from a Lab class to use as a case study in Numerical Methods class.

2. Strengthen an important learning outcome related to statistically analyzing and interpreting data. The department is currently discussing an opportunity to develop such course as an elective to CHEN students.

3. Add an important task of writing technical memoranda to other team members as well as synthesizing a large project report in the form of an executive summary to a Plant Design Course.

4. Provide more feedback on students' speaking skills and create new assessment rubric for CHEN 481: Seminar.

5. Encourage individual instructors to provide examples to facilitate discussion of the ever-increasing interdependence of global economies and how societal issues impact the chemical engineering profession. Include a lecture on the topic of societal impact of technology, contemporary issues affecting chemical engineering and life-long learning to a Seminar course.

Department's ABET and Undergraduate Curriculum Program committees discuss all recommended changes and actions are taken to improve current curriculum.

Conclusions

This paper described a process and implementation of the "course strings" approach to curriculum reform and development and in particular to assessment. The "course strings" approach developed through DLR project would serve as a basis for a continuous process of reviewing and improving course learning outcomes at the department. Course string faculty committees will function as assessment and curriculum subcommittees and will provide important information for a program review and assessment processes.

References

1. Carvajal J., Pchenitchnaia, L., Yurttas, L., & Cagin, T. (2008). *A web-based resource for teaching molecular modeling and simulation methods in chemical engineering.* Proceedings, ASEE Annual Conference and Exposition, Pittsburgh, Pennsylvania.

2. Yurttas, L., Kraus, Z., Froyd, J., Layne, J., El-Halwagi, M., and Glover, C. (2007). *A web-based Complement to Teaching Conservation of Mass in a Chemical Engineering Curriculum.* Proceedings, ASEE Annual Conference and Exposition, Honolulu, Hawaii.

112b Mechanistic Studies on the Activation of a Carbon-Carbon Single Bond Via Palladium-Catalyzed β-Alkyl Elimination

Jeffrey B. Johnson and **Valerie Winton**. Chemistry, Hope College, 35 East 12th St, Holland, MI 49422-9000

The activation and functionalization of carbon-carbon single bonds presents a significant challenge due to their significant thermodynamic and kinetic stability. Miura and coworkers have demonstrated the selective cleavage of carbon-carbon single bonds followed by functionalization through coupling with an aryl halide (*Journal of Organic Chemistry*, **2004**, *69*, 6942). Carbon-carbon bond activation is proposed to occur via palladium-catalyzed β-alkyl elimination of an aryl group from tertiary alcohols.

This work presents the results of the mechanistic study of β-alkyl elimination. The nature of the mechanism was more fully understood through kinetic studies of the palladium-catalyzed coupling of triphenylmethanol and bromobenzene. Furthermore, a series of tertiary alcohols with various substitutions were prepared and tested for the relative rate of carbon-carbon bond activation. Intermolecular and intramolecular competition reactions examined the relative propensity of cleavage of groups containing differing hybridization, electronic properties, and sterics. The results of this study promise to provide additional insight into the process of carbon-carbon bond activation and facilitate the future development of more general and selective methodology.

Poly-Acrylic Acid Functionalized Membranes for Iron Immobilization and Fenton Reaction

Alex Montague[1], Scott R. Lewis[1], and Dibakar Bhattacharyya[2]. (1) Chemical & Materials Engineering, University of Kentucky, 177 F. Paul Anderson Tower, Lexington, KY 40506, (2) Department of Chemicals & Materials Engineering, University of Kentucky, Lexington, KY 40506

The Fenton reaction is the oxidation of ferrous iron (Fe2+) by hydrogen peroxide (H2O2) for the generation of hydroxyl radicals. The hydroxyl radicals can be used to oxidize and dechlorinate toxic organics. Typical application of the Fenton reaction involves the reaction of Fe2+ and H2O2 in homogeneous solution. This study focused on synthesizing a membrane that could serve as the hydroxyl-radical generation sites and the reaction medium for the dechlorination. The model species used for the dechlorination study was pentachlorophenol (PCP). This is a common wood preservative and also has many other applications: insecticides, herbicides, etc. There have been several reported cases of PCP contamination of soil and groundwater. The membrane is especially useful because it allows for the immobilization of the Fe2+ through ion exchange with COOH functional groups in membrane pores. This could potentially allow for reuse of membranes. Also, the typical Fenton reaction can only take place at low pH (pH 3). This is because, Fe2+ in solution precipitates as a hydroxide at and above pH 6. Thus, by immobilizing the Fe2+ within a membrane, the precipitation of iron hydroxide is eliminated and Fenton's reaction can be employed at near neutral conditions. This study utilized polyacrylic acid (PAA) functionalized polyvinylidene fluoride (PVDF) microfiltration membranes. The PVDF membrane was chosen for its non-reactive nature. 5.85 mg of Fe2+ was immobilized in the membrane. This membrane system was shown to degrade H2O2 consistently up to 40%. PCP dechlorination was studied by studying chloride generation during the membrane reaction. The experimental results with membrane systems include: Fe(II) to Fe(III) conversion by H2O2, degradation rates of H202, and PCP dechlorination. This research has been supported by the NSF-REU program and by NIEHS.

112c Optimizing the Impregnation of Catalysts: Examining the Saturation of Alumina Particles In a Double Cone Blender

Dustin Kucko, Charles Radeke, Arthur W. Chester, and M. Silvina Tomassone. Chemical and Biochemical Engineering, Rutgers University, 98 Brett Road, Piscataway, NJ 08854

Catalyst impregnation is one of the most important steps in preparing industrial catalysts. In this procedure, metal salts or complexes are dissolved in a usually aqueous solution and then contacted on a porous oxide catalyst support. Here, alumina (Al2O3) is used as the porous oxide catalyst support and pure water is sprayed in a double-cone blender. Simulations were performed using DEM (Discrete Element Method) calculations between particles to determine their behavior during the absorption process. Because of the myriad number of calculations that must be done for a single simulation, we simulate one third of the actual impregnation process. A novel water transfer algorithm was implemented in the simulations to reproduce the actual distribution of water in the particle bed. We computed a number of parameters such as the residence time distribution (time that the particles spend under the spraying zone), the extent of the dead zones (zones that do not receive any water during the impregnation process) and the spatial distribution of the sprayed water throughout the powder bed in different cut planes. After examining the system under a variety of conditions, the residence time distribution seemed to be profoundly affected by any difference in spraying gun setup. However, after a given number of revolutions, the residence time distribution looked similar for increasing amount of revolutions. After characterizing the dynamics of the impregnation process, experiments were performed to substantiate our computer-generated data. Ultimately, a large number of variables affect the sorption rate of water onto these particles including: size and number of spray nozzles, distance from the center of the drum, number of rotations of the drum, and duration of the spraying. For all impregnator configurations there is a presence of dead zones and therefore the liquid is never distributed completely homogeneously throughout the powder bed. However, if we correctly manipulate the variables and the configuration of the spraying guns we can obtain a nearly homogeneous distribution.

112d Palladium Removal from Reaction Mixtures by Fixed Bed Adsorption

Caitlin A. Boyd[1], Brian G. Lefebvre[1], Sonia Berberana[1], Robert P. Hesketh[1], Laura Kuczynski[1], Dr. Stephanie Farrell[1], **Pamela Kubinski**[1], **Megerle Scherholz**[1], and Michael J. Girgis[2]. (1) Chemical Engineering, Rowan University, 201 Mullica Hill Rd., Glassboro, NJ 00828, (2) Chemical and Analytical Development, Novartis Pharmaceuticals Corporation, 1 Health Plaza, East Hanover, NJ 07936

The objective of this project, which was sponsored by Novartis Pharmaceuticals Corporation, was to remove palladium from a reaction mixture using a fixed bed adsorption column. Palladium is a heavy metal that is used frequently as a catalyst in many reactions in the pharmaceutical industry. Reaction mixtures can be contaminated up to 3000 ppm with palladium. A final drug substance can only contain two ppm palladium or less. For this project, the Heck coupling reaction of 4-bromobenzaldehyde and methyl acrylate was used, as it is representative of common reactions used in the pharmaceutical industry.

Adsorbents were evaluated by generating adsorption isotherms; Quadrapure TU was chosen as the adsorbent based on the high capacity of this resin at 60 °C. The adsorption column was packed with Quadrapure TU resin and glass beads. Palladium concentration was determined using a novel HPLC method that is still in development, and verified through outside analytical lab analysis. Several adsorption column runs at various flow rates and packing geometries have been performed, with the overall goal of operating the column under conditions that promote full utilization of the resin bed. The effect of feed concentration, flow rate, and packing geometry will be discussed.

112e Reactor Simulation for Industrial Process Using Sulfite Oxidation in Water

Ruth E. Kuilan, Chemical Engineering, University of Puerto Rico Mayaguez Campus, 320 Tous Soto St., Urb. Baldrich, San Juan, PR 00918

Abstract

Air Products and Chemicals, Inc. would like to improve a current process that uses a reactive gas to form a gas product in a Continuous Stirred Tank Reactor (CSTR). It has been determined that oxygen (O_2) and nitrogen (N_2) are two gases that can be used to simulate this reaction. Therefore, the aim of this project is to investigate the rate at which O_2 is absorbed when it reacts with a sodium sulfite (Na_2SO_3) solution while in the presence of cobalt sulfate ($CoSO_4$) as a catalyst. Under the conditions chosen, the reaction of oxygen can be considered sufficiently fast that the O_2 concentration in the reactor is practically zero. Therefore the O_2 uptake rate is affected primarily by mixing and is a direct measure of the O_2 mass transfer rate. The following equation shows the oxidation reaction that will be completed:

$Na_2SO_3 + 1/2\ O_2 \rightarrow Na_2SO_4$

This project will be realized with the intent of simulating a modified industrial process of one of the company's various products. By modifying this industrial process it is expected to obtain a better yield of the product with the same amount of raw material. This will result in the optimization of an existing process, which will eventually lead to economical savings.

115a Dependence of the Hydrolytic Degradation of Poly(ester-anhydride) Copolymers on pH, Temperature, and Time

Mathew D. Rowe, **Sarah Crosby**, and Keisha B. Walters. Swalm School of Chemical Engineering, Mississippi State University, 330 Swalm Chemical Engineering Building, Mississippi State, MS 39762

Renewable resource copolymers were developed containing both ester and anhydride linkages to allow for hydrolytic degradation with the objective of being able to control degradation time under variable temperature and pH conditions. It was expected that the anhydride bonds would hydrolyze initially, thereby reducing the molecular weight significantly, and a slower hydrolysis of the ester bonds would follow until complete degradation was achieved. Copolymers derived from 1,3-propanediol and malonic acid (PDO-MA) and 1,3-propanediol and itaconic acid (PDO-IA) were synthesized in a previous study at 155 °C using aluminum chloride. Water absorption and changes in dry weight of samples are monitored as a function of pH, time, and temperature. Changes in copolymer chemical structure as a function of pH, time, and temperature were analyzed with Fourier transform infrared (FTIR) spectroscopy by monitoring the absorbance changes corresponding to the anhydride and ester groups. The changes in molecular weight and polydispersity index are monitored by gel-permeation chromatography (GPC). The hydrolytic degradation of renewable poly(ester-anhydride) copolymers was studied as a function of pH, time, and temperature.

115b Development of a Consumer-Friendly Indoor Air Monitoring Device

Karen C. Dannemiller, Division of Engineering, Brown University, 75 Waterman St., Box 5443, Providence, RI 02912

With Americans spending 90% of their time indoors and indoor air pollutants at levels up to 100 times greater than outdoor air levels, it is important to sample indoor air. These contaminants result in thousands of cancer deaths and hundreds of billions of dollars in economics costs each year. Several design aspects of an indoor air sampling device to detect volatile organic compounds (VOCs) are examined. Challenges to design include breakthrough volume considerations, pressure drop constraints, calibration, and determination of detection limits. To avoid exceeding the breakthrough volume, the sampling tube should contain 0.210 g of Tenax TA sorbent. It is recommended that the sampling device use a pump to achieve the desired flow rate as opposed to a fan due to the pressure drop of the packed bed. In determining the calibration curve and the detection limits, the detection of formaldehyde proved to be difficult and deserves further research.

115c Design of a Laboratory Scale Pyrolysis System for Evaluating the Recycle Potential of Cca-Treated Wood Waste

Melissa Cook[1], Amy M. Parker[1], and Mark Bricka[2]. (1) Dave C. Swalm School of Chemical Engineering, Mississippi State University, P.O. Box 9595, Mississippi State, MS 39762, (2) Chemical Engineering, Mississippi State University, P.O. Box 9595, Mississippi State, MS 39762

Since the early 1970's, the most widely used wood preservative has been chromated copper arsenate (CCA), resulting in nearly 80% of all treated wood products in North America being treated with CCA. By the end of 2003 CCA-treated wood was restricted to industrial applications, resulting in a considerable increase in the volume of CCA-treated wood slated for disposal. Landfilling was considered an acceptable means of discarding CCA-treated wood products until recently, leading to approximately 24 million tons of CCA waste. Problems with metals leaching from the landfills and contaminating the surrounding soil and groundwater have generated the need for a safe and efficient disposal method for CCA-treated wood.

Pyrolysis, the heating of biomass at temperatures between 400°C and 650°C in the absence of oxygen, is a well studied technology that can be applied to CCA-impregnated wood waste. Pyrolysis of lignocellulosic material produces char, liquid condensate (bio-oil), and non-condensing gases. The focus of this research is on removing the CCA components from wood while recovering the energy value of the wood. This is accomplished by concentrating the CCA metals in the bio-oil, for possible re-use in wood preservatives, during pyrolysis. A lab-scale pyrolysis system, capable of operating in the desired temperature range under atmospheric and vacuum conditions, was designed. The system allows for the collection of each pyrolysis product so that complete mass balances on the metals can be performed, tracking the fate of the CCA metals. This paper discusses the process of designing the laboratory scale pyrolysis system and preliminary experimental results.

116a Non-Isothermal Solidification of Polymer-Drug Melts: Crystallization

Marlena Brown, Chemical and Biochemical Engineering/Food Science, Rutgers University, 65 Dudley Road, Lab 401, New Brunswick, NJ 08901 and Paul Takhistov, Food Science, Rutgers University, 65 Dudley Rd., New Brunswick, NJ 08901.

Current attempts in designing these systems have led researchers to focus on micro and nano sized structures as the foundation in the architecture for precise dosage formulation.

One of the more promising and forward thinking technologies that can be employed in the design of controlled delivery and precise dosage formulation is drop on demand technology. The design of robust platform for solid dosage manufacturing via drop on demand technology requires intimate knowledge of the phase behavior and morphology of deposited drug/excipient mixtures. We have studied solidification kinetics of melted polyethylene glycol (PEG) with various loads of ibuprofen, griseofulvin, and chlorpropamide deposited on the biopolymer film substrates. It was determined that drug type, concentration, temperature, substrate film type, and film concentration have an effect on nucleation, crystallization rate, evaporation rate, surface energy, and droplet solidification. Wetting behavior of droplets containing griseofulvin and chlorpropamide on hydroxypropyl cellulose (HPMC) was measured and controlled through engineering of surface properties of substrate film and drug loaded PEG.

Wetting behavior differs according to changes in the contact angle caused by the development of a new solid phase with the density higher than the liquid one. Research revealed that cooling rate is determining the crystallization process. Conditions that prevent crystallization and lead to formation of amorphous material have been found.

116b Evaluation of Sodium Alginate for Protein Drug Delivery

Fuyue Li, Yinyan Zhao, and Michael Harris. Chemical Engineering, Purdue University, Forney Hall of Chemical Engineering, Purdue University, West Lafayette, IN 47907

Alginate is a naturally occurring polysaccharide, composed of (1→4) linked β-D-mannuronate residues and its C-5 epimer α-L-guluronate residues. There has been increasing interests in utilizing alginate as an effective encapsulation material for the delivery of biopharmaceutical compounds. Therefore, the understanding of protein-polysaccharide interactions and the investigation of protein stability in sodium alginate solution have been important issues in this application.

In this study, the physiochemical property changes of model protein, Bovine Serum Albumin (BSA) during the interaction with sodium alginate in water solution, were monitored by Dynamic Light Scattering (DLS), Fourier Transform Infrared Spectrometer (FT-IR) equipped with Attenuated Total Reflectance (ATR) and Ultraviolet-Visible Spectrometer (UV-Vis). Parallel experiments were conducted for both native BSA-alginate and heat-denatured BSA-alginate mixture. DLS results showed that the hydrodynamic radius size of alginate did not change significantly after mixing with native BSA over 5 days of observation, indicating a thermodynamically stable system. However, the heat-denatured BSA alginate mixture had greater particle sizes than that of alginate only. A consistent turbidimetric analysis showed a sharp increase in turbidity of the heat-denatured BSA alginate mixture which was due to the formation of protein-polysaccharide complexes but a constant trend for the native BSA-alginate mixture. The reduction and addition of characteristic peaks absorbance of native BSA after mixing with alginate solution in FT-IR spectra indicated its secondary structures were changed by the electrostatic interactions between alginate chains and protein coils. Much more intense effects on the peak intensity were observed for the heat-denatured BSA. UV-Vis spectroscopy results confirmed with FT-IR. The increasing absorbance with increasing alginate concentration was caused by the formation of complexes due to interchain hydrophobic association between alginate and native/heat-denatured BSA. Our results showed that intermolecular chain associations were easily formed between BSA and alginate solution by electrostatic interactions. Not very significant conformational changes of native BSA entrapped in alginate solution suggested the potential pharmaceutical application of alginate as a carrier for protein drug delivery.

116c Experimental and Numerical Modeling of Granular Mixing in a Coater Pan

Robert Yau[1], Ekneet Sahni[2], and Bodhisattwa Chaudhuri[2]. (1) Department of Chemical, Materials, and Biomolecular Engineering, University of Connecticut, 69 North Eagleville Road, Storrs, CT 06269, (2) Department of Pharmaceutical Sciences, University of Connecticut, 69 North Eagleville Road, Storrs, CT 06269

Text Not Available.

116d pH-Triggerable Liposomes for Delivery and Release of Doxorubicin

Amey Bandekar, Shrirang Karve, and Stavroula Sofou. Othmer-Jacobs Department of Chemical and Biological Engineering, Polytechnic University, 6 Metrotech center, Brooklyn, NY 11201

Cancer is one of the leading causes of death in the US. Drug delivery systems still face the challenge of selectively increasing the availability of drug at the target tissue, while maintaining stability during circulation. pH sensitive liposomes can be used to release their contents during endocytosis. This approach relies on selective destabilization of liposomal membrane following acidification of the surrounding medium, therefore exploiting the acidic environment of the endosomal pathway to trigger content release.

In this work we developed a novel pH-sensitive lipid-based drug carrier based on lamellar-forming lipids for triggered release of encapsulated therapeutic agents. For enhanced circulation time, high Tg lipids (>70^0C) and lipids with anchored polyethylene glycol chains were used. We efficiently loaded doxorubicin on preformed liposomes. In addition, we studied the release of contents in serum supplemented media as a function of pH and time, and compared the content retention profiles to a DSPC-based liposome resembling the Dox® composition. Our results demonstrate that the proposed liposomes exhibit 80 % of content retention at physiological pH for the first 90 minutes and 35 % release of the encapsulated drug at pH 4.0 within the first one hour of incubation. The outcomes of this study indicate the potential in utilizing this newly developed pH-triggered liposome in delivering anticancer therapeutics.

116e Soluble Tissue Factor: Experiments In Silico

Erik John Haussmann, Kristin Patterson, and Coray M. Colina. Department of Materials Science and Engineering, Pennsylvania State University, 121 Steidle Building, University Park, PA 16802

Genetic disorders of blood coagulation, including hemophilia and von Willebrand's disease, affect approximately 1% of the U.S. The Tissue Factor (TF) pathway (or the extrinsic pathway) is believed to play a primary role in initiating blood coagulation during normal hemostasis as well as during many pathologic situations, including arterosclerosis and septicemia. However, there is indirect evidence that this scenario is not accurate. Human TF consists of 263 residues, the first 219 of which comprise the extracellular region. Extracellular TF (219) consists of two immunoglobulin-like domains connected by a single polypeptide linker (PRO 102 to ASN 107).

One way to help understand the motion in proteins is to use computer simulations, also called "virtual experiments". Among the several techniques available within the computer simulations, there are the so-called Molecular Dynamics. Molecular Dynamics (MD) are techniques that permit the study of the motions of flexible molecules as a function of time. In this work, we have used MD simulations to predict the solution structure of free TF (fTF) and TF bound to factor VIIa (bTF). A complete refined solution equilibrated model for TF based on available crystal structures (1BOY, 2HFT, 1DAN) using MD simulations in aqueous medium for over 15 ns is presented.

The X-ray crystal structures (1BOY, 2HFT, 1DAN) were used initially as starting points in modeling the unbound and bound structures. The complete models were subjected to aqueous-phase MD simulations, where unconstrained dynamics were performed using AMBER9/PMEMD9 programs. The solute, ions/counter-ions and crystal water molecules, were immersed in rectilinear periodic boxes of at least 20Å each side. Standards steps of energy minimization and dynamics were carried out before beginning the "production run". Finally, the post-equilibration NPT dynamics at 300 K, or production run, were completed for at least 15 ns depending of the study case.

The model sizes cover from 70,000 atoms up to 200,000 atoms, and a simulation length, for each system, up to 20 ns. All simulations were performed with (unless otherwise indicated): the ff99 and ff02, at 1 atm, 300 K, and 1fs time-step on the NPT ensemble. All proteins are solvated with TIP3P water molecules in rectangular periodic boxes, with 12, 15 or 20 Å between the protein and the box boundaries. Long range electrostatic interactions are calculated by the PME method. After a detailed study of the effect of the box's size in our simulations, we have found that these relatively large boxes are required to provide meaningful physical insight. We are using rectilinear boxes (due to the symmetry of the protein) of, e.g. 160x96x96 Å3 for a 134,422 atoms (TF-FVIIa complex). The size of the boxes is similar to the recently reported21 in the literature to model the lac repressor-DNA complex III DNA receptor with a total of 226,314 atoms.

Finally, these equilibrated solution structures for tissue factor should provide insight into the details of how tissue factor enhances the action of FVIIa. The possibility to predict these dynamic conformations could represent a crucial impact for the areas of

protein and drug design.

116f A Novel Approach to Cross-Linking Gelatin Matrices

Steven Castleberry, School of Chemical Engineering, Oklahoma State University, 423 Engineering North, Stillwater, OK 74078

The goal of this research was the optimization of a cross-linking strategy that would yield high compressive strength, as well as increased durability of gelatin-based three dimensional porous structures, useful in tissue regeneration. A novel approach of using a dual cross-linking series of Dimethyl Suberimidate (DMS) and N-Hydroxysuccinimide (NHS) to affect two different functional groups present in gelatin was chosen for this purpose. The hypothesis is that the first cross-linker (DMS) forms an unstable ester with gelatin. Subsequent addition of a second cross-linker (NHS) would react with the formed ester and stabilize it. Experiments performed showed support for the proposed hypothesis, as it was noted that the strength and durability of the structures increased when both the cross-linkers were used. Optimization was carried out through a factorial design of experiments and compression tested in hydrated conditions. A six-parameter regression model was developed from which surface plots were generated. Cross-linking density was assessed with Trinitrobenzenesulfonic acid labeling of primary amines. FTIR analysis was performed to identify cross-linking of both the functional groups. Work is currently being done to enhance the effectiveness of the secondary stabilizing reaction, most notably the use of N-Ethyl-N'-(3-dimethylaminopropyl) carbodiimide in place of NHS.

116g Bioreactive Membranes through Layer-by-Layer Assembly

Caitlyn R. Cecil, Saurav Datta, and D. Bhattacharyya. Chemical Engineering, University of Kentucky, 177 F Paul Anderson Tower, Lexington, KY 40506

Immobilizing enzymes within membrane pores is advantageous because immobilized enzymes exhibit greater stability than free enzymes, and it provides a means of directing the biocatalytic reaction. Two common methods of enzyme immobilization are through covalent bonding and electrostatic bonding. Enzymes attached electrostatically exhibit higher activities than those covalently attached because the active sites are not obstructed. The layer-by-layer (LBL) adsorption technique is a method for assembling a film of layered polyelectrolytes. An advantage of the LBL technique is that the properties of the assembled film such as its thickness and overall surface charge can be controlled. Thus, for a membrane functionalized with the LBL technique, the permeability and selectivity of the membrane can be controlled. In this study, glucose oxidase enzymes are immobilized with a layer-by-layer assembled membrane. Glucose oxidase is a dimeric protein with a molecular weight of 160 kDa and a pI of 4.2. Glucose oxidase catalyzes the reaction of b-D-glucose to hydrogen peroxide and D-glucono 1,5 lactone, which is then hydrolyzed into gluconic acid. The effects of the residence time, substrate concentration, and pH on the activity of the immobilized enzyme were considered in this study. Also, studies were conducted to prove that the enzyme could be detached from the membrane while maintaining its activity and fresh enzyme can be reattached to the same assembly.

This research was sponsored by the REU/NSF Program and NIEHS.

116h Gold Nanoparticle Based Protease-Detecting Biosensor

Kathleen A. Vermeersch[1], Craig D. Buckley[1], and Jessica O. Winter[2]. (1) Chemical and Biomolecular Engineering, The Ohio State University, Columbus, OH 43210, (2) Chemical and Biomolecular Engineering/Biomedical Engineering, Ohio State Unversity, 140 W 19th Ave, Columbus, OH 43210

Biosensors are very important in society and have many applications, including detecting biological hazards and diagnosing certain diseases. Nanoparticles, with their unique size-dependent properties, are an extremely promising technology for biosensor creation. However, some of the most promising models employ aqueous unsupported nanoparticles, which are difficult to adapt to applications outside of a laboratory. Instead, by supporting the nanoparticles in a gel, our biosensor would be portable, more adaptable to different environments, and more commercially applicable than unsupported, liquid-based biosensors.

Our biosensor consists of gold nanoparticles (AuNP's) embedded in a supporting polymer matrix, which permits the development of portable devices with potential for gas phase operation. The supporting matrix consists of poly(ethylene glycol) diacrylate (PEGDA) hydrogels, in which particles can be easily embedded. Biosensor detection is based on the color shift experienced by AuNP's, which is a function of their surface plasmon resonance. AuNP's less than 40 nm appear red, and as they grow larger the visible color shifts to violet. By bringing two AuNP close together, this same affect can be achieved as a result of proximity. If the particles are moved apart again, the absorbance shifts back resulting in a red solution. Our biosensor is designed to detect trypsin, a common protease used to decrease cell binding, and also important in diagnosis of cystic fibrosis. To produce violet particle aggregates, a peptide (N term-CGGGRGDSGGGC-COOH term) is used to bind the AuNPs together. If trypsin is present it will cleave the peptide, returning the particles to their initial red unaggregated state.

AuNP's were successfully aggregated in solution using the peptide, resulting in a color change. The absorbance spectra of the AuNP-peptide solution shifted so the system appears violet, whereas the unaggregated AuNP absorbance peak did not move. However, the system did not function properly in biological solution as a result of unwanted AuNP aggregation caused by saline and other solution contaminants. To reduce aggregation, we decided to use a stabilizing agent that would protect the AuNP's but still allow the peptide to bind. After experimenting with several agents, we are currently using tri(ethylene glycol) mono-11-mercaptoundecyl ether. This ligand is mixed together with the peptide and the two are then added simultaneously to the AuNP's. This is yielding a better passivated AuNP system than before, but the proper ratio of stabilizing agent to peptide has not yet been achieved.

Next, we plan to suspend these conjugates in PEGDA hydrogels and confirm that a color change can be observed in the presence of sensate (trypsin). This biosensor is designed to detect the extent of cell cleavage in response to trypsin. However, this biosensor could be customized to almost any enzyme-substrate system. Any substrate with thiol ends (which can be added through cysteine termination) has the ability to bind the AuNPs together, and any substrate specific enzyme can cleave the peptide bond activating the sensor. With the enzyme-substrate flexibility and the adaptability that our sensor will offer, it could be customized to fit almost any sensing need and would therefore have large commercial appeal.

116j Ribosome Kinetics and Aa-tRNA Competition Determine Rate and Fidelity of Peptide Synthesis

Aaron M. Fluitt, Elsje Pienaar, and Hendrik J. Viljoen. Department of Chemical and Biomolecular Engineering, University of Nebraska-Lincoln, 207 Othmer Hall, Lincoln, NE 68588-0643

It is well established that translation, the process by which the ribosome synthesizes proteins, proceeds at a varying rate, but several theories have been proposed for this variation. We have hypothesized that the cognate aminoacyl tRNA (aa-tRNA) faces competition for binding at the ribosome from near- and non-cognate aa-tRNAs, causing the observed delays and errors in peptide synthesis. We have calculated the competition levels, translation times, and error frequencies for all 64 RNA codons. The model sheds light on the arrangement of fast- and slow-translating regions in a particular gene and the kinetics of programmed frameshifts.

116k Production of Poly (3-Hydroxybutyric Acid) from Waste Sweet Potato Starch

James D. Winkler[1], Ratna Sharma[2], and Mari S. Chinn[2]. (1) Chemical Engineering, Rice University, 6330 Main Street, Houston, TX 77005, (2) North Carolina State University, Campus Box. 7625, Biological and Agricultural Engineering, Raleigh, NC 27695, United States

The organism Ralstonia eutropha has been reported to produce Poly (3-Hydroxy-butyric Acid) (PHB), a valuable biodegradable polymer. PHB synthesis in R. eutropha occurs during periods of transient nutrient deprivation in order to sequester carbon in anticipation of a restoration of better growth conditions. Artificial nutrient (primarily nitrogen or phosphorus) limitation is therefore used to induce large scale PHB production in the organism. The mechanical properties of PHB synthesized depend heavily on the carbon source used; the desired properties of the polymer can be controlled by carefully selecting the feedstock used as the carbon substrate for R. eutropha. This project investigated the novel use of a hydrolyzed industrial sweet potato (FTA-94) starch in an effort to determine whether PHB could be produced from the substrate and how the choice of feedstock would influence properties of the resulting polymer. Cultures incubated in FTA-94 limited media exhibited significant PHB inclusions under transient electron microscope (TEM) imaging. Gas chromatographic analysis of cell samples confirmed the presence of PHB.The results demonstrate that FTA-94 can serve as the sole carbon substrate for PHB synthesis without causing undesired inhibitory effects on bacterial growth. Further analyses are expected to reveal average molecular weight and compositional information of chemically extracted PHB.

116l Experimentally Validated Numerical Modeling Study of Mixing in a Coating Pan

Robert Yau, Department of Chemical, Materials, and Biomolecular Engineering, University of Connecticut, 8 Franklin Woods Dr., Somers, CT 06071, Ekneet Sahni, Department of Pharmaceutical Sciences, University of Connecticut, 69 North Eagleville Road, Storrs, CT 06269, and Bodhi Chaudhuri, Pharmaceutical Sciences, University of Connecticut School of Pharmacy, 69 North Eagleville Road, Storrs, CT 06269.

Coating of particles is done in the pharmaceutical industry, to mask the unpleasant taste or odor of the drug, to control the bioavailability of the drug, to protect the tablet from the environment (increase shelf-life), and sometimes to improve tablet appearance. During coating, the coating solution is sprayed at discrete locations on the cascading region of the granular bed, and only the liquid jet directly coats a small fraction of tablets. The distribution of the coating solution from these locations to the remainder of the bed occurs through mixing. Thus, the knowledge of particle flow and mixing in a pan coater is thus critical to optimize the design and operation of such equipment. Mixing is an important but poorly understood aspect of coating of pharmaceutical preparations (esp. tablets). Experimental and computational methods are developed to perform quantitative examination of mixing patterns and mixing rates in rotating coating pan. White and red non-pareils of 6-7 mesh size loaded in the ellipsoid pan coater to check the effect of initial loading (side-side and front-back), fill level, orientation of the vessel and the vessel speed on granular mixing. Video-imaging, MATLAB and discrete pocket samplers are used to quantify the mixing and to finally estimate the optimal operating condition. DEM (Discrete element method) based numerical model was also developed to study the effect of granular mixing in a coating pan. Digitally recorded mixing states from experiments are used to fine tune the model. In both the experiments and simulation, we observed slower axial dispersion than radial convection when the axis of rotation of the mixer is horizontal (no tilt). However, tilt enhances axial mixing, and faster axial mixing is seen for higher tilt angles from the horizontal. The speed of the rotating vessel has no prominent effect on the rate of mixing in a coating pan, as observed in experiments and simulations. Moreover, fill level has no significant effect on the rate of mixing. Finally effect of vessel speed on coating is simulated modeling the coating of the particles passing through prescribed spray location on the cascading zone. Higher vessel speed enhances granular coating for moderate filling and tilted orientation of the coater.

116m The Effect of Cholesterol on Lung Surfactant

Jacob M. Vestal, Chemical and Biomolecular Engineering, NC State University, 9817 Adlie Dr., Wake Forest, NC 27587, Siegfried Steltenkamp, Chemical Engineering, University of California, Santa Barbara, University of California, Santa Barbara, CA 93106, and Joesph A. Zasadzinski, Department of Chemical Engineering, University of California, Santa Barbara, CA 93106.

Lung surfactant (LS) is a complex collection of proteins and lipids found at the interface between air and the pulmonary fluid that lines the inside surfaces of the alveoli in mammalian lungs. The vital functions of lung surfactant are disrupted by Acute Respiratory Syndrome (ARDS), a condition in which serum proteins replace LS molecules on the interface following acute trauma to the abdomen. No effective treatment for ARDS exists and the crucial need for a synthetic LS-like preparation, able to restore function to serum-disabled LS, is of paramount importance to the effort to lower the 30% mortality rate of this fatal condition. An important question in the design of an ARDS treatment is weather cholesterol should be included. Although known to be present in mammalian lungs, the role that cholesterol plays in functional LS is almost totally unknown.

In this study the effect of cholesterol on bovine-derived LS was investigated with fluorescence microscopy, atomic force microscopy, Wilhelmy-plate surface force measurements, and surface viscometry. The effect of cholesterol on the phase behavior of LS during the breathing cycle is discussed, speculations are offered on the interactions between cholesterol and the surfactant proteins.

PAPER NO. **: 145**

TITLE **: "NON BACTERIAL WAY OF SYNTHESIS OF BIPFUEL FROM CELLULOSE AND OTHER POLYSACCARIDES"**

AUTHORS **: PRAVEEN KUMAR GORAKAVI, Dr. V.V. BASAVA RAO, Dr. Ch. SAILU, Dr. P. LAXMINARAYANA**

ABSTRACT:

Cellulosic or starch feedstock is the main reactant polysaccharide carbohydrate in the natural fermentation process. Out of the many products obtained, ethanol is aimed at during fermentation. So bacterium is exposed to strain that optimizes this product. Complete chemical process totally takes place in few steps. The feedstock is meant for catalytic hydrolysis which is done by treating the feedstock in acidic conditions at a high temperature and pressure. This can be done by acidulous conditions and vigorous stirring at ambient pressure. The presence of coordinate compounds with chloride on ionic sphere of complex accelerates the reaction. The obtained products take a shape of carbohydrate which can be further synthesized to smaller alcohols which is desired bio-fuel is normally meant through yeast fermentation. Some have been discovered that directly converts cellulose/starch into ethanol. However to obtain a mono product at a high purity, the hydrolyzed feedstock is meant for catalytic reduction with HI in presence of heat. This reaction converts all sugars into long chain alcohols which can be converted into smaller alcohols and hydrocarbons through anaerobic pyrolysis. Since it is recently found that butanol is a best substitute (Refer to Chart1.) for gasoline than ethanol, we can alter our pyrolysis so as to obtain butanol as main product. The other products would be a 2carbon hydrocarbon at maximum. These smaller HCs can be removed as they are gasses and they find their application as biogas. This facility of obtaining variety of products as requires is only possible by an extreme controllable reaction like this artificial synthesis. So the availability of this technology irrespective of economics is at its absolute requirement.

INRODUCTION:

Fossil fuels are generally composed as a mixture of crude oil which is typical compound containing many hydro carbons that can be combusted and water. The composition of this mixture of crude oil and water varies with ambience temperature and many other environmental parameters. The extant of crude oil availability is also lessening gradually. Moreover the market demand for fuels is rising day by day with increasing population and automobile usage. Use of fuel in fuel cell to overcome energy crisis in countries like India and Pakistan is also a reason for excess demand for the fuel. This scenario is forcing a hike in the fuel costs. Moreover the war features and other international politics abmolish fuel rates. This hike in fuel rates reached an abnormal value of 143$/barrel (1barrel~=117.34Lts.) in July 2008. All these scenario forces people to find out an alternate fuel for automobiles. Though automobiles can get another fuel, the matter of fact is that all the vehicle engines throughout the world cannot be replaced at a stroke. This at least requires a 30-50 year time for complete change. Thus there arose an absolute need for raise in interest for biofuels. Biofuels resemble the normal petrochemicals and function similar to them in automobile engines that are using petrochemical now in the market. However there arises a change in octane/cetane number and efficiency of fuel per liter. However out of these biofuels that can be synthesized we look for probable fuel that best suit as petrochemicals.

Obtaining of biofuels can be done by fermentation of the biomass (containing good amount of starch/cellulose/sugar content) like cane, corn/maize which can be generalized as the mass that is obtaining from a biological source. The process involved in getting the biofuel from biomass takes part in several unit processes which are already in practice. Out of the available techniques, the most prevailing method now is biological fermentation of the biomass using yeasts. Yeasts are a growth form of eukaryotic microorganisms classified in the kingdom fungi. Even the alcohoic beverages are also obtained through fermentation of grapes which are indeed a form of biomass in anaerobic atmosphere for a long time. Indeed, alcoholic beverages contain smaller alcohols that also substitute as petrochemical for combustion. Biofuels can also be prepared from plants with higher amount of oil content like jathropa, oil palm, soy bean, algae, etc.,. Its oil can directly be used as biodiesel after reducing its viscosity. Both the ways are benevolent. However there is an interesting competition between these two ways of extractions. Fuel from seed oil involves a time taking and a very market risk as there is every possibility for spoilage of crop in that long time. For example if we take the crop of jathropha, an ingenious plant to southern asia, takes a long time of 3-4 years to grow. Though it resist for the drought conditions, its existence is at doubt for such a long time. Its seeds have to be cared a lot for their existence as there is every possibility for it to be lost from the plant, especially in rural parts of India. So, obtaining them through another slow but effective process i.e., by yeast fermentation of biomass is opted by many industries. However, this is also catalyzed by biocatalyst and other strains are established on bacterium races meant for this reaction to increase the desired product in greater extent. To have a rapid production of biofuel an artificial fermentation reaction is needed that creates absolute production in accordance with the demand. It however also depends on availability of feedstock.

Normal production plants are now able to deliver the fuel by 3-4 weeks time. But the urgency may cause a low concentrated alcoholic product. This effects combustion of fuel and at a long run effects the engine too. So, obtaining of purity and rapidity in biofuel production is only possible through artificial synthesis of biofuel from selective feedstock.

Let us get through the various unit processes involved in this non bacterial biofuel production plant in detailed. For renting quality in understanding, let us first stay away from economical calculations and let us do the price estimation at last.

On an over look, we can just say that the feedstock is processed through the following chemical unit process step wise. Though it cannot be taken as a mere chemical flow chart, but can be compared with it.

a) The feedstock is selected so as to obtain maximum biomass from a unit raw stock.
b) The feedstock is subjected for crystallization and then is processed for catalytic hydrolysis. This reaction is called as catalysis of cellulose or cellulolysis. If the feedstock contains greater extent of cellulose composition in it, it is likely profitable to use and exploit that nature of the feedstock and thus producing cellulosic ethanol or any other cellulosic fuel. If once we are prepared with hydrolysis of cellulose, it is a known fact that the reactor takes up even hydrolysis of the few more polysaccharides like hemicelluloses, etc. However, we can understand that the hydrolysis of some compounds such as lignin is not possible.
c) After obtaining the hydrolyzed lignocelluloses, the unhydrolysed part of feedstock which mainly contains lignin and other similar carbohydrates are meant for separation to separate lignin from the hydrolysed sugars. Depending upon the case, a unit process that converts polysaccharide cellulose to disaccharide can be installed. The expected product would be sugar of different carbon number.
d) The purified sugars can then be processed to a catalytic reduction using a hydrogen iodide (HI) solution. Once, the sugar gets reduced and alcohols of various carbon numbers are obtained.
e) The obtained high carbon alcohols can then be subjected for anaerobic pyrolysis stimulating the parameters for an adjustable product yield. Product can start from a single carbon methane gas to butanol, which is now being called as biobutanol representing the source which is found as a best substitute for gasoline.

f) The obtained mixture of hydrocarbons is then subjected for distillation where we can obtain almost pure forms of different hydrocarbons and then transferring them to use them in their respective utilizing equipments.

Let us evaluate the ideas/claims of this paper in detail

Selection of the ideal feedstock for evaluation:

Selection of an ideal feedstock is at its absolute necessity. Selection is done by means of the biomass percentage the source gives which can be converted by processing through this chemical plant to biofuel. We generally select a plant that has this sort of high biomass content and also we look for the indigenous plants for their ability to resist local infections and their rapid growth at the local temperature.

If we have a look at the plants we are using as sources for biomass, the majority of indigenous biofuel sources are opted for their ingredient carbohydrate content of starch/cellulose/hemicelluloses. They are also desired to have low amount of non hydrolysable carbohydrates such as lignin. If these are the characteristics depicted by any plant, then obviously it stands in the race of plants that stays as a source for biofuel. The main idea beyond this is that there are already discovered bacterium races that release yeasts for fermentation of this feedstock for a complete conversion of hydrolysable carbohydrates to smaller alcohols which suit as petrochemicals. To increase the production of a specific compound, the parameters are set up extremely favorable for the desired compound production.

If we look at the biological sources we are currently aware of for biofuel production, we find that sugars, starch, vegetable oils, animal oils, seeds/grains such as wheat, sunflower seeds served their purpose in first generation biofuels through a natural fermentation process after hydrolysis. The microbes that convert feedstock directly into biofuel through internal hydrolysis and fermentation were not in reputation at that time. An overview on vegetable oil as biofuel can be summered to state that the vegetable oil can directly be used in diesel engine after reducing their viscosity. In fact, diesel is meant as a dilute vegetable oil. However it is only meant at higher (warm) temperatures, i.e., summer season as solidification of oils is possible in engines that effect the engine prospects. The obtained diesel is termed as biodiesel depicting the source through its name. Biodiesel is the most common biofuel in Europe. It is produced from oils or fats using transersterification and is a liquid similar in composition to mineral diesel. Its chemical name is fatty acid methyl (or ethyl) ester (FAME). Oils are mixed with sodium hydroxide and methanol (or ethanol) and the chemical reaction produces biodiesel (FAME) and glycerol. One part glycerol is produced for every 10 parts biodiesel. Feed stocks for biodiesel include animal fats, vegetable oils, soy, rapeseed, jatropha, mahua, mustard, flax, sunflower, palm oil, hemp, field pennycress, and algae. Bio alcohols are also assumed as biofuels as they suit as substitute for gasoline. Biologically produced alcohols, most commonly ethanol, and less commonly propanol and butanol, are produced by the action of microorganisms and enzymes through the fermentation of sugars or starches (easiest), or cellulose (which is more difficult). Biobutanol (also called biogasoline) is often claimed to provide a direct replacement for gasoline, because it can be used directly in a gasoline engine (in a similar way to biodiesel in diesel engines). Out of all the bioalcohols, the predominant alcohol as a biofuel is bioethanol. Ethanol fuel is the most common biofuel worldwide, particularly in Brazil. Alcohol fuels are produced by fermentation of sugars derived from wheat, corn, sugar beets, sugar cane, molasses and any sugar or starch that alcoholic beverages can be made from (like potato and fruit waste, etc.). The ethanol production methods used are enzyme digestion (to release sugars from stored starches, fermentation of the sugars, distillation and drying. Second generation biofuel utilizes non food crops as their source for biomass. It is a solution for increasing inflation due to the migration of food crops to biofuel production in first generation biofuel plants. Cellulosic biofuels are a part of second generation biofuels. These, as the name itself suggests, are obtained from cellulose which is the most abundant carbohydrate in world.

The above given discussion shows the nature of market demand and requirement for both food crops and biofuel. So, obtaining the biofuel from a non food crop or an energy crop is ideal choice of the biomass source. Let us have a look a various energy crops and their status for use as a good source for a biofuel feedstock.

Switchgrass- indigenous plant of North America:

We observe that switch grass is the most used biological source in northern parts of America. It is a native crop of the continent. Similarly when it is the case with southern Asia, jathropa seed oil is prominent as biodiesel. But there are few practical problems to establish a biodiesel plant in south Asia. Switch grass is an ideal source but extending it to a humid environment that prevails in south Asia through cultivation in artificial environment obstructs the expected output. If we look at the profile of switch grass we find some interesting aspects.

Switchgrass (*Panicum virgatum*) is a summer perennial grass that is native to North America. It is a natural component of the tall-grass prairie which covered must of the great Plains, but which also was also found on the prairie soils in the Black Belt of Alabama and Mississippi. Many people do not realize that the natural vegetation of the Black belt was grassland, and not frost like most other parts of the southeastern USA. Need for agricultural chemicals to grow switchgrass is relatively low. Switchgrass is also very tolerant of poor soils, flooding and drought, which are widespread agricultural problems in the southeast. There are two main types of switchgrass: upland types, which usually grow 5 to 6 feet tall and are adapted to well drain soils, and low land types, that grow up to 12 feet tall and which are typically found on heavy soils in bottomland sites. Although switchgrass is native, plant breeders have developed a fairly large number of improved varieties for use as forage.

'Alamo' switchgrass is a robust lowland variety of switchgrass most suited to the southern US. In Auburn University test plots, it has frequently produced over 10 tons per acre per year, but on a commercial scale, it is more reasonable to expect 6 to 8 tons per acre. This is because test plots usually have perfect establishment, but commercial plantings almost always have weak spots in the field. However, for comparison, the average annual hay yield for Alabama is about 2.5 tons per acre, and the productivity of forests is only about half that of switchgrass. Despite poorer soils than in the Midwest, switchgrass yields are higher in the Southeast because of the adaptation of more productive switchgrass varieties in our region, and because we have a longer growing season. The seed of switchgrass is very small, and much of it is dormant (will not germinate) right after it is harvested. However, aging, treating it with water and chilling temperatures (stratification) or storing it in warm conditions will break dormancy. Partly because of the small size of the seed, switchgrass seedlings tend to be slow to develop, and are susceptible to weed competition. Unfortunately, there are no herbicides approved by government for weed control during establishment of switchgrass. However, it can still be successfully established by no-till planting and other strategic approaches.

Switchgrass reaches full yield only in the third year after planting; it produces a quarter to a third of full yield in the first year, and about two thirds of full yield in the second year. When managed for energy production it can be cut once or twice a year with regular hay or silage equipment. At maturity, widely spaced switchgrass plants can measure 20 inches in diameter at ground level. Switchgrass has a huge, permanent root system that penetrates over 10 feet into the soil, and weighs as much (6-8 tons/acres) as the above-ground growth from one year. It also has many fine, temporary roots. All these roots improve the soil by adding organic matter, and by increasing soil water infiltration and nutrient-holding capacity. Switchgrass fields provide habitat and a home for many species of wildlife, including cover for deer and rabbits, and a nesting place for wild turkey and especially quail.

Switchgrass has several other environmental benefits. If it is used to produce energy, it will reduce the risk of global warming by replacing fossil fuels (coal, natural gas and oil). When fossil fuels are burnt, carbon is removed from below ground (gas and oil wells and coal mines) and release into the atmosphere as carbon dioxide (CO_2). This is a greenhouse gas that increases the risk of global warming. In contrast, switchgrass (like all other plants) removes CO_2 from the atmosphere and incorporates it into plant tissue, both above and below the ground.

The accumulation of carbon, especially below the ground, is known as carbon sequestration, and is considered to be a very important strategy for reducing atmospheric CO_2. Switchgrass is unquestionably one of the best crops for doing this. when above-ground switchgrass is harvested and burned for energy, CO_2 is once again returned to the atmosphere from where it was originally obtained by the plant, but it will have reduced the need for some fossil fuel. Therefore, CO_2 is obviously just being recycled by use of switchgrass for energy, making this process CO_2-neutral (or actually CO_2-

negative if soil carbon sequestration is considered), compare to fossil fuels that add CO_2 to the atmosphere. when compared to low grade coal, burning switchgrass for energy will probably result also in less toxic emissions, such as the oxides of sulphur and nitrogen.

Together with its energy benefits, swicthgrass offers great opportunity for farmers. Because of its perenniality, compared to annual crops switchgrass is a true conservation crop which will substantially reduce soil erosion and release of soil carbon which are related to annual tillage, and it will reduce the use of toxic chemicals.

Perhaps most important, we must recognize that fossil fuels will be our main energy base for many years, and bioenergy from switchgrass is not intended to compete with these valuable resources, but rather, to complement them by softening their environmental impact.

Artificial cultivation of the switchgrass in India and other southern asia countries is possible by extending artificial envionment to the harvesting plant. However, the practice of waiting for 3-4 years for total yield is not amicable in such an expensive environment.

Jatropha a native crop of south Asia and central America (II generation biofuel crop):

Currently the oil from Jatropha curcas seeds is used for making biodiesel fuel in Philippines, promoted by a law authored by Philippine senators Miriam Defensor-Santiago and Miguel Zubiri. Likewise, jatropha oil is being promoted as an easily grown biofuel crop in hundreds of projects throughout India and other developing countries. The rail line between Mumbai and Delhi is planted with Jatropha and the train itself runs on 15-20% biodiesel. In Africa, cultivation of Jatropha is being promoted and is grown successfully in countries such as Mali.

Estimates of Jatropha seed yield vary widely, due to a lack of research data, the genetic diversity of the crop, the range of environments in which it is grown, and Jatropha's perennial life cycle. Seed yields under cultivation can range from 1,500 to 2,000 kilograms per hectare, corresponding to extractable oil yields of 540 to 680 liters per hectare.

Jatropha can also be intercropped with other cash crops such as coffee, sugar, fruits and vegetables. Large plots of waste land have been selected for Jatropha cultivation and will provide much needed employment to the rural poor of India. Businesses are also seeing the planting of Jatropha as a good business opportunity. The Government of India has identified 400,000 square kilometres (98 million acres) of land where Jatropha can be grown, hoping it will replace 20% of India's diesel consumption by 2011.

Jatropha is also similar to switch grass in its growth rate. However unlike switchgrass, jathropa produces seeds that are utilized for biodiesel production after reduction in their viscosity. But, harvesting a crop for 3-4 years without expecting a bet gain from land is not merely possible for establishing a heavy production plant. Even if it is subsidized by government, the care that has to be initiated inorder to save the seed and ascertain its perfect growth is very high as there is every possibility in irregularity in climatic conditions. So we are in an need for a crop that gives maximum cellulose (as in II generation crop) that requires least care and be able to yield even with the case of spoiled crop. Even, complete utilization of plant is not desirable since recultivation is again a time taking process. So a crop like cotton comes into picture with the desired qualities.

Cotton crop for biodiesel production:

Cotton is a soft, staple fiber that grows around the seeds of the cotton plant (*Gossypium* sp.), a shrub native to tropical and subtropical regions around the world, including the Americas, India and Africa. The fiber most often is spun into yarn or thread and used to make a soft, breathable textile, which is the most widely used natural-fiber cloth in clothing today. Successful cultivation of cotton requires a long frost-free period, plenty of sunshine, and a moderate rainfall, usually from 600 to 1200mm (24 to 48 inches). Soils usually need to be fairly heavy, although the level of nutrients does not need to be exceptional. In general, these conditions are met within the seasonally dry tropics and subtropics in the Northern and Southern hemispheres, but a large proportion of the cotton grown today is cultivated in areas with less rainfall that obtain the water from irrigation. Production of the crop for a given year usually starts soon after harvesting the preceding autumn. Planting time in spring in the Northern hemisphere varies from the beginning of February to the beginning of June. The area of the United States known as the South Plains is the largest contiguous cotton-growing region in the world. It is heavily dependent on irrigation water drawn from the Ogallala Aquifer. Out of different varieties of cotton organic cotton is notable cotton. Organic cotton is cotton that is grown without insecticide or pesticide. Worldwide, cotton is a pesticide-intensive crop, using approximately 25% of the world's insecticides and 10% of the world's pesticides. According to the World Health Organisation (WHO), 20,000 deaths occur each year from pesticide poisoning in developing countries, many of these from cotton farming. Organic agriculture uses methods that are ecological, economical, and socially sustainable and denies the use of agrochemicals and artificial fertilizers. Instead, organic agriculture uses crop rotation, the growing of different crops than cotton in alternative years. The use of insecticides is prohibited; organic agriculture uses natural enemies to suppress harmful insects. The production of organic cotton is more expensive than the production of conventional cotton. Although toxic pollution from synthetic chemicals is eliminated, other pollution-like problems may remain, particularly run-off. Organic cotton is produced in organic agricultural systems that produce food and fiber according to clearly established standards. Organic agriculture prohibits the use of toxic and persistent chemical pesticides and fertilizers, as well as genetically modified organisms. It seeks to build biologically diverse agricultural systems, replenish and maintain soil fertility, and promote a healthy environment.

Cotton fiber, once it has been processed to remove seeds (ginning) and traces of honeydew (a secretion from aphids), protein, vegetable matter, and other impurities, consists of nearly pure cellulose, a natural polymer. Cellulose content in cotton is nearly 90%. So cotton is treated as an excellent source of cellulose. Moreover, as cotton is an established crop and is being already under use, there is no need for awareness in cultivation of crop. The major feature is that spoiled or futile cotton can be used for biofuel synthesis.

Moreover equipments and processors for cotton management are well known and existing. So there is no absolute need for equipments arising. In india, Large amount of cotton farmers are under loss due to irregularity in precipitation. All of them come to a profitable edge is utility of cotton as a source of cotton instead of usage as a fabric materializes. Thus it can be stated that in a country like India, use of cotton for synthesis of biofuel is ideal than use of switchgrass or jatropha.

2. Hydrolysis of lignocelluloses that is obtained from cotton:

CELLULOSE HYDROLYSIS: Cellulose can get hydrolyzed by normal conventional acid hydrolysis as depicted in many abstracts before. However they show an interesting stimulated hydrolysis by using an acid catalyst under ionic liquid medium. This nature can be exploited for hydrolyzing cellulose to sugar. Cellulose in ionic liquid medium when heated with stirring at 100 °C, ambient pressure a clear solution is formed. To this cellulose solution then we add quickly H2O (1.75 mole equiv. to cellulose) and appropriate amount of 98 wt % H2SO4. The reaction was then vigorously stirred. At different time intervals, samples were withdrawn, weighed, and quenched immediately with cold water for evaluation of product. The aqueous solutions were neutralized with NaOH, centrifuged at 10,000 rpm for 5 min, measured the volume and subjected to total reducing sugar (TRS) and glucose analysis. The reduced sugar yield can vary up to production of a 90-95%yield which is extremely preached as efficient.

This hydrolysis of cellulose in a ionic liquid with acid as a catalyst is the best way to obtain sugar from cellulose. Moreover, if you recall the prices of the reagents being used in the above cellulosis, it is obvious that it is a low expensive process. The obtaining can be manipulated over variety of ionic liquid solvents which is the future plan of this idea. There was an interesting paper presented by Dalian Institute of Chemical Physics and Graduate School of the Chinese Academy of Sciences in similar experiment where they have found and outstanding 95%sugar as output. Refer to fig 1.

Research was done by auburn university on dilute acid catalysis and the kinetics of normal cellulose hydrolysis using acid catalyst but in absence of ionic liquid solvent. They have stated a very interesting fact that the cellulosys is a single ordre kinetic reaction. They have extended their paper even through the probable reaction mechanism of cellulose hydrolysis.It was also shown that the crystallization ability of cellulose effects the hydrolyzation rate a lot.

HEMICELLULOSE HYDROLYSIS: When compared to cellulose both extraction and hydrolysis of hemicelluloses is very easy. In fact hemicelluloses get hydrolyzed to form xylase long before cellulose converts to sugars.

However there is another part of lignocelluloses called lignin which resists extremely to get hydrolysed for its aromatic behavior. Lignin hydrolysis is thus treated as impossible. So only cellulose and hemicelluloses gets hydrolysed to sugars and the obtained partially hydrolyzed solution is meant for separation where lignin is removed from the solution.

3. Catalytic reduction of the sugar solution to convert them to alcohols:
Though an endothermic reaction, it is not a difficult task to obtain a reduced mixture of hydrolyzed feedstock. In general, HI is employed for reducing the carbohydrates to sugars. It is ideal is we use HI in limited extent when compared with that of sugar so as to obtain varieties of partially hrdrolyses hydrocarbons. This conversion results into formation of iodine gas and a dilute solution of alcohol on product side.

4. Concentrating the alcohol and cracking:
The dilute alcohols can be then assumed to have a mixture of pentathol and hexanol as major products. The solution can be meant for distillation through various techniques that are already known to us and then the obtained product alcohols of higher concentrations are let for anaerobic pyrolysis. It is known that pyrolysis allows us to adjust the product by pressurizing reaction with varying parameters. Thus the required bioalcohol can be prepared and can be substituted for gasoline. A table containing the details on few products obtained after the chemical processing is done is as follows.

Depending on choice of fuel in accordance to the desire, parameters can be adjusted for a maximum yield of such a product.

Bibliography:

[1]. David Ramey, November 2004,light party.
[2]. E.Coli, on march 17th 2008, 'Sugar beet ethanol',natural journal.
[3]. Atsumi, S.; T. Hanai and James.C. Liao, 2008, 'Engineering Synthetic Non-Fermentative Pathways for Production of Branched-Chain Higher Alcohols as Biofuels', natural.
[4]. M. Slesser, C. Lewis, 1979, Biological Energy Resources, Halsted Press Book, London,.
[5]. Jack Bradley,1979, , Making fuel in your backyard, Biomass Resources, Wenatchee, WA.
[6]. J. Mintosh,1923, Industrial Alcohol, H.B. Stocks, F.I.C., London.
[7]. C. Bliss, D.O. Blake, 1977, Silvicultural Biomass Farms Vol. 5 Conversion Processes and Costs, The Mitre Corporation/Metrek Division, USA.
[8]. F. Collins, J. Ellis, J. Hermsen, K. Kawaoka, M. Malloy, C. Pate, B. Ryan, J. Simons, H. White, 1980, Alchohol Fuels From Biomass: Production Technology Overview, Aerospace Corporation Report ATR-80(7874)-1, MA.
[9]. T. Duncan, J. Reimer,1998, Chemical Engineering Design and Analysis, Cambridge University Press, UK.
[10]. N. Rask, 1980,Alcohol Production from Biomass in the Developing Countries, World Bank, Washington, DC.
[11]. W. Bello,1989, Brave New Third World? Strategies for Survival in the Global Economy, Food First Development Report No. 5, San Francisco, CA.
[12]. J. Childress,1981, Fuel Alcohol: An Energy Alternative for the 1980s, U.S. National Alcohol Fuels Comission, Washington, DC.
[13]. H. Bluestein, B. Greenglass, W. Holmberg,1982, Ethanol Fuels: Reference Guide, Solar Energy Research Institue, CO.
[14]. Hirokazu Nankai, Wataru Hashimoto,* Hikaru Miki, Shigeyuki Kawai, and Kousaku Murata ,1999 Microbial System for Polysaccharide Depolymerization: Enzymatic Route for Xanthan Depolymerization by Bacillus sp. Strain GL1.
[15]. Hashimoto, W., Miki, H., Tsuchiya, N., Nankai, H., Murata, K. ,2001, Polysaccharide Lyase: Molecular Cloning, Sequencing, and Overexpression of the Xanthan Lyase Gene of Bacillus sp. Strain GL1. Appl. Environ. Microbiol. 67: 713-720
[16]. Nankai, H., Hashimoto, W., Murata, K. ,2002,. Molecular Identification of Family 38 {alpha}-Mannosidase of Bacillus sp. Strain GL1, Responsible for Complete Depolymerization of Xanthan. Appl. Environ. Microbiol. 68: 2731-2736

[17]. Hashimoto, W., Nankai, H., Mikami, B., Murata, K. ,2003, Crystal Structure of Bacillus sp. GL1 Xanthan Lyase, Which Acts on the Side Chains of Xanthan. J. Biol. Chem. 278: 7663-7673
[18]. Itoh, T., Akao, S., Hashimoto, W., Mikami, B., Murata, K.,2004, Crystal Structure of Unsaturated Glucuronyl Hydrolase, Responsible for the Degradation of Glycosaminoglycan, from Bacillus sp. GL1 at 1.8 A Resolution. J. Biol. Chem. 279: 31804-31812
[19]. Atsumi, S.; T. Hanai and J.C. Liao ,2008, Non-Fermentative Pathways for Synthesis of Branched-Chain Higher Alcohols as Biofuels, Nature, 451:86-89.

[20]. Hanai, T., S. Atsumi, J. C. Liao ,2007, Engineered synthetic pathway for isopropanol production in Escherichia coli. Appl. Env. Microbiol. 73:7814–7818

[21]. Atsumi, S., A. F. Cann, M. Connor, C.R. Shen, K. M. Smith, M.P. Brynildsen, K.J.Y. Chou, T. Hanai, and J. C. Liao ,2007, Metabolic engineering of Escherichia coli for 1-butanol Production. Metabolic Eng., In print.

[22]. Bernstein, J.R., Bulter, T., Shen, C.R. and Liao, J.C. ,2007, Directed Evolution of Ribosomal Protein S1 for Enhanced Translational Efficiency of High GC Rhodopseudomonas palustris DNA in Escherichia coli. J. Biol. Chem., 282(26):18929-36.

[23]. Braatsch, S., Bernstein, J.R., Lessner, F., Morgan, J., Liao, J.C., Harwood, C.S., and Beatty J.T. ,2006, Rhodopseudomonas palustris CGA009 has two ppsR genes that each encode repressors of photosynthesis gene expression. Biochemistry, 45(48):14441-51.

[24]. Farmer, W.R. and J.C. Liao, ,2000, Improving Lycopene Production in Escherichia coli by Engineering Metabolic Control. Nature Biotechnol., 18, 533-537.

[25]. Wang, C.W. and J.C. Liao ,2001, Alteration of Product Specificity of Rhodobacter sphaeroides Phytoene Desaturase by Directed Evolution. J. Biol. Chem., 276, 41161-41164.

[26]. Farmer, W.R. and J.C. Liao, 2001, Precursor Balancing for Metabolic Engineering of Lycopene Production in Escherichia coli. Biotechnol. Prog., 17, 57-61.

[27]. Barkovich, R. and J.C. Liao ,2001, Metabolic Engineering of Isoprenoids. Metabolic Engineering, 3:27-39

[28]. SUOMALAINEN, Irma AHO, Sirpa SAARELAINEN, Ritva KRISTO,1996, PLANT XYLOSE ISOMERASE' , WIPO in 1996.
[29]. Hugh G. Lawford and Joyce D. Rousseau, 2005, 'Fuel ethanol from hardwood hemicellulose hydrolysate by genetically engineeredEscherichia coli B carrying genes fromZymomonas mobilis', springerlink.
[30]. Joseph B gonsalves in 2006 october, 'An assesment of biofuel industries in India', UNCTD
[31]. John timmer , may 24th ,2008 , ' boifuels : moving beyond food crops' ars technical.
[32]. Alderson, J., and W.C. Sharp. 1994. Grass varieties in the United States. USDA-SCS Agric. Handb. 170. U.S. Gov. Print. Office, Washington, DC.
[33]. Brejda, John J. 1996. Evaluation of arbuscular mycorrhiza populations for enhancing switchgrass yield and nutrient uptake. Ph.D. Dissertation. University of Nebraska-Lincoln, Lincoln, NE. 160p.
[34]. Brejda, John J., L.E. Moser, and K.P. Vogel, 1999. Evaluation of switchgrass rhizosphere microflora for enhancing yield and nutrient uptake. Agron. J. (In Press).
[35]. Hopkins, A.A., K.P. Vogel, K.J. Moore, K.D. Johnson, and I.T. Carlson. 1995a. Genetic effects and genotype by environment interactions for traits of elite switchgrass populations. Crop Sci. 35:125-132.

[36]. Hopkins, A.A., K.P. Vogel, K.J. Moore, K.D. Johnson, and I.T. Carlson. 1995b. Genetic variability and genotype x environment interactions among switchgrass accessions from the Midwestern USA. Crop Sci. 35: 565-571.

[37]. Hopkins, Andrew A., Charles M. Taliaferro, Christopher D. Murphy, and D'Ann Christian. 1996. Chromosome number and nuclear DNA content of several switchgrass populations. Crop Sci. 36:1192-1195.

[38]. Hultquist, Sherry J., K.P. Vogel, D.J. Lee, K. Arumuganathan, and S. Kaeppler. 1996. Chloroplast DNA and nuclear DNA content variations among cultivars of switchgrass, Panicum virgatum L. Crop Sci. 36:1049-1052.

[39]. Hultquist, Sherry J., K.P. Vogel, D.E. Lee, K. Arumuganathan, and S. Kaeppler. 1997. Nuclear DNA content and chloroplast DNA polymorphisms among accessions of Panicum virgatum L. from remnant midwestern prairies. Crop Sci. 37:595-598. Martin, A. R., R. S. Moomaw, and K. P. Vogel. 1982. Warm- season grass establishment with atrazine. Agron. J. 74:916- 920. Martinez-Reyna, Juan M. 1998. Hybridization between upland and lowland and within upland cytotypes of switchgrass. Ph.D. Dissertation, University of Nebraska, Lincoln, NE.

[40]. Martinez-Reyna, Juan M. and K.P. Vogel. 1998 Controlled hybridization technique for switchgrass.. Crop Sci. 38:876:878. Moser, L.E. and K.P. Vogel. 1995. Switchgrass, big bluestem, and indiangrass. Chapter 32. p. 409-420.

[41]. Vogel, K.P., A.A. Hopkins, K.J. Moore, K.D. Johnson, and I.T. Carlson. 1996. Registration of 'Shawnee' switchgrass. Crop Sci. 36:1713.

[42]. Vogel, K.P. 1987. Seeding rates for establishing big bluestem and switchgrass with pre-emergence atrazine applications. Agron. J. 79:509-512.

[43]. Michael D. Duffy and Virginie Y. Nanhou, 2002, 'Costs of Producing Switchgrass for Biomass in Southern Iowa',natural.

[44]. Duffy, M.D. and D. Smith. 2000. Estimated costs of crop production in Iowa—2001. Iowa State University Extension, Fm-1712. Rev. Dec. 2000.

[45]. Duffy, M.D. and V.Y. Nanhou. 2001. Costs of producing Switchgrass for Biomass in Southern Iowa. Iowa State University Extension, Pm 1866.

[46]. Edwards, W., D. Smith, B. Mc Lees, and J. Baitinger. 2000. 2000 Iowa farm custom rate survey. Iowa State Univ. Ext., FM 1698. Rev. March 2000.

[47]. Iowa State University (ISU) Extension. 1998. Switchgrass seeding recommendations for the production of biomass fuel in southern Iowa. Pm 1773.

[48]. Lemus, B.R.W. 2000. Cultivar and fertility effects on switchgrass biofuel production in southern Iowa. MS thesis. Iowa State Univ., Ames.

[49]. Park, Y.W. 1996. Economic feasibility of growing herbaceous biomass energy crops in Iowa. Ph.D. dissertation. Iowa State Univ., Ames.

[50]. Radiotis, T., J. Li, K. Goel, and R. Eisner. 1999. Fiber characteristics, pulpability, and bleachability of switchgrass. Tappi J. 82:100–105.

[51]. Teel, A., S. Barnhart, and G. Miller. 1997. Management guide for the production of switchgrass for biomass fuel in southern Iowa. Iowa State Univ. Ext., Pm 1710.

[52]. Walsh, M.E. 1994. The cost of producing switchgrass as dedicated energy crop. Biologue, 4th Qtr:35–36.

[53]. Walsh, M.E. and D. Becker. 1996. Biocost: A software program to estimate the cost of producing bioenergy crops. Proc. Bioenergy 96. Sept. 15–20, 1996, Nashville, TN.

My whole hearted thanks to all the authors, organizations and sponsors of above cited research article. It was observed not to violate any copyright of above article. Thank you all for your support.

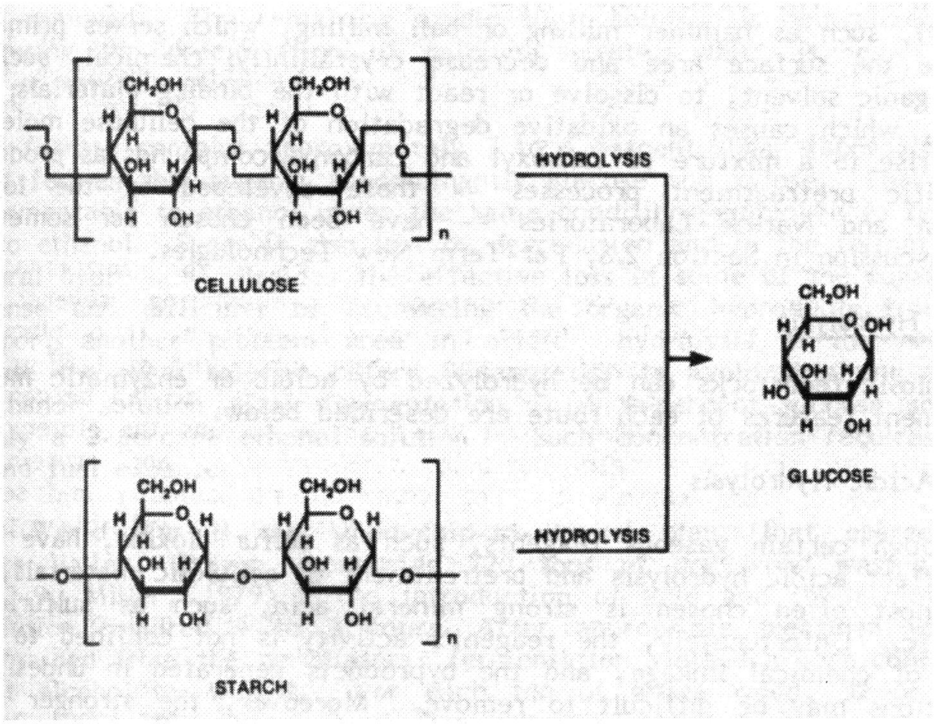

Figure1: STARCH/CELLULOSE HYDROLYSIS REACTION.

Fuel	Energy density	Air-fuel ratio	Specific energy	Heat of vaporization	RON Octane rating	MON
Gasoline and biogasoline	32 MJ/L	14.6	2.9 MJ/kg air	0.36 MJ/kg	91–99	81–89
Butanol fuel	29.2 MJ/L	11.2	3.2 MJ/kg air	0.43 MJ/kg	96	78
Ethanol fuel	19.6 MJ/L	9.0	3.0 MJ/kg air	0.92 MJ/kg	129	102
Methanol	16 MJ/L	6.5	3.1 MJ/kg air	1.2 MJ/kg	136	104

Chart1: CHART DEPICTING RELATIVE DETAILS OF ENERGY.

117b Effect of Calcium and Sodium Deposition on a Pemfc

Steven Mouton, Chemical Engineering, University of South Alabama, 5428 Henning Dr. West, Mobile, AL 36619

Even in a low temperature fuel cell such as the Proton Exchange Membrane Fuel Cell (PEMFC) contaminants from bi-polar plates, sealants, and gaskets can leach into the membrane. Here a study of two ions is presented because these ions, sodium and calcium, are present in present-day materials. This study was focused on whether the ions deposit on the sulfonic sites within the membrane electrode assembly (as would be expected through an ion-exchange mechanism) and on whether equal normality poisonings of sodium and calcium (5 mM and 2.5 mM) would produce the same percentage of performance loss.

The results confirm the hypothesis of performance loss based on ion charge. This loss for both contaminants was 45 %. However, the data show an unequal recovery after a 48 hour period of constant voltage. This recovery corresponds to 17 % degradation for sodium and 25 % for calcium. Electron microscope results illustrate cationic buildup mainly around areas of high sulfonic sites.

The higher recovery for sodium may be characteristic of a monovalent cation contaminant as compared to divalent. Additional experiments are being performed in an effort to generalize the results for all monovalent and divalent cations.

117c Techniques to Improve the Use of Biomass Fuels Used to Provide Heating to the Facilities at the Natchez Trace Greenhouses

Derek J. Rice[1], Mark Bricka[2], and Monty Singletary[2]. (1) Chemical Engineering, Bucknell University, 701 Moore Avenue Unit C0688, Lewisburg, PA 17837, (2) Dave C. Swalm School of Chemical Engineering, Mississippi State University, MS 9595, Mississippi State, MS 39762

Natchez Trace Greenhouses (NTG) uses boilers to heat water, which is then sent through the greenhouses during colder months to maintain an adequate temperature at the facility. The manager of the facility recently began to use biomass fed boilers instead of natural gas boilers to reduce his reliance on more expensive natural gas. The boiler feed currently consists of a combination of planar wood shavings, blended with corn to give the necessary higher heating value. Given that the price of corn has risen significantly, and that the facility may lose its current source of planar shavings, other potential biomass feed sources must be investigated.

This investigation aims to discover and test possible local fuel alternatives that can be easily transported to NTG at a cost lower than the current fuels. Possible options include glycerol, off-spec flour, various wood wastes and cotton wastes. Combustion experiments were performed using bomb calorimetry. These results were combined with moisture and density analysis to obtain a BTU density to determine combustion efficiency. With these results, the optimum mixture of glycerol and solid fuel source can be obtained. The boilers at NTG also produce large quantities of bottoms ash. Samples of this ash have been collected and tested for their potential use as a soil amendment for plants at the facility. Using these results, a experimental matrix has been set up. It outlines the plan to grow begonia (Begoniaceae) plants at the greenhouse using this ash. There has also been a problem with slag build-up in the boilers, as well as fly ash emissions from the boilers at NTG. Solutions to these problems will be investigated as work with NTG continues. This study aims to evaluate the impact of these problems while providing real-world solutions that will allow the NTG to lower their overall heating costs.

118a Preliminary Design of Bioethanol Process Plants

Anne Grydgaard and Camilla Berg. Department of Chemical and Biochemical Engineering, Technical University of Denmark, Soeltofts Plads, Building 229, Lyngby, DK-2800, Denmark

Lately the demand for ethanol utilized as a petrol additive for combustion engines has increased due to the escalating crude oil prices. It is thus desirable to produce ethanol from alternative renewable sources of raw materials such as corn, cereal and sugar or waste materials such as wood chips, straw, bagasse, manure. Organic matter provides a source of trapped and unutilized energy which can be exploited to supply the ever-increasing energy demand in the world.

The objective of this presentation proposal is to summarize the preliminary designs of process plants for production of bioethanol with a purity of 99 % or higher. The preliminary designs were carried out as part of a process design course at the Technical University of Denmark (DTU) in the spring 2008. The design project covers all aspects of the process design procedure from initial considerations regarding qualitative aspects of the flow sheet to detailed process simulations and economic evaluations. Base case designs were developed systematically using several known design models and procedures. Heat and mass integration and environmental impact analysis were performed in order to optimize the design. Feasible process flow sheets with unit operations were generated based on decisions made through design tools, which were later verified using commercial process simulators. Optimisation and derivation of heat recovery networks for the final processes are of importance for the design project and had a great influence on the economic feasibility of the final design. Specific features of interest regarding the development procedure will be highlighted in the presentation.

118b Physiochemical Surface Properties of Lung Tissues for Modeling Drug Deposition

Michael Lamb, Mathew D. Rowe, and Keisha B. Walters. Swalm School of Chemical Engineering, Mississippi State University, 330 Swalm Chemical Engineering Building, Mississippi State, MS 39759

The objectives of this research are to investigate and model the physical and chemical surface properties of lung tissues and model the interaction of these tissues with inhalation drugs. By studying the contact angles of saliva and drug doped saliva on lung tissue and mucin layer samples, appropriate deposition and adsorption boundary conditions can be developed with eventual implementation in fluid dynamic simulations. In order to measure the contact angle between these surfaces and the probe liquids, a contact angle goniometer was used to take digital images of the sessile drops and software was used to determine surface contact angle using several different models. Static contact angles of three probe liquids, pH 7 phosphate buffer solution, HPLC water, and artificial saliva doped with potassium iodide, salbutamol, or theophyiline, have been measured on bovine lung tissue samples. The contact angle of artificial saliva doped with potassium iodide was found to change rapidly (~ 40° to 25°) within the first 120 seconds of contact with bovine lung tissue, and the contact angle of artificial saliva doped with theophyiline demonstrated a similar behavior in the first 120 seconds (~33° to 22°). At low drug concentrations, the saliva composition was found to have a stronger influence on the contact angle than the dopant. Investigation will continue using the saliva and drug-doped saliva as the probe liquids on both lung tissue and mucin layers and development will begin on a model to account for the relationship between drug chemistry, molecular weight and shape, and concentration and the degree of deposition on lung tissues.

118c Cfd Simulation and Modeling of Conduction Heat Transfer at the Wall of a Packed Reactor Tube

Amanda K. Gurnon[1], Anthony G. Dixon[1], Michiel Nijemeisland[2], and Hugh Stitt[2]. (1) Worcester Polytechnic Institute, 100 Institute Road, Worcester, MA 01609, (2) Johnson Matthey, PO Box 1, Billingham, Cleveland, TS23 1LB, United Kingdom

Heat transfer at the reactor tube wall is especially important for highly exothermic reactions, such as partial oxidations and highly endothermic reactions such as methane ateam reforming. In such cases, flow rates tend to be high and the tube-to-particle diameter ratio (N) tends to be low, and the limiting resistance to heat transfer occurs very close to the tube wall. Empirical correlations of the wall heat transfer coefficient are extremely scattered and unreliable. An approach in which the individual heat transfer mechanisms are simulated and modeled is being tried by our research group.

In this presentation we focus on the conduction mode of heat transfer. CFD simulations under stagnant conditions and/or low flow rates are shown, to identify near-wall phenomena. Standard fixed bed reactor conduction models are evaluated using empirical and measured void fraction profiles, along with models for the stagnant thermal conductivity. Comparisons between the detailed CFD simulation temperature profiles and those obtained from the modeling approaches allow evaluation of their suitability to describe near-wall conduction heat transfer.

118d Cuo Nanospheres Based Non-Enzymatic Glucose Sensor

Michael Gentile, Eliot Reitz, Wenzhao Jia, and Yu Lei. Chemical, Materials and Biomolecular Engineering, University of Connecticut, 191 Auditorium Rd, U3222, Storrs, CT 06269

In this study, we reported the fabrication and characterization of CuO nanospheres based nonenzymatic glucose sensor. CuO nanospheres, consisting of many small and uniform nanoneedles with tips about tens of nanometers in diameter, were first synthesized by a simple, convenient, template-free, and one-step hydrothermal method, and then facilely co-deposited on a glassy carbon (GC) electrode with Nafion. The CuO modified GC (Nafion/CuO/GC) electrode shows excellent oxidation of glucose in basic solution and high sensitivity to glucose concentration change. Moreover, the presence of a high concentration of chloride ions does not affect the sensor's performance. These results demonstrate CuO nanospheres possess excellent electrocatalytic properties and are attractive in electrochemical detection.

Generalized Vapor Pressure Prediction Consistent with Cubic Equations of State

Laura L. Petrasky and Michael J. Misovich, Hope College, Holland, MI

Introduction

Equations of state may be used to calculate pure component vapor-liquid equilibrium properties such as vapor pressure, heat of vaporization, liquid density, and vapor density. The standard approach requires coupling the EOS with a phase equilibrium criterion such as free energy, chemical potential, or fugacity. The resulting equations are nonlinear and must be solved by numerical methods.

An alternative approach is applicable to cubic EOS such as the commonly used Soave-Redlich-Kwong and Peng-Robinson equations. Equilibrium properties may be explicitly expressed as power series in reduced temperature or related functions. These results are more convenient than numerical calculations, but are subject to truncation error in many practical situations.

Results from such a power series method for the SRK equation were used to generate generalized dimensionless vapor pressure equations which were extensions of the commonly used Antoine equation but valid over wider temperature ranges. The substance-specific adjustable constants of the vapor pressure equations were expressed as functions of the acentric factor of the substance and its critical temperature and pressure. The deviations between these results and the exact vapor pressure predictions from the SRK equation were quantitatively characterized.

VLE from Cubic Equations of State

Cubic equations of state of the Van der Waals type are widely used in chemical engineering practice because they provide a reasonable balance between accuracy and simplicity. Two common examples are the Soave-Redlich-Kwong [1] and Peng-Robinson [2] equations, shown below as Eqs. (1) and (2),

$$P = \frac{RT}{V - b_{SRK}} - \frac{a_{SRK}(T)}{V(V + b_{SRK})} \quad (1)$$

$$P = \frac{RT}{V - b_{PR}} - \frac{a_{PR}(T)}{V(V + b_{PR}) + b_{PR}(V - b_{PR})} \quad (2)$$

In these equations, the attractive function $a_{EQ}(T)$ has the form

$$a_{EQ}(T) = a_{c,EQ}[1 + f_{EQ}(\omega)(1 - T_r^{1/2})]^2 \quad (3)$$

with the critical value of

$$a_{c,EQ} = \frac{a_{0,EQ} R^2 T_c^2}{P_c} \tag{4}$$

and the acentric factor functions given by

$$f_{SRK}(\omega) = 0.480 + 1.574\omega - 0.176\omega^2 \tag{5}$$

$$f_{PR}(\omega) = 0.37464 + 1.54226\omega - 0.26992\omega^2 \tag{6}$$

The occupied volume parameter b is given by

$$b_{EQ} = \frac{b_{0,EQ} R T_c}{P_c} \tag{7}$$

Values of the dimensionless numerical constants in Eqs. (4) and (7) are $a_{0,SRK}$ = 0.42748, $a_{0,PR}$ = 0.45724, $b_{0,SRK}$ = 0.08664, and $b_{0,PR}$ = 0.07780.

Like many equations of state, SRK and PR may be used to calculate vapor pressure and other pure component vapor-liquid equilibrium properties. Doing this requires coupling the EOS with a phase equilibrium criterion such as free energy, chemical potential, or fugacity. Eqs. (8) and (9) give expressions for the pure component fugacity coefficient for SRK and PR.

$$\ln \phi_{SRK} = z - 1 - \ln z - \ln(1-h) - \frac{a_{SRK}}{b_{SRK} RT} \ln(1+h) \tag{8}$$

$$\ln \phi_{PR} = z - 1 - \ln z - \ln(1-h) - \frac{a_{PR}}{2\sqrt{2} b_{PR} RT} \ln \frac{1 + h(1+\sqrt{2})}{1 + h(1-\sqrt{2})} \tag{9}$$

$$z = \frac{PV}{RT} \tag{10}$$

$$h = \frac{b_{EQ}}{V} \tag{11}$$

When an equilibrium state exists, Eq. (1) gives three real roots for volume from SRK. The smallest and largest of these are liquid and vapor volumes, respectively. The fugacity coefficients calculated by Eq. (8) will be equal when the liquid volume is substituted and when the vapor volume is substituted. For PR, Eqs. (2) and (9) are used. A numerical algorithm is required to solve either of these sets of nonlinear equations.

Power Series Methods for Cubic Equation VLE

A formal procedure [3] is available to express results for phase densities and vapor pressure of coexisting liquid and vapor phases as analytic power series in temperature. The approach begins by writing reduced deviation variables for phase densities ($\Delta\rho^L$ and $\Delta\rho^V$) and temperature (ΔT) about the critical point. These variables are defined as

$$\Delta M = M_r - 1 = \frac{M - M_c}{M_c} \tag{12}$$

with M representing either ρ^L, ρ^V, or T. The resulting formal power series are

$$P_r = \sum_{j=0}^{\infty} A_j (\Delta T)^j \tag{13}$$

$$\Delta\rho^L = \sum_{i=1}^{\infty} B_i (-\Delta T)^{i/2} \tag{14}$$

$$\Delta\rho^V = \sum_{i=1}^{\infty} (-1)^i B_i (-\Delta T)^{i/2} \tag{15}$$

After equilibrium constraints are applied to these series, the coefficients A_j and B_i are observed to be functions of the acentric factor function $f_{EQ}(\omega)$ which depend upon the cubic EOS being applied. They are substance-dependent since they contain the acentric factor function. For SRK and PR, A_j is given by a polynomial of degree j in the acentric factor function, B_i is given by a polynomial of degree (i-1) in the acentric factor function for even i, and by a similar polynomial multiplied by the square root of $(1+ f_{EQ}(\omega))$ for odd i. Expressions for these polynomials have been tabulated. [4,5]

The utility of this method stems from the convenience of direct computation of coexisting phase properties without need for auxiliary equilibrium criteria such as fugacities. The weaknesses of the method arise from being based upon a series expansion about the critical point. In practice, the infinite series given by Eqs. (13) to (15) must be truncated. Few terms are required near the critical point, but this is the situation where cubic EOS are least accurate. At moderately high or moderate temperatures, where the equations have reasonable accuracy, many series terms are needed to prevent truncation error. The magnitude of truncation error also depends upon the property (phase density or vapor pressure) and varies with acentric factor. In general, truncation errors are largest for vapor density and smallest for liquid density, they are larger for the PR equation as compared to SRK, and they increase with increasing acentric factor.

Generalized Antoine Vapor Pressure Functions Derived from Power Series

In previous work, a generalized form of the Antoine vapor pressure equation was developed. In the traditional Antoine equation, dimensional constants are used to fit vapor pressure data over a temperature range.

$$\ln P = A - \frac{B}{T + C} \tag{16}$$

This was adapted by writing the Antoine equation in reduced variables as Eq. (17) where the constants are expressed as functions of the acentric factor.

$$\ln P_r = A^* - \frac{B^*}{T_r + C^*} \tag{17}$$

Results for the functions A^*, B^*, and C^* for the SRK equation are available for reduced temperatures in the range $0.6 < T_r < 1.0$, and for the PR equation in the range $0.7 < T_r < 1.0$. [6] A sample of typical results is shown in Table 1. As is the case with the traditional Antoine equation, the range of accuracy is limited -- in this example to reduced temperatures between 0.70 and 0.84. Figure 1 illustrates the accuracy of the method. Each curve in Figure 1 represents a different acentric factor. Within the specified temperature range ($0.70 < T_r < 0.84$) the vapor pressures predicted by Eq. (17) match the exact SRK vapor pressures to within less than 0.1 percent deviation for all acentric factors between -0.12 and +0.72.

Table 1. Antoine Constant Functions for the SRK Equation for $0.70 < T_r < 0.84$.

$A^* = 4.4401 + 2.2128f_{SRK}(\omega) - 0.53518f_{SRK}(\omega)^2 + 0.17368f_{SRK}(\omega)^3 - 0.018512f_{SRK}(\omega)^4$
$B^* = 5.0075 + 1.2494f_{SRK}(\omega) - 0.78155f_{SRK}(\omega)^2 + 0.32010f_{SRK}(\omega)^3 - 0.047601f_{SRK}(\omega)^4$
$C^* = 0.124652 - 0.273702f_{SRK}(\omega) + 0.0750076f_{SRK}(\omega)^2 - 0.0137818f_{SRK}(\omega)^3 + 0.00145038f_{SRK}(\omega)^4$

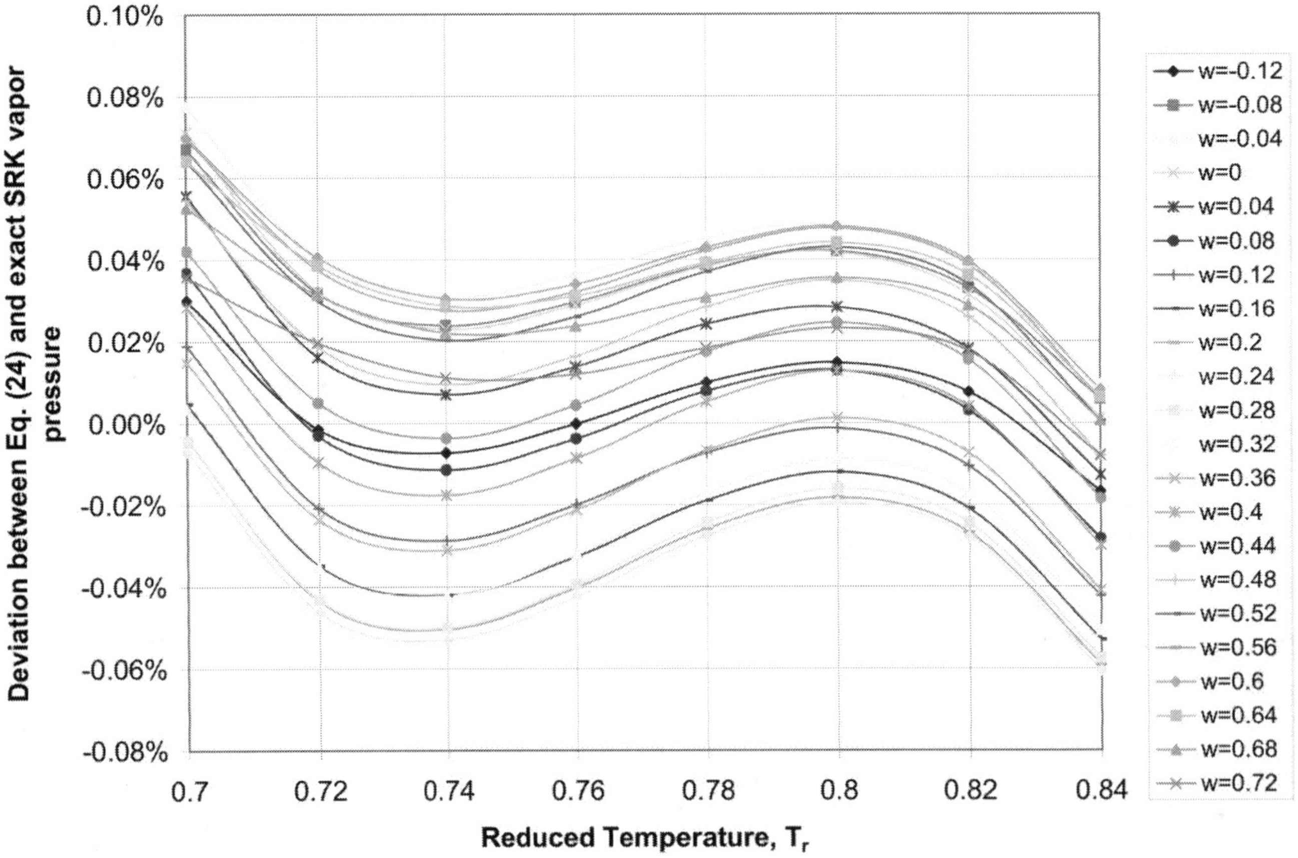

Figure 1. Percentage Deviation between Vapor Pressure Predicted by Generalized Antoine Equation, Eq. (17), and Exact SRK Vapor Pressure for Various Acentric Factors.

In the current work, the method was adapted and applied to generalized versions of extended forms of the Antoine equation, given by Eqs. (18) and (19).

$$\ln P_r = A^* - \frac{B^*}{T_r} + K_1^* \ln T_r \tag{18}$$

$$\ln P_r = A^* - \frac{B^*}{T_r + C^*} + K_4^* T_r + K_5^* \ln T_r \tag{19}$$

Analogous to Eq. (17), the constants in Eqs. (18) and (19) were polynomials of the acentric factor function $f_{SRK}(\omega)$. Some examples of these functions are shown in Tables 2 and 3.

Like Eq. (17), the generalized extended Antoine equation given by Eq. (18) was capable of estimating vapor pressures within less than 0.1 percent deviation from the exact SRK values for the reduced temperature range $0.6 < T_r < 1.0$. Also like Eq. (17), three distinct sets of Antoine constant functions were needed to span this temperature range. One advantage of Eq. (18) was that the Antoine constant functions were cubic polynomials in the acentric factor function $f_{SRK}(\omega)$, whereas Eq. (17) required quartic (4th degree) polynomials to achieve the same accuracy of prediction.

Table 2. Extended Antoine Constant Functions in Eq. (18) for the SRK Equation for the Temperature Range $0.60 < T_r < 0.72$.

$A^* = 2.8278 + 4.8255\ f_{SRK}(\omega) + 2.7682\ f_{SRK}(\omega)^2 - 0.057280\ f_{SRK}(\omega)^3$
$B^* = 2.8422 + 4.8991\ f_{SRK}(\omega) + 2.8244\ f_{SRK}(\omega)^2 - 0.069460\ f_{SRK}(\omega)^3$
$K_1^* = 1.1068 - 2.2926\ f_{SRK}(\omega) - 3.2566\ f_{SRK}(\omega)^2 + 0.13569\ f_{SRK}(\omega)^3$

Table 3. Extended Antoine Constant Functions in Eq. (19) for the SRK Equation for the Temperature Range $0.60 < T_r < 1.0$.

$A^* = 2.5713 + 3.8253\ f_{SRK}(\omega) + 0.56359\ f_{SRK}(\omega)^2 + 0.42434\ f_{SRK}(\omega)^3 - 0.011805\ f_{SRK}(\omega)^4$
$B^* = 4.2778 + 6.9822\ f_{SRK}(\omega) + 6.8127\ f_{SRK}(\omega)^2 - 0.57592\ f_{SRK}(\omega)^3 - 0.027427\ f_{SRK}(\omega)^4$
$C^* = 0.031642 + 0.00022474\ f_{SRK}(\omega) - 0.00095525\ f_{SRK}(\omega)^2 + 0.00074762\ f_{SRK}(\omega)^3$
$\quad - 0.00017628\ f_{SRK}(\omega)^4$
$K_4^* = 1.5758 + 2.9403\ f_{SRK}(\omega) + 6.0484\ f_{SRK}(\omega)^2 - 0.98555\ f_{SRK}(\omega)^3 - 0.014798\ f_{SRK}(\omega)^4$
$K_5^* = -1.5345 - 6.4179\ f_{SRK}(\omega) - 12.447\ f_{SRK}(\omega)^2 + 1.5234\ f_{SRK}(\omega)^3 + 0.036455\ f_{SRK}(\omega)^4$

Eq. (19) was more complex than Eqs. (17) and (18) because it contained five Antoine constant functions instead of three. The advantage of Eq. (19) was that a single set of Antoine constant functions, shown in Table 3, was able to estimate SRK vapor pressures within the specified deviation for the entire reduced temperature range $0.60 < T_r < 1.0$, as Figure 2 verifies.

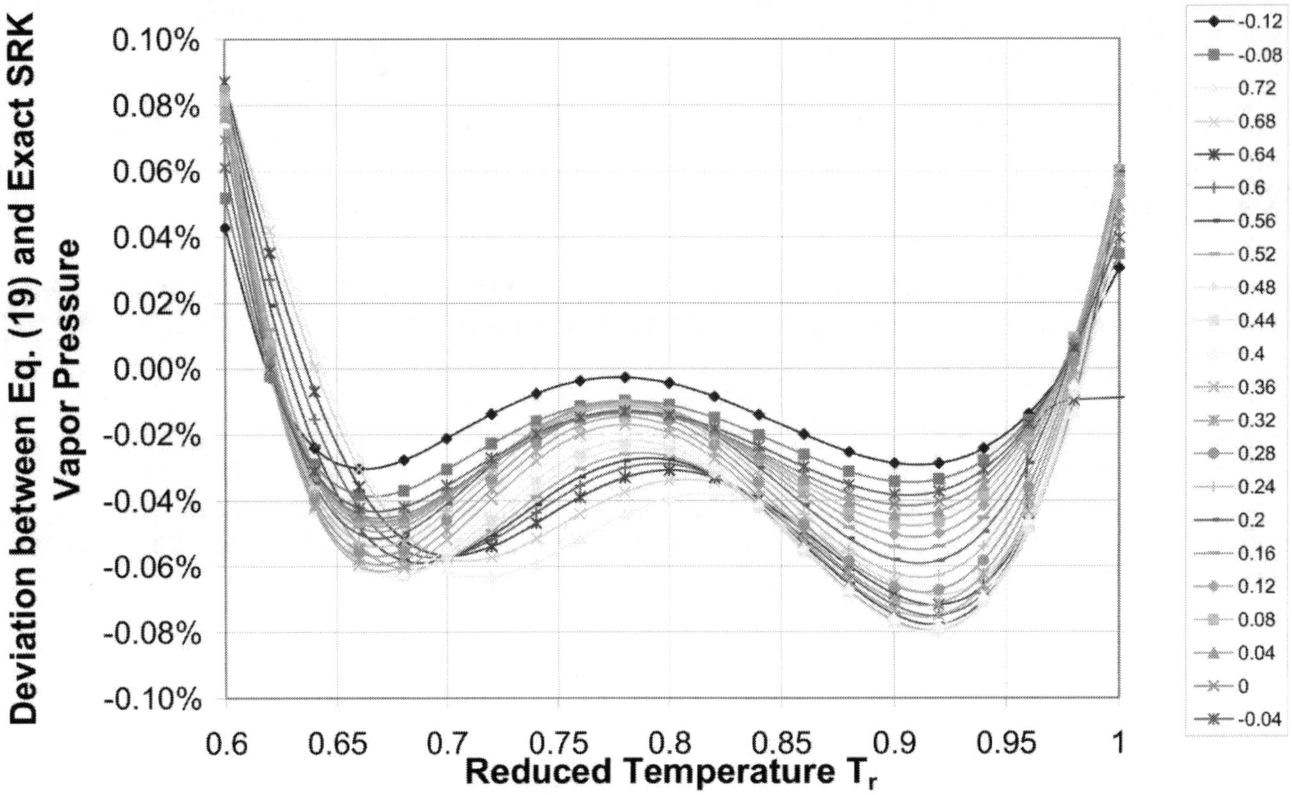

Figure 2. Percentage Deviation between Vapor Pressure Predicted by Generalized Extended Antoine Equation, Eq. (19), and Exact SRK Vapor Pressure for Various Acentric Factors.

Work continues on adapting these methods to other extended versions of the Antoine equation, especially those commonly used in process simulation software, to other cubic EOS such as PR, and to additional temperature ranges.

Conclusions

Results for a generalized dimensionless form of two versions of an extended Antoine equation were presented. These were shown to reproduce the vapor pressure predictions of the SRK cubic equation of state to within 0.1 percent deviation over the temperature range $0.60 < T_r < 1.0$ for all acentric factors between -0.12 and +0.72.

Acknowledgements

This material is based upon work supported by the Michigan Space Grant Consortium and by the Brookstra Faculty Development Fund. Any opinions, findings, and conclusions or recommendations expressed in this material are those of the authors and do not necessarily reflect the views of these institutions.

Nomenclature

A Constant in Eq. (16), Antoine vapor pressure equation
A^* Function of acentric factor in Eqs. (17) to (19), generalized Antoine vapor pressure equations
A_j temperature coefficients in Eq. (13) for reduced pressure
$a_{EQ}(T)$ energy parameter in attractive term
$a_{c,EQ}$ energy parameter in attractive term at the critical point
$a_{0,EQ}$ dimensionless numerical coefficient of energy parameter
B Constant in Eq. (16), Antoine vapor pressure equation
B^* Function of acentric factor in Eqs. (17) to (19), generalized Antoine vapor pressure equations
B_i temperature coefficients in Eqs. (14) and (15) for reduced density deviation variables
b_{EQ} excluded volume parameter in equation of state
$b_{0,EQ}$ dimensionless numerical coefficient of excluded volume parameter
C Constant in Eq. (16), Antoine vapor pressure equation
C^* Function of acentric factor in Eqs. (17) and (19), generalized Antoine vapor pressure equations
$f_{EQ}(\omega)$ quadratic function of acentric factor
h ratio of excluded volume parameter to molar volume
K_1^* Function of acentric factor in Eq. (18), generalized Antoine vapor pressure equation
K_4^* Function of acentric factor in Eq. (19), generalized Antoine vapor pressure equation
K_5^* Function of acentric factor in Eq. (19), generalized Antoine vapor pressure equation
M general thermodynamic state property (T, ρ^L or ρ^V) in Eq. (12)
P absolute pressure
R gas constant
T absolute temperature
V molar volume
z compressibility

Greek letters

ΔM reduced deviation variable of state property M (T, ρ^L or ρ^V) defined by $M_r - 1$
ρ molar density
φ_{EQ} fugacity coefficient for equation of state
ω acentric factor

Subscripts

c critical (temperature, pressure, or molar density)
EQ parameter or coefficient applying to equation of state EQ (SRK or PR)
r reduced (temperature, pressure, or molar density)

Superscripts

L liquid (molar density or molar volume)
V vapor (molar density or molar volume)

References

[1] Peng, D.-Y. D.B. Robinson. Ind. Eng. Chem. Fundam. 15. 59-64. 1976.
[2] Soave, G. Chem. Engng. Sci. 27. 1197-1203. 1972.
[3] Sengers, J.V., J.M.H. Levelt Sengers. In Croxton, C.A. (Ed.). "Progress in Liquid Physics." 103-174. Wiley: New York. 1978.
[4] Singley, C.D. I.P. Burns. M.J. Misovich. Fluid Phase Equilib. 137. 99-109. 1997.
[5] Schaich, D.A. A.R. Runge. S.A. Gangloff. M.J. Misovich, manuscript in preparation for Computers and Chemical Engineering. 2008.
[6] DeDoes, A.J. M.D. Goetz. M.J. Misovich, "A Simple Approach to Generalized Vapor Pressure Prediction." Chem. Eng. Prog. 103(1). 39-44. 2007.

Applying a Low Temperature Limit of a Cubic Equation of State to Model Pure Component Phase Equilibrium

Caitlin A. Kowalsky[1], Joshua L. Lanser[1], Emily J. Walsh[2], Kimberly Wadelton[1] and Michael J. Misovich[1], (1) Hope College, Holland, MI, (2) SmartSignal Corporation, Lisle, IL

Introduction

Equations of state may be used to calculate pure component vapor-liquid equilibrium properties such as vapor pressure, heat of vaporization, liquid density, and vapor density. The standard approach requires coupling the EOS with a phase equilibrium criterion such as free energy, chemical potential, or fugacity. The resulting equations are nonlinear and must be solved by numerical methods.

An alternative approach is applicable to cubic EOS such as the commonly used Soave-Redlich-Kwong and Peng-Robinson equations. Equilibrium properties may be explicitly expressed as power series in reduced temperature or related functions. These results are more convenient than numerical calculations, but the series diverge and become unbounded at low to moderate temperatures away from the critical point.

An alternative method was developed using a series expansion in a low temperature limit rather than the critical point limit. Although the limiting behavior itself was unphysical because it fell below the triple point of the substance, the expressions remained convergent over a very wide range of temperatures and they remained bounded over all temperatures from zero to the critical point. A second alternative method replaced the series expansions with closed form expressions which could be used up to reduced temperatures at least as high as 0.814. Dimensionless results for equilibrium properties such as vapor pressure and liquid density were determined as general expressions of the acentric factor of a substance. The deviations between these results and the exact equilibrium property predictions from the EOS were quantitatively characterized and found to be less than seven percent in one version of the second alternative method which combined a quadratic liquid density estimate with a vapor phase fugacity correction.

VLE from Cubic Equations of State

Cubic equations of state of the Van der Waals type are widely used in chemical engineering practice because they provide a reasonable balance between accuracy and simplicity. Two common examples are the Soave-Redlich-Kwong [1] and Peng-Robinson [2] equations, shown below as Eqs. (1) and (2),

$$P = \frac{RT}{V - b_{SRK}} - \frac{a_{SRK}(T)}{V(V + b_{SRK})} \qquad (1)$$

$$P = \frac{RT}{V - b_{PR}} - \frac{a_{PR}(T)}{V(V + b_{PR}) + b_{PR}(V - b_{PR})} \qquad (2)$$

In these equations, the attractive function $a_{EQ}(T)$ has the form

$$a_{EQ}(T) = a_{c,EQ}[1 + f_{EQ}(\omega)(1 - T_r^{1/2})]^2 \tag{3}$$

with the critical value of

$$a_{c,EQ} = \frac{a_{0,EQ} R^2 T_c^2}{P_c} \tag{4}$$

and the acentric factor functions given by

$$f_{SRK}(\omega) = 0.480 + 1.574\omega - 0.176\omega^2 \tag{5}$$

$$f_{PR}(\omega) = 0.37464 + 1.54226\omega - 0.26992\omega^2 \tag{6}$$

The occupied volume parameter b is given by

$$b_{EQ} = \frac{b_{0,EQ} R T_c}{P_c} \tag{7}$$

Values of the dimensionless numerical constants in Eqs. (4) and (7) are $a_{0,SRK} = 0.42748$, $a_{0,PR} = 0.45724$, $b_{0,SRK} = 0.08664$, and $b_{0,PR} = 0.07780$.

Like many equations of state, SRK and PR may be used to calculate vapor pressure and other pure component vapor-liquid equilibrium properties. Doing this requires coupling the EOS with a phase equilibrium criterion such as free energy, chemical potential, or fugacity. Eqs. (8) and (9) give expressions for the pure component fugacity coefficient for SRK and PR.

$$\ln \phi_{SRK} = z - 1 - \ln z - \ln(1 - h) - \frac{a_{SRK}}{b_{SRK} RT} \ln(1 + h) \tag{8}$$

$$\ln \phi_{PR} = z - 1 - \ln z - \ln(1 - h) - \frac{a_{PR}}{2\sqrt{2} b_{PR} RT} \ln \frac{1 + h(1 + \sqrt{2})}{1 + h(1 - \sqrt{2})} \tag{9}$$

$$z = \frac{PV}{RT} \tag{10}$$

$$h = \frac{b_{EQ}}{V} \tag{11}$$

When an equilibrium state exists, Eq. (1) gives three real roots for volume from SRK. The smallest and largest of these are liquid and vapor volumes, respectively. The fugacity coefficients calculated by Eq. (8) will be equal when the liquid volume is substituted and when the vapor volume is substituted. For PR, Eqs. (2) and (9) are used. A numerical algorithm is required to solve either of these sets of nonlinear equations.

Power Series Methods for Cubic Equation VLE

A formal procedure [3] is available to express results for phase densities and vapor pressure of coexisting liquid and vapor phases as analytic power series in temperature. The approach begins by writing reduced deviation variables for phase densities ($\Delta\rho^L$ and $\Delta\rho^V$) and temperature (ΔT) about the critical point. These variables are defined as

$$\Delta M = M_r - 1 = \frac{M - M_c}{M_c} \tag{12}$$

with M representing either ρ^L, ρ^V, or T. The resulting formal power series are

$$P_r = \sum_{j=0}^{\infty} A_j (\Delta T)^j \tag{13}$$

$$\Delta\rho^L = \sum_{i=1}^{\infty} B_i (-\Delta T)^{i/2} \tag{14}$$

$$\Delta\rho^V = \sum_{i=1}^{\infty} (-1)^i B_i (-\Delta T)^{i/2} \tag{15}$$

After equilibrium constraints are applied to these series, the coefficients A_j and B_i are observed to be functions of the acentric factor function $f_{EQ}(\omega)$ which depend upon the cubic EOS being applied. They are substance-dependent since they contain the acentric factor function. For SRK and PR, A_j is given by a polynomial of degree j in the acentric factor function, B_i is given by a polynomial of degree (i-1) in the acentric factor function for even i, and by a similar polynomial multiplied by the square root of $(1+ f_{EQ}(\omega))$ for odd i. Expressions for these polynomials have been tabulated. [4,5]

The utility of this method stems from the convenience of direct computation of coexisting phase properties without need for auxiliary equilibrium criteria such as fugacities. The weaknesses of the method arise from being based upon a series expansion about the critical point. In practice, the infinite series given by Eqs. (13) to (15) must be truncated. Few terms are required near the critical point, but this is the situation where cubic EOS are least accurate. At moderately high or moderate temperatures, where the equations have reasonable accuracy, many series terms are needed to prevent truncation error. The magnitude of truncation error also depends upon the property (phase density or vapor pressure) and varies with acentric factor. In general, truncation errors are largest for vapor density and smallest for liquid density, they are larger for the PR equation as compared to SRK, and they increase with increasing acentric factor.

Low Temperature Limit Methods for Cubic Equation VLE

An alternative approach for prediction of pure component coexisting phase properties was proposed to overcome some of the shortcomings of the critical point power series, specifically the large truncation errors and subsequent poor convergence at low to moderate temperatures. This new approach utilized expressions for density and vapor pressure that were asympotically correct in a low temperature limit yet still converged well at moderate to high temperatures. Note that the zero temperature limit used is an unphysical limit from the standpoint of VLE, as liquid behavior does not persist below the triple point which typically falls near a reduced temperature of 0.3 to 0.4. In the best case, the existing series about the critical point did not extrapolate accurately below a reduced temperature of approximately 0.58, and even that accuracy required j = 13 in the series expansion for SRK.

For the SRK equation, a relationship between reduced vapor pressure and reduced temperature was found in the mathematical (unphysical) zero temperature limit as the following truncated series in one-half powers of reduced temperature. [6]

$$\ln P_r = \frac{C_1}{T_r} + \frac{C_2}{T_r^{1/2}} + C_3 + C_4 T_r^{1/2} + C_5 T_r \tag{16}$$

$$C_1 = -\frac{a_{0,SRK}(1+f_{SRK}(\omega))^2 \ln 2}{b_{0,SRK}} \tag{17}$$

$$C_2 = \frac{2a_{0,SRK} f_{SRK}(\omega)(1+f_{SRK}(\omega))\ln 2}{b_{0,SRK}} \tag{18}$$

$$C_3 = \ln\frac{a_{0,SRK}(1+f_{SRK}(\omega))^2}{2b_{0,SRK}} - \frac{a_{0,SRK}(f_{SRK}(\omega))^2 \ln 2}{b_{0,SRK}} - \ln b_{0,SRK} \tag{19}$$

$$C_4 = -\frac{4f_{SRK}(\omega)}{1+f_{SRK}(\omega)} \tag{20}$$

$$C_5 = -\frac{2b_{0,SRK}}{a_{0,SRK}(1+f_{SRK}(\omega))^2} + \frac{3(f_{SRK}(\omega))^2}{(1+f_{SRK}(\omega))^2} \tag{21}$$

This result was derived by substituting an expression for reduced liquid density ρ_r' given by Eq. (22) into Eq. (8) to evaluate liquid fugacity coefficient. (The prime in this definition is used to emphasize that this is not the standard reduced density determined by dividing by the critical density.) A value of unity was assigned to the vapor fugacity coefficient, reflecting nearly perfect adherence to the ideal gas law at the extremely low vapor pressures corresponding to a low temperature limit, and liquid and vapor fugacity coefficients were equated to give Eqs. (16) to (21).

$$\rho_r' \equiv b_{SRK}\rho = 1 - \frac{2b_{0,SRK}}{a_{0,SRK}(1+f_{SRK}(\omega))^2} T_r - \frac{4f_{SRK}(\omega)b_{0,SRK}}{a_{0,SRK}(1+f_{SRK}(\omega))^3} T_r^{3/2} \tag{22}$$

Figure 1 shows results when vapor pressure predicted by Eq. (16) is compared to the exact result from the SRK equation. The curves plotted in Figure 3 represent various acentric factors. Since the coefficients in Eq. (16) were fit in the low temperature limit, the deviation between the predictions and the exact SRK result went to zero in that limit. Behavior at increasing temperatures was interesting. Although no additional corrections were made to the fit of the equation, the zero temperature limit equation was capable of tracking the exact vapor pressure from the SRK equation within an order of magnitude; in fact, within a factor of two, over the entire temperature range from zero to the critical point. This was unlike the behavior observed with the truncated critical point series of Eq. (13), where the deviation caused by truncation error increased by polynomial or exponential order away from the critical point.

Results analogous to Eqs. (16) to (22) were also determined for the Peng-Robinson equation. Details of those results and additional descriptions of the method used to derive these results for both SRK and PR will be given in a subsequent publication.

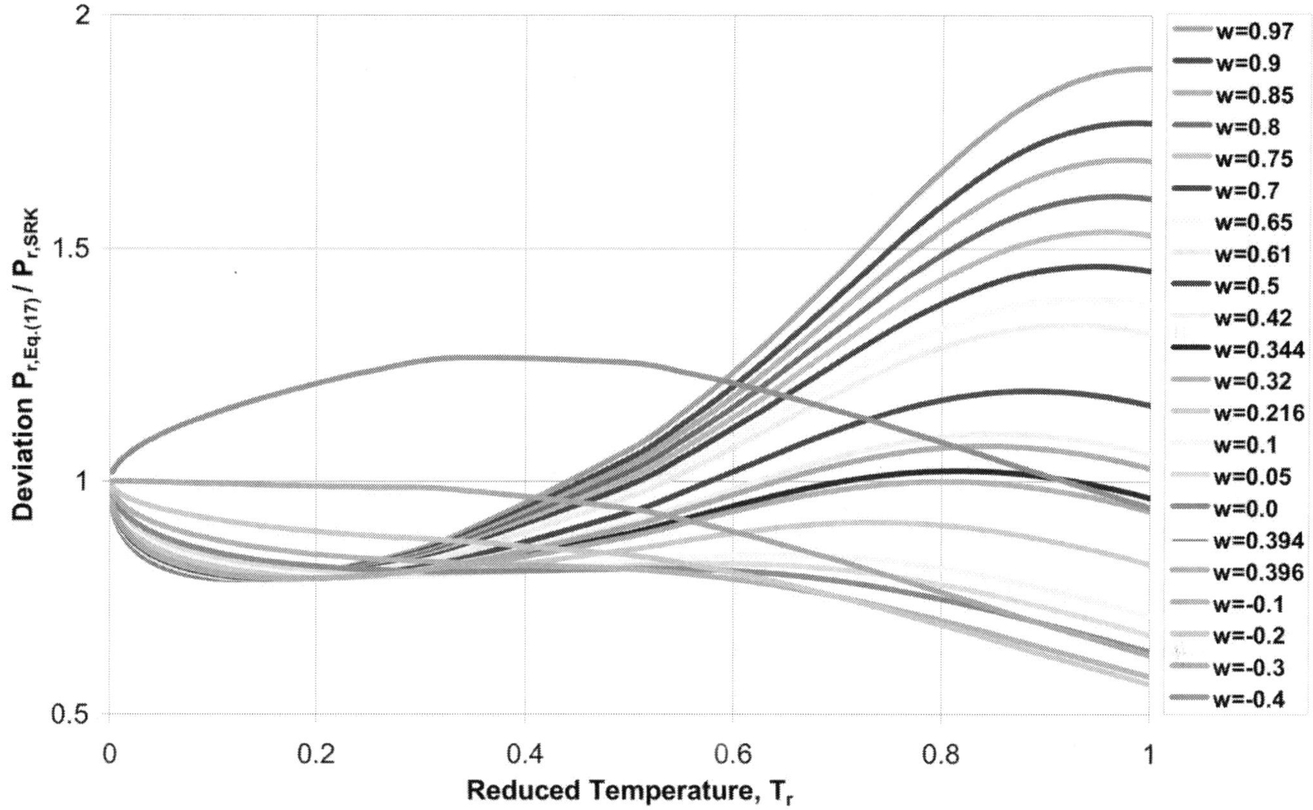

Figure 1. Deviation Ratio Between Eq. (16) and Exact SRK Vapor Pressure.

It was noted that a principal source of the deviations shown in Figure 1 was the truncation error in the prediction of liquid density by Eq. (22). To improve the predictions of the method, an alternative approach was developed for predicting liquid density. A cubic EOS such as SRK or PR was written in dimensionless variables using the critical point to nondimensionalize temperature and pressure in the standard way, but nondimensionalizing density by multiplying it by the parameter b of the EOS as was done in Eq. (22). The result for the SRK equation is shown as Eq. (23), where α_{SRK} refers to the second factor in Eq. (3) which is shown here as Eq. (24).

$$P_r - \frac{T_r}{b_{0,SRK}}\rho_r' + \left(\frac{a_{0,SRK}\alpha_{SRK}}{b_{0,SRK}^2} - \frac{T_r}{b_{0,SRK}} - P_r\right)\rho_r'^2 - \frac{a_{0,SRK}\alpha_{SRK}}{b_{0,SRK}^2}\rho_r'^3 = 0 \qquad (23)$$

$$\alpha_{SRK} = [1 + f_{SRK}(\omega)(1 - T_r^{1/2})]^2 \tag{24}$$

In the low temperature limit, it was observed that P_r approaches zero exponentially as T_r decreases. Applying this observation to Eq. (23), the P_r terms were neglected to give a cubic equation with no constant term. By factoring out the common factor of ρ_r' from each term, a low temperature limit of the cubic EOS was written as a quadratic equation, Eq. (25).

$$\frac{T_r}{b_{0,SRK}} + \left(\frac{a_{0,SRK}\alpha_{SRK}}{b_{0,SRK}^2} - \frac{T_r}{b_{0,SRK}}\right)\rho_r' - \frac{a_{0,SRK}\alpha_{SRK}}{b_{0,SRK}^2}\rho_r'^2 = 0 \tag{25}$$

$$\rho_r' = -\frac{T_r b_{0,SRK}}{2a_{0,SRK}\alpha_{SRK}} + \frac{1}{2} + \sqrt{\left(\frac{T_r b_{0,SRK}}{2a_{0,SRK}\alpha_{SRK}}\right)^2 - 3\left(\frac{T_r b_{0,SRK}}{2a_{0,SRK}\alpha_{SRK}}\right) + \frac{1}{4}} \tag{26}$$

Eq. (25) could be solved easily by the quadratic formula for two roots of reduced density ρ_r'. The larger of these was an approximation of the liquid reduced density, shown as Eq. (26) and the smaller was an approximation of the unphysical middle root from the cubic EOS. The vapor reduced density root disappeared when the common factor of ρ_r' was factored from Eq. (23). This factoring process was equivalent to identifying an additional root of ρ_r' equal to zero, which would be the ideal gas limit at the extremely low pressures of the zero temperature limit. Note that the novelty was not merely in replacing a cubic polynomial with a quadratic polynomial as both may be solved for a closed form, but rather in eliminating P_r from the equation so that only one degree of freedom remained instead of two.

The same steps used in deriving Eqs. (16) to (21) from Eq. (8) were then applied, except that Eq. (26) was used to represent reduced liquid density instead of Eq. (22). This gave a relatively simple closed form expression for P_r.

$$P_r = \frac{a_{0,SRK}\alpha_{SRK}\rho_r'^2}{e b_{0,SRK}^2 (1 + \rho_r')^{(1 + a_{0,SRK}\alpha_{SRK}/T_r b_{0,SRK})}} \tag{27}$$

The results of Eq. (27) had several interesting characteristics. They did converge asymptotically to the exact vapor pressures generated by a numerical algorithm in the low temperature limit. At higher temperatures, the deviation between Eq. (27) and the exact values was frequently less than for the more mathematically complex model given by Eqs. (16) to (21), especially for large values of the acentric factor; however, Eq. (27) could not be used beyond the reduced temperature where the discriminant (the quantity under the square root) in Eq. (26) became zero. This occurred for reduced temperatures between 0.814 (for an acentric factor of -0.4) to 0.94 (for an acentric factor of +0.97). Observed deviations were always negative, meaning that Eq. (27) underpredicted the exact SRK result, with the maximum negative deviation being 32 percent, occurring when the discriminant was zero.

One of the most interesting and unexpected observations was that the deviation between Eq. (27) and the exact SRK vapor pressure predictions, which appeared to be a function of two parameters (T_r and ω), could be written as a function of a single parameter.

This parameter was an expression containing both T_r and ω and it was identified as the discriminant of Eq. (26). This result was verified empirically and a master graph of the deviation versus the discriminant was produced, shown as Figure 2. The combination of this figure with Eqs. (26) and (27) allows essentially exact estimation of an SRK vapor pressure at any reduced temperature below the point at which the discriminant of Eq. (26) becomes zero.

Results analogous to Eqs. (23) to (27) were also determined for the Peng-Robinson equation, including the characteristics discussed above. Again, the maximum negative deviation observed was 32 percent. Details of those results and additional descriptions of the method used to derive these results for both SRK and PR will be given in a subsequent publication.

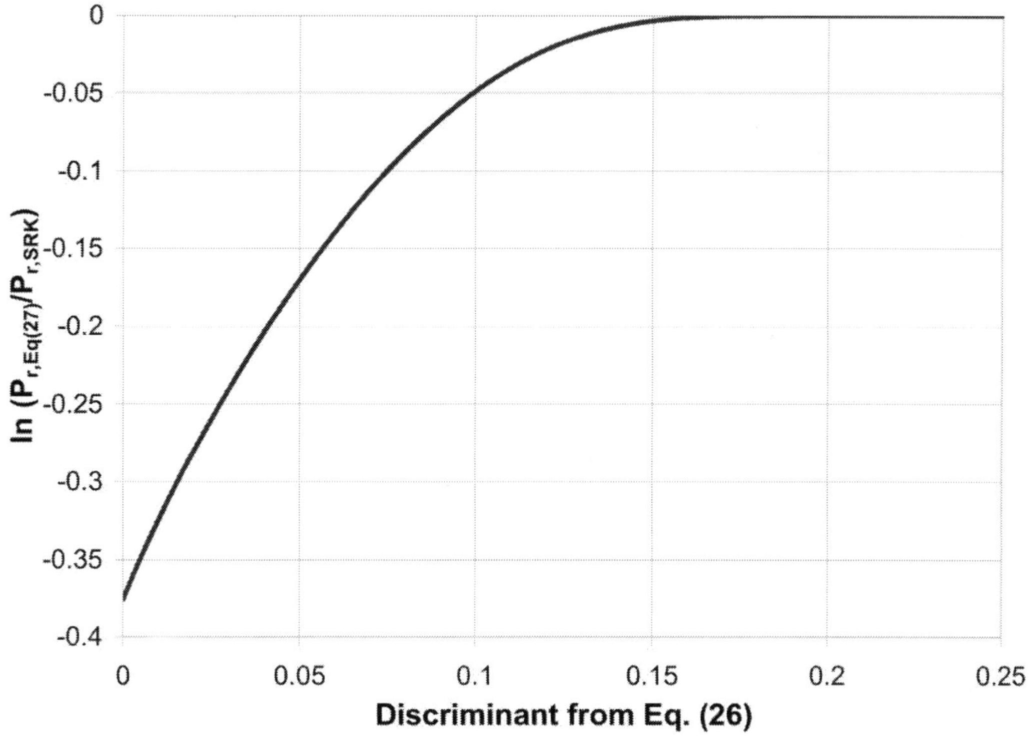

Figure 2. Log Deviation Ratio Between Eq. (27) and Exact SRK Vapor Pressure.

Work continues on characterizing the empirical function of Figure 2 with a functional expression and understanding the mathematical basis for the empirical result that the deviation between Eq. (27) and the exact SRK vapor pressure depends on the single parameter which is the discriminant of Eq. (26).

In deriving the previous results, the vapor phase fugacity coefficient was always assumed equal to unity ($\ln \varphi^V = 0$). An additional enhancement to the method was made by making a first order correction. The SRK equation was expressed in a virial expansion and the second virial coefficient term was used to correct the vapor fugacity coefficient. When the liquid and vapor fugacity coefficients were equated, the result was a quadratic equation which could be directly solved for reduced pressure P_r.

$$\left(\frac{a_{0,SRK}\alpha_{SRK}}{T_r^2}-\frac{b_{0,SRK}}{T_r}\right)P_r^2-P_r+\frac{a_{0,SRK}\alpha_{SRK}\rho_r'^2}{eb_{0,SRK}(1+\rho_r')^{(1+a_{0,SRK}\alpha_{SRK}/T_r b_{0,SRK})}}=0 \tag{28}$$

As was the case with Eq. (27) above, the value of reduced liquid density ρ_r' from Eq. (26) is used in evaluating Eq. (28). Like Eq. (27), Eq. (28) asymptotically converged in the low temperature limit to the exact SRK vapor pressure. Furthermore, the deviations between vapor pressures found by solving Eq. (28) and exact SRK vapor pressures were quite small even at higher temperatures, less than seven percent for all acentric factors between -0.4 and +0.97. As with Eq. (27), the method could not be used at reduced temperatures very close to the critical point where the discriminant in Eq. (26) became negative. Unlike Eq. (27), the deviations between Eq. (28) and exact SRK vapor pressures could not be expressed as a function of the discriminant as a single parameter, but had to be given as functions of the two parameters T_r and ω.

Results analogous to Eqs. (28) were also determined for the Peng-Robinson equation, with the maximum negative deviation observed being 1.4 percent. Details of those results and additional descriptions of the method used to derive these results for both SRK and PR will be given in a subsequent publication.

Conclusions

Methods for explicit estimation of vapor pressure from cubic EOS were presented. The methods relied upon expressions which were asymptotically correct in the zero temperature limit. These approaches were shown to reproduce, with reasonable accuracy, the vapor pressure determined by combining the EOS with a phase equilibrium constraint such as fugacity. The methods were applicable for reduced temperatures from zero up to at least 0.814 and in some cases as high as 0.94. Deviations from the exact SRK vapor pressure were less than 32 percent for an approach which used a quadratic density estimate and less than 7 percent for a modified approach which added a vapor phase fugacity correction. The methods also were applicable to the PR equation with similar accuracy.

Acknowledgements

This material is based upon work supported by the National Science Foundation under NSF-REU Grant Nos. PHY-0097578 and PHY-0452206, and by the Brookstra Faculty Development Fund. Any opinions, findings, and conclusions or recommendations expressed in this material are those of the authors and do not necessarily reflect the views of these institutions.

Nomenclature

A_j	temperature coefficients in Eq. (13) for reduced pressure
$a_{EQ}(T)$	energy parameter in attractive term
$a_{c,EQ}$	energy parameter in attractive term at the critical point
$a_{0,EQ}$	dimensionless numerical coefficient of energy parameter

B_i	temperature coefficients in Eqs. (14) and (15) for reduced density deviation variables
b_{EQ}	excluded volume parameter in equation of state
$b_{0,EQ}$	dimensionless numerical coefficient of excluded volume parameter
C_i	coefficients in Eq. (16), vapor pressure equation at low temperature limit
$f_{EQ}(\omega)$	quadratic function of acentric factor
h	ratio of excluded volume parameter to molar volume
M	general thermodynamic state property (T, ρ^L or ρ^V) in Eq. (12)
P	absolute pressure
R	gas constant
T	absolute temperature
V	molar volume
z	compressibility

Greek letters

ΔM	reduced deviation variable of state property M (T, ρ^L or ρ^V) defined by $M_r - 1$
ρ	molar density
ρ_r'	reduced molar density of liquid defined by convention of Eq. (22)
φ_{EQ}	fugacity coefficient for equation of state
ω	acentric factor

Subscripts

c	critical (temperature, pressure, or molar density)
EQ	parameter or coefficient applying to equation of state EQ (SRK or PR)
r	reduced (temperature, pressure, or molar density)

Superscripts

L	liquid (molar density or molar volume)
V	vapor (molar density or molar volume)

References

[1] Peng, D.-Y. D.B. Robinson. Ind. Eng. Chem. Fundam. 15. 59-64. 1976.
[2] Soave, G. Chem. Engng. Sci. 27. 1197-1203. 1972.
[3] Sengers, J.V., J.M.H. Levelt Sengers. In Croxton, C.A. (Ed.). "Progress in Liquid Physics." 103-174. Wiley: New York. 1978.
[4] Singley, C.D. I.P. Burns. M.J. Misovich. Fluid Phase Equilib. 137 99-109. 1997.
[5] Schaich, D.A. A.R. Runge. S.A. Gangloff. M.J. Misovich, manuscript in preparation for Computers and Chemical Engineering. 2008.
[6] Misovich, M.J., K.F. Blohm, A.J. DeDoes, M.D. Goetz, E.J. Walsh and K. Wadelton, "Explicit Phase Equilibrium Calculations from Cubic Equations of State," AIChE Ann. Mtg. CD-ROM Proc. New York: AIChE. 2007.

118g Fabrication of Microarray-Based ELISA Via Micromolding in Capillaries

Kuo-Feng Lo and Yi-Je Juang. Chemical Engineering, National Cheng Kung University, No. 1, University Road, Tainan, Taiwan

Enzyme-linked immunosorbent assays (ELISA) is one of the most commonly used immunoassays for detection and quantification of chemical and biological molecules (antigens; mainly proteins and polypeptides) in the biotechnology industry, and is becoming more and more important in clinical diagnostics, food safety testing, and environmental monitoring. Recently, research attention has been focused on developing a miniaturized and highly multiplexed microarray format. In this study, instead of using the 96-well microtiter plate, we proposed to utilize the micromolding in capillaries (MIMIC) technique for fabrication of the microarray-based ELISA. By modifying the conventional ELISA procedures, we were able to pattern the antiboby and antigen on the substrate to perform ELISA process. This technique provides large-area patterning to fabricate microarray-based ELISA at a fast, low-cost and reliable fashion. Moreover, the design of the feature size and selection of the substrate is relatively flexible.

Impact of calcium citrate precipitation on the reaction of citric acid-calcite

M.H. AlKhaldi [a,*], H.A. Nasr-El-Din [b], H. Sarma [a]

[a]The University of Adelaide, Adelaide, SA 5005, Australia
[b]Texas A&M University College Station, Texas 77843

Abstract

Mass transfer process during the reaction of citric acid with calcite was investigated using the rotating disk apparatus. The effects of disk rotational speed and temperature on the diffusion coefficient of citric acid were examined.

Using the rotating disk apparatus, the calcite dissolution in 7.5 wt% citric acid solutions was measured at different temperatures, (25, 40, 50°C). The reaction of citric acid with calcite is mainly dependent on disk rotational speed, and the temperature value. For example, at 50°C, the reaction is mass-transfer limited at 500 rpm, while it is reaction-limited below these rotational speeds. Calcium citrate formation is also dependent on the rotational speed. More calcium citrate forms at high rotational speeds.

The effective diffusion coefficient of citric acid was calculated based on the rotating disk theory. The effective diffusion coefficient of citric acid was found to be a function of the interplay between the calcium citrate precipitation and the presence of the counter calcium ions. At an initial acid concentration of 7.5 wt%, the effects of calcium citrate precipitation and counter calcium ions were significant and the reported citric acid diffusion coefficients were not comparable with its measured effective diffusion coefficients using the rotating disk. The effect of temperature on the diffusion coefficient of citric acid at a constant citric acid concentration was found to follow Arrhenius law, and the activation energy is equal to 37.9 kJ/mol.

Keywords: citric acid; calcite; dissolution rate, mass transfer, diffusion, rotating disk

Corresponding Author
***E-mail address*:** mohammed.alkhaldi@student.adelaide.edu.au

Introduction

Over the last few years, various acid systems, such as gelled acids, viscoelastic surfactants-based acids (VES), emulsified acids, and encapsulated citric acid, have been introduced to address some of the concerns with HCl such as severe corrosion problems and poor etched fracture lengths (Crowe et al., 1988). Except of citric acid system; these acids have been extensively studied and successfully employed in acid fracturing treatments (Nasr-El-Din et al., 2006; Al-Mohammad et al., 2006; Al-Al-Mutairi et al., 2007; Nasr-El-Din et al., 2008; Mutairi et al., 2008).

Citric acid was introduced because it can be used in an encapsulated form, which prevents its reaction with both the production tubulars and the formation up to 180°F (Blauch et al., 2003; Burgos et al., 2004), and hence this will protect well tubulars, and those made of corrosion resistant alloys from strong and corrosive acids, especially at high temperatures. This use of encapsulated citric acid in petroleum fields has met with mixed field results. Blauch et al. (2003) reported positive field results and that the formation damage because of calcium citrate precipitation was not a major concern at down-hole conditions. However, Burgos et al. (2004) reported that there was no improvement in the performance of an acid-fractured well after the application of encapsulated citric acid. They explained the unexpected field results in terms of calcium citrate precipitation. These mixed field results highlighted that a better understanding of the reaction of citric acid with calcite is needed.

Previous studies have investigated the reaction of citric acid with calcite. Using the rotating disk apparatus, these studies have shown that the reaction is mainly limited by the mass transfer process and the calcium citrate precipitation plays an important role in the reaction of citric acid with calcite. However, these studies have been conducted over very limited range of parameters (Alkhaldi et al., 2005; Alkhaldi et al., 2007).

In order to design an effective acid fracturing treatment, it is essential to know the effective diffusivity of the acid system at different field conditions, which is needed to determine the geometry of the fracture. Therefore, the objectives of this study are to: (1) measure the calcite dissolution rate using the rotating disk apparatus over a wide range of parameters and (2) investigate the effect of temperature on the overall mass transfer coefficient.

Chemistry of citric acid and precipitation of calcium citrate

The reaction of citric acid with calcite at pH values of 1.8 to 4 is as follows:

$$2H^+ + CaCO_3 \leftrightarrow Ca^{2+} + H_2O + CO_2 \uparrow, \ pH = 1.8 - 4 \tag{1}$$

$$H_2AOH^- + Ca^{2+} \leftrightarrow CaH_2AOH^+, \ pH = 1.8 - 4 \tag{2}$$

$$Ca^{2+} + 2(CaH_2AOH^+)(aq) \leftrightarrow Ca_3(AOH)_2\ (s) \downarrow + 4H^+, \ pH = 3.0 \tag{3}$$

According to Alkhaldi et al. (2005), precipitation of calcium citrate complexes occurs at low pH values of nearly 3.0. If calcium citrate precipitates during the reaction, it will form a layer on the calcite surface and then it will act as a diffusion barrier, which will slow the overall mass transfer process. Therefore, it is very important to examine the surface of the calcite disk at the end of each experiment to determine whether calcium citrate precipitated during the citric-calcite reaction or not.

Mass transfer into a rotating disk in reactive environments

The calcite-citric reaction can be described by the following three steps: (1) the transport of citric acid molecules to the solid surface, (2) the dissociation of citric acid and the reaction at the surface, and (3) the transport of products away from the interface. The overall reaction rate is determined by the slowest step.

In the diffusion-limited regime, the mass transfer flux of a specie *i*, J_i, can be expressed in terms of the concentration of the transferring specie, and the mass transfer coefficient, k_{mti}, as follows:

$$J_i = k_{mt_i}(C_{Bi} - C_{Si}) \tag{4}$$

where C_{Bi} is the concentration of specie *i* in the bulk solution, and C_{Si} is its concentration at the surface of the disk. The mass transfer coefficient in Newtonian fluids, is a function of the effective diffusion coefficient of specie i, physical properties of the solution, and the square root of the rotational speed. It can be determined under laminar flow conditions ($Re \leq 3*10^5$ and $Sc > 100$) using **Equation 5** (Levich, 1962; Newman, 1966):

$$k_{mt} = \frac{(0.62048)\ (Sc^{-2/3})\ (\sqrt{\omega\ \nu})}{1 + 0.2980\ Sc^{-1/3} + 0.14514\ Sc^{-2/3}} \tag{5}$$

where *Sc* is Schmidt number (*v/D*), *Re* is Reynolds number $(\omega^*R^2)/v$, *v* is the kinematic viscosity of specie i solution, *D* is the effective diffusion coefficient of specie i, ω is the disk rotational speed, and *R* is the calcite disk radius.

Experimental studies

The experimental procedure reported by Alkhaldi et al. (2007) was used to conduct this study.

Results and discussion

Calcium citrate precipitation

Calcite disks reacted with different citric acid solutions for 50 minutes at 50°C were examined. The reacted disks were analyzed by scraping the surface with a spatula and running XRD analysis on the collected solids. The results showed that the scrapped solids contained a mixture of calcium citrate tetra-hydrate and calcium carbonate. Due to calcium citrate precipitation, it was expected that the pH value of the bulk solution to be above 3.0, based on **Equations 1-3**. However, the maximum increase in the pH value of the bulk solution was measured at 7.5 w% at 50°C, where the pH increased from 1.8 to 2.3 due to the reaction, but no precipitate was noted in the solution.

The pH at the reaction interface is expected to exceed 3.0 based on the fact that calcium citrate precipitate was noted at the reacted calcite disk.

Calcite dissolution in citric acid solutions

Figure 1 shows the dissolution rate of calcite in citric acid solutions of initial concentrations of 7.5 wt% as a function of the reaction temperature at rotational speeds of 100 to 1,000 rpm and 1,000 psi. The dissolution rate of calcite is proportional to the square root of the rotational speed from 100 to nearly 500 rpm. Hence, the reaction of citric acid with calcite is mass-transfer limited. The plateau portion, from 500 to 1000 rpm, suggests that the dissolution rate is limited by the surface reaction rate because the dissolution rate remained constant when the disk rotational speed was doubled (Lund et al., 1975). In addition, it was found that the increase in temperature resulted in higher dissolution rate, as expected. However, it is evident from the data that temperature has a significant effect on the calcite dissolution in the reaction-limited regime when it is compared to its effect in the mass-transfer-limited regime. This is mainly due to the strong dependence of the reaction rate constant on temperature (Schmidt, 1998). In general, log of the rate constant of any chemical reaction has a linear relationship with the inverse value of the reaction temperature.

The effective diffusion coefficient of citric acid

The diffusion coefficient of citric acid at 7.5 wt% is 5.9E − 6 cm^2/s (Levien, 1955) at 50◦C. The pH value under these conditions was measured in the present study and found to be 1.8. The diffusion coefficient calculated using the data given in **Figure 1** was found to be 4.5E−6 cm^2/s at 50◦C and $pH_{solution}$= 2.3. The diffusion coefficient of citric acid determined using the RDA is 24% lower than that given by Levien (1955). This is probably due to following reasons: (1) calcium citrate precipitation on the disk surface, which acted as a diffusion barrier, (2) the diffusion coefficient of the citric acid was measured in solutions that contained more than 1,900 mg/l calcium ions. The calcium ions may act as counter ions, diffusing from the reaction boundary to the bulk solution (Conway et al., 1999), and hence decrease the diffusion coefficient of citric acid, and (3) the diffusion of the H^+ from the bulk solution to the disk was neglected. This may also affect the diffusion coefficient measured by the RDA.

Effect of temperature on the effective diffusion coefficient of citric acid

Figure 2 and **Table 1** show the effect of temperature on the effective diffusion coefficients of citric acid in its reaction solutions of initial concentration of 7.5 wt% at 1,000 psi. The temperature dependence of diffusion coefficients is usually quantified using the Arrhenius equation:

$$D = D_0 \, exp\left(\frac{-E_a}{RT}\right) \tag{6}$$

where D_0 is the maximum diffusion coefficient (at infinite temperature), E_a is the diffusion activation energy (J/mol), R is the universal gas constant (8.314 J/mol/K), and T is temperature (K).

From **Figure 2**, it is evident that the effect of temperature on the effective diffusion coefficient of citric acid follows Arrhenius law. The activation energy was found to be 37.9 kJ/mol.

Conclusions

1. The dissolution rate was found to be dependent on reaction temperature, and the disk rotational speed.
2. The presence of counter calcium ions and the calcium citrate precipitation had significant effect on the effective diffusion coefficient of citric acid.
3. The effect of temperature on the effective diffusion coefficient of citric acid was found to follow Arrhenius law.

Acknowledgments

The authors would like to thank the Operation and Services Division of Saudi Aramco EXPEC ARC for lending their Rotating Disk Apparatus for conducting this study. Ali Al-Aamri is acknowledged for his help in conducting the Rotating Disk experiments. In addition, thanks go to M. Al-Fahad, of Saudi Aramco R&D Center for measuring the concentration of calcium ions.

References

Akhaldi, M.H., Nasr-El-Din, H.A., Blauch, M.E., Funkhouser, G.P.: "New findings on damage potential, geochemical reaction mechanisms, and production enhancement applications for citric acid," *SPEJ* 10 (3) (2005) 267–275.

Alkhaldi, M., Nasr-El-Din, H., Mehta, S., Alaamri, A.: "Reaction of citric acid with calcite," *Chemical Engineering Science* 62 (2007) 5880-5896.

Al-Mohammad, A.M., Nasr-El-Din H.A., Al-Aamri, A.M., and Al-Fuwaires, O.: "Reaction of calcite with surfactant-based acids," paper SPE 102838 presented at the 2006 SPE Annual Technical Conference and Exhibition, San Antonio, TX, 24-27 September.

Al-Mutairi, S.H., Hill, A.D., Nasr-El-Din, H.A.: "Effect of droplet size and acid volume fraction on the rheological properties and stability of emulsified acid," accepted for publication, SPE Production and Operations, 2008.

Al-Mutairi, S.H., Nasr-El-Din, H.A., Hill, A.D., Al-Aamri, A.: "Effect of droplet size on the reaction kinetics of emulsified acid With Calcite," paper SPE 112454 presented at the 2008 SPE International Symposium and Exhibition on Formation Damage Control ,Lafayette, Louisiana, U.S.A., 13–15 February.

Blauch, M.E., Cheng, A., Rispler, K.: "Novel carbonate well production enhancement application for encapsulated acid technology: first-use case history," paper SPE 84131 presented at the 2003 SPE Annual Technical Conference and Exhibition, Denver, CO, 5–8 October.

Burgos, G., Birch, G., Buijse, M.: "Acid fracturing with encapsulated citric acid," paper SPE 86484 presented at the 2004 SPE International Symposium and Exhibition on Formation Damage Control, Lafayette, LA, 18–20 February.

Conway, M.W., Asadi, M., Penny, G., Chang, F.: "A comparative study of straight/gelled/emulsified hydrochloric acid diffusivity coefficient using diaphram cell and rotating disk," paper SPE 56532 presented at the 1999 SPE Annual Technical Conference and Exhibition, Houston, TX, 3–6 October.

Crowe, C.W., McGowan, G.R., Baranet, S.E.: "Investigation of retarded acids provides better understanding of their effectiveness and potential benefits," paper SPE 18222 presented at the 1988 SPE Annual Technical Conference and Exhibition, Houston, TX, 2–5 October.

Levien, B.J.: "A physicochemical study of aqueous citric acid solutions," *J. Physical Chemistry* 59 (1955) 640–644.

Levich,V.: "*Physicochemical Hydro-dynamics*," Prentice-Hall, Englewood Cliffs, NJ (1962) 100–130.

Lund, K., Fogler, H.S., McCune, C.C, Ault, J.W.: "Acidization II. The dissolution of calcite in hydrochloric acid," Chemical Engineering Science 30 (1975) 825–835.

Newman, J.: "Schmidt number correction for the rotating disk," *J. Physical Chemistry* 70 (1966) 1327.

Nasr-El-Din H.A., Al-Mohammad, A.M., Al-Aamri, A.D., and Al-Fahad, M.A.: "Quantitative analysis of reaction rate retardation in surfactant-based acids," paper SPE 107451 presented at the 2007 SPE European Formation Damage, Scheveningen, Netherlands, May 30 - June 2.

Nasr-El-Din H.A., Al-Mohammad, A.M., Al-Aamri, A.M., and Al-Fuwaires, O.: "Reaction kinetics of gelled acids with calcite," accepted for publication, SPE Production and Operations, 2008.

Schmidt, L.: "The *engineering of chemical reactions*," Oxford University Press, Oxford (1998) 31.

Table 1: Effect of temperature on the diffusion coefficient of citric acid at initial acid concentration of 7.5 wt% and 1,000 psi.

D, cm^2/s	ln (D)	T, K	1/T, K^{-1}
1.3634E -6	-13.505	298	0.003356
2.68E -6	-12.830	313	0.003195
4.5E -6	-12.311	323	0.003096

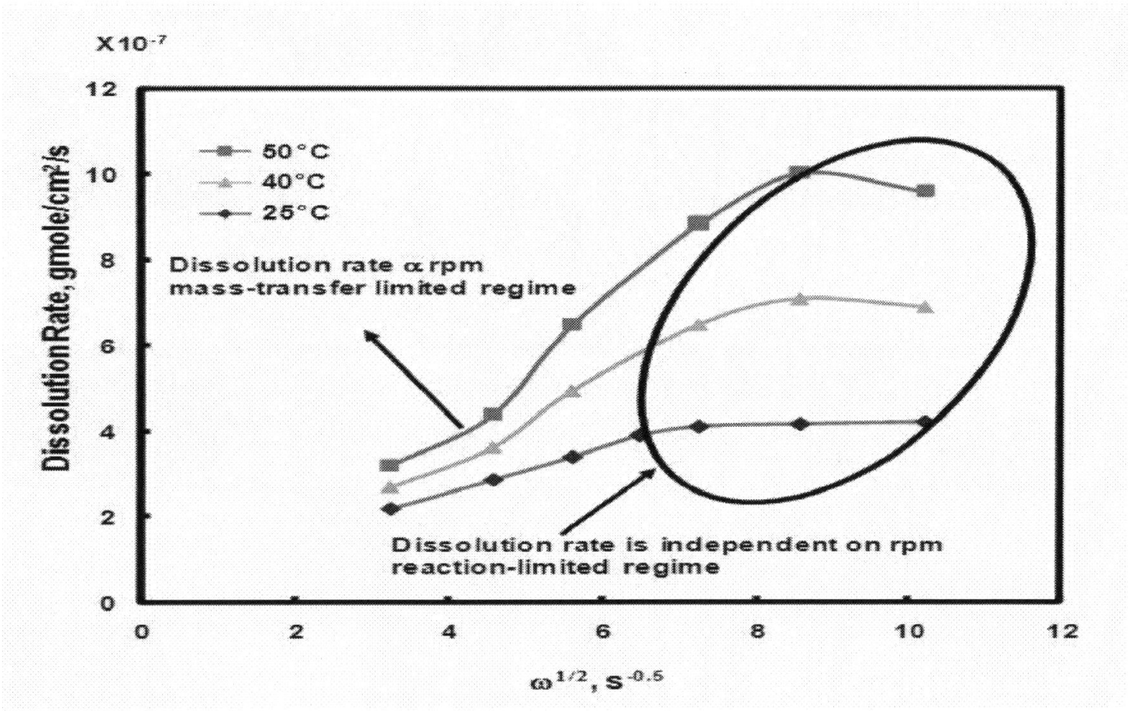

Figure 1: Dissolution rate of calcite in citric acid solutions of initial concentration of 7.5 wt% as a function of reaction temperature at rotational speeds of 100 to 1,000 rpm and 1,000 psi.

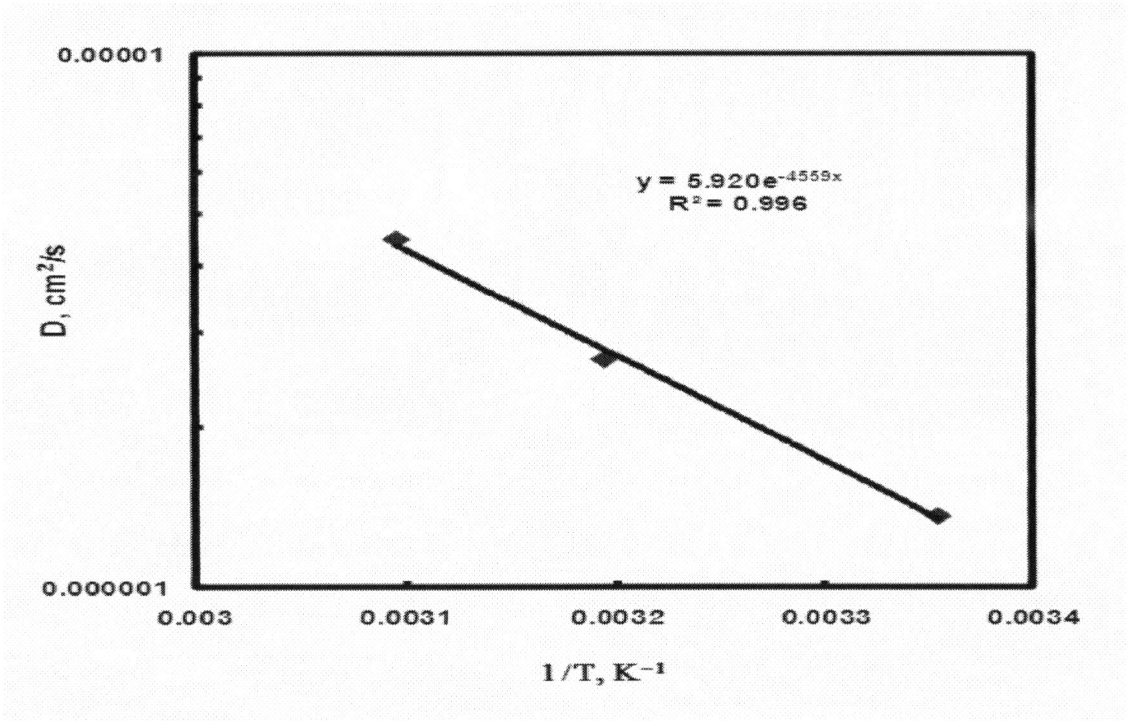

Figure 2: Diffusion coefficient of 7.5 wt% citric acid as a function of temperature at 1,000 psi.

118i CdS/CdTe Thin Film Solar Modules and Atmospheric Pressure Physical Vapor Deposition

Brian Ashenfelter, The McKetta Department of Chemical & Bioprocess Engineering, Trine University, 1 University Ave., Attn: Chemical Engineering, Angola, IN 46703

With the rising cost of energy and the increasing demand to make it more green, solar power has become an increasingly popular market. The new technology of thin film CdTe solar modules has been proven as a cheap affective way to generate power. In this competitive field, many different applications have been developed which relay on physical and chemical vapor transport systems. The objective of this presentation is to discuss the principles behind CdTe thin film solar energy and also film deposition. This includes the viability of atmospheric physical vapor deposition (APPVD) as an industrial application. Summarized, will be a general process in current use by CalyxoUSA (Perrysburg, OH) which is proof of this concept.

Web Page: www.calyxo.com

119a Metal Oxide Nanofiber and Its Application in Gas Sensing

Ashley Schempf, Wenzhao Jia, and Yu Lei. Chemical, Materials and Biomolecular Engineering, University of Connecticut, 191 Auditorium Rd, U3222, Storrs, CT 06269

Electrospinning is a process in which polymers can be spun into nanofiber mats by applying high voltages. Electrospun nanofibers have been widely applied in tissue engineering, chemical sensors, filters, and drug delivery. In this study the electrospinning process is used to produce metal oxide nanofibers. The morphology of the as-prepared metal oxide nanofibers is characterized by scanning electron microscope (SEM), transmission electron microscope (TEM) and X-ray diffraction (XRD). The application of the as-prepared metal oxide nanofibers in gas sensing will also be demonstrated.

119b The Effect of Collector Geometries on the Nanofiber Alignment

Matthew Cagnetta, Wenzhao Jia, and Yu Lei. Department of Chemical, Materials and Biomolecular Engineering, University of Connecticut, 191 Auditorium Rd, U-3222, Storrs, CT 06269

The ability to produce electrospun polymer nanofibers in a repeatable fashion and with a variety of desired morphological characteristics is integral to their use in many nanotechnoloical applications. In this study, the ability of collectors with different geometries to influence the deposition characteristics and in particular, the alignment of polymer fibers was investigated with poly(acrylic acid). It has been seen that fiber deposition is driven by the shape of the electric field while alignment occurs to minimize electrostatic forces on the fibers. Both preferential deposition and alignment can be obtained by manipulating the collector geometry.

119c Polyhedral Oligomeric Silsesquioxane Coatings Stimulate Osteoblast Attachment to Cp-Ti

Jacob G. Dickinson[1], **Jacob P. Lyons**[1], Michael J. Misovich[1], Andre Y. Lee[2], and Melissa J. Baumann[2]. (1) Engineering, Hope College, 27 Graves Pl, Holland, MI 49422-9000, (2) Chemical Engineering & Materials Science, Michigan State University, 2527 Engineering Building, East Lansing, MI 48824-1226

Given that the number of total joint replacements will increase with an aging population, along with a concomitant rise in joint revision surgeries that is expected to double within the next 20 years, there is an immediate need for improved bonding and fixation between bone and orthopaedic implant surfaces. Polyhedral oligomeric silsesquioxane (POSS) trisilanols have been used as a dental restorative material and for contact lenses but have yet to be utilized in conjunction with implants in contact with bone and soft tissue. We hypothesize that osteoblast (OB) attachment to commercially pure titanium (cp-Ti) will be enhanced by a structurally robust trisilanol POSS surface coating. The dip coating method was used to apply the trisilanol POSS coatings onto commercially pure Ti (cp-Ti) surfaces that had received an industry standard surface cleaning. Osteoblast attachment and Live/Dead assay tests were used to evaluate the effectiveness of the trisilanol POSS coating. It was found that cp-Ti coated with trisilanol POSS showed a statistically significant increase in OB attachment over uncoated cp-Ti controls as a function of POSS coating exposure times of 1, 10 and 100 minutes and attachment times of 1, 2 and 4 hours. This increased attachment is also statistically significantly greater than the cell attachment data reported by the manufacturer for trisilanol POSS coated cp-Ti which did not receive the industry standard cleaning treatment. Live/Dead assay results showed no statistically significant increase in cell death with trisilanol POSS coating.

119d Photopatterned Nanoporous Polymer Monoliths as Passive Mixers to Enhance Mixing Efficiency for on-Chip Labeling Reaction

Thomas Schwei, Chemical Engineering, University of California - Berkeley, 2562 LeConte Avenue, Berkeley, CA 94709

Efficient mixing is an important component of micro total analytical systems that rely on chemical reactions such as DNA sequencing, cell lysis and protein analysis. However, laminar flow in these devices causes mixing to be diffusion-limited and slow. In order to increase the extent of reaction for on-chip fluorescent labeling of proteins a passive mixer was prepared by using UV light to photopattern a periodic arrangement of nanoporous polymer monolith (nPPM) structures directly within the channel of a plastic microfluidic chip. By optimizing the composition of the polymerization solution and irradiation time we demonstrate the ability to pattern nPPM in regular 100 μ m segments. The mixing efficiency of this patterned array was evaluated by monitoring the laser-induced fluorescence intensity of the reaction product of fluorescamine and lysine introduced into a tee-junction

119e Determining Applicability of Known Material Equations to Biomaterial Models and Basic Characterization of Blood Vessel Performance

Kevin P. Timms and Robert J. Wilkens. Department of Chemical and Materials Engineering, University of Dayton, 309 Stonemill Rd., Dayton, OH 45409

In order for biological materials and organs to be developed using tissue engineering, data needs to be obtained on the properties and performance capabilities of their healthy natural counterparts. Many proven and well-known models have been characterized for many materials. However, certain aspects of biomaterials are unique to this field of materials and their performance needs to be characterized, either with current models such as Young's Modulus and Poisson's Ratio, or with newly developed models. The cardiovascular system is being studied in particular because its main attribute, fluid flow, has been well characterized in non-biological systems.

The purpose of this experiment is to measure fluid mechanics performance of an elastic conduit under various amounts of tension with pulsatile flow. Specifically, the pressure drop and degree of pulse amplitude dampening will be measured for pulsatile flow across the tubing under axial tension. The data obtained will quantify the effect of tension on these two characteristics of fluid performance.

This data will reveal relationships for an analog system, blood vessels. The information obtained will provide better knowledge of the performance of the tubular section of the mammalian cardiovascular system and will provide data that can be specifically applied to blood vessels (which are contrastingly not fully cylindrical and vary in size, shape, and micro-structural properties from specimen to specimen). Furthermore, the experiment could provide relationships that suggest a mode of objective vessel abnormality detection in living biological organisms.

Future work will include quantifying these relationships using porcine arteries and non-Newtonian blood analogs.

119f Effect of Diameter on Band Pattern Formation of Silica Nanoparticles by Convective Assembly

Kayla L. Reibel[1], Alex Lee[2], and Michael Tsapatsis[2]. (1) McKetta Department of Chemical & Bioprocess Engineering, Trine University, 1 University Avenue, Angola, IN 46703, (2) Chemical Engineering & Material Science, University of Minnesota, 151 Amundson Hall, 421 Washington Avenue SE, Minneapolis, MN 55455

Films of silica nanoparticles have many applications in diverse fields such as molecular separations and electronics. In the process of producing coatings of silica nanoparticles by convective assembly, an interesting phenomenon occurs where the silica nanoparticles form a periodic band pattern instead of covering the glass substrate. The band pattern will be studied to see the effect of particle diameter on the spacing between the bands. Silica nanoparticles of different sizes will be grown and coated onto a glass substrate. The spacing of the bands will be observed with the scanning electron microscope. After producing and analyzing the particles and coatings, the spacing of the bands will be able to be predicted based on particle size.

119g Characterization of the Nafion Fuel Cell Membrane Via Observation of Surface Wetting Properties

Sharonmoyee Goswami, Chemical Engineering, Princeton University, Princeton, NJ 08544, Shannon Klaus, Chemical Engineering, University of Missouri, Columbia, MO 65201, and Jay B. Benziger, Department of Chemical Engineering, Princeton University, Princeton, NJ 08544.

A primary difficulty in using fuel cells as a viable energy alternative is that the Nafion membrane has dynamic physical and mechanical properties that depend upon hydration. Nafion undergoes microphase separation due to energy differences between the hydrophobic backbone and the hydrophilic sulfonic acid side chains. We examine here the effect of microphase separation on the surface wetting of Nafion and Nafion composite films. A combination of Wilhelmy plate and sessile drop measurements were carried out with Nafion membranes of varying thickness (0.002" to 0.010"), amount of composite material (0,3,and 10% TiO2), and pre-strain levels. These sessile drops on Nafion membranes were monitored through time-lapse photography to observe physical changes at the Nafion-water interface.

Wilhelmy plats measurements showed the Nafion was hydrophobic (contact angle of ~110°) in the descending mode but was hydrophilic (contact angle of ~0°) in the ascending mode. Sessile drops were hydrophobic on Nafion, but remained pinned in position when the drop evaporated, indicating the Nafion surface under the drop must be hydrophilic. Water droplets on the membrane immediately induced local swelling; the maximum state of protrusion increased with membrane thickness, and drop size. After complete liquid evaporation, the membrane reverted to its pre-wetted state.

Wetting of Nafion was also examined with octane and ethanol. Octane wet the Nafion surface with a contact angle ~0°, but the octane did not cause any measurable deformation of the Nafion. Ethanol induced the most dramatic response; it resulted in enormous film deformation.

Nafion deformation due to sessile drops of water did not show any significant effect due to pre-straining, or adding 3% and 10% TiO2 nanoparticles to the membrane.

The results suggest that Nafion surfaces can be switched from hydrophobic to hydrophilic due to exposure to hydrogen bonding liquids: water and ethanol. Swelling properties originate from the solvent penetration into the membrane causing a differential expansion of the membrane in contact with the liquid. Even though octane wets the Nafion surface it does not appear to penetrate into the membrane suggesting that the solvent interactions are with the hydrophilic sulfonic acid domains.

Formation of Magnetite Clusters using a Confined Impinging Jet Mixer

H. L. Sewell[1], R. Mejia-Ariza[1], W. C. Miles[2], P. Huffstetler[3], J. Goff[3], R. M. Davis[1,2], J. S. Riffle[1,3]
[1]Macromolecular Science and Engineering, [2]Department of Chemical Engineering, [3]Department of Chemistry, Virginia Polytechnic Institute and State University, Blacksburg, VA 24061.

Superparamagnetic nanoparticles (SPM NP) are commonly employed as magnetic resonance imaging contrast agents. Magnetite (Fe_3O_4) is of increasing interest because of its high magnetic susceptibility in an applied magnetic field. It has been shown in the literature that an increase in size of transverse NMR relaxivity agents leads to a significant decrease in T2 relaxation times and thus higher contrast in an MRI. It is therefore important to develop methods and conditions for controlling the size of polymer-stabilized magnetite clusters. Our strategy is to elucidate critical coagulation concentrations of water for magnetite particles that have been coated with an amphiphilic poly(propylene oxide-*b*-ethylene oxide)-OH (PPO-*b*-PEO) copolymer to control clustering size through the use of a confined impinging jet (CIJ) mixer. Using the critical coagulation concentrations, operating conditions for the CIJ mixer have been found where micellization of the copolymer occurs simultaneously with magnetite clustering. Preliminary results have shown that control over the size distribution may be possible by changing the concentration of water in the mixing chamber.

Introduction

The study of superparamagnetic nanoparticles (SPM NP) has led to the potential employment of polymer-coated nanoclusters in the areas of drug delivery (1), magnetic resonance imaging (MRI) relaxivity agents (2), bio-catalysis, and bio-separations (3). Gadolinium, cobalt, and magnetite are common SPM NPs used in biological systems. Magnetite (Fe_3O_4), being an iron oxide, is well suited for oxygen-rich environments and will not lose its superparamagnetic properties over time due to oxidation (4; 5). For particle sizes less than 25 nm, the thermal energy is stronger than the magnetic coupling forces so there is no longer a net magnetic moment (6; 7; 8; 9; 10). When a magnetic field is applied, the magnetic coupling forces once again dominate and all particles align with the magnetic field (11).

MRI contrast agents represent a particularly useful application of polymer-stabilized magnetite clusters. Magnetite is interesting for MRI as it has a larger magnetic moment in an applied image field (12). In MR imaging, a magnetic field is applied in the presence of time-varying radiofrequency fields. At equilibrium, the hydrogen atoms in the body are aligned with a net magnetization in the direction of the applied magnetic field. When the radiofrequency fields are applied, the alignment of the atoms is disturbed. The time for the atoms in the body to return to equilibrium is called the relaxation time. There are two types of relaxation of hydrogen atoms measured in MRI, T1 and T2. T1 is the spin-lattice or longitudinal relaxation time and T2 is the spin-spin or transverse relaxation time. T1 and T2 occur and are measured simultaneously. Relaxivity is the inverse of relaxation time, so a smaller T1 or T2, leads to a higher relaxation, R1 or R2, respectively.

SPM NP, including magnetite, can be used as T2 contrast agents (13). It has been found that an increase in the size of MRI T2 contrast agents leads to a significant decrease in the T2 relaxation time (5). Magnetite particles greater than 25 nm in diameter do not exhibit superparamagnetic behavior, a prerequisite for functioning as a contrast agent. However, by forming clusters of smaller particles, superparamagnetism may still be maintained while also allowing for a larger size of the contrast agent.

In the presence of a magnetic field the magnetic moments of the particles align with the field. As water molecules diffuse randomly around the SPM NP they experience a range of field conditions (15). Polymer stabilization of the SPM NP is crucial for proper design of contrast agents. However, before polymer-stabilized magnetite nanoclusters can be employed for use in biological systems the conditions for optimal cluster formation must be elucidated.

To form polymer-stabilized magnetite nanoclusters we investigated the use of oleic acid versus poly(butylene oxide) (PBO) stabilized magnetite nanoparticles to determine which coating would allow for the better control of cluster size. Once the best coating for the magnetite was found, the critical water concentrations for magnetite nanoparticles and a diblock copolymer needed for cluster formation were examined and an operating region for confined impinging jet mixing was found for forming stabilized magnetite clusters.

Methods and Materials

Synthesis of Coated Magnetite Nanoparticles

Oleic-acid coated magnetite nanoparticles with a narrow particle size distribution were synthesized in the Riffle group by the method outlined by Pinna et al. and adapted by Nikorn Pothayee [16]. Iron(III) acetylacetonate is added to deoxygenated benzyl alcohol and slowly heated over the course of two hours to 205°C and then kept stable for 24 hours. The resultant black particles are washed with acetone, dried, and then redispersed in chloroform with oleic acid. The chloroform is evaporated and the particles are washed to remove any remaining oleic acid. Particle sizes are measured to be on the average diameter of 5.1 nm with dynamic light scattering (DLS) and average radius of 3.8 nm with transmission electron microscopy (TEM) [17].

Poly(butylene oxide), PBO, coated magnetite was synthesized in the Riffle group. PBO was made by reacting 1,2-epoxybutane initiated with 3-hydroxypropyl trivinylsilyl in a pressure reactor. The vinyls on the PBO fluid were then reacted with mercapto acetic acid in chloroform to functionalize the PBO with carboxylic acids. The magnetite was formed via coprecipitation of Fe_2Cl_3 and $FeCl_3$ with 50% ammonium hydroxide while mixed with an ultra-high speed homogenizer. The functionalized polymer was then dispersed in chloroform and added to the magnetite. The PBO adsorbs onto the magnetite particles via the carboxylic acid

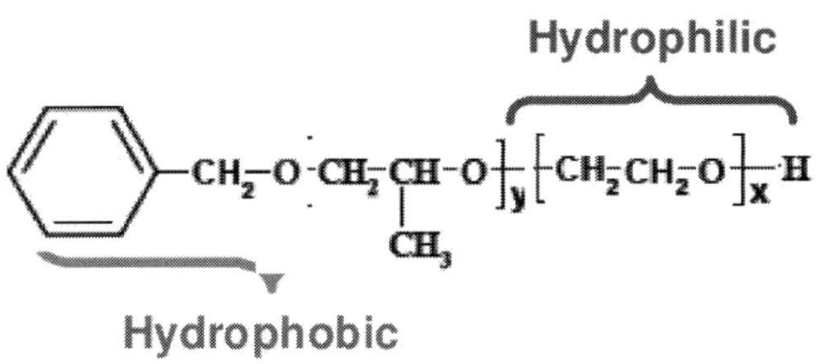

Figure 1: PPO-b-PEO

(x=34, y= 9)
MW = 2,000 g/mole

groups. The complex is then placed in a separation funnel and allowed to separate, washing with chloroform. The chloroform is then evaporated off with a rotational evaporator to leave a PBO ferrofluid.

Amphiphilic Diblock Copolymer

The amphiphilic diblock copolymer stabilizer is a benzyl alcohol initiated poly(propylene oxide-*b*-ethylene oxide)-OH (Bz-PPO-*b*-PEO-OH) as shown in Figure 1. PPO has been used safely in cell studies and PEO has been approved by the FDA [18]. The PPO is hydrophobic so it interacts with the hydrophobic magnetite coating. The PEO is hydrophilic, providing stabilization in the aqueous antisolvent.

Derjaguin, Landau, Verwey, Overbeck (DLVO) Theory

Derjaguin, Landau, Verwey, Overbeck (DLVO) theory proposes that the stability of colloidal systems can be described by the following:

$$V_T = V_{VdW} + V_{ES} + V_S + V_M \qquad \text{Equation (1)}$$

V_T is the total potential, V_{Vdw} is the van der Waals attraction, V_{ES} and V_S are electrostatic and steric repulsions, and V_M is the magnetic attraction [19; 20]. DLVO theory can be used to predict the particle size needed for the magnetic forces to have a large enough range that steric repulsion is overcome in the presence of a magnetic field [19; 20]. Using DLVO theory, conditions such as desired diblock molecular weight and necessary polymer loading for a specific cluster size can be estimated.

Critical Coagulation Concentration

To find the amount of water needed for cluster/micelle formation 1mL coated-magnetite solutions (magnetite-oleic acid or PBO-magnetite) or diblock copolymer was placed in a cuvette. Microliter amounts of water were added and dynamic light scattering (DLS) was used to determine when clustering or micellization occurred.

Synthesis of Magnetite Clusters

Magnetite clusters contained in polymer micelles are formed by the use of the Flash NanoPrecipitaion technique (21; 22). The Flash NanoPrecipitation technique requires rapid mixing of the magnetite and an amphiphilic polymer suspended in an organic solvent with an appropriate antisolvent, which in this case is deionized (DI) water.

The Flash NanoPrecipitation method is carried out in a confined impinging jet (CIJ) mixer. For the CIJ process the contents of two syringes – one holding the magnetite and copolymer stabilizer suspended in the organic solvent, and the other holding the aqueous solvent – are injected into a mixing chamber as shown in Figure 2.

CIJ mixing is more effective at forming polymer-stabilized clusters than traditional mixing because it allows for very rapid mixing at a controlled and maintained weight percent of antisolvent in the mixing chamber (17).

The CIJ mixer works on the basis of mixing on three different scales: macromixing, mesomixing, and micromixing (21).

Macromixing occurs on the scale of the mixing vessel which, in the case of the present CIJ mixer is 0.109 inches, the mixing chamber diameter. Mesomixing occurs on the scale of turbulent eddies caused by the mixing of the two fluids. Micromixing occurs on the molecular level by diffusion of molecules across laminar regions. The equation for the momentum diffusivity (kinematic viscosity) of the liquid in the mixing chamber, i.e. the effluent from the mixer, v_3, is shown below:

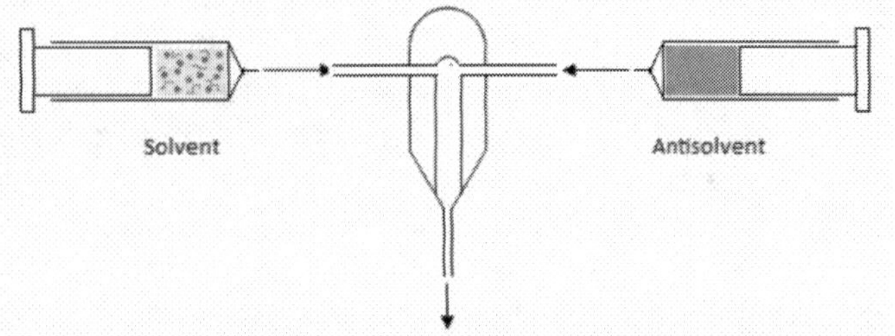

Figure 2: Confined Impinging Jet Mixer

$$v_3 = \frac{(0.5\lambda_K)^2}{\tau_{diffusion}} \qquad \text{Equation (2)}$$

where $\tau_{diffusion}$ is the time for diffusion of molecules from one fluid to the other. λ_K is the Kolmogorov length, the smallest length scale for turbulent flow, described by (5):

$$\lambda_K = \left[\frac{v_3}{\varepsilon}\right]^{1/4} \qquad \text{Equation (3)}$$

ε is the energy dissipation rate, described as the rate of energy that is put into a system, P, divided by the mass over which the energy is dissipated. The mass is the the product of the density of the effluent stream and its volume so that:

$$\varepsilon = \frac{P}{\rho_3 V_m} \qquad \text{Equation (4)}$$

$$P = \left(\frac{1}{2}m_1u_1 + \frac{1}{2}m_2u_2\right) \qquad \text{Equation (5)}$$

The mixing time is proportional to the time for diffusion (21).

$$\tau_m \propto \tau_{diffusion} = \frac{(0.5\lambda_K)^2}{v_3} \qquad \text{Equation (6)}$$

Combining the above Equations (2) through (5) and substituting into Equation (6) gives a new relationship for the mixing time.

$$\tau_m \propto \frac{1}{4}\left(\frac{v_3}{\varepsilon}\right)^{1/2} = \frac{v_3^{1/2}\rho_3^{1/2}V_m^{1/2}}{4\left(\frac{1}{2}m_1u_1 + \frac{1}{2}m_2u_2\right)^{1/2}}$$

Equation (7)

In order for the streams to meet in the center of the chamber, the momenta of the streams must be matched.

$$\dot{m}_1 u_1 = \dot{m}_2 u_2 \qquad \text{Equation (8)}$$

The mass flowrate (kg/s) of the fluid is m and the linear velocity is u (m/s).

The mass flowrate of the fluid can be found by multiplying the volumetric flowrate, Q, by the density of the fluid, ρ.

$$\dot{m}_1 = Q_1 \rho_1 \qquad \text{Equation (9a)}$$

$$\dot{m}_2 = Q_2 \rho_2 \qquad \text{Equation (9b)}$$

The volumetric flowrate is the linear velocity multiplied by the cross sectional area, A, of the needle.

$$Q_1 = u_1 A_1 \qquad \text{Equation (10a)}$$

$$Q_2 = u_2 A_2 \qquad \text{Equation (10b)}$$

The equation for the cross sectional area is shown below, where R is the radius of the needle.

$$A_1 = \pi R_1^2 \qquad \text{Equation (11a)}$$

$$A_2 = \pi R_2^2 \qquad \text{Equation (11b)}$$

Combining equations (10) and (11) and solving for the linear velocity gives equation (12) where d is the diameter of the needle.

$$u_1 = \frac{4Q_1}{\pi d_1^2} \qquad \text{Equation (12a)}$$

$$u_2 = \frac{4Q_2}{\pi d_2^2} \qquad \text{Equation (12b)}$$

Substituting equation (9) into equation (8) gives a new equation for matched momentum.

$$Q_1 \rho_1 u_1 = Q_2 \rho_2 u_2 \qquad \text{Equation (13)}$$

Substituting equation (12) into the above and rearranging gives a new equation relating volumetric flowrate and densities to the ratio of the needle diameters.

$$\frac{Q_1^2 \rho_1}{Q_2^2 \rho_2} = \frac{d_1^2}{d_2^2} \qquad \text{Equation (14)}$$

A simple rearrangement gives the equation for the volumetric flowrate of fluid 1.

$$Q_1 = Q_2 \frac{d_1}{d_2} \sqrt{\frac{\rho_2}{\rho_1}} \qquad \text{Equation (15)}$$

If the diameters of the needles are the same for both fluids ($d_1 = d_2$) then equation (15) can be simplified.

$$Q_1 = Q_2 \sqrt{\frac{\rho_2}{\rho_1}}$$

Equation (16)

If fluid 1 is THF with a density of 881.4 kg/ m^3 at 25°C and fluid 2 is water with a density of 997.1 kg/m^3 25°C then,

$$Q_1 = 1.06 Q_2 \qquad \text{Equation (17)}$$

However, as there are only two streams, in order to vary the weight percent of water in the mixer, fluid 2 has to be a mixture of THF and water.

To find the flowrate of the aqueous stream required for the matched momentum case, a series of simultaneous equations must be solved.

First the density of the fluid inside the mixer is found by:

$$\rho_{mix} = w_{H_2O} \rho_{H_2O} + (1 - w_{H_2O}) \rho_{THF}$$

Equation (18)

w_{H_2O} is the weight percent of water in the mixer and ρ_{H_2O} and ρ_{THF} are the densities of water and THF, respectively.

The total volume, V_{Total} of fluid that enters the mixer is the sum of the volume of the organic stream – containing the THF, magnetite and diblock – and the aqueous stream containing the water/THF mixture, $V_{organic}$ and $V_{aqueous}$.

$$V_{Total} = V_{organic} + V_{aqueous} \qquad \text{Equation (19)}$$

The volume of the organic stream is pre-determined as the product of the volumetric flowrate and the time for which the experiment will be run.

$$V_{organic} = Q_1 t \qquad \text{Equation (20)}$$

The volume of the aqueous solution is determined by:

$$V_{aqueous} = \frac{V_{organic}^{\ 2} \rho_{THF}}{m_{H_2O} + m_{THF(aq)}}$$

Equation (21)

m_{H_2O} is the mass of water in the mixer and $m_{THF(aq)}$ is the mass of THF in the aqueous stream.

m_{H_2O} is simply found by the product of the weight percent of water in the mixer, m_{mix}, and the total mass in the mixer as there is no water entering in the organic stream.

$$m_{H_2O} = w_{H_2O} m_{mix} \qquad \text{Equation (22)}$$

$m_{THF(aq)}$ is found by subtracting the mass of THF in the mixer coming from the aqueous stream and m_{H_2O} from the total mass of fluid that enters the mixer m_{mix}.

$$m_{THF(aq)} = m_{mix} - m_{THF(org)} - m_{H_2O}$$

Equation (23)

The mass of fluid in the mixer is found by multiplying the density of the fluid in the mixer by the total volume.

$$m_{mix} = \rho_{mix} V_{Total} \qquad \text{Equation (24a)}$$

Similarly, the mass of THF from the organic stream is found by the density of THF multiplied by the volume of the organic stream.

$$m_{THF} = \rho_{THF} V_{organic} \qquad \text{Equation (24b)}$$

To find the volume of water and THF required for the aqueous stream the following equations were used:

$$V_{H_2O} = \frac{m_{H_2O}}{\rho_{H_2O}} \qquad \text{Equation (25a)}$$

$$V_{THF(aq)} = V_{aqueous} - V_{H_2O} \qquad \text{Equation (25b)}$$

Results and Discussion

Magnetite Nanoparticles

Determination of the water concentration at which

Figure 3:
magnetite-oleic acid (0.02mg/mL) versus PBO-magnetite (0.02mg/mL) clustering for determination of best magnetite particle for CIJ mixing trials.

Magnetite-oleic acid clusters before 1wt% water in THF but PBO-magnetite does not start clustering until 15 wt% water.
Magnetite-oleic acid clusters in the region of diblock micellization. However micellization occurs in the entire PBO-magnetite clustering region.

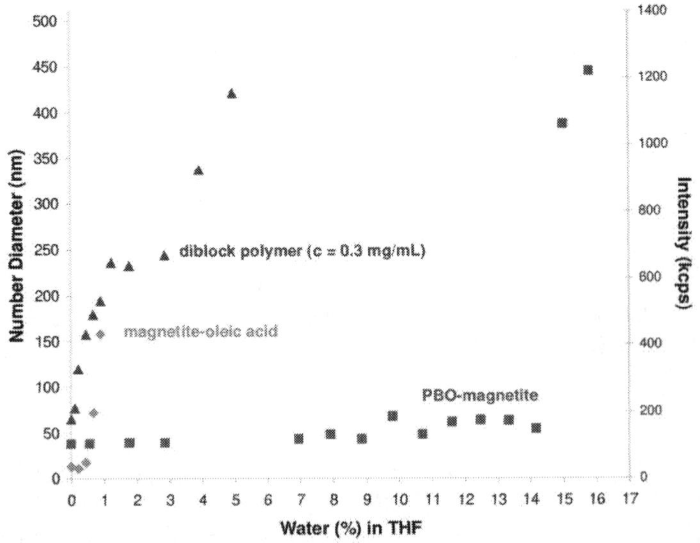

Figure 4: Critical Water Concentrations

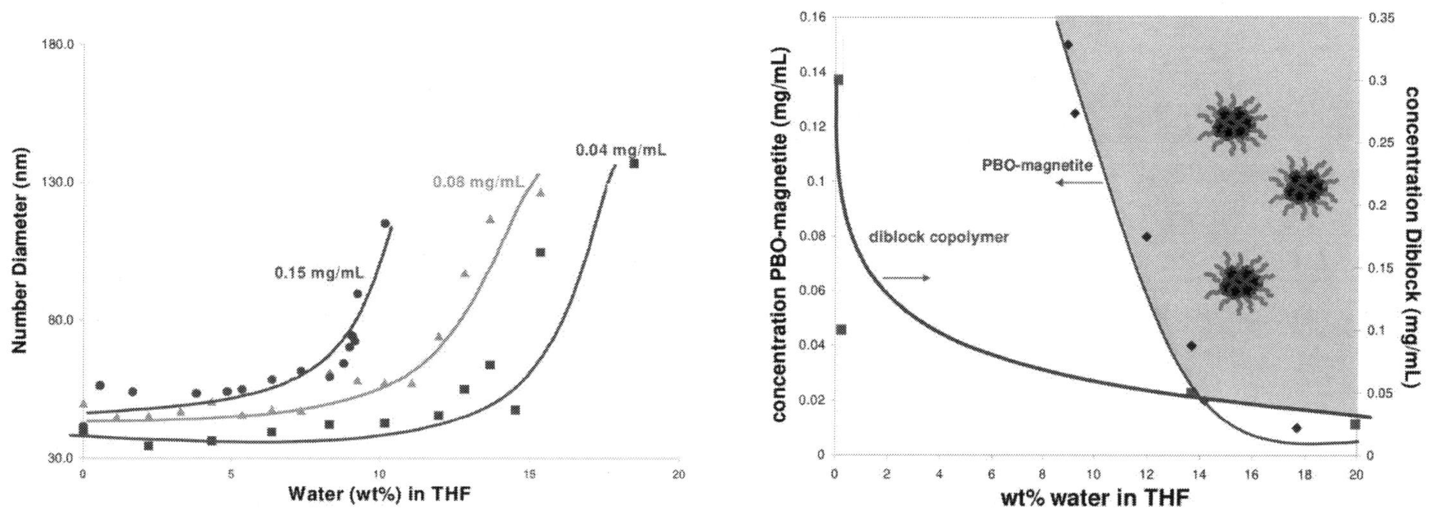

A) Critical Water concentration for PBO-magnetite

B) Operating Region for CIJ mixer

both the hydrophobic magnetite particles clustered and the amphiphilic diblock copolymer micellized, both in THF, was essential to determining how to synthesize uniform, stable clusters. Figure 3 shows a plot of the change in cluster size of magnetite-oleic acid (0.02 mg/mL) and PBO-magnetite oleic acid (0.02 mg/mL) with increasing weight percent water in THF. Magnetite-oleic acid clusters appreciably between 0.04 and 0.07 weight percent water whereas PBO-magnetite clusters around 15 weight percent. If the clustering sizes of the magnetite particles are compared to a plot of the intensity of the diblock copolymer (0.3 mg/mL) versus weight percent water, it can be seen that micellization of the diblock copolymer occurs at a weight percent water similar to that at which clustering of the magnetite-oleic acid occurs. However, full micellization of the diblock copolymer occurs at water concentrations well below that needed for clustering of PBO-magnetite.

If magnetite-oleic acid is used for CIJ mixing experiments, difficulty would be expected in finding the optimal weight percent water needed for simultaneous clustering and micellization. However, by using PBO-magnetite, micellization can occur around single particles or clusters of particles, thereby eliminating concern for micellization kinetics.

It was also noticed that over time the magnetite oleic-acid would aggregate without the addition of water. Both suspension in THF as well as storage of the sample under nitrogen were attempted. It was proposed that the oleic acid coating may be degrading over time with exposure to light, removing the steric stability and causing uncoated magnetite particles to aggregate.

No similar observations with stability of the PBO-magnetite were noted and thus many cluster formation experiments focused on this system.

Critical Water Concentrations

Critical water concentrations were found for both PBO-magnetite and the Bz-PPO-*b*-PBO-OH diblock copolymer. Figure 4A shows the general trend that increasing weight percent water increases cluster size. Plotting the weight percent water required to form clusters or micelles at varying concentrations of the PBO-magnetite and the diblock copolymer gives a plot of operating conditions for the CIJ mixer as seen in Figure 4B. Operating the mixer to the right of the diblock

Table 1: **Particle/cluster sizes for PBO-magnetite: PPO-*b*-PEO with varying wt% water**

wt% water	$Q_{aqueous}$ (mL/min)	Dn (nm)
5	39.7	39.6
10	39.5	32.6
15	39.2	161.3
20	39.0	207.0

Figure 5: TEM images of CIJ mixer experiment

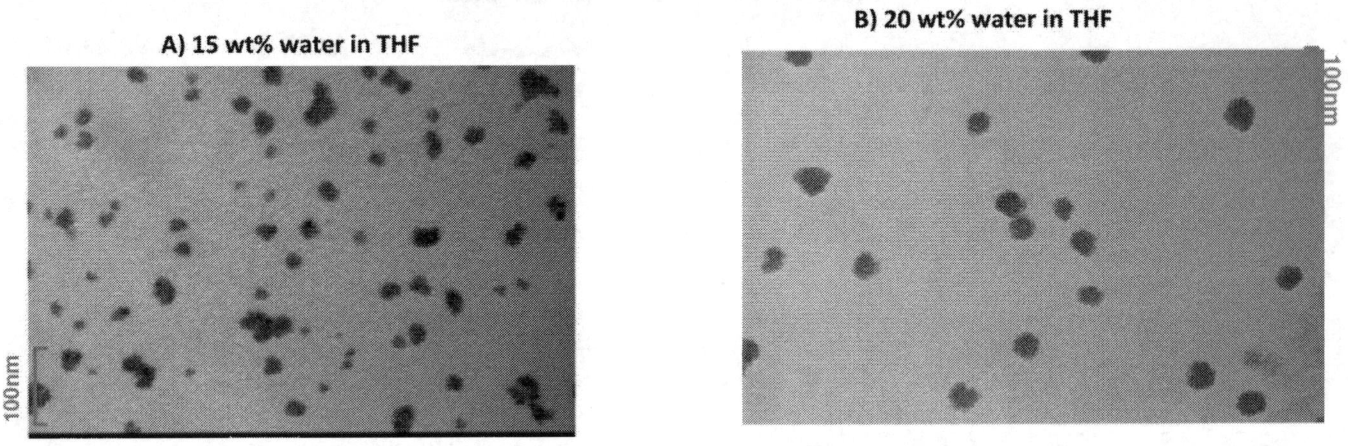

copolymer line will result in micellization of the polymer. Operating to the right of the PBO-magnetite line would result in clustering of the particles. Operating the mixer in the shaded region should result in formation of the desired polymer stabilized magnetite clusters.

Using the CIJ Mixer

The CIJ mixer was operated within the operating region as illustrated in Figure 4A. The PBO-magnetite concentration was 0.02mg/mL and the diblock copolymer concentration was 0.03 mg/mL in the mixing chamber. The CIJ mixer was run at an organic flowrate of 40 mL/min for 30 seconds. The weight percent water in THF was varied from 5-20% at 5% increments. From Figure 3 it can be seen that clustering starts to occur at about 15 wt% water in THF. Table 1 shows the calculated aqueous flowrates and the cluster sizes measured by DLS. As expected, at 5 and 10 wt% water, there appear to be single particles with a micelle coating. At 15 and 20 wt% water, the number diameter (D_n) as measured by DLS is much larger, indicating that clustering may have occurred.

To verify the DLS measurements, transmission electron microscopy (TEM) was performed. The results for 15 and 20 wt% are shown in Figure 5 A and B, respectively.

Both images verify that clusters were formed but the most promising result is that, while at 15 wt% water the clusters are somewhat polydisperse, those clusters made at 20 wt% are fairly uniform.

Conclusions and Future Work

Polymer-stabilized magnetite clusters have successfully been synthesized using a confined impinging jet mixer. Using an operating region developed by the use of critical water concentrations for PBO-magnetite and Bz-PPO-***b***-PBO-OH, conditions for cluster formation were elucidated. Preliminary results appear to show that forming uniformly sized clusters may be possible.

Future work will include variation of PBO-magnetite and diblock copolymer concentrations as well as the weight percent water in THF to determine further effects of these components on cluster formation.

Experiments were also conducted at higher water concentrations but neither DLS nor TEM showed cluster formation. This may be due to the required mixing time not being achieved in the mixer. Mixing times should be calculated and flowrates should be adjusted accordingly if these experiments are to be repeated.

Acknowledgements: The following people and organizations provided invaluable support: The staff of the Virginia Polytechnic and State University Summer Undergraduate Research Program especially Maggie Bump and Angie Flynn, NSF, ICTAS, MII, and John Boyd.

Works Cited

1. *Antitumor Immunity Induction by Intracellular Hyperthermia Using Magnetite Cationic Liposomes.* **Yanase, Mitsugu, et al.** July 1998, Jpn. J. Cancer Res., pp. 775–782.

2. *Magnetite-Loaded Carrier Erythrocytes as Contrast Agents for Magnetic Resonance Imaging.* **Brahler, M, et al.** 11, 2006, Nano Letters, Vol. 6, pp. 2505-2509.

3. *Magnetic Nanotubes for Magnetic-Field-Assisted Bioseparation, Biointeraction, and Drug Delivery.* **Son, Sang Ju, et al.** 20, April 29, 2005, JACS, Vol. 127, pp. 7316-7317.

4. *Detection of magnetic nanoparticles in tissue using magneto-motive ultrasound.* **Oh, Junghwan, et al.**, May 17 2006, Nanotechnology, Vol. 17, pp. 4183-4190.

5. *Characterization of Superparamagnetic Nanoparticle Interactions with Extracellular Matrix in an in Vitro System.* **Khun, Sam J., et al.**, August 20 2005, Annals of Biomedical Engineering, Vol. 34, pp. 51-58.

6. *Elaboration of PLLA-based superparamagnetic nanoparticles: Characterization, magnetic behaviour study and in vitro relaxivity evaluation.* **Hamoudeh, M. et al.,** International Journal of Pharmaceutics *2007,* 338, (1-2), pp. 248-257.

7. *Magnetic-Properties of Ultrafine Ferrite Particles.* **Sato, T., et al.**, 1987. Journal of Magnetism and Magnetic Materials Vol. 65, 252-256.

8. *Antibiofouling polymer-coated superparamagnetic iron oxide nanoparticles as potential magnetic resonance contrast agents for in vivo cancer imaging.* **Lee, H. et al.**, 2006. Journal of the American Chemical Society Vol. 128, pp. 7383-7389.

9. *Generation of superparamagnetic liposomes revealed as highly efficient MRI contrast agents for in vivo imaging.* **Martina, M. S. et al.**, 2005. JACS Vol. 127, pp. 10676-10685.

10. *Stability of dispersions of colloidal nickel ferrite spheres.* **Plaza, R. C., et. al.**, 2001. Journal of Colloid and Interface Science, Vol. 242, pp. 306-313.

11. **Calister, W D and Rethwisch, D.** *Fundamentals of Material Science and Engineering: An Integrated Approach, Third Edition.* New York : John Wiley & Sons, Inc., 2008.

12. *Physical and Chemical Properties of Superparamagnetic Iron Oxide MR Contrast Agents: Ferumoxides, Ferumoxtran, Ferumoxsil.* **Jung, Chu W. et al.,**1995. Magnetic Resonance Imaging, Vol. 13, No. 5, pp. 661-674.

13. *Synthesis of Two 3,5-Disubstituted Sulfonamide Catechol Ligands and Evaluation of Their Iron(III) Complexes for Use as MRI Contrast Agents.* **Schwert, Daniel D. et al.,** 2005. J. Med. Chem., Vol. 48, pp. 7482-7485

14. *Controlled Clustering of Superparamagnetic Nanoparticles Using Block Copolymers: Design of New Contrast Agents for Magnetic Resonance Imaging.* **Berret, Jean-Francois, et al.** 5, 2006, J. Am. Chem. Soc., Vol. 128, pp. 1755 -1761.

15. **Hornak, Joseph P.** The Basics of MRI. [Online] RIT, 1996. [Cited: June 18, 2008.] http://www.cis.rit.edu/htbooks/mri/inside.htm.

16. *Magnetite nanocrystals: Nonaqueous synthesis, characterization, and solubility.* **Pinna, N., et. al.** 11, 2005, Chemistry of Materials, Vol. 17, pp. 3044-3049.

17. **Caba, Beth Lynn.** *Solution and Adsorption Characterization of Novel Water-Soluble Ionic Block Copolymers for Stabilization of Magnetite Nanoparticles.* Macromolecular Science and Engineering, Virginia Polytechnic Institute and State University. 2007. pp. 193-194.

18. *DOCKET NO. 2004P-0139: SUPPLEMENT TO CITIZEN PETITION.* 17 September 2004.

19. **Israelachvili, Jacob.** *Intermolecular and Surface Forces, Second Edition.* San Diego: Academic Press Limited, 1992.

20. **Hiemenz, Paul C. and Rajagopalan, Raj.** *Principles of Colloid and Surface Chemistry, Third Edition.* New York: Marcel Dekker, Inc, 1997.

21. *Chemical Processing and Micromixing in Confined Impinging Jets.* **Johnson, Brian K and Prud'homme, Robert K.** 9, 2003, AICHE Journal, Vol. 49, pp. 2264-2282.

22. Flash NanoPrecipitation of Organic Actives and Block Copolymers using a Confined Impinging Jets Mixer. **Johnson, Brian K and Prud'homme, Robert K.** 2003, Aust. J. Chem, Vol. 56, pp. 1021-1024.

a Roadmap for Sustainable Development

Jeshua D. Gonzalez, Chemical Engineering, Lafayette College, Box 7799 Farinon Center, Easton, PA 18042

Text Not Available.

APPLYING NEW TECHNOLOGIES TO THE CLASSROOM – WHAT HAVE WE LEARNED FROM PAST EXPERIENCE

Mordechai Shacham, Ben-Gurion University, Beer-Sheva, Israel

Introduction

Application of new technologies in the classroom in the modern era has probably started with the introduction of the concept of Computer-Based Instruction in the nineteen seventies (see, for example, Shacham and Cutlip, 1981), where interactive computer terminals with touch sensitive screens were used to enhance the learning process. Since those early days considerable amount of experience has been accumulated in successful and less successful applications of new technologies in the classroom.

A well knows success story is the Process Control laboratory where "virtual" experiments have replaced some of the hands-on experiments. The use of virtual (simulated) experiments can reduce considerably the cost of a laboratory course, increase the number of experiments included and enable carrying out experiments that otherwise would require working with dangerous materials and/or working in dangerous conditions. In the nineteen eighties the Advanced Control System (ACS) of IBM was used in several Universities for carrying out simulated laboratory experiments (Corripio, 1987). In the nineteen nineties PC based simulation programs have replaced the main frame based ACS program (Cooper, 1993). Nowadays, the virtual laboratory experiments are widely used in the process control laboratory and there are attempts to introduce them into additional laboratories, such as the Unit Operations laboratory (Wiesner and Lan, 2003).

There are also indications of success of the application of new technologies to enhance "distance learning". Distance learning is directed mainly toward a target audience of employed professionals. The use of distance education in Bachelor engineering studies has been very limited so far, however there are several successful distance learning based Masters' programs (Bourne et al. 2005). Distance learning has been also successfully utilized for delivering various courses to practicing engineers in programs called "Continuing Education" and "Lifelong Learning."

In courses that teach technical skills, such as the use of various computer-based tools for problem solving, preparation of documentations and presentations, and even computer programming, there is considerable amount of evidence to show that in such courses classroom instruction can be successfully replaced by e-learning even in on-campus study environments. The textbook by Seider et al. (2004), for example, contains a CD-ROM that provides multi-media instructional materials to teach the usage of the process simulation programs ASPEN PLUS and HYSYS. These software packages are studied only in a self-paced manner, without any classroom instruction.

In contrast to these success stories there is also evidence of failures in the use of instructional technology. Felder and Brent (2005), for example, caution in their article "Death by PowerPoint" against the excessive use of pre-prepared PowerPoint visuals for teaching. The undesired effects of such extensive use may include the following: 1) the attendance in the class session may drop as students have access to copies of the presentation in the course web site; 2) the class may become mind-numbing for students who have studied the material, and the pace may be too fast for students who did not; 3) the students may not appreciate the knowledge of the instructor as he/she only "repeats what is already written on the slides".

Roskowski et al. (2002) report on student resistance to use an instructional and computational software package that had no particular assignment attached to it. To overcome the initial resistance, the package was introduced into homework problems and tests. After the students learned to use the new package they found it beneficial to continue using it in assignments where the use was not requested explicitly. Thus, students may often disregard instructional material if it does not seem to affect the final grade.

Incorporating new technologies into a course may require tremendous amount of effort while taking the risk of failure to reach the desired goals. However, successful incorporation of new technologies is often an evolutionary process. The initial development of the instructional material should be followed by application-feedback-revision cycles, until the results are satisfactory. In the following sections our experience in fifteen years of evolutional refinement and use of Computer Based Exams (CBE) in the Chemical Engineering Department of the Ben-Gurion University of the Negev are briefly described.

Use of CBE in a Self Paced, Mastery Oriented Framework

The development of the early versions of CBEs was motivated by the introduction of the "Keller Plan" of Personalized System of Instruction (PSI, Keller, 1968). The key concepts of the PSI are the following:

1. *Individually Paced*, divided into 15 to 20 units and detailed study guides are provided
2. *Mastery Oriented*
3. *No Punishment for Failure*
4. *Proctors Deliver Course*: repeated testing and immediate scoring
5. *Stress upon Written Materials*: lectures are optional

CBE can be a preferable choice for self paced learning with requirement for repeated testing and immediate scoring. Indeed CBE has been widely used in disciplines such as social sciences and medical studies. The exams given in those disciplines are fairly simple; contain multiple choice and true/false type questions. Such type of questions can be used in exams where the students are tested for knowledge, recollection and understanding of the course material. In engineering the student is requested to do more than that. He must also demonstrate ability to use the knowledge for problem solving. Thus the CBE in engineering must include problem solving.

Following this premise CBEs were developed in the Chemical Engineering Department in the early 1990, in several courses ("Introduction to Personal Computers", "Modelling and Simulation of Chemical Processes", "Optimization" and "Fluid Mechanics"). Detailed descriptions of some CBEs are provided by Shacham (1998). A brief discussion of the subject is included here.

According to the principles of the mastery learning most of the material learned must be included in the exams. To ensure inclusion of as much material as possible five to six mastery oriented quizzes and an evaluative final exam were given during the semester.

The key components for the CBE were the exam grading programs. These programs were typically Visual Basic programs that checked the credentials of the student, carried out the computations required for the solution of a particular problem, presented questions related to the problem being solved, provided feedback on the answers and, finally, recorded the grade which was obtained based on the fraction of the questions that were answered correctly. The tests were typically delivered in a classroom with networked PC's. Numerical software packages such as POLYMATH[1], Excel[2], or MATLAB[3] were used to solve the exam questions. All the questions required some numerical data that could be varied so that to yield different solutions. Every student received three different sets of data for his exam. At first he had to solve the problem using the first set of the data. The solution files were saved and some of the numerical results were copied into the exam sheet. Next, the exam's grading program was opened and questions related to the solved problems are presented. For a correct answer the feedback "correct" was provided. For an incorrect answer the student was given one chance to correct his answer. This was done in order to allow him to correct technical mistakes, such as those originating from typing errors, accidental key-press or insufficient significant digits in the answer. If the second answer was also wrong, the correct answer was shown and points were deducted from the grade.

If all the answers were correct, the student had to answer only a few key questions. If not, more questions related to the intermediate stages of the solution are presented. This, together with the presentation of the correct answer, allowed the student to find out where the mistake was made. If there was enough time left, he could correct his mistakes, resolve the problem (using a different set of data) and improve his grade. After finishing the exam the student had to turn in the entire set of Polymath, Excel or MATLAB solution files and the exam form, where his particular data and key results are written. The solution files, that contain a complete documentation of the solution process, could be used to verify that the grade given by the computer represents correctly the student's achievements in the exam, or additional reviews of the solution file were required.

Practical stumbling blocks in the implementation of the PSI

In spite of the proven pedagogical advantages of the "Keller Plan" (Keller 1968), its practical implementation using CBE run into several stumbling blocks. The most serious difficulties were the following.

"Self Pacing" dictates that the students can take quizzes, test or exams whenever they feel ready. What actually happened is that the best students took the tests first and they disclosed the information regarding the questions presented and the correct method of solution to the rest of the students. Thus, most of the students did not have to learn all the course material but only the parts which were actually covered in the quizzes and exams.

Mastery learning requires covering in quizzes or exams as much of the course material as possible, so that many quizzes should be scheduled during the semester. The supervision in the rooms where the exams or quizzes are carried out is expensive; therefore, the University Administration limited the number of quizzes that could be given. Furthermore, mastery learning enables most of the students to obtain grades which are significantly higher than in regular courses. Grades that do not reflect the ranking of the students according to their abilities are unacceptable by the administration, either.

Because of these practical considerations it was decided to abandon self pacing and mastery learning. Unfortunately, when CBEs (of the form that was described in the previous section) are used not in a mastery learning mode, they are not easier to grade than the traditional exams. The grading program will detect the first incorrect numerical answer, but most probably all the rest of the answers down the solution path, from this point on, will be

---

[1] POLYMATH is copyrighted by M. Shacham, M. B. Cutlip and M. Elly, http://www.polymath-software.com/

[2] Excel is trademark of Microsoft Corporation, http://www.microsoft.com

[3] MATLAB is trademark of The Math Works, Inc., http://www.mathworks.com

judged incorrect. Thus, the program itself cannot detect how many errors were actually made, and manual grading is required. With manual grading the extra effort in preparation of CBEs is not justified and a different approach must be developed.

The evolution of CBE in Chemical Engineering Courses

In light of the difficulties described in the previous section it was essential to modify the requirements in the courses where CBE's were used. All the quizzes were converted to Internet based homework assignments. Details of such assignments (self tests) are available in the publication Shacham (2005) and also on the website: http://echem.bgu.ac.il/staff/Mordechai%20Shacham/introduction.htm. The benefits of the homework assignments are the same as of the CBEs provided that the students complete all the assignments and also that they work them out independently on their own. To ensure that all the assignments are completed their grades are included to the extent of 10 % in the course grade. To ensure that the students solve the problems on their own, they are requested to use a software package that requires programming, (such as MATLAB) to reach the solution. In the midterm and final exams the students are requested to use the same problem solving tool. We have found that a student, who does not solve the homework assignment on his own, does not acquire enough proficiency in the use of the problem solving tool and have no chance to pass the exams. The importance of working out the homework assignments on their own, is explained to the students in the beginning of the course.

A Currently Used CBE in Midterm and Final Exams

The midterm and the final exams of the course are held in a computer laboratory, under supervision, where the students have to present their identity card to take the exam. The duration of the midterm exam is two hours and the final exam is three hours.

A typical midterm exam which was recently given in a "Mathematical Modeling and Numerical Methods" course is presented in Appendix A. The exam questions are based on problem 12.3 in the book of Cutlip and Shacham (2008). There are two questions in the exam. The first one involves the calculation of the Wilson equation coefficients for a binary system which includes ethyl alcohol and a another randomly assigned organic compound. The Wilson equation represents activity coefficients for non-ideal systems and in this question the students should use azeotropic point data to calculate the coefficients. This requires the solution of a system of two nonlinear algebraic equations. The students should specify the mathematical model of the problem, use MATLABs' symbolic manipulation capabilities to derive the partial derivatives of the functions and solve the problem iteratively using the Newton-Raphson method. All the steps of the solution are implemented in MATLAB programs. The second question involves the calculation of the dew point temperature for the same non-ideal binary system that was used in question 1. The method of solution is similar to the solution of question 1, except that in this case there are three simultaneous nonlinear algebraic equations and the partial derivatives are calculated using finite differences.

After finishing the exam the students turn in the exam form, where their individual data are specified and all the MATLAB files that were used for the solution. The MATLAB programs provide clear and precise documentation of all the solution steps. Thus, the programs are the best means to assess the knowledge level of the student and to grade the exam. No self grading is used now in the midterm and final exams.

Conclusions

Using CBE as an example it has been demonstrated that application of new technologies to the classroom is a lengthy, evolutionary process. At first, educational theory can be used to determine the best way of application of the new technologies; however cycles of class test-feedback-revision are often needed in order to reach the most beneficial combination.

In the case of CBE the initial intention was to use it in order to enable self pacing and mastery learning in chemical engineering courses. Class tests have shown that traditional universities (based mainly on classroom learning) are not ready yet to accept these forms of learning. Evolutionary change of the concepts, over a period of about ten years, enabled reaching a satisfactory combination where computer grading and immediate feedback are used to enhance the effectiveness of the homework assignments. The presentation of the solution in a form of a computer program (in the midterm and final exams) ensures complete understanding of the course material.

References

1. Bourne, J., Harris, D. and Mayadas, A.," Online Engineering Education: Learning Anywhere, Anytime", *Journal of Engineering Education*, Vol. 94. No. 1, 2005, p. 131

2. Cooper, D. J., "Picles: A Simulator for Teaching the Real World of Process Control," *Chemical Engineering Education*, Vol. 27, 1993, p. 176.

3. Corripio, A. B., "Automatic Process Control: Using IBM's Advanced Control System(ACS)," *Academic Computing*, Vol 1, No. 2, 1987, pp. 30-31, 60-62.

4. Cutlip, M. B. and Shacham, M., *Problem Solving In Chemical and Biochemical Engineering with Polymath, Excel and MATLAB*, Prentice-Hall, Upper Saddle River, New-Jersey, 2008.

5. Felder, R. M. and R. Brent, "Death by PowerPoint", *Chemical Engineering Education*, Vol. 39, No. 1, 2005, pp. 28-29.

6. Keller, F. S., "Good-bye Teacher", Journal of Applied Behavioral Analysis, Vol. 1, 1968, p. 78

7. Roskowski, A. M., Felder, R. M. and Bullard, L.G., "Student Use (and Non-Use) of Instructional Software", *Journal of SMET Education*, Vol 2, 2002, pp. 41-45

8. Seider, W. D., Seader,J. D. and Lewin, D. R., *Product and Process Design Principles: Synthesis, Analyis, and Evaluation*, 2nd Ed., Wiley, New York, 2004.

9. Shacham, M., and Cutlip, M.B., "Computer-based Instruction: Is There a Future in ChE Education?" *Chemical Engineering Education*, Vol. 15, No. 2, 1981, p. 78.

10. Shacham, M., "Computer Based Exams in Undergraduate Engineering Courses", *Computer Applications in Engineering Education*, Vol. 6, No. 3, 1998, pp. 201-209

11. Shacham, M., "An Introductory Course of Modeling and Computation for Chemical Engineers", *Computer Applications in Engineering Education*, Vol. 13, 2005, pp. 137 – 145

12. Wiesner, T. F. and Lan, W., "Comparison of Student Learning in Physical and Simulated Unit Operations Experiments, "*Journal of Engineering Education*, 2003, pp. 195-204

Appendix A

Problem Statement for the Midterm Exam of the Course: "Mathematical Modeling and Numerical Methods in Chemical Engineering"[†]

The Wilson[*]1 equations are used to correlate the activity coefficients of strongly non-ideal, but miscible systems. The Wilson equations for activity coefficients in binary system are:

$$\ln\gamma_1 = -\ln(\theta_1) + \frac{x_2(G_{12}\theta_2 - G_{21}\theta_1)}{\theta_1\theta_2}$$
$$\ln\gamma_2 = -\ln(\theta_2) - \frac{x_1(G_{12}\theta_2 - G_{21}\theta_1)}{\theta_1\theta_2} \quad \text{(A-1)}$$

where $\theta_1 = x_1 + x_2 G_{12}$ and $\theta_2 = x_2 + x_1 G_{21}$, G_{12} and G_{21} are adjustable parameters which must be determined from experimental data, x_1 and x_2 are mole fractions of components 1 and 2 respectively.

The vapor liquid equilibrium ratios can be calculated assuming that the modified Raoult's law apply, thus $k_i = \gamma_i P_i / P$, where γ_i is the activity coefficient of component *i* in the liquid phase, P_i is the vapor pressure of component *i*, and *P* is the total pressure.
Vapor pressures at various temperatures can be obtained from the Antoine equation

$$\log P_i = A_i - \frac{B_i}{C_i + T} \quad \text{(A-2)}$$

where *T* is the temperature in °C , P_i is the vapor pressure of component *i* in mmHg and A_i, B_i and C_i are the Antoine equation constants given in Table A-1.

Azeotropes represent a condition where vapor and liquid phases have identical composition, meaning that all the vapor liquid equilibrium coefficients: $k_i = \gamma_i P_i / P = 1$, at the temperature, pressure and composition of the azeotrope. Thus at the temperature of the azeotrope $\gamma_i = P / P_i$. This allows the coefficients of the Wilson equations to be calculated from one set of azeotropic data.

1. For the binary system ethyl alcohol (No.1) and ___________ (No. 2) calculate the Wilson equation coefficients using the binary azeotrope data in Table A-2. For the solution rewrite equations (A-1) in the form of: $f_1(G_{12}, G_{21}) = 0;\ f_2(G_{12}, G_{21}) = 0$. Derive the expressions for the partial derivatives of the functions and solve this system of equations using the Newton-Raphson method. Start from the initial guess: G_{21}=G_{12}=0.1. Stop the iterations when $\|f\| < 10^{-5}$ or the number of iterations ≥20. Solve with MATLAB.
2. For the same binary system of Question 1, calculate the dew point temperature (T_d) and liquid composition at the dew point (x_1, x_2) for the case where the ethyl alcohol mole fraction in the vapor phase is y_1 = 0.4. For calculation of activity coefficients use the Wilson equations with the coefficients calculated in question 1. Solve by the Newton-Raphson method using finite difference approximations for the partial derivatives. Start from the initial

[†]Based on Problem 12.3 in Cutlip and Shacham (2008)
[*]Wilson, G. M. J. Amer. Chem. Soc. 1964, 86, 127-130.

guess: T_D = 70 °C, x_1 = 0.5. Stop the iterations when $\|f\|<10^{-5}$ or the number of iterations ≥20. Solve with MATLAB.

Table A-1 Antoine Equation Constants for Vapor Pressure of Various Substances‡

Name	Formula	*A*	*B*	*C*
Ethyl Alcohol	C_2H_6O	8.04494	1554.3	222.65
Methyl acetate	$C_3H_6O_2$	7.20211	1232.83	228.0
Methyl propionate	$C_4H_8O_2$	7.12841	1257.14	216.4
Ethyl propionate	$C_5H_{10}O_2$	7.07293	1298.30	210.7
n-Propyl formate	$C_4H_8O_2$	7.04006	1235.00	216.1
n-Propyl acetate	$C_5H_{10}O_2$	7.06665	1304.10	210.0
Benzene	C_6H_6	6.90565	1211.033	220.79
Toluene	C_7H_8	6.95464	1344.80	219.482
n-Pentane	C_5H_{12}	6.85221	1064.63	232.00
n-Hexane	C_6H_{14}	6.87776	1171.530	224.366
n-Heptane	C_7H_{16}	6.90240	1268.115	216.900
n-Octane	C_8H_{18}	6.92374	1355.126	209.517
Carbon Teterachloride	CCl_4	6.9339	1242.43	230.0

Table A-2 Binary Azeotropes Containing Ethyl Alcohol‡

Ethyl Alcohol (B.P. 78.3 °C)	B.P. @ 760 mm (°C)		% by weight	
	Other component	Azeotrope	Alcohol	Other component
Methyl acetate	57.0	56.9	3	97
Methyl propionate	79.7	72.0	33	67
Ethyl propionate	99.2	78	75	25
n-Propyl formate	80.8	71.8	38	62
n-Propyl acetate	101.6	78.2	85	15
Benzene	80.2	68.2	32.4	67.6
Toluene	110.8	76.7	68	32
n-Pentane	36.2	34.3	5	95
n-Hexane	68.9	58.7	21	79
n-Heptane	98.5	70.9	49	51
n-Octane	125.6	77.0	78	22
Carbon teterachloride	76.8	65.1	15.8	84.2

‡Dean. A. (Ed.), *Lange's Handbook of Chemistry*, McGraw Hill: New York, 1973

Concurrently Collaborative Spreadsheets in the Chemical Engineering Classroom

David L. Silverstein, Ph.D., P.E.
PJC Engineering Professor and Associate Professor of Chemical & Materials Engineering
University of Kentucky
P.O. Box 7380
4810 Alben Barkley Drive
Paducah, KY 42002-7380

Abstract

A project investigating the viability of a concurrently collaborative online spreadsheet to improve the effectiveness of student teams when solving chemical engineering problems is described. Students in two classes representing sophomores and seniors were assigned a problem to be solved using a spreadsheet on Google Docs, an online browser-based suite of productivity applications. The unique feature of this spreadsheet is that multiple users on multiple machines can edit the same spreadsheet simultaneously, with changes appearing on all users screen within about one second. Assessment was performed to determine whether use of this spreadsheet was technically viable, suitable for students not in the same room, and useful for improving the effectiveness and efficiency of students working in teams. From a limited sample size assessment, the Google spreadsheet does appear to be viable, to allow effective communication amongst participants, and to contribute to a more efficient and effective team problem-solving experience.

Keywords: spreadsheets, computing, collaboration, teamwork

Introduction

To prepare students for practice in the modern industrial world, graduates of chemical engineering degree programs are expected to function effectively in teams. At the same time, problem-solving skills are a focus, typically involving some computation. Often, these computations are completed with some computer software, with the most common type of software package used being a spreadsheet.

Collaboration on solving computational problems involving software has typically followed one of two models: participants gathered around a single computer with one individual interacting with the software; or a single computational file shared amongst multiple users, either from a common storage location or revision sharing via e-mail or other file transfer method. Neither method is efficient due to the need for reconciliation amongst edited versions or a limit of one concurrent editing session.

A new spreadsheet software, currently in beta status, is available from Google as part of its free Google Docs service. Google Docs is a web-browser based collection of office software (word processor, spreadsheet, and presentations) which is not operating system dependent, using Java to provide a rich user interface. The spreadsheet contains the key inline functions required for most chemical engineering problems, though it does lack other capabilities engineers frequently use in Microsoft Excel, such as Goal Seek, Solver, and advanced graphing functionality. Perhaps the most interesting feature of Google Docs is the ability to share a single online document amongst multiple users, and when configured appropriately, to enable simultaneous editing by multiple users.

Students in two chemical engineering courses were each assigned a different group problem for which they were expected to use the Google spreadsheet to solve. To prepare them for this process, they were given basic instruction in group problem solving, focusing on planning a solution and task distribution. Students then were placed in separate rooms and asked to create a spreadsheet solution using the online spreadsheet, using the built-in messaging software to communicate with teammates as needed. Students were observed by the instructor developing their solution, and the session was recorded from the perspective of an editor. The assessments of both the students and the instructor regarding the use of a concurrently edited spreadsheet will be presented, along with an overview of alternative approaches to collaborative computations using spreadsheets.

This paper was previously presented at the 2008 Annual Meeting of the American Society for Engineering Education.[1]

Background

The need to collaborate when using calculations to solve engineering problems is not new. Typically, collaborative problem solving in engineering courses means one of two things: students prepare a solution by huddling around a single computer while the person at the keyboard does most of the problem-solving; or students prepare portions of a solution and then gather to try to reconcile those contributions either in person or by sharing multiple computer files.

Many vendors of computational software have been working to address this need for teamwork by enhancements to existing software. Most attempts to add collaborative capability have followed one of two models.

- Library model. A single file is maintained with all calculations and associated documentation. When an individual accesses the document for editing from a central database, it is marked as unavailable. Multiple users can have simultaneous read-only access.
- Revision marking model. The software maintains an original document while incorporating changes the document into the file. An editor later has the option of accepting each change made by each author.

Each method has strengths and weaknesses. The library model is restricted to a single concurrent user. If a user fails to "return" the "borrowed" document, perhaps by failing to close the document, it is not available for editing without administrative action. The process of teamwork via this model forces a significant lag in the incorporation of ideas, since the review process is linear (write→review→revise→review→revise…) and often depends on the editing of a lead participant. The approach has the advantages of security (no changes will be lost under normal circumstances) and rapid accessibility of edited documents. It requires a server hosting the document library and managing access.

The revision marking approach is most common because it is decentralized (does not require a server host). Using this approach, changes are added to the document without deleting any information. Changes are coded to correspond to an individual editor. At some point, an individual may go through the document reviewing, accepting, and rejecting changes to the document. The final result may be the consolidation of the contributions of many individuals, but requires significant management by a single editor before the document is finalized. One key advantage is that all team members may edit the document simultaneously. The key disadvantage is that most team members will never see the suggestions of those whose contributions are rejected by an editor before they see a consolidated file. Another disadvantage to this approach it that at any given time there are multiple versions of a document file in existence, making it difficult to track which is the most current version.

Some approaches combine these methods, using revision marking on files managed by a central server.

This review focuses on the current state of collaborative functionality offered by companies offering computational software used by chemical engineers.

Desktop Applications

Microsoft. Microsoft offers several mechanisms to facilitate collaboration in its ubiquitous spreadsheet, Excel. It allows users to "Track Changes" (the revision marking approach) and add comments to facilitate contributions from multiple users. Additionally, Microsoft offers a server product known as SharePoint, implementing the library model. In addition to document management, it also provides services to enhance communication amongst users including "Wiki" style document writing, "blogs", and persistent discussion forums. Some versions of Office (including the Enterprise version available through the campus licensing agreements at many universities) include Groove, a client-based program which offers some similar functionality to SharePoint but decentralized with less administrative effort. Finally, Microsoft has recently introduced its Live Office suite, which is a collection of web-browser based applications which mimic members of its office suite. Its spreadsheet equivalent, however, is currently not intended as a calculation tool.[2]

MathWorks. MATLAB, one of the most common math packages in use in chemical engineering departments, does not offer integrated collaborative functionality. However, since it is modeled after traditional development software, it does interface with industry standard source control software provided from other vendors. This is a library management approach with some revision tracking handled by the server.[3]

PTC. Mathcad offers user the capability to share worksheets including an edit-protected mode of "live" worksheets, but does not allow multiple users to edit a file in any non-trivial mode (other than providing a copy worksheet file to another user).[4]

Wolfram. Mathematica does not offer integrated collaborative functionality. The company offers a companion product, Wolfram Workbench, that serves as an integrated development environment (IDE) allowing multiple users to work on a development project. Individual files in that workspace may be edited by only a single user at a time.[5]

Maplesoft. There are no advertised collaborative features in Maple.[6]

OpenOffice.org. OpenOffice is a collection of open source projects oriented toward competing with the functionality of Microsoft Office. The Calc spreadsheet module does not offer native collaborative capabilities.[7]

Internet Based Options

Online Storage. There are multiple options available for engineers to use desktop applications on their desktop and store those files online for broader access. This includes services such as Xdrive (www.xdrive.com), Windows Live Folders (skydrive.live.com), Basecamp (www.basecamphq.com), WebOffice (http://www.weboffice.com), and Central Desktop (www.centraldesktop.com).[8] Some of these resources offer collaborative features using the library model, but the primary emphasis is that files are available from any location a user has network access.

Google. Google Docs (docs.google.com) offers a suite of web-browser based productivity applications (spreadsheet, word processor, presentations) that offer many of the standard capabilities of spreadsheets. For engineering purposes, it contains basic graphing capabilities, all standard functions and calculation capabilities. It does lack some features of particular use to engineers, including a "goal-seek" capability and circular (iterative) calculations. Another downside is a limit of 1MB for a single spreadsheet file. Documents may be imported from and exported to desktop applications, including Excel. The distinguishing feature of the offering is the ability to not only share documents with other users, but for all of those users to simultaneously edit the document with all changes appearing on the spreadsheet in nearly real-time. The application uses subtle outlines and color changes to indicate a cell is currently locked for editing by another user. Google currently does not charge individual users for access to the application.

Zoho. Google is not the only company to offer an online collaborative spreadsheet. Zoho (www.zoho.com) offers similar capabilities in what is arguably a more attractive package. A significant functionality recently added to Zoho is the ability to incorporate macro programs written in Microsoft's Visual Basic for Applications syntax. Zoho is also free to individual users. The greater likelihood of a student already having a Google account led to the decision to use Google Docs in this project.

Objectives

This project was intended to investigate the practicality and effectiveness of using an online collaborative spreadsheets for small groups of chemical engineering students to solve problems.

In particular, we examined the following questions:

- Are online spreadsheets adequate for solving problems not requiring advanced spreadsheet capabilities?
- Are the communication capabilities of the online spreadsheet application sufficient for students unable to speak to one another to complete the solution?
- Does the requirement for network connectivity and use of a browser-based application significant impact the usability of a spreadsheet?
- Does the collaborative nature of the online spreadsheet contribute to training students to function effectively as a team?

Methods

To address these questions, students in two chemical engineering classes at the University of Kentucky Extended Campus in Paducah were each assigned problems to be solved as a team. One course consisted of a group of 5 sophomores in a material and energy balances class, and the other three students in a senior level engineering economy course. These sample sizes represent the total enrollment in this program at those levels. Students were given pre- and post-assessment surveys, with selected questions common to both surveys. The post-project survey included free-answer questions to illuminate student perceptions of their experience. Just prior to students being released to complete their assignments, they were given a 15-minute lesson on how to function as a team. Prior team training varied by student as indicated in the results. Students were observed moving to different locations in the engineering facility and were assumed to have followed the assignment requirement to not communicate outside of the application's instant messaging function.

The students in the sophomore course were given a problem requiring completion of a spreadsheet to calculate the compressibility of a mixture using the Peng-Robinson equation of state. Since registration is required to use Google Docs, students registered for the site a week in advance and added their name to a shared spreadsheet to confirm they had access to the spreadsheet. Students were to complete selected portions of the spreadsheet based on equations provided on the assignment sheet. The spreadsheet was color-coded to indicate cells the students should edit, and those cells representing inputs to the problem. A brief explanation of how to use the spreadsheet was given in the context of the spreadsheet they were going to edit, and the correspondence of the equations on the assignment sheet to the spreadsheet was explained. Finally, students were instructed to use computers in multiple locations in the building and not to speak with each other, relying solely on the instant messaging system included with the spreadsheet to communicate. The chat traffic was consistent with individuals unable to otherwise communicate. The instructor was also logged in as a user and recorded portions of the solution process. A screen capture of the sheet in progress is presented as Figure 1. The sophomores spent about an hour on the problem.

The seniors in the engineering economy course were expected to develop a spreadsheet to enable a user to compare the total costs of living for purchasing a home and for renting a home. No template was used for this assignment. Following a brief explanation of the project requirements as given on the assignment sheet, students were left to discuss their plan briefly before heading to computers in different locations as before. Chat traffic was again consistent with a group of students not able to speak with one another. A recording was also made of their solution process, although it was more complicated because this class (appropriately) used multiple sheets in the workbook to isolate inputs and outputs from computations. A screen capture of the student's efforts near completion is given in Figure 2. The seniors spent about 45 minutes completing the project.

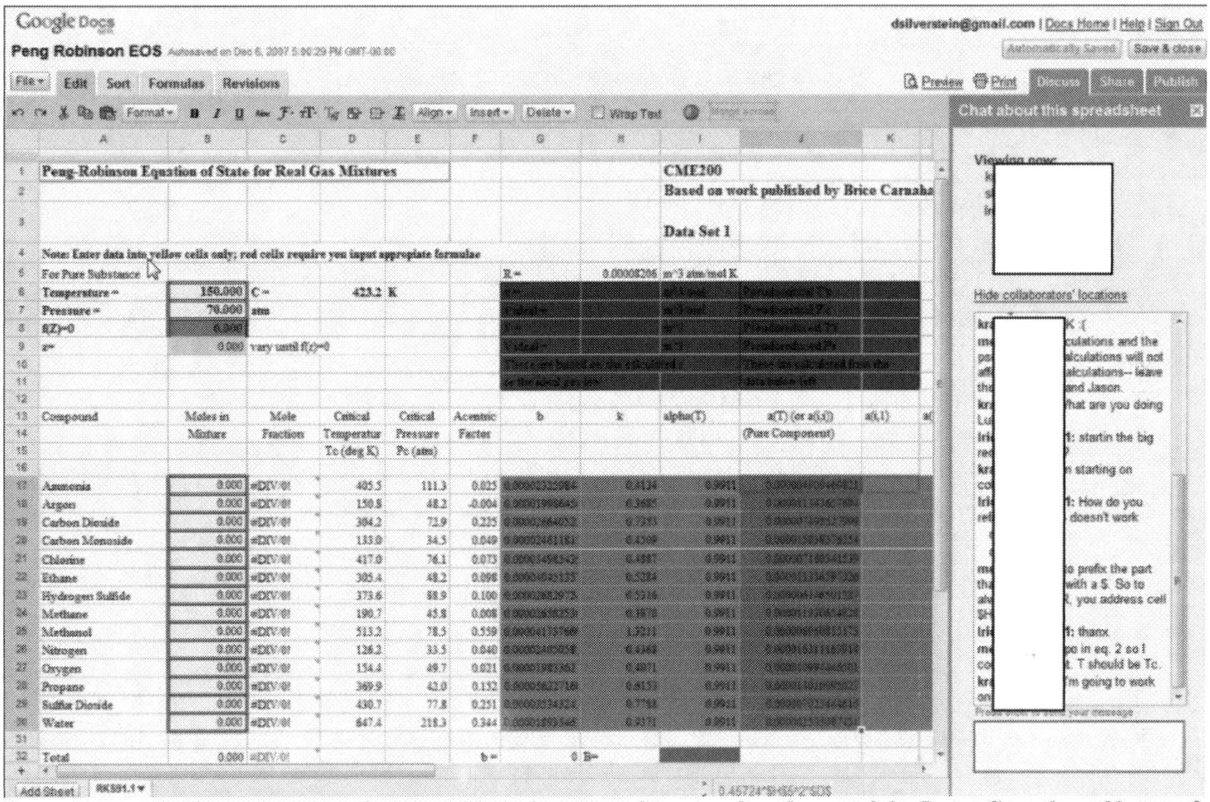

Figure 1. Peng-Robinson equation of state problem given to sophomores, based on work by Bryce Carnahan. Names of particpants have been obscured.

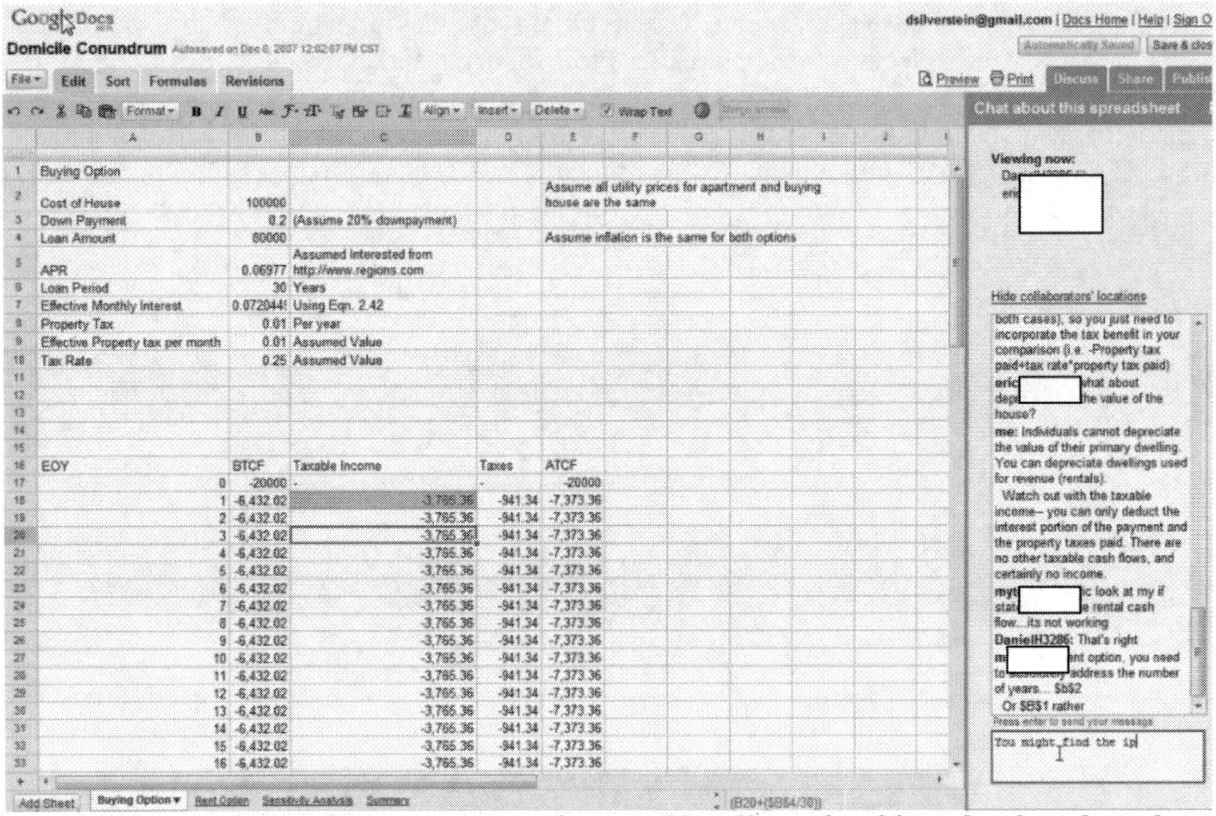

Figure 2. Seniors completing their rent vs. own comparison spreadsheet. Names of participants have been obscured.

Results and Discussion

In both cases, students were able to complete the solutions of their problems on a timely basis. Suggestions were infrequently offered by the instructor via instant messaging to limit the time spent on this project.

All surveys asked students to respond to questions and answer them using a 5-point Likert scale (1=Strongly Disagree, 5= Strongly Agree). Sample size for sophomores was 5 on the pre-assessment and 4 on the post-assessment. For the seniors, the sample size was 3 on both assessments. Standard deviations are not given due to the small sample size.

The first group of questions that students were asked were used to establish the prior experience of the students as presented in Table 1. Students indicated they had not previously used Google spreadsheets, but almost all had experience with instant messaging. More advanced students were far more likely to have had teamwork training, which is a component of our curriculum.

	Sophomores	Seniors	Combined
Used Google spreadsheets previously	1.0	1.7	1.25
Used instant messaging previously	3.0	5.0	3.75
Had prior teamwork training	2.8	4.7	3.5

Table 1. Results from questions regarding prior experience asked prior to the project.

Students were satisfied with the capabilities of the online spreadsheet, though some were uncomfortable with the lack of familiar features which might have been useful for the assigned problems. Table 2 shows the results from relevant questions. In particular, the sophomore class problem would have been solved more elegantly with a goal-seek function. Perceptions changed notably before and after using the spreadsheet. Perhaps the most interesting result was that students indicated they were not as likely to use the online spreadsheet for an individual assignment was they were for a team project.

	Sophomores		Seniors		Combined	
	Pre-	Post-	Pre-	Post-	Pre-	Post-
Google spreadsheets (GS) are as easy to use as Excel	1.8	4.3	3.0	2.7	2.3	3.6
GS contain all the functionality I need for engineering problems	1.8	3.0	3.0	4.0	2.3	3.4
We never lost data when developing the spreadsheet		4.5		4.3		4.4
I would use GS for suitable individual problems		2.5		3.0		2.7
I would use GS for suitable team problems		4.8		4.0		4.4

Table 2. Results from questions associated with the usability of Google spreadsheets.

Teamwork was the final topic addressed by student surveys as presented in Table 3. Sophomore students appeared to develop a greater appreciation for the value of planning and organization in team projects, while seniors indicated no change in their perceptions. The spreadsheet itself seemed to facilitate communication amongst team members adequately, both through an effective instant messaging applet and through adequate indication of where other team members were working within the spreadsheet. The collaborative features of the spreadsheets were perceived to have required less time to solve the problem compared to other collaborative methods.

	Sophomores		Seniors		Combined	
	Pre-	Post-	Pre-	Post-	Pre-	Post-
We have planned how to execute this task	1.6		2.7			2.0
We have/had determined how to solve the problem prior to beginning calculation	1.6	3.0	2.7	3.3	2.0	3.1
We planned sufficiently to execute the assigned task		4.5		2.7		3.7
Planning is more important when working on a project simultaneously	2.8	4.8	4.7	4.3	3.5	4.6
It was less time-consuming to work simultaneously on the same spreadsheet		4.5		4.0		4.3
Instant messaging was an effective means of communication		4.5		4.3		4.4
I coud tell what my team members were doing		4.5		4.0		4.3
I worked more than I would have if we had gathered around a single computer		3.5		2.7		3.1
Teamwork instruction was important to the project's timely completion		4.3		4.0		4.1

Table 3. Teamwork assessment questions.

Students were also asked some free-answer questions regarding the spreadsheet application and the project. Students commonly indicated that the collaborative nature of the spreadsheet was appealing, with the problem being solved rapidly with everybody contributing simultaneously. Shortcomings of the spreadsheet primarily involved usability, including lack of common shortcuts (F4 for toggling absolute cell references), poor visibility of cell contents, and lack of prompts for function arguments. The best feature appeared to be the integration of the instant messaging system.

Students indicated that more time to prepare for the project would have been useful, or perhaps setting a time limit to make certain students plan ahead. More general comments included:

> "I really enjoyed it. I really like working in teams."
> "The project was very interesting due to the fact that the spreadsheet can be compiled so quickly."
> "It seemed to fly by once everything started clicking."
> "fun yet challenging"
> "interesting possibilities"
> "That was fun"
> "Best homework all semester"
> "Got to chat with each other"

All students participating received full credit for the assignment, which counted as a homework grade in each course.

The instructor noted that students had greater interest in this assignment than they have had for essentially the same assignment in previous terms. Much of this excitement can likely be attributed to the novelty of the browser-based spreadsheet. Individual contributions could be observed by the instructor, and those students with more developed spreadsheet skills completed a great proportion of the tasks required by the assignment. The spreadsheet was judged by the instructor to be sufficient for problems that do not require iterative calculations or goal-seek functions, and will likely be included in future course offerings.

There is no additional instructional overhead required for an assignment involving this collaborative spreadsheet beyond a brief introduction to Google Docs, nor is there need to monitor student activities in the manner performed for the purposes of this project. Any assignment involving use of a spreadsheet can involve use of this application, provided that the calculations are not subject to the computational limitations of the application (no iterative calculations, Goal Seek, or Solver). The application offers no additional benefit in monitoring for academic integrity, and is potentially more susceptible than Excel since Excel offers internal tracking data which can be extracted by the skilled instructor. Training requirements are the same as for Excel with minor additions for operational details specific to Google Docs.

Conclusions

This project investigated the utility of an online collaborative spreadsheet for small groups of chemical engineering students to solve problems. The software tools were judged adequate by both the students and the instructor. Despite the requirements of a browser-based online spreadsheet, it was responsive and stable on multiple classes of system running different browsers. The calculation capability is adequate for many engineering problems, but will not replace the desktop spreadsheet in the near term. The communication capabilities of the software make the spreadsheet very appealing for group projects, enabling the social element of instant messaging to group discussions. Team training enhanced the experience of collaborative problem-solving for students, and appeared to lead to a better appreciation of the importance of planning to successful group projects.

References

[1] Silverstein, David L., "Using a Concurrently Collaborative Spreadsheet to Improve Teamwork", *Proceedings of the 2008 Annual Meeting of the American Society for Engineering Education*, American Society for Engineering Education, June, 2008.
[2] http://www.microsoft.com, accessed August 2008.
[3] http://www.mathworks.com, accessed August 2008.
[4] http://www.ptc.com/appserver/mkt/products/home.jsp?k=3901, accessed August 2008.
[5] http://www.wolfram.com/, accessed August 2008.
[6] http://www.maplesoft.com/, accessed August 2008.
[7] http://www.openoffice.org/, accessed August 2008.
[8] Null, Christopher, "Your Data Anywhere", *PC World*, p. 110, Sept. 2007.

Wiki Technology as a Design Tool for a Capstone Design Course

Kevin R. Hadley and Kenneth A. Debelak
Department of Chemical Engineering, Vanderbilt University,
Nashville TN 37235-1604

Abstract

Technology is a key component to the net generations social interaction, study skills, and professional practice. It is prudent to investigate the potential benefits of new technology in education. In this study, we incorporated a wiki into a senior design course, and investigated its benefits and downfalls with respect to implementing and learning design. We found it improved organization and interaction amongst the students and between the students and the professor and teaching assistant. We also discovered the potential wikis can contribute towards pedagogical research. For example, a wiki can provide insight into the thought process of a student with respect to how they approach design.

Introduction

Web 2.0 technologies, like wikis and blogs, are becoming an integral part of the net generation's social and academic life[1]. Educators and students, alike, benefit from incorporating these technologies into the classroom[2]. With wiki technology, student interaction, idea collaboration, and organization of information can be improved upon compared to traditional ways of teaching[3].

A wiki is a website where users edit and add content as needed. Different users can add content, and review material added from other users. A wiki provides potential for collaboration within groups.

According to a survey conducted by the Educause Center for Applied Research (ECAR), the use of information technology and web 2.0 technologies is astonishing[1]. Out of the 20,000+ students surveyed, engineers spent more time online (an average of 21.9 hrs/week) than any other discipline. Specifically pertaining to wiki usage, 41.7% of all of those surveyed access or use wikis on a weekly basis. According to the conductors of the study, this number may be understated because the students may not know what a wiki is or realize their internet searches direct them to a wiki site. Another note is the survey does not distinguish between access and contribution. Another part of the survey reported 32.6% of the students liked learning through contribution to wikis and blogs. Again, this number may be skewed due to the ignorance of what constitutes a wiki.

Even though technologies like these are invading all facets of the net generation's life, doesn't mean educators should rush to implement them all into every facet of a class's curriculum. It is up to the teacher to decide the most appropriate use of this technology. The main point to keep in mind when deciding to include new technology (or a new approach in general) is what will benefit the students. According to Oblinger and Oblinger, even though the net generation values what older generations consider new technology, wikis and podcasts, what they value most is interaction[4]. Professors can't replace interaction with technology, but must augment and enhance interaction with technology. Interaction and learning are the keys when bringing something new into your classroom.

Of course, it is hard to know whether or not something new will enhance interaction or learning, which is why we conducted this study. We wanted to understand how to introduce wikis to students and what their value was as a design tool. In this article, we present a description of wikis, the details of our study, what we learned from the study, and suggestions for further wiki use in the engineering classroom.

Wikis and their Features

Wiki is a Hawaiian word for quick, but in the context of this study it is a type of website any user can view and edit like any word processor without the knowledge of html or similar programming languages. A nice illustration of wikis and their possibilities can be found online: http://www.youtube.com/watch?v=-dnL00TdmLY[5]. There are many wiki hosts on the web, but we chose to use www.pbwiki.com, because it was very user-friendly and it was free.

Wikis have a number of features appealing to engineering and engineering education. First, there is a complete revision history for each page in the wiki. From the student's perspective, they can go back to any version and not only see what has been changed, but they have access to the version in order to make it the current version if mistakes were made. From an educator's perspective, a professor can track the progression of the project through the student's eyes. Also, the teacher can observe specific changes between versions.

Every saved change to a page can be tracked to a specific user and that change is time-stamped. This means a professor can really enforce accountability with respect to each team member. If Jean and Tom say Billy isn't working, the professor can go to the wiki and confirm Billy made four minor additions to the wiki. In addition, a professor has verification if a group is being lazy or if they are procrastinating.

A final feature is the ability to add comments. If a professor doesn't feel comfortable editing a student's work, they can leave a comment on a page of interest. Instead of a meeting at key points in the semester, a professor can go to the wiki and look at how things are going. If something of concern exists or if the students have questions or concerns, the professor can address those concerns or notify the group of his/her concerns. The ability to do this ties directly into the goal of integrating technology into the classroom, while promoting interaction between students and the faculty.

Senior Design Project and Wiki Study Details

The participants of our study were the seniors enrolled in our Chemical Engineering Process Design course at Vanderbilt University. Their final project was the 2008 AIChE National Student Design Competition project. In short, the project was to design a process to convert coal into methanol and perform a complete economic analysis.

Ten groups of three students were given 30 days to complete their design and present their results to the rest of the class. At the start of the project, the students were introduced to the use and potential of wikis, and were encouraged (although not required) to add content to the wiki as part of the project. Each week, weekly reports were received, some from the wiki, and the group's progress was discussed in a weekly meeting. In addition, throughout the project, the authors of this article would monitor the wiki of each group and add comments/questions/concerns when necessary and reply to comments/questions/concerns expressed by the group within their wiki.

At the end of the project, 25 out of the 30 students were given a survey asking about their experience with the wiki, their opinions of the wiki's use, and their suggestions for further use. Within the survey, there were three sections: positive and negative statements requiring a numerical allocation from the student on a 6-point Likert scale, a list asking which project items were included in the wiki, and open-ended questions about the good and bad points of the wiki and its implementation.

Survey Results

In the first section of the survey, there were twelve statements, and the students were asked to circle a number between 1 and 6 to describe how much they agreed with the statement where a choice of 1 indicated strong disagreement and a choice of 6 indicated strong agreement. To prevent a neutral response, an even number was chosen for the maximum. Also, a mix of positive and negative responses was included to ensure valid results from those surveyed.

The scores were analyzed and (for the most part) students liked the use of wikis in the design course. One group didn't add any content to their wiki in any form, and their responses were negative with respect to the wiki. Their open-ended comments were helpful, but their responses were excluded from the numerical analysis of the survey. We decided the absence of their participation didn't qualify them for a valid opinion about the implementation and general opinion of the wiki.

The first step in evaluating wiki use in a design course was taking the average score for each statement. If the score was greater than 4.0 or less than 3.0, we considered that score to have a significant positive or negative agreement. The statements with values between 3 and 4 were regarded as a neutral response. Table 1 lists the average score for each statement and you can see, the students agreed with six statements:

- (I.) They will tell others about wiki technology for collaboration
- (II.) They would like to use a wiki in their future career
- (III.) They recommended use of the wiki for other senior design courses
- (V.) They used the wiki only because it was required
- (VI.) The wiki helped organize their work and findings
- (XII.) There was more interaction from the professor and the TA in this project than others in the past.

In addition, the students disagreed with three out of the twelve statements:

- (IV.) Adding to the wiki took more time than it saved
- (IX.) The wiki overcomplicated the project
- (XI.) The wiki was confusing, and it made the project more difficult.

There was not a definitive agreement as to whether the wiki was a key component to finishing the project (VIII.), if the wiki helped them finish the project more efficiently (VII.), or if the student had better understanding of their team member's progress because of the contributions to the wiki (X.).

Table 1. Average Score for Survey Statements

I.	I will tell others about wiki technology for collaboration	4.18
II.	I would like to use a wiki in my future career	4.02
III.	I believe the wiki should be implemented in next year's senior design course	4.52
IV.	Adding to the wiki took more time than it saved	2.91
V.	If it wasn't required, I wouldn't have used the wiki	4.32
VI.	The wiki helped organize our work and our findings	4.00
VII.	This design project was finished more efficiently than other school projects	3.59
VIII.	The wiki was a key component to finishing this project quickly and thoroughly	3.18
IX.	The wiki overcomplicated the project	2.80
X.	I had a better understanding of my team member's progress because of their individual contributions to the wiki	3.36
XI.	The wiki was confusing, and it made the project more difficult	2.25
XII.	Dr. Debelak's and Kevin's interaction/involvement in this project was more productive to my progress than other professor/TA's involvement in the past.	4.82

Taking these numbers into account, statement V. was the only one expressing a negative opinion towards wikis. From the rest of the statements, the benefits outweigh the shortcomings. The students may have realized this if the project was longer term and/or involved more members per group who didn't have a history of working with each other. In addition, the neutral statements may have shifted towards a positive response if the project was changed with respect to the two factors mentioned in the previous sentence.

Another criteria for acceptance of the wiki was average individual and group scores for each statement. For the negative statements, the scores were adjusted by reflecting their value across the median of the Likert scale. The score for each statement was summed. We looked at the average score for all of the statements to evaluate if an individual/group had a positive response to the implementation of the wiki, as shown in Table 2. To ensure anonymity, chemical names are given to the group names and their members. On an individual basis, three students had a negative response, eight had a neutral response, but the rest had a positive response with four individuals having an average score of 5 or greater. On a group basis, half of the groups had a positive response, four had a neutral response, and the group who didn't use the wiki had a negative response.

Table 2. Average Positive Response of Each Student/Group

Student	Score	Group	Score
Hydrogen 1	1.58	Hydrogen	1.86
Hydrogen 2	1.83	Oxygen	3.08
Hydrogen 3	2.17	Uranium	3.29
Uranium 1	2.58	Aluminum	3.46
Oxygen 1	2.67	Fluorine	3.53
Oxygen 2	2.83	Sodium	4.08
Aluminum 1	3.08	Sulfur	4.22
Fluorine 1	3.08	Nitrogen	4.44
Nitrogen 1	3.33	Carbon	4.61
Sulfur 1	3.58	Lithium	4.83
Oxygen 3	3.75		
Fluorine 2	3.75		
Fluorine 3	3.75		
Aluminum 2	3.83		
Carbon 1	4.00		
Uranium 2	4.00		
Sodium 1	4.08		
Sulfur 2	4.42		
Lithium 1	4.58		
Carbon 2	4.67		
Sulfur 3	4.67		
Nitrogen 2	5.00		
Nitrogen 3	5.00		
Lithium 2	5.08		
Carbon 3	5.17		

The frequency, volume, and quality of content added to a group's wiki correlates with the average opinion of the group. In other words, the groups who utilized their wiki liked using the wiki, but it is difficult to evaluate cause and effect with respect to recommendation of the wiki and wiki contribution. However, if the wikis was implemented throughout the entire senior design course versus a 30 day project, the students might begin to see the appeal of the technology and have a more positive reaction. We have other ideas as to what might have increased the positive opinion of the use of wikis in design, but those are discussed in a later section.

Table 3 summarizes student use of the wiki. From the open-ended responses, the ability to have a central hub for shared files like Excel© files or Aspen© files was a very appealing feature of the wiki. The students also used the wiki to organize their meetings and update the specifics of the group's timeline and task allocation. Finally, most of the students utilized the capability of the wiki to quickly make links to important references.

Table 3. Summary of how many groups included each item in their wiki.

Task Allocation	6
Coordinate Meeting Times	6
Timeline/Calendar	8
Attaching files (I.e. Aspen)	8
Premeeting Agenda	3
Meeting Notes	8
Links to References	7
Group/Professor Discussion	7

Positive Reactions to Wikis from the Student's and Educator's Perspective

The last section of the survey asked open-ended questions regarding their likes and dislikes of the wiki, in general, and its integration into senior design. A lot of what the students liked about wikis didn't surprise us. The main thing commented on was the wiki serving as a central hub for files and information. They said it depleted inconsistencies in the content of files (i.e. weekly reports) and everybody had easy access to the most-up-date files. Finally, in cases where certain individual tasks of the project overlapped, a student could find the necessary details in their partner's added content to help steer the progression of their portion of the project.

Another appealing feature to the students was the dynamics of the wiki. Instead of sending multiple emails, it was much easier to come to a consensus on meeting times or have discussions without having to schedule a formal group meeting. Like instant messaging, the students could add little pieces to a discussion or make slight alterations to a plan (meeting schedule) until the group was satisfied with the final result. Unlike instant messaging, the whole discussion is automatically recorded and archived.

The other thing related to the appeal of the dynamics of the wiki was the interaction from the authors of this article. Students felt their questions and concerns were addressed frequently and timely. The wiki provided more interaction on this project compared to other projects. It is critical to reiterate what was said in the introduction about new technology in the classroom. Students perceive enhancement of interaction as a main requirement when deciding to integrate technology into a class[4]. From the responses to the survey, wiki use seems to have met this requirement.

A lot of the appealing qualities of the wiki are shared between the educator and the student. With respect to the wiki acting as a central hub for information, it was easy for us to find what we wanted/needed and evaluate it. With the students adding content, we could see what options the students were exploring, what decisions were being made, and sometimes see why those decisions were made. Also, in terms of correspondence with the students, it was easy to find what information was shared versus hunting through a slew of emails.

From a pedagogical standpoint, wikis provide a great potential for study. Wikis allow easy sharing of information amongst a group. A professor may get a lot more information about what went on throughout the semester, than they might get within a weekly or final report. Also, because of its revision history, we can observe the dynamics of the design process from the student's point of view. If the students utilize the wiki and add content as information is gathered and decisions are made, an outside observer can start to see the thought process of the designers. Another appealing piece of the revision history is the record of who added what and when. As observed by Heys, individual accountability can really be enforced[6]. Early in a project, if there is a lack of content added or participation by an individual, the group or teacher can take steps to prevent further laziness or problematic procrastination.

Another reason we believe makes wikis a great tool for students, is the faculty interaction with the wiki better simulates the interaction they'll receive in practice. In industry, an engineer doesn't collaborate solely by writing a report every month or at the end of a project. The supervisor keeps constant tabs on a group's progress, so the project gets completed on time and the results are valid. With respect to the students' comments about the frequency of interaction from the professor and the teaching assistant (TA), it was easy to address concerns and questions of the students. If a change had been made to a wiki (whether it was a question or addition of information), it was easy to go to the site and see what content was added since the last time it was observed. Specifically, pbwiki will highlight the differences between two versions of a page. If a student/group had a question or uncertainty about their design, it took no more than 15 minutes to see the question and answer it in a place all of the members could see it. Also, the response would be on record and be easily accessible to reference for further viewings by either the student or the teacher.

Negative Response to the Wiki

Not every tool you expose to students is 100% well received. However, this study was to investigate what was good and bad about the wiki. From the opinions pointing out the flaws of the wiki and its implementation, we could make constructive decisions about what to change in the future and how.

The students' three main arguments against the use of the wiki involved the preference of email, the small size of the groups, and the small scale of the project (amount of work required and time to finish). A large percentage of the students commented on how they "prefer to use email." They thought the wiki was more work and it was easier to use email. In addition, they thought it was easier to use email because of the size of the groups. It was much easier to meet up with two other people or email two other people, then to add their content to the wiki. Another factor related to the size of the groups was the familiarity of the group members. Each group member shared at least two (if not more) classes with their teammate, they socialized in their personal time, and they saw each other outside of group meeting times very often.

With the project lasting only 30 days, the students didn't think it was worth adding content to the wiki. One student is quoted as saying, "...given more time than four otherwise busy weeks with graduating and major life changes approaching, we would have had time to use it for effective group and time management." Because there was no requirement for the content added, some students minimized the content added to save them time.

There were other hurdles preventing or discouraging the students from adding to their wikis. The students began the study with minimal familiarity with wiki technology. Some embraced the new technology, but others strayed away from it because it was new. This is consistent with what was seen in the ECAR study. They found students who considered themselves early adopters of new technology had a greater affinity to using wikis in the classroom, than those who utilize technology at the same rate as the average population.

Other hurdles were the organization of the wiki, the limited amount of file storage, and full group participation. Some students thought the wiki could be a great tool if the information was organized, but it took a lot of time to organize to a point where it would be useful. With respect to the amount of storage, the free pbwiki account only allows a maximum of 15.0 MB worth of files to be uploaded. There is no limit on the amount of content added directly to the wiki, but pictures and actual files saved to the wiki count towards the maximum. Finally, there was at least one group where one of the members didn't attempt to contribute to the wiki, discouraging the rest of the group from also adding to it.

Suggestions for Future Use

In general, we believe the wiki is a good design tool for students and recommend it to all design groups in education and in industry. The authors have been communicating with the design team at pbwiki.com in order to improve what the wiki has to offer. Since the beginning of this study, some of our suggestions for changing pbwiki have been implemented into the newest version, or the feature is being tested as a beta version.

Our suggestions for change try to address all of the things that prevented students from embracing the wiki. To address the problems with organization, we suggest having a preconstructed skeleton structure for the wiki. The designers have come up with a way to make this very easy for an educator. First off, any previously made wiki page can serve as a template for future pages made. Another feature in beta is the ability to "clone" a wiki. In this fashion, not only does the structure of one page get copied (as in a page template), but all of the links and the whole structure of the wiki can be carbon-copied. Utilizing these two new features, we plan on making one wiki with headings and space for new additions in a manner we, as supervisors, prefer. We will clone this wiki to make each group's beginning wiki exactly the same.

Being able to give the students a skeleton structure of the wiki helps alleviate a lot of problems with how the wiki was initially implemented. The biggest benefit is organization. The students can appreciate this, because they don't have to spend as much time organizing, and can spend more time adding content. The teacher can utilize this organization to allow them to find exactly what they want and where, i.e. the results of a decision matrix. The teacher won't need to go through page after page looking for the justification for the use of a piece of technology. Another helpful aspect is how it will take away the intimidation of the wiki and deemphasize the lack of familiarity with it by giving them a head start on the project.

Changing the logistics of the project could solve the other issues with the use of the wiki in the class. To address the issues discovered from this study, we are planning on doing the following:

- Increase the number of members in each group
- Randomize the members of the group to deplete familiarity
- Increase the scale of the project and assign it the first day of class
- Require specific entries into the wiki (not just weekly reports)
- Require equal contributions to the wiki by all members
- Incorporate other major's for an interdisciplinary design project

Considering what the student said prevented their wiki contribution, we think the above will alleviate those problems.

Being digital immigrants, we didn't enforce using wikis over email. Throughout the project, some students would email us with their concerns and questions (versus putting them on the wiki) and we would reply using email. The main advantage of wikis over email is the centralization of data and its organization. By responding to the students over email, we decentralize correspondences and add to the disarray of information. In short, a project manager or professor needs to be consistent about adding to wikis if all group members are expected to use wikis over email.

Conclusions

Web 2.0 technologies like wikis and blogs have great potential in the classroom for the net generation. However, these technologies should be used with caution. We as educators can't integrate these technologies into our classes because we want to seem novel and up-to-date, but we should integrate them to improve student learning. By doing studies like the one from this article, we can decide the best way to involve technology in lectures and teaching design.

Wikis have a lot of potential in the classroom. Heys used wiki technology for a class project to improve the learning of his Mass & Energy Balances class[6]. Some educators are using wikis as a replacement for traditional textbooks, where the students add problems and edit the educational content[3]. In this study, we utilized wikis as a design tool.

Overall, the students liked using the wiki and recommend it for further use. They liked how the wiki improved interaction amongst the group members, the professor, and the TA. In addition, they utilized how wikis centralize their findings and its dynamic nature for collaboration. They didn't like taking the time to organize their wiki and prefer using emails for a variety of reasons. Email can be used for collaborating, but for a large design project, we think organized wikis are more beneficial. As a result, we have suggested changes for further usage of the wiki in a design course. Finally, we think wikis have great potential for pedagogical research. If the students properly add content to their wikis, we can delve into how students approach and implement a design project. Also, research can explore what factors affect group productivity and design quality.

Acknowledgements

The authors would like to thank the chemical engineering senior design class of 2008 for participating in this study. We also appreciate the helpful input put forth by the teaching certificate working group of the Vanderbilt Center for Teaching.

References

1. Salaway, G., J.B. Caruso, and M.R. Nelson. *The ECAR Study of Undergraduate Students and Information Technology, 2006.* Boulder, CO: EDUCAUSE Center for Applied Research, 2007.
2. Chubin, D., K. Donaldson, B. Olds, and L. Fleming. "Educating Generation Net – Can U.S. Engineering Woo and Win the Competition for Talent?" *J. Eng. Educ.*, v.97 no.3: 245-258 (2008).
3. Richardson, W. Blogs, Wikis, Podcasts, and Other Powerful Web Tools for Classrooms. 2nd ed. Thousand Oaks, CA: Corwin Press (2006).
4. Oblinger, D.G. and J.L. Oblinger. "Is it Age or IT: First Steps Toward Understanding the Net Generation." In Educating the Net Generation, ed. D.G. Oblinger and J.L Oblinger, 2.1-2.20. Boulder, CO: EDUCAUSE (2005).
5. Commoncraft. "Wikis in Plain English." Available at http://www.youtube.com/watch?v=-dnL00TdmLY: accessed March 15, 2008.
6. Heys, JJ. "Group Projects in Chemical Engineering Using a Wiki." Chem. Eng. Educ., v.42 no.2: 91-95 (2008).

133d Software Programming Features That Facilitate Process Control Education

Ricardo Dunia, LabVIEW R&D, National Instruments, 11500 N. Mopac Expwy, Austin, TX 78759, Thomas F. Edgar, University of Texas at Austin, 1 University Station C0400, Austin, TX 78712, and **Ivan Castillo**, Chemical Engineering, The University of Texas in Austin, 1 University Station C0400, Austin, TX 78712.

Process control classes and laboratories have adopted affordable control systems from industry with easy to use software/hardware configurations and interactive design tools in order to make process control more practical and accessible to students. Consequently, the implementation of control algorithms in real time systems has become an important stage in the understanding of different control design topics. Many process control classes have adopted a practical component in which students are exposed to a laboratory project or an interactive software in the classroom.

Engineering departments are always in search of better tools for teaching in classrooms. Among these tools, computer software and process hardware have been recently considered a necessity in control courses. The more such tools resemble the commercial software used for industrial applications, the more valuable the course is. At the same time, some academic software packages have adopted features from industrial software applications to close the gap between theory and practice. Among the different commercial features that have been adopted in technical institutions to configure and implement a control system in an academic environment are:

-Interactive design of controller and process parameters; Mathematical expressions or static plots are not enough to visualize the effect of a parameter in the design of a closed loop system. Interactive class demos are very important to illustrate the effect of the classical tuning parameters for PID controllers, as well as the adjustment of penalty weights applied to the objective functions in optimal control algorithms. Moreover, instructors should be capable of adjusting a variable as the simulation executes to determine its effect and illustrate its most desirable value to the classroom audience.

-Software/Hardware configuration should be friendly; In many cases the configuration of software on hardware that is manufactured by a different vendor brings data conflicts and user interface restrictions that limits the visual teaching and frustrates students in their effort to write comprehensive reports. Having a smooth transition between design and implementation indicates a straight forward configuration of software executed in standard hardware.

-There should be easy transfer of information to and from process and controller hardware; signals should travel without protocol conversion from the physical system to the controller, and viceversa. These signals include measured disturbances, controlled variables and manipulated variables. The more the number of conversions or devices between the controller and the process, the more the risk of adding unnecessary delays and signal saturation in the closed loop system. Such delays and lags are detrimental to the controller closed loop performance.

-The simulation and control programming language should provide tools for easy debugging; a great amount of time is spent when trying to determine the source of programming errors. Students learning a new subject get frustrated when they are not capable of running simple cases because of not knowing the reason of the failed execution. Manuals tend to be more self-contained, and error messages should properly describe the source of the problem. Features like breaking points, high-light execution and variable probes are a must.

-Code portions should be easy to fragment and reuse for later projects; the instructor may require the project to be finished in several stages. Therefore, the programming language should be flexible enough to be fragmented such that code portions can be reused in future assignments. Also, students may divide the work in several tasks or coding routines that could be written in separate files. Such a divide and conquer strategy is common in computer science applications, and it is getting popular in control courses. Different control strategies are implemented by replacing blocks of code made by different developers with a common programming interface.

-The user should remain in the same software framework during the different design stages; storing and loading information used in different software packages may required more effort from the student to learn different user interfaces and programming languages. Some parameters may have similar names among software packages, but these parameters could be used in different ways according to user manual instructions. A well-known example, the tuning of parameters used for PID controllers, could be applied in different ways depending on the control algorithm placed in practice.

-Affordable price and scalable when the application requires it};the same piece of code should be reusable for different hardware platforms. Therefore, software used in an initial prototype with inexpensive hardware should be compatible with the one implemented in sophisticated equipment. The programming code use for proof of concept in a lab can be brought to the actual industrial application without major modifications. In this manner slow and inexpensive academic equipment can be replaced by a full demonstration and fast execution hardware ready for industrial use.

The success of a control package in academia depends on how well the features enumerated above are implemented in academic environments. Software packages that brings all these features to educational institutions provide user-friendly software that communicates to hardware. Therefore, the same piece of programming code could be used for data acquisition, system identification, model creation and validation, model-based control design, closed-loop simulation and controller deployment on different hardware sites. A friendly user interface and interactive functionality are also crucial in the success of an educational programming environment.

In this work we present the desired educational features of a programming framework, facilitating the learning process and the instructor communication to students either in the classroom or in academic labs. There are few packages that provide these features. We focus our attention in using LabVIEW as an example of on one educational programming language that has made control design and simulation classes more friendly and practical in recent years.

Virtual Learning Environment as a Conceptual Tool in Phase Equilibrium Course at Universidad de Los Andes.

Camargo G[1]., Tirano J.E[1]. , Altamar A[2].

1. Department of Chemical Engineering. Universidad de Los Andes, Bogotá – Colombia. 2. Department of environmental Engineering. Universidad libre. Bogota - Colombia.

Topics reviewed in phase equilibrium, may seem complex and difficult for many students. The mathematical background adds an abstraction level that makes difficult to observe the practical implementation and interpretation of phase equilibrium. Inconveniences such as these have been overtaken by developing a virtual teaching tool, designed project between the Chemical Engineering Department and Teaching Assistant laboratory at the Universidad of Los Andes. With this tool, students can see different fields of application of concepts such as free energy, partial molar properties, relation between phase equilibrium and chemical equilibrium. This virtual learning environment is an additional tool that allows the student to review their progress, reviewing specific topics such as pure compounds, thermodynamic properties of solutions, criteria of equilibrium. In addition to reading and study materials, invites the students to perceive the phase equilibrium as something real from the virtual reality and reflect on some examples of specific application linking the themes of the ongoing process with applications in chemical engineering. The results about acceptance of the course have been favored with this tool implementation.

1. Introduction

One aim of the project "Virtual Environments as the Higher Education Support Attendance at Universidad de los Andes", is to implement a system of monitoring and evaluation to ascertain the impact of Virtual Learning Environments – VLE - developed for undergraduate courses that have participated in the project.

The proposed Virtual Learning Environments is a proposal led by the academic rectory at Universidad of the Andes. The project seeks to enhance educational processes VLE unconventional, to invite students to investigate, explore, learn, learn and expand on the subject of study. Similarly, the intention is to exploit the possibilities of multimedia and Hypermedia technology to support academic activities were present. Additionally, the project will seek to implement processes of planning, design, installation, implementation and evaluation of learning environments no physical presence. Finally, the VLE project seeks to promote relations between innovative content-student, student-teacher-student and student.

The implementation of the project is through a step-by-step methodology: Analysis of education, educational design, definition of the virtual learning environment, graphic design and computer, editing, programming, implementation, testing and adjustments and finally assessing the level of environmental impact virtual.

The evaluation of conducted process was inspired by some criteria:: Being a contextualized evaluation, giving voice to the participants to look beyond the numbers, look for an improvement in practice. To achieve this purpose the monitoring and evaluation system focuses both on the results generated in the VLE designed, as in the processes and conditions that enable the achievement of objectives.

2. Methodology.

In order to have a vision of the project, has designed an evaluation to identify, first, an overview of the process and the conditions of implementation and, secondly, an assessment of the results from measurements of central components of the course. This will define a time of 2 measurements, which constitute the basis of analysis to realize the potential changes generated in the courses. This will structure a formative assessment and summative which covers two academic semesters:

Base Line (LB): aims to record the development of courses and identifying information as the particular need identified in the analysis phase of education. This information corresponds to the course without the intervention of the virtual environment.

Line of Departure (LS) is focused on the valuation of the courses with the VLE running. The information is analyzed in relation to the earlier time, with the aim of identifying changes generated by the project in terms of educational need for the course.

From the analysis and design for the course of Thermochemistry II, it was determined that the students are difficult to abstract the cycle that occurs on the equilibrium of substances which are part of a natural situation, mathematical tools are used to calculate equilibriums and obtained an interpretation of the results in optimal conditions of operation.

Therefore, the objective of the VLE would strengthen the students in the abstract and evidence of the cycle that occurs on the equilibrium of substances from the approach of natural situations and the use of tools for calculating equilibriums through the demonstration in the learning environment. In that sense, it is clear that the VLE is a conceptual and that the theories of learning concepts share three basic assumptions (Bolton 1978): a.los concepts are formed through the recognition of similarities between objects, b. progress in the formation of concepts ranging from the particular to the general and c. the concepts are raw concrete, as they form the basis for the acquisition of more abstract concepts.

Thermochemistry II is a course that is issued to students from sixth semester of chemical engineering, whose goal is the conceptual understanding of the equilibrium of substances and how it can be used in industrial processes and transformations, the teacher explained that the course was not focuses on studying the separation of substances in them as conditions for the equilibrium of the same (one subject later shows them how is this separation). For him, that it can arise in terms of skills: to transform these processes into equations that students can handle, setting conditions for the changes they want can be made. At the starting line professor adds that the skill is expected in students refers to the ability of interpretation of the calculations made, integrating them together, "have to interpret that this equilibrium is demarcated in a series of steps that transforms a raw materials into products. "

For Professor at this time (baseline), it is important that the aid would provide the VLE to show the students that processes need to be very visual abstraction, so that they can draw conclusions from there: "generate a full cycle of abstraction, where this model applies, what can I interpret this model, and feedback from that. "

3. ***Results***

The VLE of Thermochemistry II is composed of 1 space organization (Resource Center), 1 Information (Help), and 4 training (Board Room, Applications Equilibrium, multicomponent systems, and systems from a single component). The following describes the meaning of each of these areas:

1. One component systems: In this space the student starts the game of experimentation and exercises about three types of equilibriums: equilibrium Liquid - Liquid Liquid equilibrium - Steam, liquid equilibrium - Solid

   At the student is proposing a challenge or problem to solve, if the student does not answer so successful, the tool gives feedback and offers the possibility to solve other challenges, to understand the conditions of a pure substance. In answering the questions correctly will continue with the experimentation of the case.

2. Multicomponent systems: In this space the student will do an analysis of the conditions of the substances, and the questions it will lead to perform the procedure to be able to abstract in terms of calculating balances. (Calculations made by those outside the tool.) What's interesting is the process that makes the student to take into account different types of substances and mixtures and solutions to the challenges proposed. As in the space above the tool gives you the ability to respond various types of questions if you do not hit it correctly solve the challenges continue with the case.

3. Applications balances: In this space the student experience from a real situation or event. This situation gives you an overview of how, from a physical situation, there is the analysis of the balance of substances, either pure or composite, this analysis lead the student to determine conditions.

4. Resource Center: Space for publishing documents to support the kinds of equilibrium liquid - liquid, liquid - vapor, liquid - solid, as well as links to other sites and all the resources necessary to achieve the objectives of the course.

Las animaciones en el espacio de "Aplicaciones de equilibrios" le permitió analizar el equilibrio de sustancias puras o compuestas, a partir de las demostraciones que allí se le presentaban.

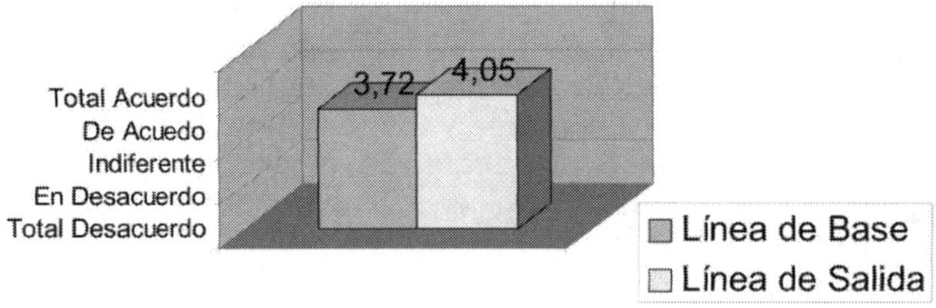

Space Applications Balance, composed of various kinds of animation, had a very good welcome from the students interviewed who entered this space, mainly because it seemed fun and useful to understand some content. One of the students said that initially this kind of spaces he questioned many things that led him to ask the teacher: "Things that I was not clear. But I realized that with the animation. " The professor mentions that the sessions of listening to appoint this space laboratory by students, and how useful it was to understand the content. This corresponds with the results showing the above chart, which compares the importance of a certain type of resources in line with the basic line VLE in output, to analyze the balance of pure substances or complex: it looks like There is a trend the agreement (X = 4.05).

Integrating virtual presence

The initial presentation was held in a room of computers in which each student had his computer and witnessed how they sailed by the VLEs. For the observation was made of it, you can verify that it met all the requirements to make a presentation of quality: Deal introduced the charge of the spaces in its entirety, how to access them, clarifying the doubts techniques were taken, the professor for his part gave a presentation on the general sense was that each of these areas in order to achieve the educational goals of the course, making some students, chosen at random-resolved any of the exercises appeared in the exercised.

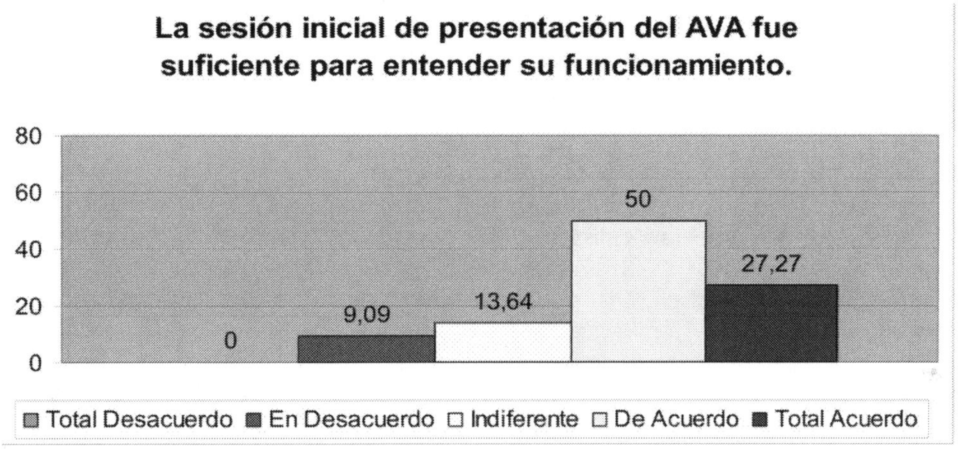

Relationship between the VLE and educational need identified.

Bearing in mind that the educational need to be found in this course was that the students are difficult to abstract the cycle that occurred in the balance of substance, the professor believes that the VLE served as a source of questions, such as a space for The exercises they could get to see if wanted.

For its part the students believe that while the VLEs can not replace the experience of a laboratory presence, the area of the animations on the lot, "where one sees the mixture of moles combining with each other, it makes sense for the course." However, some students think they should put the focus not only on micro processes (in the interview refer to chemical levels) but also the industrial level.

Evaluation of VLEs

Regarding the feedback that was online basic, and its importance to student learning, the contribution was significant in the VLE starting line, as shown in the chart below, which shows a greater tendency to agree that he had at baseline:

This is somewhat at odds with one of the problems that arose not because of a technical but management tool, and that has to do with the bank of questions that had been practicing for such spaces: the students stated that when got to answer the questions, they ended up going out after them repeat a cycle of four years.

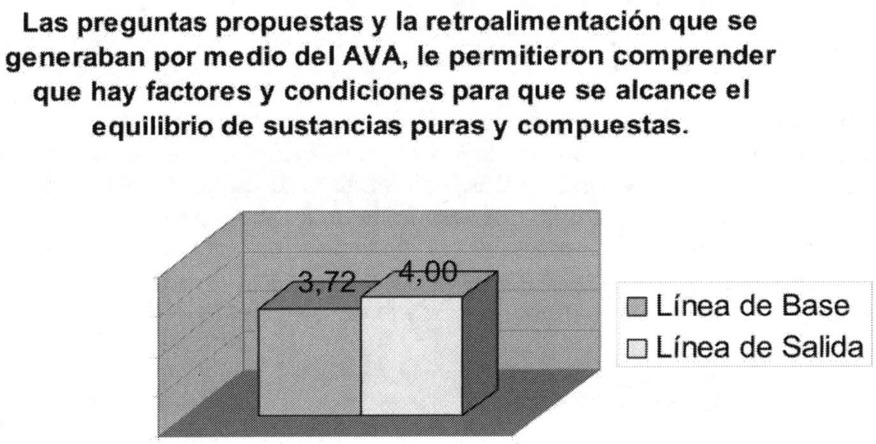

It can be said that both the feedback wrong, as the lack of questions may have varied disheartened those who were trying to find the VLE so frequent, which is understandable: the first generates a lack of credibility in the tool, the latter a feeling that no worthy of review by this means but to seek other remedies to go with the highest number of exercises, "a bored and was already" and answered all because I have nothing to learn "" At first I used it for the part but he knew that Gabriel was never going to get a question from there for the part. " This corresponds to the level of motivation that the students had to enter the VLE, and that as shown in the chart below shows a high percentage of students with indifference to the question (36.36%), while the rest tends to Total Agreement.

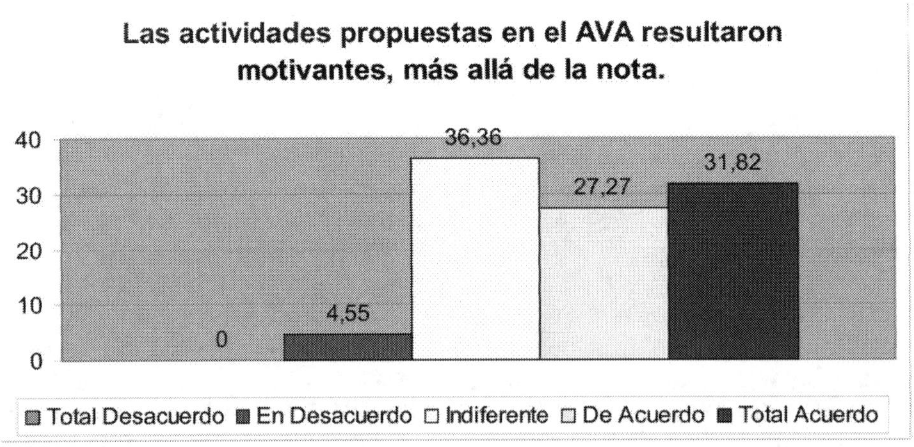

Conclusions

For some interviewed students, there is a huge difference between the use and utility of a course with Sicua®, and the virtual environment that was for this course in terms of learning: "There are very few subjects at the University who are as such a virtual environment in which one can try to measure the knowledge that one has acquired over the matter. " So this VLE managed to become some of the students in an area of significant exercise for the area.

Equally valuable is for students to have a tool that not only were to enter links to other pages (mention that this happens in other courses), but could interact with the VLE, to know its own course, as well as watching their own process throughout the semester to measure their level of knowledge. As can be seen in Annex 9.4, all students admitted at least once to see or do a test drill in spaces and some of them did so many times (one student showed up to 123 attempts).

Bearing in mind that the specific objectives that had this VLE, they had to do with giving students activities and problems to help them strengthen their capacity for analysis on the processes involved in balancing the separation of substances and compounds, it can be said that the was achieved successfully.

Bibliography

1. ALVARADO, S. V. La educación superior y la aplicación de las tecnologías. De la información y la comunicación. Síntesis y conclusiones. En: Programa Nacional De Formación De Profesores De La Educación Superior. Tercera Versión Cátedra ICFES. Agustín Nieto Caballero. Bogotá, Octubre 11 de 2002.

2.BARBERO, M., GARCÍA, J, (et al.) (1998). Mapas nocturnos. Diálogos con la obra de Jesús Martín Barbero. Universidad Central- DIUC: Siglo del Hombre Editores, p 202.

3.GALVIS, A. (1997) Ingeniería de Software Educativo. Universidad de los Andes, Bogotá - Colombia.

4.MORENO, M.G “El desarrollo de habilidades como objetivo educativo. Una aproximación conceptual” En Revista Educar. Revista de Educación - Estrategias Cognitivas No. 6 Julio – Septiembre 1998.

5. Keith Ward, Jim Hartley. Using a virtual learning environment to address one problem with problem based learning. Nurse Education in Practice, , Volume 6, Issue 4, July 2006, Pages 185-191.

6. Suzanne M. Byron, Jon I. Young. Information seeking in a virtual learning environment.Research Strategies, , Volume 17, Issue 4, 4th Quarter 2000, Pages 257-267.

7. David Shulman. D.S. Preston, Virtual learning and higher education, The Internet and Higher Education, , Volume 8, Issue 2, 2nd Quarter 2005, Pages 175-179.

133f Application of Multimodal Software Tools to Teach Problem Solving Skills

Paul Blowers, Chemical and Environmental Engineering, The University of Arizona, PO Box 210011, Tucson, AZ 85721-0011

Many faculty members have attempted to apply new technological advances in classroom settings to improve pedagogical approaches, increase student learning, and to run classrooms more effectively. Unfortunately, many of the approaches of applying these new tools do not accomplish these goals. This work investigates the use of multimodal (spoken verbal, written verbal, and visual) approaches integrated through Microsoft's OneNote program to change student access to problem solving frameworks in the context of a sophomore and junior level courses.

One of the challenges in engineering is getting students to move beyond plug-and-chug type solutions and frameworks to having them fully integrate new knowledge into an independently organized thought process for bringing complex information to bear in applied ways. Students have a difficult time developing these skills from textbooks because the texts are not interactive and because they often do not present information in more than one way (written verbal). Lectures can engage students in active learning techniques where students can apply information with guidance. However, the fast pace of some lecture materials will leave weaker students with little recourse other than to extensively rely on the textbook. Archived materials from lectures recorded with OneNote accompanied by verbal spoken explanations timed with the flow of the note development allows students the ability to review notes at their leisure and to capture the nuances of explanation they may have missed during the actual lecture. Examples of materials generated for the sample course and a brief tutorial of how to apply the software will be discussed.

133g Technology In Computer Lab Environments: Challenges and Opportunities

Eric L. Maase, Chemical Engineering Department, Lafayette College, Acopian Engineering Center Rm 268, 740 High St., Easton, PA 18042

As instructional environments gradually move closer to each student having access in the classroom to a computers available on the desks new opportunities and new challenges for both the instructors and the students become evident. In a specific lab environment, during a course in chemical engineering computing, students' computers are connected to the classrooms' overhead projection system with the specific student or instructor's computer projected according to the instructor's direction. The challenges and unique aspects of implementing such an environment are discussed in the context of teaching and engaging students with the materials of a chemical engineering computational methods course.

A COURSE MODULE ON CHEMICAL PRODUCT DESIGN

James C. Watters
Chemical Engineering Department
University of Louisville
Louisville, KY 40292

The changing nature of the chemical engineering profession has promoted interest over the past decade in introducing students to product design concepts in addition to traditional chemical process design. The author began teaching design at Louisville in fall 2004 following the retirement of a senior colleague who had taught it for many years, and decided to incorporate a module on product design into the second course of a two-course sequence. The students survey the text "Chemical Product Design" by Cussler and Moggridge (2001) and then are required to spend about one month identifying a needed product and performing a preliminary product design for it, including financial analysis, following the Cussler-Moggridge procedure.

Some products proposed and analyzed over the past four years include:

- A swizzle-stick capable of detecting a date-rape drug in a mixed drink
- Chewing gum to whiten teeth
- Nanofluid production using induction heating, and
- Design and production of a better golfball.

The paper discusses these student projects, as well as some common issues students have with such an open-ended exercise. Student projects are generally of acceptable quality and show that graduates can think and rise to a challenge, even as they are about to be loosed on an unsuspecting world!

INTRODUCTION

Chemical Engineering Design at University of Louisville consists of a two course sequence offered in the 4th year of the program. The present author began teaching the design sequence in 2004. CHE 471, Strategy of Design, is a basic "how to do design" course, comprising flow-sheet development, component design, use of process simulators (Aspenplus) and process economics including profitability analysis. Over the course of the semester each student develop a design process to synthesize his or her own assigned commodity chemical. The second course, CHE 572, Plant, Process, and Project Design, focuses on teamwork. In the first eight weeks of the semester (from start of semester to Spring Break) students prepare a team design report with a topical focus, for example biodiesel synthesis, or the AIChE design contest problem. For the four to five week period between Spring Break and the end of semester, a module on Chemical Product Design was introduced, and is the focus of the remainder of this paper.

Chemical Product Design

Several authors of the past decade have proposed that today's Chemical Engineering design courses need to include some instruction on Chemical Product Design (see for example: Moggridge and Cussler 2000, 2003; Westerberg, 2000; Shaeiwitz and Turton, 2001; Grossmann, 2003; Wei, 2004; Cussler, 2006; Moggridge et al, 2008). This has been reinforced by the changing nature of the chemical industry in the US. Large scale new production of commodity chemicals is being off-shored while the domestic focus is often on specialty chemicals and new products. 2001 saw the publication of the monograph "Chemical Product Design" by Cussler and Moggridge which presented an easy avenue to introduce the topic into a design course.

The Cussler-Moggridge text is easily surveyed in a half dozen classroom lectures and presents a solid framework for students to develop a product design study. Topics presented in the text and in class include:

- Why students ought to be interested in product design
- Assessment of customer needs
    - Market analysis
- Generation of ideas to meet the needs, through brainstorming sessions
- Culling and selection of ideas for further study
    - Patents and trade secrets (Presentation by an alumna who is a patent attorney)
- Product manufacture and how it compares to traditional process design
- Economic issues, again compared to process design

The text material is interspersed with a wealth of examples of successful products, ideas for new products and "quick and dirty" analysis approaches appropriate to rapid selection of workable ideas for further study.

Class Mechanics

The 4th year class at Louisville in recent years has ranged in size from 15 to 30 students; class sizes projected for the next several years are closer to 30 students per year. Students are given considerable latitude in how they approach the product design project. They may work individually, in teams of 3 or 4, or even as a whole class, though no group has even taken this option. This project comes in what may be the last month of formal education for many of the students and often "senioritis" is rampant among even the more dedicated members of the class. The product design assignment is presented to them as a project that ought to be undertaken as a challenge to be creative and, within, some reasonable bounds, should be viewed as a "fun" exercise. This is their opportunity to be innovative and inventive, and obviously the effort they put into the exercise will reflect what they get out of it. This approach seems to work with most students and leads to their putting an appropriate level of effort into the project.

Figure 1 shows the problem statement as presented in spring semester 2008. Figure 2 shows the outline format provided for the final report; some leeway is allowed here as the exact product chosen will often dictate how specific some areas of the report can get. Generally the textbook material occupies the three weeks or so before spring break, while the class is working on that semester's first project which is due at last class before break. The assignment for the product design project is distributed at one of the classes in the week before spring break with the instructions that they decide know how they will work on the project (individually, in teams with members specified, etc.) by the class immediately following the break.

Selected Projects from Past Three Years

Given the constraints on how the course is organized the quality of the final projects has varied somewhat each year. However the vast majority of the students have taken the product design project quite seriously and produced creditable work. Table 1 lists the products designed by these four student groups. While some of the ideas may be somewhat far-fetched, and a few others show insufficient research to see that they copied existing products or patents, there were surprising few completely sophomoric proposals. Alcohol-related concepts often featured in the initial product list generated by brainstorming ("boys will be boys!") but relatively few of these made it to final selection. Many of the ideas came from the students' own life experiences by asking "Wouldn't it be nice if ... ". Brief discussion of some of the more interesting ideas follows the table.

TABLE 1: SUMMARY OF PRODUCTS DESIGNED BY STUDENTS IN CHE 572, 2005-2008

Spring 2005
Date-Rape-Drug Detection Device
Ale-No-More: The Beer That Prevents Hangovers
Biodiesel Fuel (before this became a commodity)
Teeth Whitening and Calcium Enriched Chewing Gum
Spring 2006
Photochromatic Automobile Windows
"Alledrex" – A Proposed Anti-Allergy Ocular Cream
Digital Price Display System for Stores
"Fizz" – A Sodium Bicarbonate-Citric Acid Tablet to Reinvigorate Partly-Consumed Bottles of Soft Drinks
Transdermal Patch for Delivery of Vitamins
Nanofluid Production by an Induction Heating Deposition Process
Spring 2007
Coated Dog Food for Outdoor Dogs
Aerosol Spray for Killing Salmonella
Transdermal Patch for Delivery of Thrombolytic Drugs
"Breath Assure" – A Litmus Test to Detect Bad Breath
Spring 2008
Refrigerator Cartridge for Absorption of Odors and Ethylene Gas
Biodegradable Cigarette Filter
"Vitamin Gum" – Get Your Daily Dose While You Chew!
Design of a Seamless Golf Ball
A Bite-size, Healthy, High-Protein Snack – An Alternative to Cereal and Protein Bars

1. DATE-RAPE-DRUG DETECTION DEVICE

(Donna H., Haymanot M, Cathy R., Erin W.)

This all-female team, obviously well attuned to the potential dangers lurking in singles bars, sought to design a cheap, effective and unobtrusive device to detect the presence of one or more date-rape drugs surreptitiously slipped into an alcoholic beverage. Research elicited that there are three often used such drugs in the US – Rohypnol, γ-Hydroxybutyrate (GHB) and Ketamine (US Department of Health and Human Services). Further research identified a number of chemicals that could detect one or more of these drugs by means of a color change or fluorescence generating reaction (Guerra et al., 2004).

Though some drug detection kits using some of these chemicals were currently marketed, they were neither inexpensive nor unobtrusive to use.

The students proposed designing a stirrer or "swizzle" stick that could be coated with the detector and sold cheaply, or even given away, by bars as a service to their customers. These could be customized by individual bars or clubs in the same manner that swizzle sticks are currently customized and indeed would differ from existing sticks only in their extra external coating. How to apply such a coating to sticks was investigated and seemed to be feasible using either dip-coating or spray-coating. Initial calculations showed favorable process economics with the cost of a single stick being in the 5 to 15 cent range, an amount easily absorbed into the price of a drink.

2. Teeth Whitening Chewing Gum

(Andrew B., Kyoung L., Min K., Gordon S., Randy S.)

This enterprising all-male team focused on the rapidly growing field of teeth whitening to capitalize on the vanities associated with an aging baby boomer generation. Research showed a 6% annual growth rate in the confectionary chewing gum industry in the early 2000's with a steady growth of specialty gum products in the same period. This growth has continued to increase. Retail sales of gum are up by about 10% in the past year, while sugarless gum sales are up 18% (National Confectioners Association). In the early 2000's sales of tooth whitening aids had tripled, auguring well for the future of a marriage between the two products. They proposed using Xylitol as a low calorie alternative to sugar. Some studies have shown that Xylitol fights cavities and plaque in teeth and may also counteract some symptoms of osteoporosis and diabetes, other concerns of aging baby boomers.

Gum manufacture is well documented and includes steps for the insertion of various additives and flavorings (see for example Wikipedia, 2008). A very dilute solution of hydrogen peroxide (less than 1% by volume) was proposed as the whitening ingredient. Current over-the-counter whitening products use as much as 2-3% hydrogen peroxide, but such concentrations may lead to an offensive odor problem. The team suggests that users of their product can be exposed to the same net quantity of hydrogen peroxide by chewing more gum over an extended period. They estimate that a plant to produce 10,000 tons per year of gum would pay back in about 2.5 years and yield an internal rate of return of about 25%.

3. Nanofluid Production by an Induction Heating Deposition Process

(Matthew C.)

The premise behind this project is that properties of materials may be enhanced by the addition of nanoparticles. Ethylene glycol impregnated with copper nanoparticles has been shown to have improved heat exchange capacity (US Department of Energy). Ethylene glycol with improved heat transfer characteristics would reduce the quantity of anti-freeze needed in an automobile radiator permitting smaller lighter radiators to do the same cooling as today's standard units. This in turn lightens the weight of the car, with the end result of improving fuel efficiency.

Building on the work of Kostic (2008) the student proposed a continuous process to vaporize copper into nanosized particles using an induction heater and then disperse these particles in ethylene glycol. Figure 3 illustrates the proposed design. The design showed reasonably good economics though some of the data seem unrealistic. However the project showed an impressively high level of critical thinking on the part of a student who had elected to perform the exercise on his own and who, up to then, had shown decidedly average performance in most of his coursework.

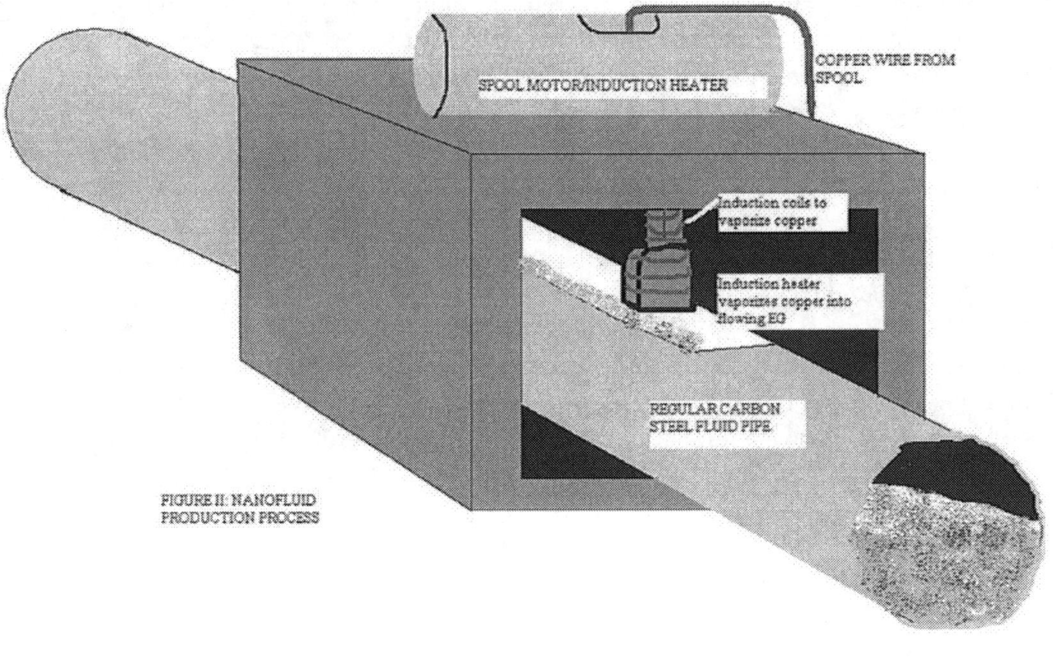

FIGURE 3: NANOFLUID PRODUCTION PROCESS

4. Design of a Seamless Golf Ball

(Nick C., Boris C., Dustin C., Devin C.)

This team of avid golfing enthusiasts, "The Four C's", proposed to design a better golf ball. Currently the top-selling golf ball on the market is the Titleist Pro V-1. It consists of a double layer core surrounded by a thin polyeurethane cover which is applied by encasing the core in twin hemispherical molds containing the liquid polymer. The mold surface is designed to form the polyeurathane into the conventional dimpled surface characteristic of a golf ball. However, the process creates a seam at the equator of the ball where no dimples are present and which causes the ball to have different flight characteristics depending on where it is struck relative to this seam. This concerns many professional golfers as it can mean a difference of up to 6 feet in a 250 foot drive.

Team 4C proposed an improved injection molding process whereby polyeurethane is injected into the mold via multiple ports (as opposed to a single port in the Titleist process) in such a manner that each half of the mold forms a fraction of a dimple at the points of contact between the top and bottom half of the ball. The result would be a complete set of whole dimples being formed where the seam formerly existed. The design is based loosely on a patent filed by Miller (1993). Miller's design does not seem to have been put into production though he still owns the intellectual property and may have rights on aspects of the design proposed.

Preliminary economic analysis of a plant to produce 40 million golf balls per year shows a fixed capital investment of about $10 million, an ROI in the range of 50%, and a payback period of about 18 months, premised on making a golf ball for under $2 and selling it for about $4.50.

Discussion

The course has been conducted in the manner described for four years now. The projects performed by the students meet the level expected of senior chemical engineers in terms of the degree of rigor exhibited in the needs assessment, product selection, analysis of manufacturing criteria and economic projections. While some of the economic data seem overly optimistic many of the cost factors must be estimated from minimal available information. The students are forced to move beyond their comfort zone of readily available chemical or equipment cost data in publications such as ICIS Chemical Business (2008) or Peters et al (2001), the textbook currently used in CHE 471 and 572, and make informed estimates based on limited information. In addition most of the students find themselves dealing with synthesis and processing equipment that is widely different from what they normally encounter in traditional process design. This broadens their appreciation for what they may be called upon to work with in real life, and also builds their confidence that they can indeed work with the "unknown".

Students have generally viewed the exercise favorably. There is an initial apprehension on the part of some students to seek out new ideas but usually one member of a team is adventurous enough to take on a leadership role in the early stages. Often this turns out to be one of the less academically strong students who perhaps was a Boy or Girl Scout in younger days and developed a desire to meet a challenge head-on, or an older student who has had broader life experiences than the norm. Once the idea for a new product has crystallized then the academically stronger students may take the lead in the actual design and "number-crunching" phase. Through this exercise all members of the team see the value of having different strengths represented when undertaking a project.

It is indeed very heartening to see the full maturing of students following four or more years of nurturing as they undertake and master a truly open ended challenge.

Conclusions

The author has successfully introduced a one-month module on Chemical Product Design based on the methodology proposed by Cussler and Moggridge (2001) into the second course of the design sequence at University of Louisville. Projects tend to be generally of acceptable quality and most students enjoy the challenge of a completely open-ended problem. After four years of refining the module, the concept is working to bring out a high level of critical thinking on the part of the students and to enhance their confidence in their technical ability as they move towards the world of work or graduate study.

Acknowledgement

JCW acknowledges the hard work of the University of Louisville BSChE graduates of 2005 to 2008. Without their willingness to take on the new challenge of Chemical Product Design this paper would not have been possible. May they have long, fruitful and productive lives and careers and may they never lose their willingness to meet a challenge head-on!

CHE 572 Plant, Process and Project Design SPRING 2008

Design Problem 2 Introduction to Product Design

You may do this problem individually, in groups of 3 or 4 or as a whole class. The goal is to produce a report on the design of a new or modified product that we will propose to management. Let me know on March 20 how people will work on the project. The steps in the process are as follows:

1. Identify a need. In the real world this would be done by interviewing customers and potential end users. It will also include a market analysis to determine potential number of end users, etc. In a class exercise it is up to you to come up with the need from your own devices and project the market needs. I have provided some ideas in the class and there are several in the Cussler book. You are also encouraged to suggest some of your own. Ideally what you propose should have some relationship to chemical engineering, engineering in general or chemistry.
2. Convert that need to Specifications for a product, chemical or device. Revise those specifications so that they are realistic, i.e., feasible at a "reasonable" cost, which is related to a price you would expect people would be willing to pay for your product.
3. Brainstorm ideas for your product. Ideally you should generate 100 +/- ideas. Sort and screen the ideas to about 5 viable ones.
4. Select one or two ideas for future research. Use scientific principles (thermodynamics, kinetics), subjective criteria and risk analysis to guide your selection.
5. Propose a manufacturing process for your product. Be aware of issues such as potential for patents or trade secrets, how to get "missing information", setting final specifications, and the need to be "first to market".
6. If your product involves the manufacture of a specialty chemical, be aware of the differences between commodity and specialty chemical manufacture – generic equipment, FDA/USDA issues for pharmaceuticals and food products, scale-up issues, etc.
7. Economics considerations – differences between commodity and specialty products such as generic equipment, shorter product sales life, etc.

Prepare a report that addresses the above issues, making a recommendation to proceed (or not) with manufacturing the product. A suggested report format is on page 2. Be aware that some of the items requested in the report may require estimation from few or no data – use your best engineering guesses.

Report is due by 9.30 am on Thursday April 17, 2008, last day of class for CHE 572.

FIGURE 1: CHEMICAL PRODUCT DESIGN ASSIGNMENT AS PRESENTED IN SPRING 2008

Format for Design Report 2

OUTLINE		Points
	Letter of Transmittal	5
	Title Page	5
	ABSTRACT (one page)	15
	SUMMARY (two or three pages)	30
	Table of Contents	
(1)	SCOPE: Identify the product, dimensions, raw materials, production rate,	10
(2)	Identification of Needs for your product	10
(3)	Ideas – lists of ideas generated by brainstorming; screening, sorting	10
(4)	Selection – selection process using objective criteria such as chemistry, thermo, etc; subjective criteria and how subjectivity was minimized using various selection matrices, etc.; risk management	10
(5)	Manufacture – filling in missing info; setting final specs; scale-up needed10	
(6)	IP issues (Patent, trade secret, other)	10
(7)	EQUIPMENT	10
	Equipment List	
(8)	ECONOMICS: Best estimate of as much as possible of	
	a. CAPITAL INVESTMENT	10
	Fixed Capital Working Capital Total Capital	
	b. Manufacturing COSTS	10
	Direct Manufacturing Costs	
	Indirect Manufacturing Costs	
	c. PROFITABILITY	10
	NPW	
(9)	CONCLUSIONS AND RECOMMENDATIONS	20
	Conclusions	
	Recommendations	
(10)	REFERENCES & BIBLIOGRAPHY	10
(11)	APPENDICES	15
	Supporting Documentation	
	Copies of most important references, etc.	
	Detailed calculations (MEB, Component Design, Financial)	
TOTAL		200

FIGURE 2: SUGGESTED OUTLINE AND POINT ASSIGNMENT FOR FINAL REPORT ON CHEMICAL PRODUCT DESIGN

REFERENCES

Cussler, EL. and GD Moggridge, *Chemical Product Design*, Cambridge University Press, New York, NY (2001)

Cussler, EL, "Teaching Chemical Product Design", presented at Annual Meeting of AIChE, paper # 132e, San Francisco, CA (2006)

Grossmann, IE, *"Challenges in the new millennium: product discovery and design, enterprise and supply chain optimization, global life cycle assessment",* Proceedings of the 8th International Symposium on Process Systems Engineering, PSE 2003, 28-47 (2003)

Guerra, FJ, BR Glover, and JST McCafferty, US Patent Application number 20040146429, filed July 29, 2004.

ICIS Chemical Business, ICIS Student Pages, http://www.icis.com/StaticPages/a-e.htm , visited on 19 September 2008

Kostic, M, *"Nanofluids: Advanced Flow and Heat Transfer Fluids"*, www.kostic.niu.edu/DRnanofluids/nanofluids-Kostic.ppt , viewed on 19 September 2008

Miller, SS, "*Molded Seamless Golf Ball*", US Patent number 5201523, April 13, 1993

Moggridge, GD and EL Cussler, *"An introduction to chemical product design",* Trans IChemE-Part A, 78, 5-11 (2000)

Moggridge, GD and EL Cussler, *"Teaching chemical product design",* Proceedings of the 8th International Symposium on Process Systems Engineering, PSE 2003, 1460-1465 (2003)

Moggridge, GD, R Costa and P Saraiva, *"Chemical Product Design Teaching",* http://www.engsc.ac.uk/an/mini_projects/cpd/index.html, viewed on 19 September 2008

National Confectioners Association, http://www.ecandy.com/ , visited on 19 September 2008

Peters, MS, KD Timmerhaus and RE West, *Plant Design and Economics for Chemical Engineers*, 5th edition, McGraw-Hill, New York, NY, 2003

Shaeiwitz, JA and R Turton, *"Chemical product design",* Chem. Eng. Edu., 35, 280-285 (2001)

US Department of Energy, "*Taking the heat off: Nanofluids promise efficient heat transfer*", http://www.eurekalert.org/features/doe/2001-10/dnl-tth060302.php , visited on 19 September 2008

US Department of Health and Human Services, http://www.4woman.gov/FAQ/rohypnol.htm , visited on 19 September 2008

Wei , J, *"The integration of process and product design",* Proceedings of the 6th International Conference on Foundations of Computer-Aided Process Design, FOCAPD 2004, 7-18, (2004)

Westerberg, AW, and E Subrahmanian, *"Product design"*, Comp. & Chem. Eng., 24, 959-966 (2000)

Wikipedia, http://en.wikipedia.org/wiki/Chewing_gum , visited on 19 September 2008

Modular Teaching and Open Ended Design Projects

Yaşar Demirel
Department of chemical and Biomolecular Engineering, University of Nebraska Lincoln,
ydemirel@unl.edu

ABSTRACT

Some latest employment surveys show that chemical engineers are working in more and more diverse industries manufacturing specialty and commodity products; they need to understand property-structure relationships using chemistry, biology, and physics. Therefore, modular teaching with open ended design projects can provide a means of responding to diverse and fast changing course contents and learning/teaching objectives. This study presents the experience on modular teaching integrated with the Aspen Plus simulator and discusses open ended plant design projects in the Department of Chemical Engineering at University of Nebraska Lincoln. The projects discussed is the conversion of available carbon dioxide and hydrogen into methanol or/and ammonia. The hydrogen comes from electrolysis of water using the electricity produced by wind power, while the carbon dioxide comes from power plants as well as ethanol plants. The feasibility of the plants at some assumed production capacities with the available technologies are discussed using the discounted cash flow diagrams of the plants. The economic data used, assumed capacities, and the cost of electrolytic hydrogen have all compounded effects on the feasibility of the plants.

Keywords: Modular teaching, capstone design, open ended design projects

1. MODULAR TEACHING

Chemical Engineering discipline has evolved to embrace biological engineering, and its graduates may work as product and process engineer to define products and processes, understand property-structure relationships using chemistry, biology, and physics. This reemphasizes the constant need for revisions of teaching strategies and instructional materials in capstone design. One of the ways of achieving this goal may be modular teaching technique [1-3]. A module is a well-organized, high quality student text based on clear level, prerequisites and learning objectives of a topic. Modules mainly consist of title page, main text, and end materials; they can accommodate institutional and temporal variations and response to diverse and fast changing learning objectives and to the dynamics of accreditation in engineering education. They also follow new practices, the advancements in technology, and departmental curricular needs, and contain design practice problems and relevant references. The main text and the practice problems help students to transfer and synthesize knowledge across the previous courses, provide them with some examples of open-ended problems, and develop strategies in making engineering decisions under uncertainty. However, the effectiveness of modular teaching technique still remains to be assessed properly [1]. The Modules used at the Department of Chemical and Biomolecular Engineering at University of Nebraska Lincon are: M1-Introduction: Design Considerations, M2-Engineering Economics, M-3 Separation Systems, M-4 Heat Transfer and Fluid Flow, M-5 Chemical Reaction Engineering, M-6 Thermodynamics in Design, M-7 Process Synthesis, M-8 Heuristics in Process Design, M-9 Safety, M-10 Green Engineering. The Aspen Plus simulator practices are incorporated into the Modules starting with M-3 Separation Systems. The modules are detailed elsewhere [3].

2. OPEN ENDED DESIGN PROJECTS

The purpose of this design project is to explore the use of the electrolytic hydrogen together with the available carbon dioxide to produce high value chemicals such as methanol and/or ammonia. Electricity produced by wind farms is used in hydrogen production by the electrolysis of water. The carbon dioxide comes from power plants and methanol plants. Using the electrolytic hydrogen and the available carbon dioxide it may be possible to convert and store renewable energy and make use of carbon dioxide if methanol and/or ammonia plant is feasible under current economic conditions and at assumed production capacities with existing technologies. So the open ended design problem is a search of feasible use of the electrolytic hydrogen and carbon dioxide to produce valuable chemicals. We designed and simulated the methanol and ammonia plants with Aspen Plus, and carried out economic analysis by discounted cash flow diagrams for the feasibility of the plants.

2.1 METHANOL PLANT

Process Description

We used the ASPEN PLUS simulator to design and simulate the methanol plant with the RK-SOAVE equation of state property method. The methanol plant uses 786.2 kg/hr hydrogen and 5721.3 kg/hr carbon dioxide, and produces 4169.2 kg/hr and 99-wt% of methanol. The plant operates for 8520 hours/year. We assume that under high pressure (50 bar) and temperature (250°C), there are no side reactions beside the following elementary reaction [4,5]: $CO_2 + 3H_2 = CH_3OH + H_2O$. This reaction is reversible. The forward reaction's conversion is 95% of carbon dioxide. The reaction takes place over a catalyst of $Cu/ZnO/Al_2O_3$ [4]. The cost of major utilities in the form of cooling water, steam, and electricity is around \$1.4 10^6/year. Figure 1 shows the conceptual PDF for the methanol plant with major equipments. The feedstock is at 25°C with a mole ratio of hydrogen to carbon dioxide of 1:3. Stream S6 is the output of the reactor R201 and contains the unused hydrogen and carbon dioxide, as well as the produced methanol and water. This effluent is cooled down to 84°C in E202. The unreacted carbon dioxide and hydrogen are recycled to the reactor in stream S8. The liquid output of F301, at 70°C, is introduced into the stage 24 of the distillation tower T301. The distillate stream 'METHANOL' contains 99.7-wt% of methanol, while the bottoms flow 'WATER' contains 99.7 wt% of water. Table 1 presents the component balance as well as the overall mass and energy balances.

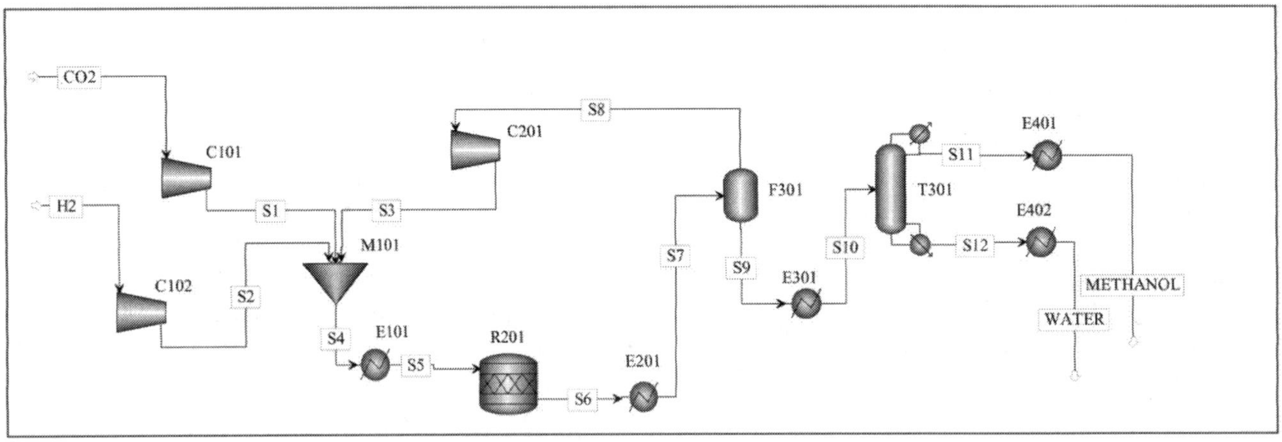

Figure 1. Process flow diagram for methanol plant.

Table 1 Mass and energy balances for the methanol plant

Component balance, kmol/hr	In	Out	Generation
H2	390.0	0.002	-389.8
CO2	130.0	0.05	-129.9
METHANOL	0.0	129.9	129.9
H2O	0.0	129.9	129.9
Mass, kg/hr	6507.5	6507.32	
Enthalpy, Gcal/hr	-12.2209	-16.1622	

Discounted Cash Flow Diagrams (DCFD)

Discounted cash flow diagrams (DCFD) provide the annual cash flows with the time value of the money taken into account [6]. Within the DCFDs it is assumed that the construction time for the plant is one year, the tax is zero, while the interest rate is 5.25%. The useful operation time is 10 years, and the depreciation is based on seven years. The selling price of the primary product methanol is assumed as \$1000.0/MT, while the economic analysis finds the cost of methanol production is \$1400/MT. This result is based on the unit cost of hydrogen by wind power assumed as \$3.5/kg H_2 and at the assumed capacity and using currently available technology for the methanol production. Table 2 displays the data used in the economic analysis.

The DCFD yields a net present value of -\$119.0 10^6 at the end of operation. The sum of working capital, land, and salvage value is fully recovered. The value of net present value (NPV) makes the operation unfeasible. If we can lower the cost of hydrogen from \$3.5/kg to \$1.5/kg and hence the cost of methanol production from \$49.0 10^6 to \$33.0 10^6 the net present value becomes zero, which may make the investment acceptable. Figure 2 shows the DCFD with the new cost of hydrogen.

Table 2 Data used in the economic analysis of methanol plant

Economic data		Economic data	
Bare Module Cost, \$	7.00 10^6	Methanol Production Cost, \$/MT	1400
Fixed Capital Investment, \$	35.00 10^6	Labor, \$/hr each labor	25
Working Capital, \$	7.00 10^6	Number of Labor	15
Land, \$	2.00 10^6	Cost of Labor, \$	3.20 10^6
Salvage, \$	1.40 10^6	Cost of Waste Treatment, \$	0.25 10^6
Revenue, \$	38 10^6	Cost of Raw Material, \$	23.44 10^6
Cost of Production, \$	50 10^6	Cost of Utilities, \$	1.375 10^6
Useful Life of Operation, n , years	10	Carbon Credit, \$/MT CO_2	3.75
Years for MACRS Depreciation, years	7	Revenue for Methanol, \$/MT	1000
Cost of Electrolytic Hydrogen, \$/kg	3.5	Revenue of Water, \$/MT	100
Cost of Carbon dioxide, \$/MT	8.0	Revenue of Carbon credit, \$/year	0.18 10^6

MT: Metric ton

Recommendation

One possible recommendation would be the co-production of methanol with other product which uses methanol as feedstock such as biodiesel, or dimethyl ether, or dimethyl carbonate.

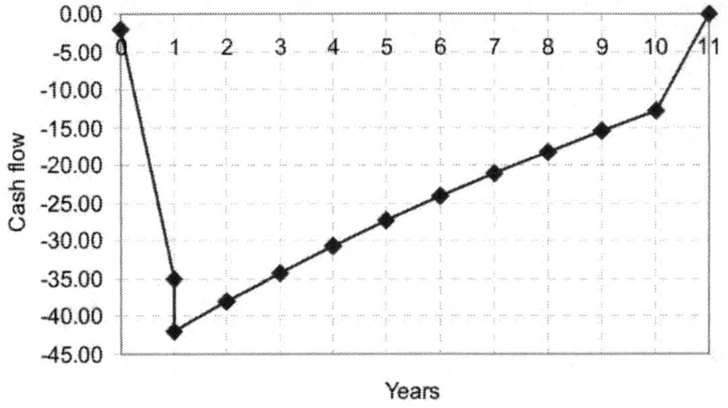

Figure 2. Discounted cash flow diagram for the methanol plant with electrolytic H_2 cost of \$1.5/ kg H_2.

2.3 AMMONIA PLANT

Process Description

The ammonia process is simulated by the Aspen Plus simulator with RK-SOAVE equation of state property method. The ammonia plant uses 9071.5 kg/hr hydrogen and 42.48.3 kg/hr nitrogen, and produces 50111.1 kg/hr 99.9 wt % anhydrous ammonia. The plant operates for 8520 hours per year. Air is separated in SEP 101, and the feeds of nitrogen and hydrogen at 20.27 bar are mixed in M101. This mixture is compressed to about 212 bar in compressors C101 and C102. In reactor R201 the ammonia synthesis $3H_2 + N_2 \rightarrow 2NH_3$ takes place at around 556°C and 212 bar with a platinum group metal such as ruthenium. The reactor R201 is a RGIBBS reactor and estimates the equilibrium composition of the reactor by Gibbs free energy minimization. The output of the reactor is conditioned in heat exchangers E202 and E203 and sent to adiabatic flash drums FL301 and FL302, which operate at 203 atm and at 12 bar, respectively. The bottom flow of FL302 is the product ammonia at -26°C and 12.4 bar. Figure 3 shows the conceptual PFD for the ammonia plant with major equipments. The ammonia plant requires electricity, cooling water, steam, and refrigeration as utilities. The flow rate of ammonia is maximized to be 2942.7 kmol/hr and its composition to be 0.99-wt% within the optimization block with constraints. Tables 3 presents the overall mass and energy balances for the ammonia plant.

Table 3 Overall mass and energy balances for the ammonia plant

Component balance, kmol/hr	In	Out	Generation
N2	1501.0	29.5	-1471.4
H2	4500.0	85.5	-4414.4
NH3	0.0	2943.0	2942.9
Overall Balance, kmol/hr	6400.00	3457.0	-2942.9
Mass, kg/hr	63887.2	3887.8	
Enthalpy, Gcal/hr	0.1770	-50.3657	

Discounted Cash Flow Diagrams (DCFD)

The DCFDs are prepared for a ten-year of operation time and a one-year of construction time for the ammonia plant. The tax is zero, while the interest rate is assumed to be 5.25%. The depreciation is based on seven years. Table 4 displays the data used in the economic analysis. The revenue of ammonia is assumed to be \$600.0/MT. The cost of ammonia production is estimated as \$985.0/MT. This result is based on the unit cost of hydrogen by wind power assumed as \$3.5/kg H_2 at the assumed production capacity and using the current technology.

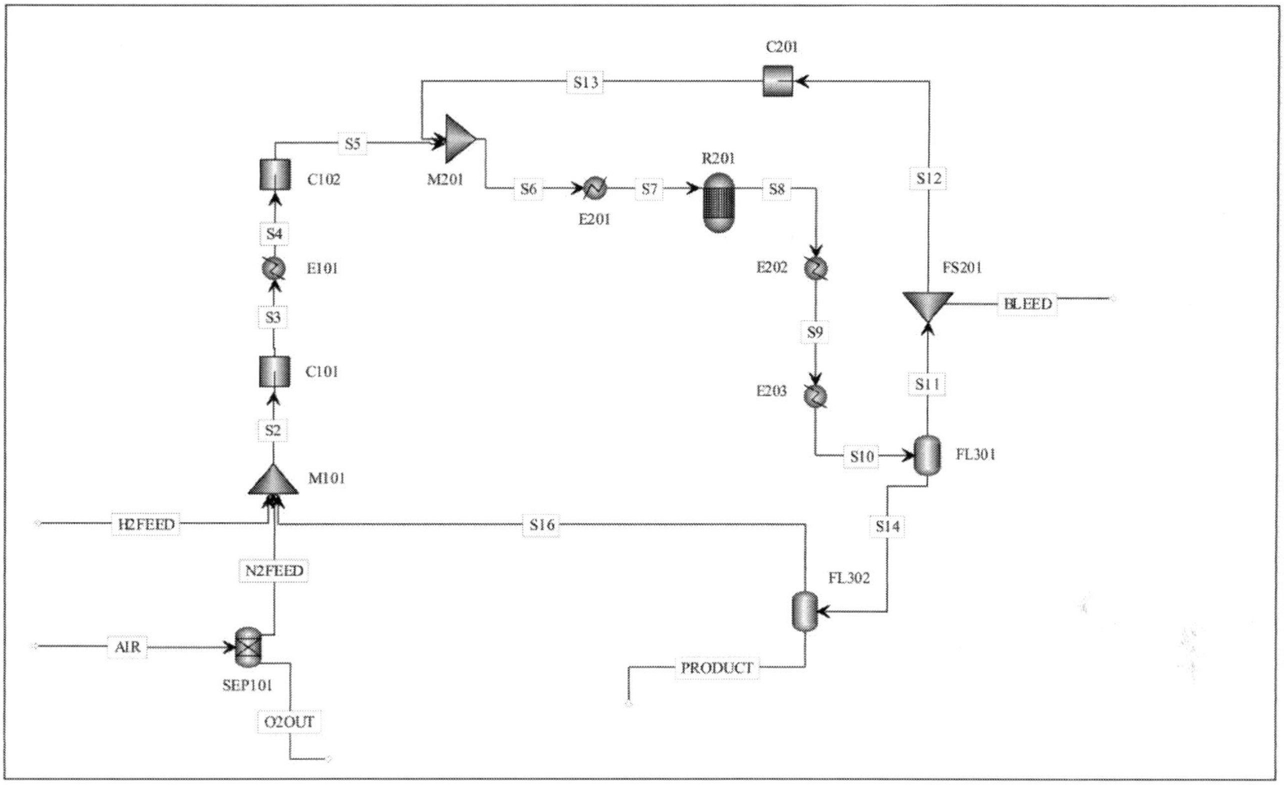

Figure 3 Process flow diagram for the ammonia plant.

Table 4 Data used in the economic analysis of the ammonia plant

Economic data		Economic data	
Bare Module Cost, \$	37,131,000	Cost of Labor, \$	4,260,000
Fixed Capital Investment, \$	185,655,000	Cost of Waste Treatment, \$	1,000,000
Working Capital, \$	37,131,000	Cost of Raw Material, \$	270,690,062
Land, \$	3,000,000	Cost of Utilities, \$	21,301,862
Salvage, \$	7,000,000	Cost of Labor,\$	25
Revenue, \$	310,557,680	Number of Labor	20
Cost of Production,	423,993,268	Cost of Hydrogen, \$/kg H_2	3.5
Revenue of Oxygen, \$/kg O_2	0.05	Cost of Nitrogen, \$/kg N_2	0.5
Revenue of Ammonia, \$/kg NH_3	0.60	Output of Oxygen , kg O_2/hr	12767

The net present value at the end of operation is -\$989.0 10^6. The sum of working capital, land, and salvage value is fully recovered. With the current economic data (Table 4), the ammonia plant is unfeasible. If we can reduce the cost of hydrogen production from \$3.5/kg to \$2.0/kg the NPV becomes zero. Figure 4 shows the DCFD with the new cost of hydrogen production of \$2.0/kg. Also, if the revenue of ammonia increases from \$0.60/kg to \$0.92/kg then the NPV becomes zero with the hydrogen cost of \$3.5/kg.

Recommendation

One of the possible recommendations would be the co-production of ammonia and urea as fertilizers.

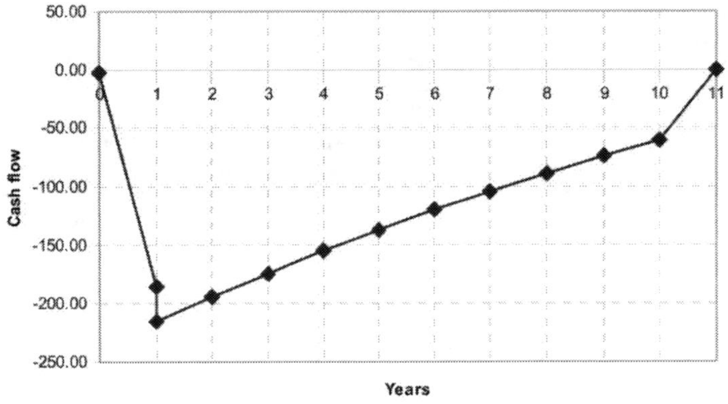

Figure 4 Discounted cash flow diagram for the ammonia plant with electrolytic H_2 cost of \$2.0/kg H_2.

Table 5 summarizes the effects of the cost of electrolytic hydrogen production and the revenue of methanol and ammonia on the NPV. The net present value will be equal to zero if the hydrogen production cost is reduced from \$3.5/kg H_2 to \$1.5/kg H_2, or the methanol revenue is increased from \$1.0/kg to \$1.46/kg at the assumed production capacity. On the other hand, the net present value will be equal to zero if the hydrogen production cost is reduced from \$3.5/kg H_2 to \$2.0/kg H_2, or the ammonia revenue is increased from \$0.6/kg to \$0.92/kg at the assumed production capacity for the ammonia plant.

Table 5 Effects of the cost of electrolytic hydrogen and revenues of methanol and ammonia on the NPV.

Cost of hydrogen, \$/kg	Revenue of methanol, \$/kg	Cost of methanol production, \$	Methanol plant Net present value, \$
3.50	1.00	50 10^6	-119.0 10^6
1.5	1.00	34 10^6	0.0
3.50	1.46	50 10^6	0.0
	Revenue of ammonia, \$/kg	Cost of ammonia production, \$	Ammonia plant Net present value, \$
3.50	0.60	424 10^6	-989.0 10^6
2.00	0.60	288 10^6	0.0
3.50	0.92	424 10^6	0.0

3. OTHER USES OF CARBON DIOXIDE AND HYDROGEN

Carbon dioxide is a very stable molecule and hence energy is generally necessary to drive the desired conversion. Thus high temperatures, extremely reactive reagents, electricity, or the energy from photons may be necessary to carry out carbon dioxide reactions. Reactions of carbon dioxide are dominated by nucleophilic attacks at the carbon, which result in bending of the O-C-O angle to 120°; i.e. hydroxide attack on carbon dioxide to form bicarbonate. Carbon dioxide may be converted to: (a) ammonia/urea $C(O)(NH_2)_2$, and (b) sodium bicarbonate ($NaHCO_3$). Carbon dioxide is also an intermediate in organic syntheses for melamine and urea resins productions. Supercritical carbon dioxide is a hydrophobic solvent which can replace organic solvents in some applications; solvent costs may be reduced and emission of toxic organics can be reduced. Fuel cells convert hydrogen to electricity, heat, and water.

4 DISCUSSION AND CONCLUSIONS

Modular teaching and open ended design projects may enhance student's skill of transferring and synthesizing knowledge across courses, and decision making under uncertainties. Modular teaching can accommodate diverse and fast changing course contents and learning/teaching objectives.

Open ended design projects considered in this study analyze the use of electrolytic hydrogen with the available carbon dioxide to produce valuable chemical products, such as methanol, ammonia, and/or sodium bicarbonate. The cost of electrolytic hydrogen strongly affects the economics of the methanol and ammonia plants at the capacities assumed in this study and using currently available technologies. The methanol plant is economically feasible if the electrolytic hydrogen cost is less than $1.5/kg hydrogen. The ammonia plant is not favorable under current economic conditions. A feasibility study is needed for sodium bicarbonate production. The analyses at some assumed production capacities with the available technologies show that the economic data, assumed capacities, and the cost of electrolytic hydrogen have all compounded effects on the feasibility of the methanol and ammonia plants. However, changing economic conditions, ongoing research and technological developments could lead to improved efficiencies and cost-effectiveness for the production of hydrogen, methanol, ammonia, and hence the analyses may lead to open ended results.

Worldwide, we use only 110 million tons to produce other chemicals (mainly urea) of the 3500 million tons carbon dioxide we add annually. Therefore it is critical to use carbon dioxide as feedstock for production of other valuable chemicals.

REFERENCES

1. Brusic, S.A., LaPorte, J.E., The Status of Modular Technology Education, Int. Tech, Ed. Assoc. Conf., 1999, Indianapolis, IN.
2. Demirel, Y., Modular Teaching Experience in Capstone Design, AIChE's Annual Meeting, November 7-12, 2004, Austin, TX.
3. Demirel, Y., Modular Teaching and Product Design Projects in Capstone Design ASEE SE Section Annual Conference, April 3-5, 2005, University of Tennessee at Chattanooga, TN.
4. M. Lachowska, J. Skrzypek, Methanol synthesis from carbon dioxide and hydrogen over Mn-promoted copper/zinc/zirconia catalysts. Reaction Kinetics Catalysis Letters, 83 (2004) 269-273.
5. D. Mignard, M. Sahibzada, J.M. Duthie, H.W. Whittington, Methanol synthesis from flue-gas CO2 and renewable electricity: a feasibility study. International Journal of Hydrogen Energy. 28 (2003) 455-464.
6. R. Turton, R.C. Bailie, W.B. Whiting, J.A. Shaeiwitz, Analysis, Synthesis, and Design of Chemical Processes, 2nd edition, Prentice Hall, Upper Saddle River, 2003.

173c Alternative Fuel Feasibility Study as the Process Engineering Senior Design Course

Katsuyuki Wakabayashi, James E. Maneval, William J. Snyder, and Michael E. Hanyak. Department of Chemical Engineering, Bucknell University, Lewisburg, PA 17837

As the chemical engineering discipline continues to transform and expand, there is a need to introduce problems that have a scope beyond the traditional chemical engineering. This is especially true for the senior design courses. Bucknell University offers senior design in two parts, where the first semester "process design" deals with simulation and plant design, while the second semester "project design" offers real-world chemical engineering problems. Both courses are based on open-ended problem teaching, co-instructed by three faculty members with diverse background and expertise. .

The process design course at Bucknell has recently focused on a feasibility study on the production of an alternative fuel that can serve as a primary energy source in our economy. The design course structure on this topical issue of renewable, sustainable fuel gives a good overview of the design/problem-formulation process, and allows the students to see the big picture of "alternative" fuels production, leading to a better comprehension of the surrounding studies and debates.

Unlike most institutions where the capstone senior design course is solving one common design problem with all the students or working from a prescribed list of industry-consulted/ sponsored design problems, we offer a truly open-ended problem; the students select their own fuel types, processes, and assessment metrics virtually from scratch. We set the production goals based on the most recent national and state initiatives for developing alternative sources of energy, and partner with a regional development agency as our client to oversee the work of the students. The course puts major emphasis on technical communication (both oral and written) as well as teamwork skills.

Although this new format of senior design is overall beneficial, the effort does have its shortcomings, especially on the ineffectiveness of process simulation software on non-petroleum chemical processes. Both the positive and the negative outcomes will be discussed, from the educational and logistical perspectives.

173d Project Engineering as Senior Design: Moving from Paper to Practice

Margot A.-S. Vigeant, James E. Maneval, William Snyder, Michael J. Prince, and Michael Hanyak. Chemical Engineering, Bucknell University, Dana Engineering Building, Lewisburg, PA 17837

This presentation describes the eight year evolution at Bucknell University towards a hybrid paper/ practical senior design sequence where each team must physically implement the solution to a real chemical engineering problem. These problems are supplied by local industry as well as individuals and non-profit groups. Solutions to the problems must be demonstrated experimentally. Problem areas have ranged from developing and operating bench- and pilot-scale processes to the design and development of novel process equipment to developing novel process conditions to make better products with existing equipment. Students have produced coatings for wrestling mats, tested filtration parameters for pharmaceutical production, and created a novel process for "hopping" beer. Both survey and direct assessment results demonstrate positive student outcomes from this version of the course sequence. We will also reflect upon both the plusses and minuses of this approach from the faculty perspective.

A STUDY OF THE EFFECTS ON DISSOLUTION CONSTANTS AND ACTIVATION ENERGY IN A STIRRED TANK

Gustavo A. Orozco, Universidad de los Andes, Bogotá, Colombia.
Clara T. Gonzalez, Universidad Nacional, Bogotá, Colombia.
Fabio E. Fajardo, Universidad Nacional, Bogotá, Colombia.

Constant kinetics dissolution and activation energy for solids with spherical geometry were calculated. Also the influence of the stirring rate and agitator geometry were studied. Three types of sails arrangement were chosen: The first agitator was composed of two sails, the second agitator was composed of four sails sloping 45 degrees, and the last agitator was composed of six sails with a ring.

In the first part of the experiment it was possible to determine the dissolution constant (k), by means Hixon-Crowell model (eq-1) and experimental data of weight variation through time at constant temperature.

$$\mathbf{M^{1/3}(t) = Mo^{1/3} - kt} \quad (1)$$

Where M(t): weight variation with the time, t: time, Mo: weight at t = 0, k: dissolution constant.

Furthermore, on the dissolution constant, the stirrer effects of angular velocity were studied, finding a direct relationship between 200-650 RPM, but for highest velocities we found that there is a negative effect on the dissolution constant.

For the second part, the experimental equipment was immersed into a thermal bath (Thermohakee Phoenix II-C41P) and the agitator's velocity was maintained at 630 RPM. The temperature was modified between 10°C and 80°C determining the constant dissolution at many temperatures. Afterwards supposing a similar behaviour to Arrhenius equation, it was possible to estimate the activation energy for each type of agitator. These results are shown in Figure 1.

Figure 1.Determination of activation energy in a dissolution process for the three types of agitators.

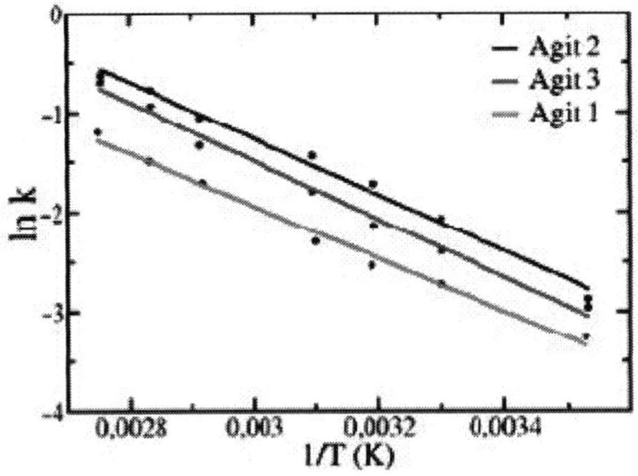

The activation energy in the dissolution process was similar for each type of agitator because the straight lines had the same slope (Figure 1). It was also possible to observe the effect of the system geometry over dissolution constant, for this case the straight lines displayed a pre-exponential factor similar for the second a third agitators meanwhile in the first one it was approximately three times lower than in the second and third agitators.

REFERENCES

[1] Hixsson A. and Crowell J, (1931), J. Ind. Eng. Chem. 1931. Vol (23) pp 923-928.
[2] Beauchamp G. (2001), "Dissolution kinetics of solids: Application with spherical Candy". J. Chem. Ed. Vol 78 (4) pp 523-524.
[3] JChemEd.chem.wisc.edu.

173f Application of Heat and Mass Transfer Analogies In the Undergraduate Engineering Laboratory to Estimate Time of Death

Jimmy L. Smart, Chemical & Materials Engineering, University of Kentucky, 4810 Alben Barkley Drive, Paducah, KY 42001

Historically, estimation of time of death has always been a valuable forensic tool. In England, 925 AD, St. John of Beverly was appointed "keeper of the pleaser of the crown", later to be shortened to "crowner", then finally to "coroner". At that time, only crude "forensic" tools of rigor mortis and corpse warmth were used to gage time of death. It was not until 1600, that the first criminal autopsy was conducted by Ambrose Pare (French physician) [1,2].

By 1811, an English physician named John Davey had formulated his famous law based upon the principle of algor mortis (slow cooling of a warm-blooded corpse). Davey's Law states that 1.5 oF is lost per hour as measured in the armpit. Therefore, at least in theory, it should be possible to record the temperature of a corpse and having knowledge of the heat loss rate curve, be able to estimate time of death. In reality, the rate curve is complicated by many factors and it is often difficult to pinpoint time of death.

The rate of heat loss from the human body can be calculated from the fundamental equation of heat transfer: $Q = UA\ \Delta T$, where Q is energy lost per area, U is the overall heat transfer coefficient, A is the area of the body, and ΔT is the temperature difference between the body and its surroundings. It is difficult to estimate a reliable value for the heat transfer coefficient because of the complicated surface area geometry of the human body. In this laboratory exercise, a heat - mass transfer analogy: the Chilton-Colburn Analogy, is used to estimate the heat transfer coefficient governing loss of heat from the human body. The naphthalene sublimation technique [3] is used to relate mass transfer coefficients to analogous heat transfer coefficients.

Initially, simple geometries of a sphere and cylinder are cast in naphthalene and suspended in a heated duct. Following procedures of earlier investigators [4], loss of weight of these simple shapes over time is monitored and used to calculate an average mass transfer coefficient. Using the Chilton-Colburn Analogy, an average heat transfer coefficient can be determined. These results are compared to existing classical correlations to give students confidence in their technique.

Similarly, a small doll in the shape of a human body is cast from naphthalene and its mass loss monitored over time at a variety of temperatures. An average heat transfer coefficient for the doll can be determined. The surface area of the human body is estimated by modeling the body as a combination of simple geometries of a sphere and various size cylinders.

1. Sachs, Jessica S., Corpse: Nature, Forensics, and the Struggle to Pinpoint Time of Death, Perseus Books Group: Basic Books, NY, October 2002.

2. Miller, Hugh, What the Corpse Revealed: Murder and the Science of Forensic Detection, St. Martin's Press, NY, June 1999.

3. Goldstein, Leonardo and E.M. Sparrow, Heat/Mass Transfer Characteristics for Flow in a Corrugated Wall Channel, J. Heat Transfer (99), May 1977, 187 - 195.

4. Perez, J.S., and E.M. Sparrow, Determination of Shell-Side Heat Transfer Coefficients by the Naphthalene Sublimation Technique, Heat Transfer Engineering (vol 6, no. 2), 1985, 19 - 29.

225a New Laboratory Course for Today's Chemical Engineering Students

Mark T. Aronson and Robert J. Davis. Chemical Engineering, University of Virginia, 102 Engineers' Way, Charlottesville, VA 22980

A new laboratory course has been developed at the University of Virginia for senior-level chemical engineering students. The new laboratory course is based on three 4-week long experiments that demonstrate the relevancy of chemical engineering in a range of technologies including bioprocess engineering, energy conversion and catalysis, and polymer synthesis and characterization. The emphasis of the new laboratory course is to provide the students with opportunities to learn skills needed to succeed in today's changing world such as the integration of process steps, product development, and teamwork. The overall concept of the new laboratory course is discussed along with an overview of each experiment and student feedback.

225b Teaching Material and Energy Balances to First-Year Students Using Cooperative Team-Based Projects and Labs

Timothy M. Raymond, Bucknell University, Chemical Engineering Dept., Lewisburg, PA 17837 and Michael Hanyak, Chemical Engineering, Bucknell University, Chemical Engineering Dept., Lewisburg, PA 17837.

A team-based cooperative learning environment for teaching Principles of Chemical Engineering (the material and energy balances course) has been used at Bucknell University for several years. This course has been carefully designed to include a variety of "best practices" to help prepare chemical engineering students in their first course in the curriculum. The course involves five two-week projects where students work in teams to complete problems covering a range of materials and, at the same time, practice team work and professional skills. Additionally, each project involves a complex laboratory experiment and use of process simulation software (HYSYS) problems.

Assessment for this course has been ongoing and involves a range of data from team self-reports, before and after project concept inventories, individual surveys, team surveys, and final course evaluations. This presentation will explain the details of the course setup, the use and evaluation of various "best practices" used in the course, and evaluation of the numerous benefits of the cooperative learning environment.

COMMON ERRORS IN NUMERICAL PROBLEM SOLVING – HOW CAN THEY BE DETECTED AND PREVENTED

Mordechai Shacham, Ben-Gurion University, Beer-Sheva, Israel

Michael B. Cutlip, University of Connecticut, Storrs, CT 06269, USA

Michael Elly, Intel Corp., Qiryat Gat, Israel

Introduction

The introduction of mathematical software packages such as Excel[1], MATLAB[2] and Polymath[3] into the field of engineering problem solving has brought the benefits of higher speed and precision in obtaining the results. However, numerical solution techniques have also introduced some new sources of errors. These errors may often pass undetected because of the lack of experience in the use of the new computational tools. A few examples of errors originating from numerical sources follow.

1. *Regression and Analysis of Data*

The most common errors in regression of data are originated from the use of regression models with too many or too few parameters[1], fitting polynomial models without proper scaling (standardization) of the temperature data, correlation of data when the model equations are improperly linearized[1]or when experimental design for obtaining the data is not satisfactory[3].

2. *Systems of Nonlinear Algebraic Equations (NLE)*

There are many examples in the literature showing that identification of good initial guesses and formulating the problem correctly for enabling convergence to a solution of an NLE system may represent a major challenge[4]. Even if the solution is found it may be infeasible in the physical sense (solution with negative concentrations, for example) or a false solution caused by improper variable and/or function scaling.

3. *Ordinary Differential Equations (ODE)*

Indiscriminate use of default error tolerances of the ODE solver tools is the most common source of errors in solving ordinary differential equations[5]. Failure to use the proper integration algorithm (stiff vs. non-stiff), carelessly rounding numbers in the model equations, using the model outside the domain of its validity and the use of inadequate resolution in presenting the results have been documented[1],[2] as additional common sources of errors in solving ODEs.

Being aware of potential causes to errors in numerical problem solving is a preliminary requirement to detection and prevention of such errors. In this paper three potential causes of errors are described and demonstrated. Means to prevent these errors are proposed.

No Consideration Given to Problem Parameter Uncertainties

The importance of the analysis of the sensitivity of the problem solution to small changes in the parameter values has been recognized, long ego, in the field of optimization. Edgar et al. (2001), for example, wrote "You should always be aware of the sensitivity of the optimal answer ... Parameter values usually contain errors or uncertainties. Information concerning the sensitivity of the optimum to changes or variations in a parameter is therefore very important in optimal process design." However, sensitivity analysis is rarely carried for other types of

---

[1] Excel is a trademark of Microsoft Corporation (http://www.microsoft.com)

[2] MATLAB is a trademark of The Math Works, Inc. (http://www.mathworks.com)

[3] POLYMATH is a product of Polymath Software (http://www.polymath-software.com)

problems. In this section the importance of the sensitivity analysis in solving systems of nonlinear algebraic equations will be demonstrated.

The example considered was originally published by Froment and Bishoff (1990) and its solution using Mathcad is presented by Parulekar (2006). The reaction considered is the catalytic hydrogenation of olefins (component A) in an isothermal CSTR. Differential mass balance on component A yields

$$\frac{dC_A}{dt} = \frac{C_{A0} - C_A}{\tau} - r_A \tag{1}$$

where C_{A0} is the inlet concentration of A (mol/L), $\tau = V / v_0$ is the space time (s), V is the reactor volume (L), v_0 is the inlet flow rate (L/s) and r_A is the reaction rate per unit reactor volume (mol/L·s):

$$r_A = \frac{C_A}{(1 - C_A)^2} \tag{2}$$

The concentration of A at steady-state is sought for the operating conditions: C_{A0} = 13 mol/L; V = 10 L; and v_0 = 0.2 L/s. Finding the steady-state solution requires solving the nonlinear algebraic equation $f(C_A) = 0$ where

$$f(C_A) = \frac{C_{A0} - C_A}{\tau} - r_A \tag{3}$$

In Figure 1 $f(C_A)$ is plotted versus C_A in the region $0 \le C_A \le 10$. Observe that in this region there are three real solutions. The solutions as obtained by the Polymath 6.1 software package are: C_A^* = [0.7515358, 2.130933, and 8.117532]. Polymath displays the results showing 7 significant digits. Mathcad displays the same results with 20 significant digits (Parulecar, 2006). The discrepancy between the number of significant digits in the CSTR parameters (one or two) and the number of those digits reported in the solution (up to twenty) cannot be missed. The determination of the number of significant digits to be shown in the documentation of the problem solution should be based, in this particular case, on the uncertainty associated with the parameter values.

In Table 1 the values of C_A at the solution are presented for various levels of uncertainty in the flow rate v_0. Assuming uncertainty of 10^{-4} L/s (thus flow rate of 0.2001 L/s instead of 0.2 L/s) yields solution values which match the base solution only in the first two digits. For uncertainty of 10^{-3} L/s there is only one correct digit in the first two roots and two correct digits in the largest root. For uncertainty of 0.01 L/s there is not even a single correct digit in the first two roots and one in the largest root. Thus, any realistic estimate for the uncertainty in the value of the flow rate leads to the conclusion that the solutions for C_A should be rounded up to two significant digits as all the additional digits are uncertain. It should be emphasized that this analysis applies only to physical quantities which obtained as the final results of the computation. There are other cases where rounding the numbers may lead to significant errors. This will be demonstrated in one of the following examples.

As the value of the flow rate is specified by one significant digit only one possible assumption regarding the uncertainty could be that the flow rate was measured (or controlled) with such an accuracy, thus its value can vary in the range of 0.15 L/s $\leq v_0 \leq 0.24$ L/s. In the last two rows of Table 1 the solutions for the two extreme values of v_0 are shown. For $v_0 = 0.24$ L/s there is only one real root at $C_A = 9.36922$ mol/L while for $v_0 = 0.15$ L/s there is one real root at $C_A = 0.3419$ mol/L. Thus, in these cases there is only one root instead of the three roots for the base value of $v_0 = 0.2$ L/s, and the roots are completely different.

Use of Inappropriate Statistics and Graphical Representations in Analysis of Regression Models

The use of inappropriate statistics and graphical representation may often lead to the acceptance unsatisfactory regression models. This phenomenon will be demonstrated by fitting the Clausius-Clapeyron equation to vapor pressure (P_V) data of ethane. The vapor pressure data, containing $N = 107$ points, are available[8] in the temperature range of $T = 92$ K ($P_V = 1.7$ Pa) through $T = 304$ K ($P_V = 4.738*10^6$ Pa). This range covers essentially the liquid phase region between the triple point temperature (= 90.352 K) and the critical temperature (T_C = 305.32 K) of ethane. The Clausius-Clapeyron equation is considered inappropriate for modeling vapor pressure for such a wide range of temperatures and we want to examine what kind of statistics and/or graphical representation can reveal this inappropriateness.
The Clausius-Clapeyron equation is given by

$$\ln(P_V) = A + \frac{B}{T} \tag{4}$$

where A and B are coefficients calculated by regression of experimental data.
In Figure 2 $\ln(P_V)_{exp}$ is plotted versus $1/T$. The straight line fitted to the data is also shown. The linear equation obtained is

$$\hat{y} = 21.816 + 1919.5x \tag{5}$$

where $x = 1/T$ and $\hat{y} = \ln(P_V)calc$. The correlation coefficient R^2 is defined

$$R^2 = \frac{\sum_{i=1}^{N}(\hat{y}_i - \bar{y})^2}{\sum_{i=1}^{N}(y_i - \bar{y})^2} \tag{6}$$

where $\bar{y}$ is the sample mean of the dependent variable. The value of R is bounded: $0 \leq R \leq 1$. If R is close to 1, there is a strong correlation between the variables, whereas a value close to zero indicates a weak or no correlation. The R^2 value in this case (as shown in Figure 2) is 0.999, very close to 1. Thus, the plot of $\ln(P_V)_{exp}$ versus $1/T$ and the statistics R^2 both indicate that the Clausius-Clapeyron equation represents the vapor pressure data adequately over the entire range from the triple point to T_C.

Another widely used graphical representation for checking adequacy of a correlation is the plot of the calculated value of the dependent variable $\hat{y}$ versus the experimental value y. In case of a good fit a straight line should be obtained with the slope ~1.0. In Figure 3 $\ln(P_V)_{calc}$ (as calculated by the Clausius-Clapeyron equation) is plotted versus $\ln(P_V)_{exp}$. Observe that the data points are aligned along a straight line with the slope of 0.9999 with only small deviations. Thus, this type of representation indicates also good fit between the Clausius-Clapeyron equation and the vapor pressure data over the entire range. However, comparing $\ln(P_V)_{calc}$ and $\ln(P_V)_{exp}$ at T = 92 K, for example, yields $\ln(P_V)_{exp} = 0.53$; $\ln(P_V)_{calc} = 0.95$, thus ~ 80 %

difference. This definitely cannot be defined as adequate representation of the experimental data.

The proper graphical means to test the goodness of fit between a regression model and the data is the "residual" plot. The residual $\hat{\varepsilon}_i$ is defined

$$\hat{\varepsilon}_i = y_i - \hat{y}_i \tag{7}$$

A random distribution of the residuals around zero indicates that the regression model represents the data correctly. A definite trend or pattern in the residual plot may indicate either lack of fit of the model or that the assumed error distribution for the data (i.e. random error distribution in y) is incorrect. In Figure 4 the residual: $\ln(P_V)_{exp}$ - $\ln(P_V)_{calc}$ is plotted versus $\ln(P_V)_{exp}$. Observe that there is definite trend (curvature) in the residuals, indicating that the Clausius-Clapeyron equation cannot represent the data adequately.

Careless rounding of model parameters.

In the DIPPR database[9] the Riedel equation is recommended for modeling vapor pressure data. The Riedel equation can be written

$$\ln(P_V) = A + \frac{B}{T} + C\ln T + DT^E \tag{8}$$

where A,B,C,D and E are the parameters of the regression model. Usually the value of E is set at 2 to enable calculating the rest of the parameters by linear regression. Using the same vapor pressure data that were discussed in the previous section the DIPPR staff obtained the following parameter values: A = 5.1857E+01; B = -2.5987E+03; C = -5.1283E+00; and D = 1.4913E-05. In Figure 5 the residual plot of $\ln(P_V)$ calculated by the Riedel equation using the DIPPR parameter values is shown. Observe that the residuals in this case are smaller by at least one order of magnitude than the residuals obtained using the Clausius-Clapeyron equation. There are two separate regions. In the first one (up to $\ln(P_V) \sim 12$) the residuals are in the range of -0.04 to 0.02, not showing any particular trend. In the second region ($\ln(P_V) > 12$) there is an order of magnitude reduction in the residual values (all smaller in absolute value than 0.003), however there is some trend in the residual values in this region. To obtain perfectly random distribution of the residuals a different correlation should be used (the Wagner equation, for example). Improvement of the correlation by use of a different type of equation, however, is outside of the scope of the present paper.

The "copy and paste" facility is a very convenient and time saving option when transferring numerical and other data between various data-bases and software packages. We, for example, copied the Riedel equation constants for various compounds into an Excel spreadsheet. After pasting the parameter values into Excel they are stored with the exact same value as they appear in the DIPPR database, however they are displayed rounded to two three decimal digits: A = 5.19E+01; B = -2.60E+03; C = -5.13E+00; and D = 1.49E-05, if the default number display format is used. Copying these values from Excel and pasting it into a document or a different software package will results in losing the additional digits which were included in the DIPPR published values.

Calculating the vapor pressure using the rounded values of the parameters leads to significant errors in the calculated pressure values. The residual plots of P_V calculated by the Riedel equation using exact and rounded parameter values are shown in Figure 6. Observe that the residual (error) in the highest pressure range is of the order of 2.5% when rounded parameter values are used in comparison to 0.3 % with exact parameter values. Thus, careless rounding of parameter values causes the prediction error to increase by almost one order of magnitude.

Conclusions

Potential causes to errors in numerical problem solving have been demonstrated and means to detect and prevent such errors are suggested.

It was demonstrated that complete definition of a problem requires not only specification of the parameter values but also the specification of the uncertainty levels for these parameters. Specification of the uncertainty levels of the parameters enables determining the uncertainty of the solution, provides justification to number of significant digits to be used in documentation of the results and sometimes (as in the example presented) it may reveal that the proposed process cannot be implemented in practice because of too high levels of uncertainty.

Vapor pressure correlation with the Clausius-Clapeyron equation was used to demonstrate that the linear correlation coefficient (R^2) statistics, the plot of $\ln(P_V)_{exp}$ versus $1/T$, and the plot of calculated values versus experimental values can be misleading in determining the goodness of the fit between the data and the regression model. The residual plot is the most reliable means for examining the goodness of the fit.

Transferring numerical data between various databases, documents and programs by "copy and paste" may result in losing significant digits of the numbers transferred. It was demonstrated that rounding the parameter values of the Riedel equation (by using "copy and paste") caused considerable increase of the vapor pressure prediction errors, especially in the higher temperature range.

Paying attention to the potential causes of errors that were presented here and in previous publications[1],[2],[3],[4],[5] can improve considerably the reliability of the results in numerical problem solving and in the modeling of data by regression.

References

1. N. Brauner, M. Shacham and M. B. Cutlip, ``Computational Results: How Reliable Are They? - A Systematic Approach to Modal Validation", *Chemical Engineering Education*, **30**(1), 20-25 (1996).

2. Shacham, M., N. Brauner and M. Pozin, "Potential Pitfalls in Using General Purpose Software for Interactive Solution of Ordinary Differential Equations", *Acta Chimica Slovenica*, **42**(1), 119-124 (1995)

3. Shacham, M. and N. Brauner, "Correlation and Over-correlation of Heterogeneous Reaction Rate Data", *Chemical Engineering Education*, **29**(1) 22-25, 45 (1995).

4. Shacham, M., N. Brauner and M. B. Cutlip, "A Web-based Library for Testing Performance of Numerical Software for Solving Nonlinear Algebraic Equations", *Computers chem. Engng.* **26**(4-5), 547-554(2002)

5. Shacham, M., N. Brauner, W. R. Ashurst, and M. B. Cutlip, "Can I Trust this Software Package? – An Exercise in Validation of Computational Results ", *Chemical Engineering Education*, **42**(1), 53-59 (2008)

6. Edgar, T. F., Himmelblau, D. M., and Lasdon, L. S., *Optimization of Chemical Processes*, 2nd Ed., McGraw-Hill, New York, 2001

7. Froment, G.F., and Bishoff, K. B., *Chemical Reactor Analysis and Design*, 2nd Ed., Wiley, New York ,1990.

7. Parulekar, S. J., "Numerical Problem Solving Using Mathcad in Undergraduate Reaction Engineering", *Chemical Engineering Education*, **40** (1), 14(2006)
8. Ingham, H.; Friend, D.G.; Ely, J.F.; "Thermophysical Properties of Ethane"; *J. Phys. Ref. Data*, *20*, 275(1991)
9. Rowley, R.L.; Wilding, W.V.; Oscarson, J.L.; Yang, Y.; Zundel, N.A. *DIPPR Data Compilation of Pure Chemical Properties Design Institute for Physical Properties*, (http//dippr.byu.edu), Brigham Young University Provo Utah, 2006.

Table 1. Isothermal CSTR solutions for various levels of uncertainty in the flow rate v_0.

	Feed flow rate (v_0) L/s					
	0.2	**0.2001**	**0.201**	**0.21**	**0.24**	**0.15**
C_A, 1st Root	**0.751536**	**0.75**33915†	**0.7**707207	1.162332	-	0.341902
C_A, 2nd Root	**2.130933**	**2.1**24481	**2.**066326	1.311824	-	-
C_A, 3rd Root	**8.117532**	**8.1**22128	**8.1**62953	**8.**525843	9.369219	-

†Correct digits are shown in **bold**

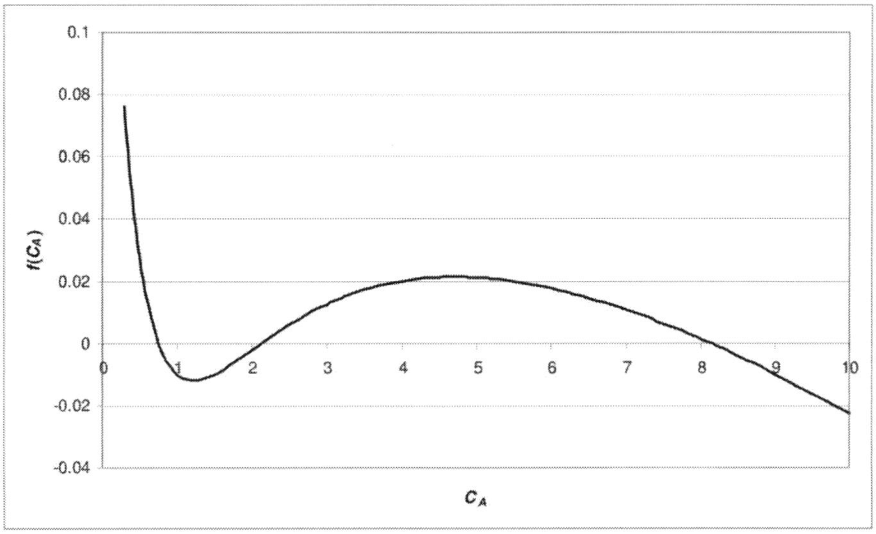

Figure 1. Plot of $f(C_A)$ versus C_A for v_0 = 0.2 L/s

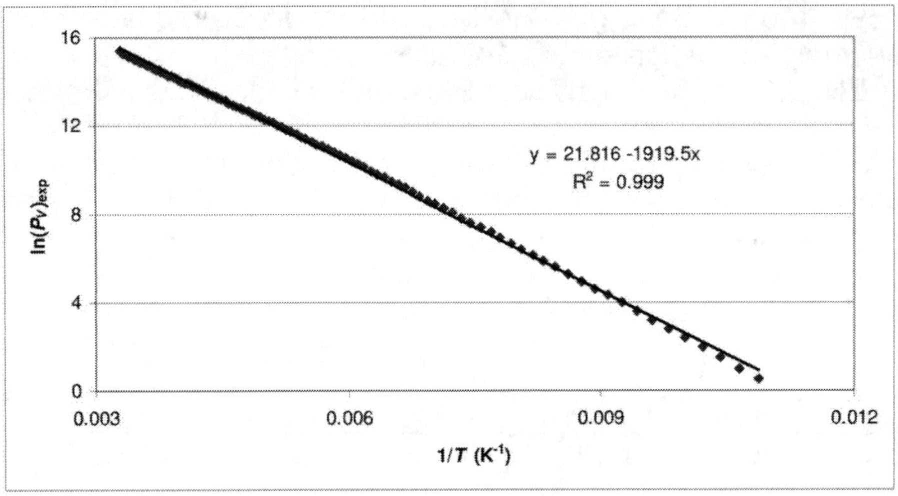

Figure 2. Plot of $\ln(P_V)_{exp}$ versus $1/T$ for ethane between the triple point and T_C

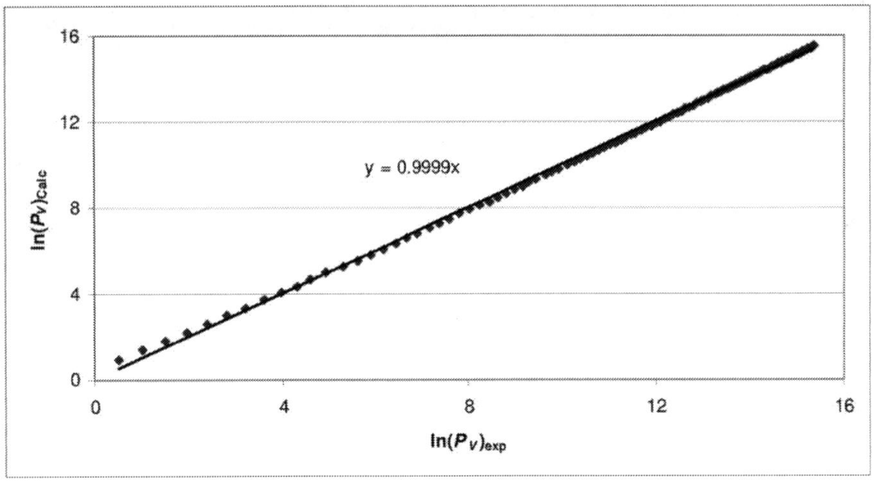

Figure 3. Plot of $\ln(P_V)_{calc}$, calculated by the Clausius-Clapeyron equation, versus $\ln(P_V)_{exp}$ for ethane between the triple point and T_C

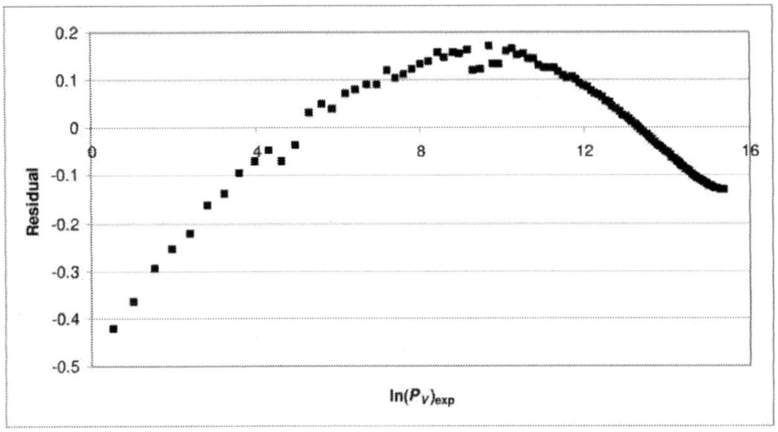

Figure 4. Residual plot for $\ln(P_V)$ calculated by the Clausius - Clapeyron equation for ethane between the triple point T_C

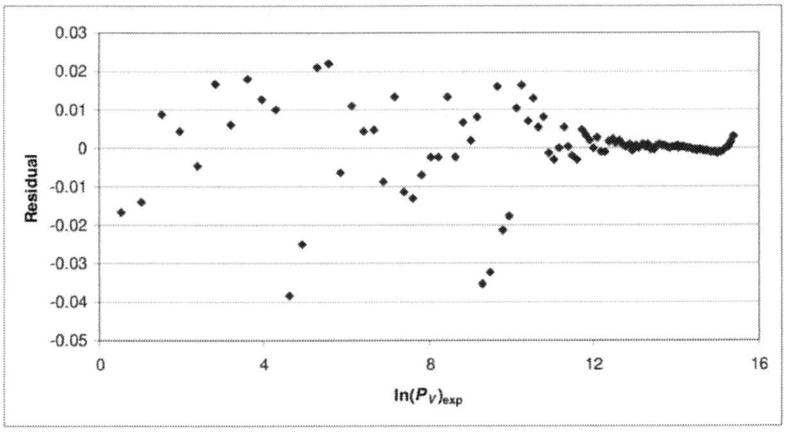

Figure 5. Residual plot for ln(P_V) calculated by the Riedel equation for ethane between the triple point and T_C

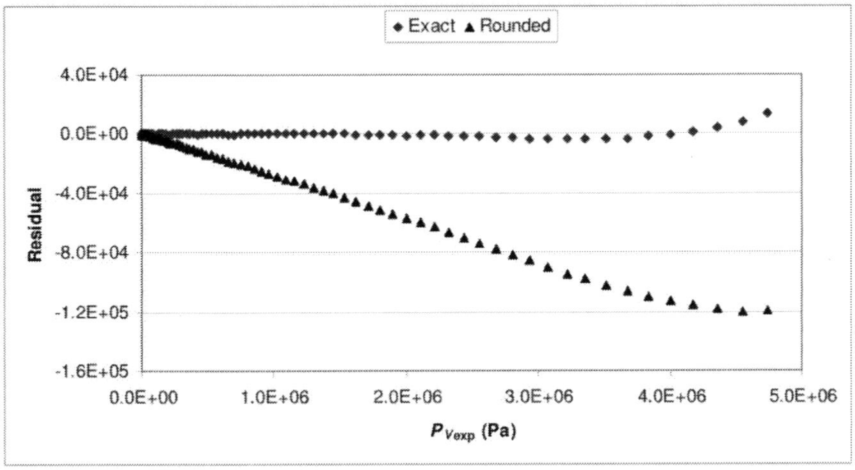

Figure 6. Residual plot for P_V, calculated by the Riedel equation, for ethane between the triple point and the T_C

225d Why Is the Fundamental of Engineering Exam (FE) Important to Chemical Engineers and the Use of the Fe as One Assessment Tool for the Chemical Engineering Programs

R. Mark Bricka, Chemical Engineering, Mississippi State University, Box 9595, Mississippi State, MS 39762

The number of chemical engineers which take the FE test is proportionally small when compared to the other engineering disciplines. This is a trend which should be changed. The FE is the first step in securing professional registration. Professional registration should be strongly encouraged by all academic institutions, and the appropriate courses should be targeted to convey this fundamental knowledge to chemical engineering students.

In addition to the need for professional registration the FE provides a useful tool for academic assessment. Details will be discussed on how to use the FE test results as an assessment tool for Chemical engineering programs.

For several years the chemical engineering department at Mississippi State University (MSU) has been using the results of the Fundamentals of Engineering (FE) Exam as one of the many tools to access the performance of students in meeting the educational objectives as part of the yearly ABET assessment process. In the past s part of the yearly process the faculty at MSU has offered informal reviews session to assist the students in preparing for the exam.

This year MSU the faculty and have formalized the FE review course and is offering this course to the all graduating chemical engineers for course credit. The students were provide a practice test prior to taking the course and will be tested once again after completing the course. The results of this assessment will be available for inclusion at this presentation.

The purpose of this presentation will be to discuss how MSU has used and will be using the FE results as an assessment tool. Previous changes in the undergraduate curricula as a result of this assessment process will be discussed. In addition the evolution and effects of the review process on testing scores and it impact will also be presented.

225e A Degree-Project Approach to Engineering Education

C. Ted Lee Jr., Chemical Engineering and Materials Science, University of Southern California, Los Angeles, CA 90089-1211 and Gisele Ragusa, Rossier School of Education, University of Southern California, Los Angeles, CA 90089.

Chemical engineering education is facing a growing disconnect between a curriculum focused primarily on "unit operations" (e.g., heat exchangers and distillation columns) and faculty research that has increasingly emphasized nano- and bio-technology. This discrepancy was recognized by an NSF-sponsored Frontiers in Chemical Engineering Education initiative, recommending a move from the macroscopic, unit-operations educational approach to instead teach from the molecular point of view in a bottom-up fashion. The challenge, however, is to continue to serve the more conventional chemical and petroleum industries while instituting this change. At USC we have developed the two-pronged approach of utilizing (1) a recently-created nanotechnology course-work emphasis within the Department of Chemical Engineering and Materials Science, and (2) vertically- and horizontally-integrated "degree projects" consisting of nano or bio laboratory modules in successive chemical engineering courses that build upon a student's growing knowledge in their chosen emphasis, while at the same time relating the degree project to traditional areas of chemical engineering. Students in the Nanotechnology Emphasis, for example, synthesize nanoparticles in the Mass Balance course, examine the interaction strength between these nanoparticles in Thermodynamics, size-fractionate these nanoparticles in Separations, investigate nanoparticle catalyst in Kinetics, and examine the thermal conductivity of nanocolloids in Heat Transfer, all culminating with an independent research project in the senior year. A comprehensive assessment strategy is utilized to study these changes to the chemical engineering curriculum, a collaboration between faculty in USC's Viterbi School of Engineering and Rossier School of Education. Three assessment measures are utilized, including an observational rubric, a chemical engineering efficacy scale, and a chemical engineering success scale. This allows robust evaluation of how the merger of traditional chemical engineering subjects with advanced nanotechnology and biotechnology topics using a degree-project approach may better prepare students for today's increasingly molecular-oriented workplace. Preliminary results of year one data reveals chemical engineering efficacy is affected by a variety of factors including gender, experiences, perspective on collaboration, and area of specialization within chemical engineering.

Targeted Infusion Project: Implementation of a Bioengineering Concentration in the Department of Chemical Engineering at Prairie View A&M University

Felecia M. Nave, Michael Gyamerah, and Irvin W. Osborne-Lee
Prairie View A&M University, Prairie View, TX

ABSTRACT

With the emergence of newer technologies, many of which are steeped deeply in chemical engineering principles, the chemical engineering profession has witnessed an increased decline in the number of students choosing it as a field of study. To address this issue, many chemical engineering programs have updated their curricula to reflect the emergence of biotechnology and nanotechnology, as chemical engineering is a natural fit. Hence, a number of chemical engineering departments have changed their names and curricula to reflect a shared focus on biology and chemistry, while others see fit to maintain their original name and have "bio-"concentrations/tracks or offer a biochemical degree within the chemical engineering department [1]. Consistent with these trends, the Department of Chemical Engineering at Prairie View A&M University has initiated steps to implement a bioengineering concentration as a option within its program. These efforts were supported using funding from the National Science Foundation HBCU-UP Targeted Infusion Program. This paper presents an overview of the accomplishment-to date of this project, future work and lessons learned.

INTRODUCTION & BACKGROUND

The engineering profession is currently facing an unprecedented array of external and internal pressures. It is becoming more complex, with traditional disciplinary boundaries blurring or disappearing; thus, graduates in any engineering discipline may find themselves working in emerging fields e.g. biotechnology and nanotechnology. As a result, graduates are under increasing pressure to have knowledge in these emerging areas. Engineering programs are in the precarious position of adapting to meet the continuously changing workforce needs in order to graduate students whose knowledge is relevant to the 21st century marketplace.

The demand for a continuous supply of highly trained scientists and engineers continues to serve as one of the most formidable issues facing our nation. Despite the increasing demand, the number of engineering degrees awarded over the next decade is projected to remain relatively stagnant [2], thus, severely hampering national efforts to meet the needs of the US technological workforce. Nationally, there is a growing consensus among leaders in the scientific and engineering communities that if we are to meet the ever increasing demand for a highly skilled workforce, it is necessary to tap the underutilized pipeline and increase the numbers of minorities who are interested in pursuing and graduating with degrees in engineering and technology.

With the daily advances in biotechnology, the increasing importance of nanotechnology and exponential demand for a more skilled biotech/nanotech workforce, many universities are updating their curriculum in order to graduate students better prepared to enter and compete in the new global marketplace. With the emergence of newer technologies, many of which are steeped deeply in chemical engineering principles, the chemical engineering profession has witnessed a decline in the number of students choosing it as a field of study [1-3]. Whereas there are a number of factors that contribute to the enrollment decline, the shift of students who would normally pursue careers in chemical engineering degrees to bio-related departments (i.e. biomedical, biological, bioengineering, etc.) has had a significant impact [3]. To address this issue, many chemical engineering programs have changed their names and updated their curriculum to reflect the shared focus on biology and engineering. Hence, a number of prominent chemical engineering departments have changed their names and curricula, while others see fit to maintain their original name and have "bio-"concentrations/tracks or offer a biochemical degree within the chemical engineering department [1]. Previously, this list contained only one Historically Black College or University (HBCU) with a department name change and one with a bio-related track.

Worldwide, the fastest growing global biotechnology marketplace includes approximately 4300 companies in 25 nations with revenues estimated at over $40 billion [5]. The biotechnology industry clusters have identified workforce development as the second or third largest hurdle to commercialization and economic success [6]. Hence, the survival, maturation and success of the biotechnology industry is ultimately linked to the ability to not only hire and retain a workforce that possesses skills and fundamental knowledge in traditional areas of biochemistry, molecular biology, pharmacology, engineering and chemistry, but also the ability to apply this knowledge to solve fundamental nano-scaled and bioengineering problems. Given the daily advances in biotechnology, the increasing importance of nanotechnology and exponential demand for a more skilled biotech/nanotech workforce, it is imperative that university curricula are reflective of market shifts (while maintaining fundamental principles) in order to graduate students better prepared to enter and compete in the new global marketplace.

HBCU's continue to contribute significantly to the overall number of minority students exiting the engineering pipeline despite limited federal support [7]. Often, they are unable to respond quickly and implement curriculum changes that focus on emerging technologies. Further, faculty at many HBCUs developed their research focus before the evolution of biotechnology and transitioning to new research areas, particularly with the high teaching loads and lack of facility infrastructure, can be quite challenging if not impossible. Currently, there are twelve (12) HBCUs that have engineering programs accredited by the Accreditation Board for Engineering and Technology (ABET), yet only six (6) have chemical engineering programs. With the ever growing emphasis on the synergy between biology and engineering and the recognized significance of HBCU's to the engineering pipeline, increasing the number of bioengineering programs at minority serving institutions is necessary if the science and technology community are to meet the ever growing needs of the biotechnology and nanotechnology workforce.

The National Science Foundation (NSF) has developed a cadre of programs to assist minority serving institutions to meet the scientific and technological needs of the US workforce. One such program is the Historically Black Colleges and Universities Undergraduate Program (HBCU-UP). "This program provides awards to enhance the quality of undergraduate science, technology, engineering, and mathematics (STEM) education and research at HBCUs as a means to broaden participation in the Nation's STEM workforce" [8] In 2005, The NSF expanded the HBCU-UP solicitation to include two additional funding opportunities: Education Research Project and Targeted Infusion Project (TIP). The TIP program provides awards to "achieve a short-term, well-defined goal to improve the quality of undergraduate STEM education such as specialized accreditation or certifications, establishing new programs or concentrations, establishing collaborations between STEM disciplines and teacher education programs, and updating programs to reflect advances in the field and workforce requirements" [8].

The Department of Chemical Engineering at PVAMU is one of only six chemical engineering programs at an HBCU. The department offered a Bachelor of Science in Chemical Engineering (B.S.Ch.E.) degree. Of the six institutions, only two programs have updated their curricula to reflect the shared focus on biology and engineering. Florida A&M University (FAMU), a public institution, has changed its name to the Department of Chemical and Biomedical Engineering and Tuskegee University, a private institution, has added concentrations to its curriculum. For more than 20 years, PVAMU Chemical Engineering (ChE) has graduated students well-trained and prepared to enter the engineering workforce and/or pursue further studies in graduate school in the traditional areas of ChE. With the ever growing emphasis on the synergy between biology and engineering, it is paramount that PVAMU also update its curriculum in order to remain competitive and be a major contributor to the biotechnology workforce. Based on thoughtful consideration of the future technological and scientific needs of the department, US workforce, and the profession in general, the faculty concluded that it is paramount our educational efforts align with the skills and knowledge relevant to our nation's needs. If bio-nanotechnology is the way of the future then, the need to be prepared cannot be over emphasized.

Over the past seven years, the department has made strategic efforts to expand its capabilities in the bioengineering area. Prior to this period, the department was comprised of five (5) faculty members, whose area of expertise was tertiary oil recovery, engineering optimization, and process safety. In 2001 and 2003, two new faculty members were hired to support the department's shift in focus and expand the academic and research capabilities in biotechnology. With the addition of the new faculty members, the department was able to initiate curriculum updates and increase student interest in course and research opportunities in this area. However, to make the required academic modifications, the faculty quickly recognized that financial support was necessary to achieve any significant changes. In 2006, the department applied for and was subsequently awarded the NSF Targeted Infusion Grant to develop a bioengineering concentration in the Department of Chemical Engineering. The objective of this paper is to discuss the steps that guided the implementation of this award, developing a bioengineering concentration in the Department of Chemical Engineering.

GOALS AND OBJECTIVES

The overarching mission of the TIP is to: to provide an intellectually rigorous undergraduate educational program that emphasizes fundamental engineering and life sciences and that will prepare students to pursue further education in bioengineering or successful careers in businesses related to various bioengineering industries. To achieve our mission, the **goal** of this project is to *transform the chemical engineering program at Prairie View A & M University (PVAMU) in order to better prepare students for careers and/or graduate study in nanoscaled bioengineering.* The specific *objective* of the project is to: establish a bioengineering concentration within the Department of Chemical Engineering.

PROJECT ACTIVITIES

Prior Works

In the five years prior to the TIP funding, the department had initiated several key steps to facilitate the transition into this new emerging technology. The initiatives included: (1) strategic recruiting of faculty whose area of expertise was in biotechnology, (2) providing start-up funds, although very limited, to support their research activities, and (3) introduction of a new course under the "Special Topics" label aimed at increasing student interest. The facility infrastructure to conduct biotechnology research was limited. To increase research capabilities in this area, the faculty applied for external financial support through the National Science Foundation Major Research Instrumentation Program. In 2005, a the department was awarded a MRI: Award Abstract #0421287 entitled: *Acquisition of Research Instrumentation for Applied Research and Training in Biotechnology and Bioprocess Engineering* , A special topics course, Introduction to Biotechnology and Bioprocess Engineering, was developed and first offered as a technical elective in 2002. Table 1 outlines the growth in student interest as reflected in the continued growth in student enrollment in the semesters the course was offered.

Table 1. Course Enrollments

Activity	Course Number	2007	2006	2005	2004	2003	2002
Intro to Biotechnology & Bioprocess Engineering*	CHEG 4103	-	-	13	9	-	10
Bioengineering	CHEG 4153	10	N/A	N/A	N/A	N/A	N/A

The equipment purchased with the MRI grant allowed the department to increase the number of undergraduate research opportunities and thus increase student interest in bio-related topics as demonstrated in Table 2. Additionally, the increased infrastructure positioned the department to partner with other larger institutions (i.e. University of Kansas, University of Iowa, University of California-Berkeley, Harvard, etc.) and successfully secure NSF funding of two (2) Engineering Research Centers: the Center for the Environmentally Beneficial Catalysis (CEBC) and the Synthetic Biology Engineering Research Center (SynBERC).

Given these strategic moves, the department began to examine the chemical engineering curriculum to assess how to implement further expansion of the curriculum in support of biotechnology. The following descriptions show how the department was able to build on prior success coupled with the TIP award to successfully implement a "new" bioengineering option within the Department of Chemical Engineering.

Proposed TIP Activities

The project activities that were proposed for the TIP project were designed to allow the department to strategically analyze its current curriculum and craft a bioengineering concentration that would provide the theoretical background students would need in order to be successful and to compete in the current biotechnology marketplace. The project activities included the following.

1. Petitioning for course and curriculum approvals.

2. Restructuring of the chemical engineering curriculum to reflect the bioengineering option.
3. Development of new courses and laboratory experiments.
4. Development supplemental materials for existing chemical engineering course.
5. Increased research opportunities and exposure to bioengineering principles for undergraduate.

Restructuring of Chemical Engineering Curriculum

The current chemical engineering curriculum is designed to prepare engineers who are well qualified to design and operate chemical processes. The undergraduate baccalaureate degree required that students complete 135 hours, of which 20 hours of chemistry and 41 hours of chemical engineering specific courses. The bioengineering option requires that students complete 20 hours of course materials, of which 17 hours are integrated into the current curriculum, specific to bioengineering concepts, which are spread amongst the current chemistry and chemical engineering requirements, as detailed in Table 3 below. The courses were selected to build the students knowledge, skills, and expertise in basic fundamental bioengineering concepts. Students choosing the bioengineering option will thus complete 138 hours for the B.S.Ch.E degree. Additionally, bioengineering concepts are being vertically and horizontally integrated throughout the chemical engineering curriculum. The concentration requirements were constructed by examining other programs with bio-related options [9-10]. Hence, the requirements are aligned with programs at other national and regional universities of similar and larger size and comparable mission statements. The course requirements are as follow (Table 3).

Table 3. Course Requirements for Bioengineering Option

Courses	Hours
Biology	**3**
Biochemistry	**3**
Biochemistry Lab	**2**
Bioengineering Technical Elective (2 courses)	**6**
Senior Design I & II (Bioengineering project)	**6**
Total Hours	**20**

Concentration Approval

Once the above curriculum requirements were outlined and received approval from its entire faculty, the department moved forward to secure approval of the concentration through the process outlined by the university. To establish a concentration at PVAMU, the department must first receive approval of any curricula changes from the requesting College Curriculum Review Committee (CRC). Once approved by the CRC, the department must then submit necessary paperwork for review by the University Academic Council (UAC). Once this process has been approved, paperwork is submitted to the Texas A&M University System Coordinating Board for final approval. The department was able to successfully navigate the outlined procedures and receive subsequent approval of the concentration. Effective fall 2008, the Department of Chemical Engineering will offer a concentration in bioengineering as an option under the B.S.Ch.E. degree program.

Course Approval

Given the increased interest among students for topics in area of biotechnology and to support the "new" concentration, two courses were updated. The technical elective CHEG 3153, Special Topics: Introduction to Biotechnology and Bioprocess Engineering was originally introduced as a Special Topics course in 2002 and has experienced growth in enrollment during the three semesters offered. CHEG 4153, Bioengineering, is a course that is in the course inventory but had not been taught for several years. Enrollment figures for both courses are listed above in Table 2. These courses status were modified using the same approval process outlined above. The CHEG 3153 course was renamed, its description updated, and it was requested to be added as a stand-alone course to support the new bioengineering option. The CHEG 4153 course description was also updated. Effective fall 2008, CHEG 3153 (formerly CHEG 4103), Introduction to Biotechnology and Nano-Scaled Processes was successfully added to the curriculum and will be offered for the first time under the new name and description in the spring 2009 semester.

Increased opportunities for Undergraduate Research

Research opportunities for both graduate and undergraduate students have increased with the hiring of the new faculty, improved academic and facility infrastructure, and financial support from the TIP as shown in Table 4. Notably, the students conducting undergraduate bio-related research represent approximately 80% of the undergraduate students conducting research in the Department of Chemical Engineering.

Table 4. Research Participation in Biotechnology

Activity	2008	2007	2006	2005	2004	2003	2002
Undergraduate Research in Bioengineering	10	8	12	12	8	4	1
Graduate Research in Bioengineering	1	1	2	2	1	1	-

The students have been actively involved in conducting authentic, publishable research. These research accomplishments have been presented at both regional and national conferences. Two students have received partial travel support to the National Organization for the Professional Advancement of Black Chemists and Chemical Engineers (NOBCCHE) annual conference. Two supplemental awards have been approved to support a Faculty and Student (FaST) research team. One faculty and 5 student researchers spent 10 weeks at Oak Ridge National Laboratories in Knoxville, TN in summer 2007 and 2008 in the FaST program. Additionally, one of the undergraduate research students received an internship with Eli Lilly during summer 2007, a first for the department with this company; the student was subsequently hired by a company to do bio-related work. Also, four (4) of undergraduate research students received a Research Experience for Undergraduates at Texas A&M University (College Station) during summer 2007.

Development of Laboratory Experiments and Supplemental Materials

During the spring 2008 semester, a lab manual was developed for implementation of the new experiments in the Chemical Engineering Lab 1 and Chemical Engineering Lab 2 courses. To support these experiments, the undergraduate lab was furnished with state-of-the-art **new instrumentation**. Bio-related material and problems have been introduced in the following courses: (1) Equilibrium Staged Separation Processes, (2) Kinetics and Reactor Design, (3) Heat, Mass, Momentum Transport, (4) Introduction to Engineering Course (freshman level), and (5) Senior Design & Professionalism I and II. The faculty anticipates developing formal supplemental documentation that can be used by all faculty for the above listed courses and other courses.

CONCLUSION AND DISCUSSION

The Department of Chemical Engineering was successful in their efforts to develop a bioengineering option. Awarding of the Targeted Infusion Project by NSF provided the last integral piece in order for the department to move forward and expand its academic and research capabilities in this area. The department is moving forward to recruit students, develop additional courses, and seek research funding.

Lessons Learned

One of the most important lessons learned was that we did not adequately anticipate the amount of time and human resources that would be needed in order to implement the bioengineering concentration. For this reason, it took longer than originally estimated to accomplish the measures that have been completed to date. PVAMU is primarily a teaching institution, hence our normal teaching load is four course (or 12 credit hours) per semester. However, there is a focus on increasing the amount of research conducted by faculty within the College of Engineering. Additionally, maintaining student interest and being able to recruit new students has become challenging. Although we are able to introduce the concepts in new courses and existing courses, it is often difficult to keep up with the expanding interest of the students and very challenging to do any active recruiting of new students. Given these factors combined with the fact that the new faculty hired to develop the bioengineering program are Assistant-level faculty, additional human resources were needed to meet the tasks associated with the new concentration.

ACKNOWLEDGEMENTS

The authors would like to acknowledge the National Science Foundation for their support through grants # 0636409 and # 0421287.

REFERENCES

1. Engineering Trends: Undergraduate Enrollments (Total Full-Time and First-Year) Provide Degree Trends in Engineering Disciplines through AY 2006-07. Retrieved from http://www.engtrends.com/IEE/0404C.php on January 4, 2008.
2. Rhinehart, R. R., "Analysis of Enrollment Cycling in Chemical Engineering", Chemical Engineering Education, Vol. 35, No. 1,Winter 2001, pp 50-57.
3. ASEE Prism: Pursuing New Paths. Retrieved from http://www.prism-magazine.org/nov03/pursuing_paths.cfm on February 13, 2008.
4. Retrieved from: http://www.doleta.gov/BRG/Indprof/Biotech.cfm on February 8, 2006.
5. Retrieved from http://www.bioorg/speeches/pubs/er/statistics.asp on February 8, 2006.

6. Holtzclaw, David. J, et. al (2006). "Incorporating a New Bioinformatics Component into Genetics at a Historically Black College: Outcomes and Lessons", The American Society for Cell Biology, Vol. 5, 52-6
7. Retrieved from http://www.doleta.gov/BRG/Indprof/Biotech.cfm on November 3, 2006.
8. Retrieved from http://www.nsf.gov on September 22, 2008.
9. Retrieved from http://www.bioorg/speeches/pubs/er/statistics.asp on November 3, 2006.
10. Retrieved from http://www.aiche.org on November 3, 2006.

Implementing Concepts of Pharmaceutical Engineering Into High School Science Classrooms

Howard Kimmel, New Jersey Institute of Technology, Newark, NJ
Linda S. Hirsch, New Jersey Institute of Technology, Newark, NJ
Laurent Simon, New Jersey Institute of Technology, Newark, NJ
Levelle Burr-Alexander, New Jersey Institute of Technology, Newark, NJ
Rajesh Dave, New Jersey Institute of Technology, Newark, NJ

Abstract

The New Jersey Institute of Technology in conjunction with the Center for Pre-College Programs and research groups within the Engineering Research Center for Structured Organic Particulate Systems (ERC-SOPS) established a summer Research Experience for Teachers (RET). The goal of the RET was to educate high-school science teachers in the opportunities and challenges involved with manufacturing of pharmaceutical products, and thus help educate the future generation of students, helping create a strong pipe-line of talented students interested in pursuing careers in engineering and science. Nine teachers representing chemistry, biology, and physics, were recruited from schools in local urban districts, and were able to develop skills and knowledge in science and engineering with a focus on the area of pharmaceutical particulate and composite systems from which they created instructional modules to be integrated into their teaching practices. The teachers worked in teams of two along with a faculty mentor on different projects that evolved from current research in C-SOPS which included: a) control and characterization of flowability of cohesive powders; b) The Electromagnetic Spectrum and Raman Scattering Spectroscopy; c) The Mixing Efficiency of Dissolution Testing for Pharmaceutics and Engineering; and d) Crystallization of Ultrafine (Nano and Micro) Particles of Active Pharmaceutical Ingredients. The evaluation process focused on the summer experience and the impact it had on the teachers and their classroom practices. During the summer research experience and through the following school year teachers completed survey instruments to gauge: 1) the effectiveness of the research experience and the supportive activities that took place, 2) changes in the teachers' concerns about integrating engineering skills into classroom practice, 3) their preparedness to teach the engineering skills they learned and 4) changes in their attitudes toward engineering and knowledge about careers in engineering. Results indicate that teachers felt the experience was useful and that their concerns about implementing engineering skills in their classroom changed from needing more information to considering the impact the change(s) in their teaching would have on their students. After completing the program teachers felt they were prepared to teach the engineering skills they learned, had more self-efficacy for helping students who might want to study engineering and teachers knew more about careers in engineering and what engineers do.

Introduction

Engineering plays a major role in shaping the world today. The application of science, mathematics and technology into engineering benefits people and makes the world we live in possible. Most students are unaware of the benefits that engineering provides people in their daily lives, from developing consumer goods, to creating artificial devices such as knees or hearts and are not interested in pursuing careers in engineering. Many bright, capable high school students choose not to pursue studies in science and mathematics and do not prepare academically for college degree programs that would allow them to pursue careers in engineering and technology (1, 2). As a result, the United States currently has a shortage of qualified workers in the science, technology, engineering and mathematics (STEM) fields that will continue at least into the next decade (3, 4). For the long-term economic health of this country it is important that more students undertake studies and pursue careers in these fields to meet the current needs of the American work force (5).

One of the more critical reasons most students, particularly those from underrepresented populations in urban school districts, are not interested in pursuing careers in engineering is that they are not exposed to topics in engineering during their K-12 studies. Most K-12 teachers have not been trained to incorporate engineering and technology topics into their classroom lessons and there is a lack of high-quality curricular materials in these areas (6). As a result, students are not exposed to the engineering and technology resources used to develop strategies for solving real-world problems.

There has been a growing interest by higher education institutions to bring engineering principles and applications to the secondary school classrooms. Comprehensive professional development programs are needed for teachers to address the new skills and knowledge needed for improved classroom teaching and learning (7, 8) if we expect them to integrate engineering concepts into their classroom practice (9-11). Some of the identified factors that should be included in successful professional development programs include: Long term effort, Technical assistance, as well as support networks, Collegial atmosphere in which teachers share views and experiences, Opportunities for reflection on one's own practice, Focus on teaching for understanding through personal learning experiences, and Professional development grounded in classroom practice.

A focus is needed on content included in currently available curriculum materials that creates connections between the science used in engineering applications in the real world and science curriculum standards for which teachers and administrators are held accountable (6, 12, 13). Most existing engineering curricula lack an appropriate translation into standards-achieving lessons for enriching the science curriculum. Hence, teachers need to be able to translate state content standards into effective practices, while integrating engineering concepts into standards-achieving lessons for enriching the science curriculum. However, many teachers lack knowledge about standards-based lesson planning, and the availability of resources for standards-based lesson planning. While substantial energy has been devoted to the identification of standard-based curriculum materials and achievement tests, little is known about new lesson planning, teaching, and student activities that may be needed in a standards-based classroom. O'Shea & Kimmel (14) have developed a protocol for standards-based lesson planning that allows teachers to systematically assess learning outcomes that are aligned with state content standards.

Research Experiences for Teachers (RET) are seen as a vehicle for introducing engineering into secondary school curricula and as a strategy for increasing student interest in engineering, and ultimately increasing enrollment of qualified students in engineering degree programs (15-17). Miller & Winter (15) described a 5-week summer experience for middle and high school science, mathematics, and technology teachers. Follow up experiences were difficult due to geographical distances between teachers and the University. No evaluation data were included. Orlich, Zollars & Thomson (16) provided a 6-week research experience for high school science teachers and were able to maintain communication with the teachers and their students through technology. Instructional modules were developed as a result of the research experiences. There were plans to develop an instrument to measure students' attitudes towards engineering. Conrad, Conrad & Auerbvach (17) provided engineering research experiences for physics teachers, with a focus on modern physics concepts and applications to microelectronics packaging. Assessments included pre- and post-surveys and surveys of teachers' activities during the school year.

An RET at New Jersey Institute of Technology (NJIT) has been designed to provide high school science teachers with a professional development program that enhances their research skills and their knowledge of science and engineering concepts that enables them to incorporate real-world applications (e.g., pharmaceutical engineering) into high school science curricula. As part of the program teachers developed instructional modules they could use to integrate engineering principles into their classroom teaching. The project also focused on helping the teachers refine their instructional planning skills and providing them with an effective protocol for developing standards-based lesson plans.

The Setting

The RET program at NJIT is a collaboration of the Engineering Research Center for Structured Organic Particulate Systems (ERC-SOPS) and the University's Center for Pre-College Programs initiated under an NSF-sponsored four-university project. The goal of the program is to educate high-school teachers in the opportunities and challenges involved with manufacturing of pharmaceutical products, and thus help educate the future generation of students, helping create a strong pipe-line of talented students interested in pursuing careers in engineering and science.

Over the past several years, a number of research and education initiatives have been developed at NJIT in areas related to particulate systems and multiphase systems, and pharmaceutical processing and engineering. These initiative have resulted in the development of a strong research program in engineered particulates (especially particulates of various types of nano-structured composites), and in educational programs including; novel graduate-level degrees, research programs for undergraduates, outreach activities targeting K-12 teachers and students; all of which have helped to strengthen NJIT's research and education role in these critical areas.

The Center for Pre-College Programs (CPCP) at NJIT has almost 40 years experience working with the public school systems in Newark and others across the state of New Jersey (18). The mission of the Center is to improve the quality of education in elementary and secondary New Jersey school districts, especially urban districts, by; 1) Providing leadership in the planning, development and assessment of STEM education programs; 2) Developing and coordinating academic programs to serve elementary and secondary school teachers; and 3) Conducting outreach programs across the state for students, parents, teachers and school administrators.

Among the many successful programs at CPCP is Pre-IOP, the Pre-Engineering Instructional and Outreach Program, established to work with the public school systems of Newark and the surrounding cities to raise awareness about the importance of pre-engineering concepts in science and mathematics curricula (10, 19). Pre-IOP included the development of pre-engineering curriculum modules (aligned with the New Jersey Core Curriculum Content Standards) for use in middle and high school mathematics and science classrooms. Teacher professional development programs were established to train teachers how to integrate the pre-engineering curriculum into their classroom teaching as a way for their students to apply their classroom lessons to real-life problems. Outreach efforts were found to increase awareness of careers in engineering and knowledge about what engineers actually do. Use of the pre-engineering curriculum in science, mathematics and technology classroom was found to increase students' and teachers' attitudes toward engineering and knowledge of careers in engineering (20, 21). The RET program at NJIT continues the work of PrE-IOP by incorporating pharmaceutical concepts into the high school science curriculum.

The Research Experience

The 2007 NJIT Research Experiences for Teachers (RET) program provided the opportunity for nine high school science teachers (chemistry, biology, and physics) to engage in a six-week experience in a research group of the Center for Structured Organic Particulate Systems (C-SOPS). Working side-by-side with university research faculty, graduate students, and undergraduate students (participating in a parallel REU-Site program), in discovery based, hands-on research projects, teachers developed basic knowledge and skills in the area of pharmaceutical particulate and composite systems that can be incorporated into their teaching practice. Implicit was the opportunity for intellectual professional growth for the teachers.

The first week was an orientation program that included an introduction to NJIT and to ERC-SOPS and its research activities, methodologies, instrumentation, and safety procedures. Teachers were introduced to the scientific tools, protocols, and equipment necessary for gaining meaningful hands-on experience in the laboratory, from which they could develop basic knowledge and skills in the area of pharmaceutical particulate and composite systems that they can incorporate into their teaching practice. The teachers were introduced to the basic knowledge and skills they would need to be a contributing member of the research teams, and to be able to develop standards-based lessons/modules that they would share with their colleagues and bring back to their classrooms. Presentations were made by research faculty on research projects in which the teachers could be involved. A discussion among NJIT faculty and teachers considered the broader impacts of the program on healthcare and the role of the high school science teachers. Teachers were also provided with an introduction to the technical literature and methodologies for searching the web to support their research activities, as part of the development of the teachers' information literacy and related skills and knowledge. On-going discussion during the summer experience focused on the development of lessons and modules.

RET projects were small sub-projects within many of the research at ERD-SOPS, recognizing that much of the research deals with concepts that can be difficult to translate into laboratory and instructional activities for high school classrooms. Simplified versions of the basic concepts in a research project were developed. For example, dissolution of particles can be related to basic concepts of solubility, equilibrium, and rates of processes by developing simple experiments that involve observing dissolving of sugar crystals of varying size, with or without stirring or agitations. Teachers worked in teams of two that also involved at least one graduate student and one undergraduate REU student. The REU students will have had several weeks of experience by the time the RET program begins, and hence the team consisting of one graduate student and one REU student will be well-versed in the research project that the RET participants will participate in.

For example, in one project, a method for dry particle coating was utilized to deposit a very small amount of nano-sized additives with high degree of precision onto drug or excipient particles in order to change their flow and other properties. RET participants examined the application of this technique on improvement, control and characterization of flowability of cohesive powders in a predictive manner through dry particle coating. Another project focused on crystallization, the most common method used in the pharmaceutical industry for generating particles of active substances or intermediates. The teachers examined the role of agitation on crystal size as part of the research study the hydrodynamics of a stirred tank-impeller assembly, with particular attention being paid to solid dispersion and the determination of the minimum agitation speed for off-bottom solid suspension, both in the presence and the absence of an impinging jet apparatus. Other teachers examined the feasibility of using some water soluble edible polymer and FDA approved surfactant to disperse water insoluble drugs molecules in aqueous medium. Detailed characterization of materials was conducted using dynamic light scattering, FT-infrared, Raman and solid-state NMR spectroscopy. The materials were also characterized for film thickness, controlled release, dissolution rate testing, kinetics and selectivity. Another project involved the mixing efficiency of dissolution testing for pharmaceuticals.

Development of the instructional modules were critical to the RET program. Teachers and their mentors met frequently to develop a simple topic that is closely related to the pharmaceutical industry as well as the research they were conducting. To be effective, the modules had to address important issues including:

- What are the real-life implications of the research?
- What experiments will best relate the information to students in an exciting, insightful way?
- What are the materials and methods required to perform these experiments? Will there be insurmountable safety issues in planning such an experiment?
- The step-by-step procedure for disseminating the information to students in a logical way.
- Are the necessary tools accessible in the HS laboratories?
- What assessment will be used to show that students have internalized the information?

Because there were an odd number of teachers, one of the teachers served as a "swing teacher" working jointly with each team to monitor progress of the teams and communicate with the mentors. The swing teacher developed an instructional module that encompassed the research projects of the other teachers, "A Step toward Discovery: Inquiry Skills in Science", designed to help students think like engineers and scientists, while connecting relevant mathematics and science skills.

Standards-Based Lesson Planning

Curricular materials in support of the integration of engineering into science instruction have been made available through organizations such as NASA, ASME and IEEE, as well as through universities and teacher developed lesson plans. However, only concepts included in state content standards are taught in the classroom, as teachers believe they will only be accountable for what is in the standards (14). As a result, the only curriculum materials usually considered, let alone implemented, are those that reinforce state content standards, since student achievement (and schools' and districts' achievement) is measured largely by student performance on the statewide assessment tests (22). So, if teachers are to make engineering principles a part of their instruction for student learning, then engineering principles and design must be a part of the state science standards. Unfortunately, most existing engineering curricula lack an appropriate translation into standards-achieving lessons for enriching the science curriculum. Translation into standards-achieving lessons is critical (6).

The use of standards-aligned curriculum materials is necessary, but not sufficient for students to achieve the standards (14, 23). Curriculum with topics aligned to engineering standards is also not sufficient. Alignment with standards must also include the assessment of student achievement of the skills and knowledge defined by the standards. Criteria must be established that allow teachers to determine whether their students have achieved the standard. Several curriculum efforts have reported developing procedures for relating learning objectives with standards and assessment of student performance (24, 25). However, the reported alignment of standards is mostly a referencing to standards only, and assessments are generally missing or do not measure student achievement of the learning objectives.

Research suggests that lesson and unit plans are essential and powerful tools for instructional improvement and increased student achievement (23). When teachers prepare truly standards-based lessons, their teaching is focused on student achievement in relation to specific standards (24). Aligning curriculum, instruction, and assessment with goals for student learning (i.e., learning objectives) is an essential first step (25). A protocol for the creation and implementation of standards-based lesson plans has been developed at CPCP and utilized in previous and current professional development programs (14). The protocol starts with the concept that is to be taught. One or more measurable learning objectives are identified for the lesson. For each learning objective the corresponding statement from the content standards is then specified. The activity for the lesson is examined to be sure that it provides the student the opportunity to acquire the skill and/or knowledge specified by the learning objective and the appropriate statement of the standards. Finally the expected student performance is described (a student behavior or work product) that will provide the evidence that the student has acquired the skill and/or knowledge of the learning objective and the statement of the standard. The RET participants were introduced to the protocol and a template was developed for use in the development of their instructional modules.

Evaluation

Teachers' Concerns about Integrating Engineering skills into Classroom Teaching: Teachers concerns about integrating engineering skills into their classroom teaching were measured using The Teachers' Concerns Questionnaire (TCQ) adapted from the Concerns Based Assessment Model (CBAM) [26]. Repeated administrations of the TCQ are used to identify teachers' concerns and track changes in their concerns as they engage in educational reforms, focusing on how they progress through seven stages of concern: Awareness, informational, personal, management, consequences, collaboration and refocusing. Teachers completed the TCQ at the beginning and the end of the RET program and again a few months into the school year after they had some time in their classrooms. All three sets of responses were examined and indicate that initially, some teachers showed low levels of awareness and\or some were not very interested. By the end of the program most teachers appeared to have increased their awareness and many had moved into the information gathering stage. Not until a few months into the school year did some of the teachers begun shifting toward whether the new curriculum would help their students learn math and/or science. Three teachers completed the TCQ toward the end of the school year, expressing fewer personal and management concerns about the time commitments required to implement their new instruction modules. The teachers were focused on how the implementation may have impacted their students and appeared to be shifting into the collaboration stage.

Teachers' Readiness to Teach: At the end of the RET program teachers completed a Readiness to Teach Questionnaire (RTQ). The RTQ (20, 21) requires teachers to indicate how ready they feel they are to teach lessons on new topics and\or skills they have learned on a scale from 1 to 4 where 1='I would have to start from scratch', 2='I would need more training to teach this topic', 3='I would have to look at my notes to do this' and 4='I can teach a lesson on this topic tomorrow'. For example, one item asks 'How ready are you to teach the concept of steady state?' Teachers were asked to complete the Readiness to Teach again a few months into the school year after they had some time in their classrooms. At end of the summer program average scores for the 13 topics ranged from 2.8 to 3.8 indicating that most of the responses were 3='I would have to look at my notes' or 4='I can teach a lesson on this topic tomorrow'. Only one teacher gave any responses that indicated 1='I would have to start from scratch'. Less than 20% of the responses indicated 2='I would need more training to teach this topic'. For many topics the percentage of teachers that indicated 4=' I can teach a lesson on this topic tomorrow' was over 50%. Average scores for most of the 13 topics increased slightly a few months into the school year; ranging from 3.2 to 3.8. The average scores for two of the topics did not change and only one topic 'Drug release from a Lozenge' showed a decrease in the average response from 3.1 to 2.8. This was due mostly to a few teachers indicating 3='I would have to look at my notes' the second time rather than their initial response of 4='I can teach a lesson on this topic tomorrow' at the end of the program. Again, three of the teachers completed the RTQ a third time toward the end of the school year. The average scores for the 13 topics ranged from 3.5 to 4.0 indicating that at least these three teachers could teach all of the topics even if they had to look at their notes.

Attitudes to Engineering: Teachers completed the Attitudes to Engineering Survey (TATE) at the beginning of the RET program and again a few months into the school year after they had completed the program and had some time in their classrooms. The TATE (20, 21) measures teachers overall attitudes toward engineers and engineering as a career as well as their knowledge of careers in engineering and their self-efficacy for assisting students who might be interested in studying engineering. Teachers' attitudes toward engineers and engineering as a career was found to be fairly high, even at the beginning of the program. All nine teachers agreed that 'skills learned in engineering are useful in everyday life' and disagreed with the statement 'I would not like any of my students to be engineers'. Their average Attitudes to Engineering scores before the beginning of the summer program was 3.9 which increased to 4.2 during the school year.

Most teachers were somewhat informed about how to help prepare students that might be interested in studying engineering. Most agreed they knew 'where to find the necessary information to help their students if they wanted to become engineers' but most disagreed with the statement 'I have all the information I need to help prepare any of my students who may want to be an engineer'. Only a few indicated they knew of summer programs to help students learn more about careers in engineering. Average scores on the items that assess teachers' self-efficacy for helping students who might want to study engineering were low, only 3.0 at the beginning of the program but increased to 4.3 during the school year.

Knowledge of engineers and engineering as a career is measured using a multiple-part open-ended question which asks teachers to 'Name five different types of engineers' and to 'give an example of the work done by each type'. Each type of engineer is coded '1' for correct and '0' for incorrect. Possible total scores range from zero to five. Each example of the work they do is coded '2' for completely correct, '1' for partly correct, and '0' for incorrect. Possible total scores range from zero to ten. At the beginning of the program only five of the nine teachers were able to correctly name five different types of engineers and two of them were only able to name two types correctly. Only one of the teachers was able to give correct or partly correct examples of the work done by all five types of engineers receiving 7 points. One teacher did not give any examples and the rest were only able to give one, two or three partly correct examples. When the teachers completed the survey again a few months later results showed that teachers' knowledge of engineers and engineer as a career had increased. Six of the teachers were able to correctly name five different types of engineers, two teachers named four types and the last teacher named three. All of the teachers were able to give at least some partly correct examples of the work done by the types of engineers they named most scoring at least 5 or 5 points, a few scored 8 or 9 points.

Teachers' feedback on program effectiveness: Periodically during the program teachers were asked to provide written feedback on how they felt the program was progressing. Teachers were asked to rate each activity or learning experience by indicating how useful they felt it was to them as a teacher (2= very useful, 1=somewhat useful or 0=not useful) and the value they felt it had for student learning (2=high value, 1=some value or 0=no value). The average usefulness rating for a majority of the activities was at least 1.5. Two activities, a poster presentation session to share their research experience with others and the mentoring process, had an average rating of 1. Many of the teachers just did not find the poster presentation very useful. Two of the teachers rated the mentoring process as not useful. Unfortunately one of these two teachers reported in additional comments that their mentor had "not been available" during the program. The teachers also found a majority of the activities to have a high value for student learning in their classrooms. Most activities had an average rating of at least 1.6. The activities that teachers did not find useful for their students, scoring only an average of 1 or less were things like tours of laboratories, poster presentations and discussions of on-going research.

Conclusions

Teachers found the RET program useful to them as teachers and found a lot of value in the experience for their students. Many of the teachers expressed an interest in repeating such an experience. Participation in the RET program increased teachers' attitudes towards engineering, knowledge of engineering careers and their self-efficacy for helping students who might be interested in studying engineering. The program is being repeated in 2008 with a greater focus on collecting student data to help evaluate the impact of the program on students as well as teachers.

Acknowledgments

This project is based upon work supported by a grant from the National Science Foundation, ERC Supplement Award for an RET Site, EEC-0540855, and is gratefully acknowledged. Any opinions, findings and conclusions or recommendations expressed in this material are those of the authors and do not necessarily reflect the views of the National Science Foundation (NSF).

References

1. American Society of Engineering Education (2004), Engineering in the K-12 Classroom: An Analysis of Current Practices & Guidelines for the Future. ASEE. Washington, D.C.
2. American Society of Engineering Education. Engineering Education and the Science & Engineering Workforce. Public Policy Internet Website. http://www.asee.org.
3. National Academy of Engineering (2002), Raising Public Awareness of Engineering. The National Academies Press. Washington, D.C.
4. National Science Board (2004), Science and engineering indicators 2004, Arlington, VA: National Science Foundation, Washington, D.C. (NSB-00-1).
5. New Jersey Commission on Higher Education (2001), Higher education outcomes and high-tech workforce demands. The fifth annual system wide accountability report. NJCHE, Trenton, NJ.
6. Kimmel, H., John Carpinelli, J., Burr-Alexander, L. & Rockland, R. (2006), "Bringing Engineering into K-12 Schools: A Problem Looking for Solutions?," Proceedings of the 2006 ASEE Annual Conference. Chicago, IL,. June .
7. Guskey, T.R. (1986) "Staff Development and the Process of Teacher Change". Educational Researcher, Vol. 15, pp. 5-12.
8. Joyce, B. & Showers, B. (1988) Student Achievement Through Staff Development. Longman. New York.
9. Zarske, M., Sullivan, J., Carlson, L. & Yowell, J. (2004), "Teachers Teaching Teachers: Linking K-12 Engineering Curricula with Teacher Professional Development". Proceedings of the 2004 ASEE Annual Conference, Salt Lake City, UT, June 20-23,.
10. Kimmel, H. & Rockland, R. (2002), "Incorporation Of Pre-Engineering Lessons Into Secondary Science Classrooms". Proceedings of the 32nd ASEE/IEEE Frontiers in Education Conference, Boston, MA, November 6-9,
11. Kimmel, H. & O'Shea, M. (1999) "Professional Development and the Implementation of Standards". Proceedings of the 29th ASEE/IEEE Frontiers in Education Conference, San Juan, PR, November 10-14.

12. Anderson-Rowland, M., Baker, D.R., Secola, P.M., Smiley, B.A., Evans, D.L. & Middleton, J.A. (2002), "Integrating Engineering Concepts Under Current K-12 State and National Standards." Proceedings of the 2002 ASEE Annual Conference, Montreal, PQ, Canada, June 16-19.
13. Schaefer, M., Sullivan, J. & Yowell, J. (2003), "Standards-Based Engineering Curricula as a Vehicle for K-12 Science and Math Integration." Proceedings of the 33rd ASEE/IEEE Frontiers in Education Conference, Boulder, CO, November 5-8.
14. O'Shea, M. & Kimmel, H. (2003) "Preparing Teachers for Content Standards: A field Study of Implementation Problems". Proceedings of the American Association for Colleges of Teacher Education, New Orleans, LA, January 25.
15. Miller, B. and Winter, R.M. (2006), "RET Site: Inspiring Educators in Rural America through Research." CD 2006 AIChE Annual Meeting Conference Proceedings. New York, NY, AIChE.
16. Orlich, D, Zollers, R, and Thomson, W. (2006). "Introducing Engineering at the Middle School and High School Level." Proceedings of the 2006 ASEE Annual Conference, Chicago, IL,. June 2006.
17. Conrad, L., Conrad, E, and Auerbach, J. (2007). "The Development, Implementation and Assessment of an Engineering Research Experience for Physics Teachers." Proceedings of the 2007 ASEE Annual Conference, Honolulu, HI,. June 2007.
18. Kimmel, H. & Cano, R.M. (2003) "Model for a K-12 Engineering Pipeline", Proceedings of the 2003 ASEE Annual Conference, Nashville, TN, June 22-25.
19. Carpinelli, J., Burr-Alexander, L., Henesian, D., Kimmel, H. & Sodhi, R. "The Pre-Engineering Instructional and Outreach Program at the New Jersey Institute of Technology." Proceedings of the International Conference on Engineering Education (ICEE 2004), Gainesville, FL, October 16-21 2004.
20. Hirsch, L., S., Kimmel, H., Rockland, R. and Bloom, J. (2005). "Implementing Pre-engineering Curricula in High School Science and Mathematics". Proceedings of the 35th ASEE/IEEE Frontiers in Education Conference, Indianapolis, IN., October 2005.
21. Hirsch, L., S., Kimmel, H., Rockland, R. and Bloom, J. (2006). "Using Pre-engineering Curricula in High School Science and Mathematics: A Follow-up Study". Proceedings of the 36th ASEE/IEEE Frontiers in Education Conference, San Diego, CA, October 2006.
22. National Research Council. Systems for State Science Assessment, 2005. National Academic Press. Washington, DC.
23. Tell, C.A., Bodone, F.M. & Addie, K.L. "A Framework of Teacher Knowledge and Skills Necessary in a Standards-Based System: Lessons from High School and University Faculty". Presented at the Annual Meeting of the American Educational Research Association, New Orleans, LA, April 24-28, 2000.
24. Rothman, R., Slattery, J. B., Vranek, J. L., and Resnick, L. B. Benchmarking and Alignment of Standards and Testing. CSE Technical Report 566. National Center for Research on Evaluation, Standards, and Student Testing, UCLA, Los Angeles, CA, 2002.
25. Rutherford, F.J. and Ahlgen. Science for All Americans. 1991. Oxford University Press, New York, NY
26. Hall, G., E., George, A., A., and Rutherford, W., L. (1980). "Stages of Concern about the Innovation: The Concept, Verification, and Implications". Distributed by Southwest Educational Development Laboratory, Austin, Texas.

In Search of the Active Site of pMMO Enzyme: Partnership between a K-12 Teacher, a Graduate K-12 Teaching Fellow, and a Research Mentor

Katherine K. Bearden[1], Tanya Culligan[2], and Daniela S. Mainardi[1]

[1]Louisiana Tech University, Ruston, LA
[2]Caddo Parish Magnet High School, Shreveport, LA

Abstract

In this work, we describe the partnership between a K-12 teacher, a GK-12 graduate student, and a Louisiana Tech faculty member collaborating in a research and education project for 6 weeks. Tanya Culligan (NanoResearcher), pairs with Katherine Bearden(Teaching Fellow), a doctoral student in Chemical Engineering, and Dr. Daniela Mainardi (Research Mentor), an assistant professor of Chemical Engineering and Nanosystems Engineering at Louisiana Tech. Culligan is assigned the task of exploring the location of the active site of a very important enzyme: Particulate Methane Monooxygenase (pMMO), which relates to environmental bio-catalysis and involves atmospheric methane consumption (oxidation) for the production of fuel (methanol). With the guidance of the Research Mentor, the help of the Teaching Fellow, and the use of scientific software with great visualization capabilities, Culligan is driven through the different steps needed to take to accomplish her goals. By the end of the 6-week program, the K-12 teacher develops a module, with the aid of her Research Mentor, to take and implement in her biology classes to explain the concept of the research she conducted on the modeling of enzymes. A description of the program, research project, and collaboration benefits is provided.

Introduction

There are many sources available for the implementation of outreach at Louisiana Tech University (LA Tech) and in the surrounding community. Within the College of Engineering and Science, Louisiana Tech has two National Science Foundation (NSF)-funded programs: the GK-12 Creating Connections (NSF grant 0638730) program, and the Nanoscience Educational Research Outreach (NERO) program's Research Experience for Teachers (RET) (NSF grant 0602029). These programs create an environment for graduate students, university faculty, and teachers in the surrounding community to (1) develop inquiry-based science laboratories for K-12 grades, (2) expose K-12 teachers to nanotechnology principles, equipment, and research, and (3) engage K-12 teachers in specific research experiences spanning six weeks in summer where they are mentored by university faculty.

Through the implementation of the NERO RET program at LA Tech since 2007, sixteen local teachers have participated in a 6-week summer program; which contains professional development and research components. The RET program provides the unique opportunity for collaboration between a K-12 teacher (NanoResearcher), a GK-12 graduate student (Teaching Fellow) and a Louisiana Tech faculty member (Research Mentor) in science, technology, engineering and math disciplines (STEM).

As part of their professional development, the RET participants are first introduced to concepts of nano-scale science and engineering through a series of seminars focusing on scientific literacy. Then they are engaged in hands-on experiences that aid them in understanding protocols, running simple experiments, collecting data, and analyzing the corresponding results.

Combined with professional development activities, RET participants are exposed to independent research work under a STEM faculty member with guidance from a Teaching

Fellow. Research projects available for the RET participants focus on various branches of nano-scale science, such as the fabrication of cellular capsules for regenerative medicine, analysis of varying L-arginine concentrations on platelet adhesion, and modeling of enzyme docking for environmental applications.

At the end of the 6-week program, the RET participants orally present their independent research work, as well as education activities they have designed together with their research mentors on how to take back to their classroom the research concepts they have learned. Explaining nano-scale research to students in the K-12 grade levels is truly challenging; however exposure to concepts of nano-scale science creates a foundation for student inquiry and provides students with extensive applications of the abstract science concepts they learn within their curriculum.

Research Initiative Overview

Tanya Culligan (NanoResearcher), a ninth grade biology teacher at Caddo Parish Magnet High School in Shreveport, Louisiana, has participated in the 2007 and 2008 NERO RET programs at LA Tech. In both opportunities, she paired with Daniela S. Mainardi (Research Mentor), an assistant professor of Chemical Engineering and Nanosystems Engineering at LA Tech, and Katherine K. Bearden (Teaching Fellow), a doctoral student in Chemical Engineering under Mainardi's supervision. In 2008, Culligan worked during her RET summer experience in the Mainardi computer laboratory on the modeling of enzymes docking with potential environmental technology applications involving methane (natural gas; chemical formula = CH_4) bioremediation.

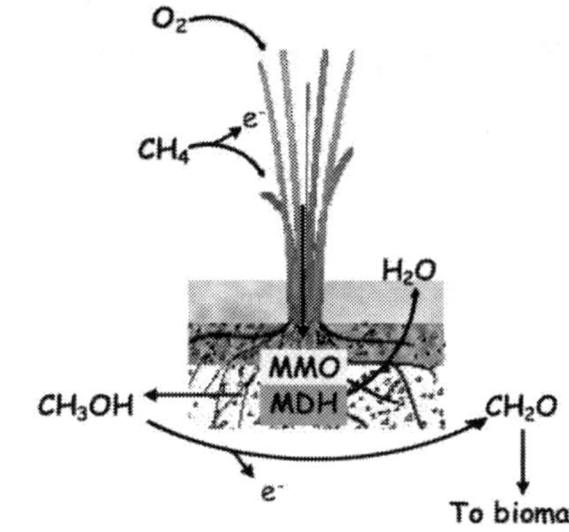

Figure 1. MDH cycle of sustainable methanol production and oxidation from atmospheric CH_4.

Methane is well known to be a relatively potent greenhouse gas with a high global warming potential. CH_4 emissions in the atmosphere are approximately 600 teragrams (Tg) per year[1], 10% of which is produced by paddy plants. On a scale similar to that of methane are methanol emissions at 82 to 273 Tg/yr with living plants as the major source[2, 3].

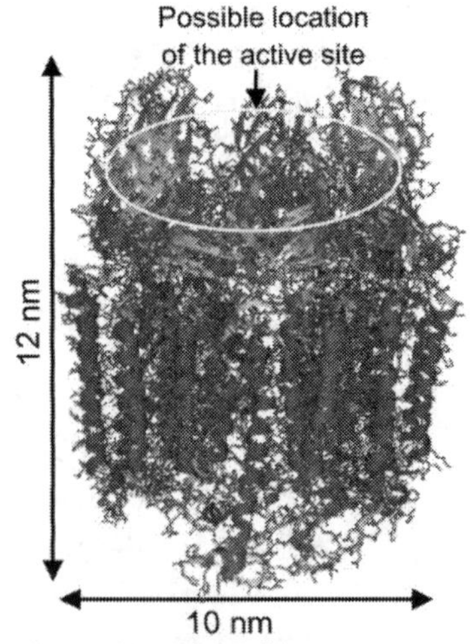

Figure 2. pMMO X-ray structure from the entry 1YEW of the Protein Data Bank.

Methane and other one-carbon compounds, including methanol, formaldehyde, and formate, are oxidized by methylotrophic microorganisms, which are associated with the rhizosphere of paddy plants[4, 5]. In particular, methane oxidation by *Methanotrophic* bacteria, a subgroup of the methylotrophs, is considered as an important sink for CH_4[6, 7]. These bacteria use methane as their sole carbon and energy source, and therefore are of significant interest in terms of playing a key role in the cycling of carbon in the biosphere[8]. Particularly, the role of methanotrophs in the reduction of global emissions of methane, their potential commercial use for the biotransformation of numerous organic chemicals into valuable products, and their capacity for the bioremediation of toxic pollutants have been well recognized[9, 10].

Methane Monooxygenase (MMO) and Methanol Dehydrogenase (MDH) enzymes are found in *Methanotrophic* bacteria. While MMO is known to exhibit the unique catalytic capacity for converting methane to methanol under ambient conditions using dioxygen as the oxidant[11], MDH is well known to catalyze the oxidation of methanol to formaldehyde, which is assimilated into biomass (Figure 1)[5, 12].

Methane Monooxygenase enzymes exist in two distinct forms: a soluble (sMMO) and a membrane-associated (pMMO) form, depending on the level of copper ions in the growth medium. When the growth medium contains copper at sufficient levels (>5μM) under high copper/biomass ratios, the particulate pMMO form is expressed[9, 11].

Particulate Methane Monooxygenase is the predominant methane oxidation catalyst in nature[6]. Lieberman et al.[13] have determined its crystal structure from *Methylococcus capsulatus* (Bath) at 2.8 Å resolution, and reported that this enzyme consists of a 300 kDa trimer with an $\alpha_3\beta_3\gamma_3$ polypeptide arrangement (Figure 2). There are two copper centers located in soluble regions of each α subunit, which resembles cytochrome *c* oxidase subunit II, and there is a third metal center occupied by zinc, which is located within the membrane[13]. These authors also suggested that the active site of pMMO is constructed partially (with 50% of the copper) of a multi-copper complex; although the nuclearity of the Cu cluster in pMMO is still unknown[13]. Even less is known about the natural electron donor(s) and electron-transfer pathway during the methane oxidation (methanol production) by pMMO, and therefore the mechanism by which this enzyme performs this unique conversion is not understood.

There is evidence that Methanol Dehydrogenase (MDH) is co-localized within the intracytoplasmic membranes network in the organism when *Methanotrophic* bacteria are expressing pMMO[14]; and recently, Myronova et al.[15] have shown that the reductase form of the pMMO complex is in fact MDH. Hence, this is an indication that pMMO and MDH not only cooperatively work together but also are in close proximity with each other in *Methanotrophic* bacteria.

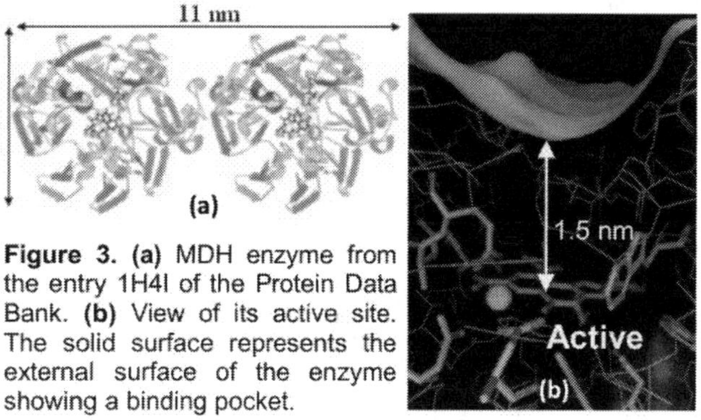

Figure 3. **(a)** MDH enzyme from the entry 1H4I of the Protein Data Bank. **(b)** View of its active site. The solid surface represents the external surface of the enzyme showing a binding pocket.

Methanol Dehydrogenase (Figure 3a) is a NAD(P)-independent enzyme of broad substrate specificity. The crystal structure of MDH from bacteria *Methlylobacterium extorquens*[16, 17] and from *Methylophilus W3A1*[18, 19] has been characterized and it has been determined that the enzyme has an $\alpha_2\beta_2$ tetrameric structure. Moreover, it was also reported that its active center (site) contains a Ca^{2+} ion, a pyrrolo-quinoline quinine (PQQ) molecule, which serves as its redox cofactor[16, 18-20], and amino acids (Figure 3b). What is not understood, however, is (1) how the two enzymes work together in the bacteria and (2) the specific location of the active site of pMMO.

Since methanol dehydrogenase enzyme is co-localized within the membranes network in the organism when *Methanotrophic* bacteria are expressing pMMO, Culligan (NanoResearcher) was assigned the investigation of different docking situations for these enzymes during her 2008 summer research experience in the Mainardi group. Hence, Culligan's overall goal was to determine the most likely configuration of the two enzymes in order to gain insight into the actual location of the pMMO enzyme's active site. In order to achieve her goal, Culligan needed to (1) create appropriately sized models of the MDH and pMMO enzymes (2) optimize the geometries of the enzymes individually, and (3) explore different MDH/pMMO arrangements by performing geometry optimizations to find the lowest energy configuration leading to the most likely position for the interacting enzymes (as they would within the *Methanotrophic* bacteria were they are found).

Methodology

The computational modeling technique utilized to obtain results for Culligan's simulations is Molecular Mechanics[21]. The principle behind Molecular Mechanics is that it uses an energy equation to describe bonded interatomic interactions including bond lengths, angles, and dihedrals, and also non-bonded interactions such as electrostatic and van der Waals. Such an equation is known as the Force Field, and it is trying to provide a fitting to the real Potential Energy surface of the molecular system[21]. During the simulation, this function is minimized, using energy minimization methods to find a minimum (equilibrium structure) of the molecular system, which represents a stable conformation.

The Universal force field was utilized by Culligan in her research because it was the best available to describe atomic interaction involving the zinc and copper metal ions in her molecular system. Upon geometry optimization using molecular mechanics, as implemented in the module Forcite of the Materials Studio software by Accelrys, Inc.[22], the lowest energy configurations for the MDH/pMMO model system was found and fully structural characterization performed.

Results

Culligan first built a model for the MDH enzyme based on the current state of knowledge on the location of its active site. The MDH enzyme,,entry 1H4I of the Protein Data Bank, was imported and its active site was found and highlighted. Then, Culligan added an amino acid shell surrounding the entire MDH active site thus creating a 1,300 atom model to represent MDH (Figure 4a). The MDH model was then geometry minimized using Molecular Mechanics.

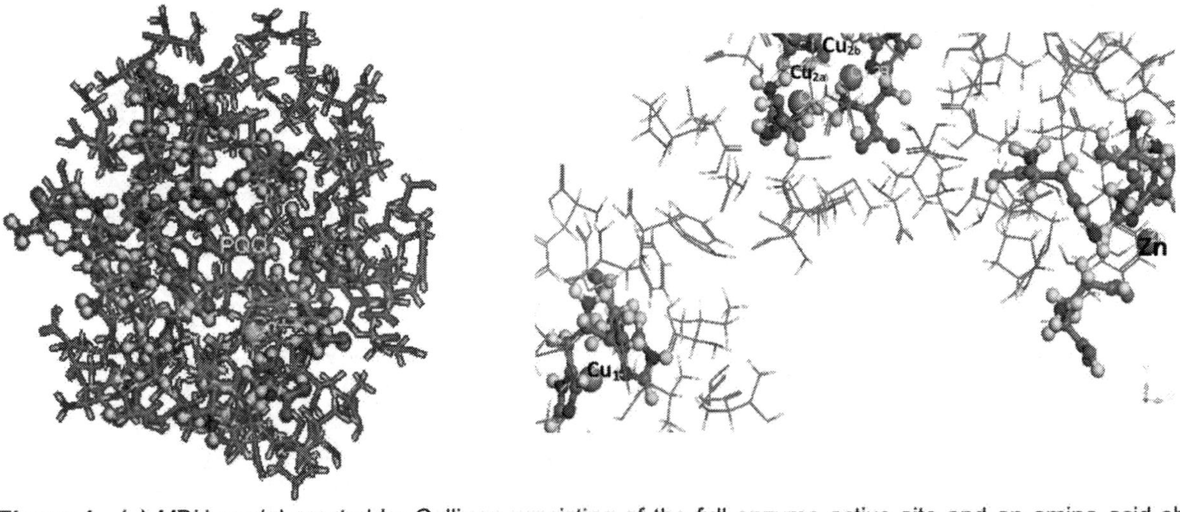

Figure 4. (a) MDH model created by Culligan consisting of the full enzyme active site and an amino acid shell surrounding it. **(b)** pMMO model created by Culligan based on the current state of knowledge on its active site location and contents. Note the three distinct areas: a single copper ion on the left (Cu_1), a two copper ion complex (Cu_{2a} and Cu_{2b}) in the center and a single zinc ion on the right.

A second model, for the pMMO enzyme, was created based on the current state of knowledge on its active site location and contents using the entry 1YEW of the Protein Data Bank. The active site of pMMO is not fully understood and is still being explored; however current research seems to indicate that it is part of a complex involving four ions and some amino acids. The four ions are metallic and make up three distinct sections of the site. The first is a single copper ion (Cu_1), the second is a double copper ion (Cu_{2a} and Cu_{2b}), and the third consists of a zinc ion (Figure 4b). The pMMO model was then geometry minimized using Molecular Mechanics.

Once the optimizations of the models were complete, they were paired together in ten different configurations to determine a likely position for the interaction (docking) of these two enzymes within the *Methanotrophic* bacteria. All cases were geometry minimized using Molecular Mechanics simulations (Figure 5).

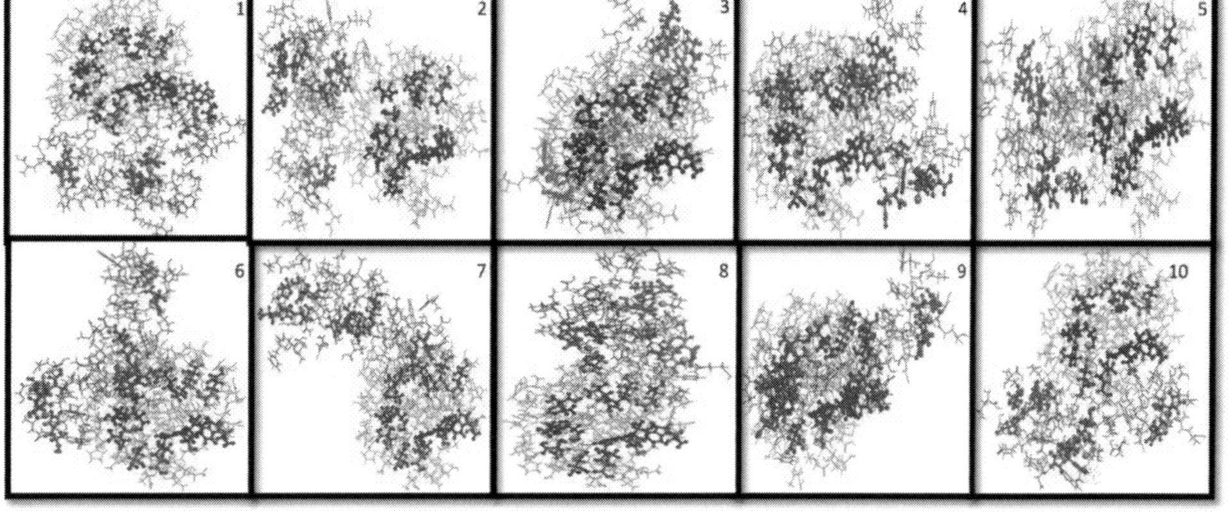

Figure 5: Ten different geometry minimized conformations for the MDH/pMMO models docking (Left to right orientation with single copper as a reference point). Color Key: MDH (blue), PQQ (yellow), calcium ion (bright green) pMMO (green), three copper ions (orange), and the zinc ion (grey).

After completing each simulation, Culligan recorded relevant bond lengths, angles between atoms of particular interest (the copper and zinc ions) as well as the overall distance between the Ca^{2+} ion in MDH active site and other metal ions in the pMMO model under investigation. This preliminary data provides insight in determining the optimum arrangement for pMMO and MDH. The information gathered provides a baseline of investigation and indicates and region of particular interest to concentrate on in future simulations in the Mainardi research group. Moreover, Culligan found the most stable MDH/pMMO configuration showed the shortest distance between the calcium ion of MDH and the zinc Ion of pMMO, making the zinc ion closest to the active site of MDH (Figure 6).

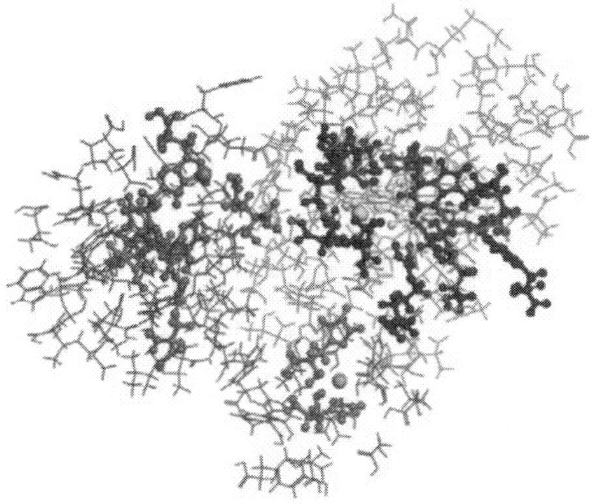

Figure 6. Minimum energy configuration of pMMO model interacting with MDH model (Case 5 from Figure 5). Color Key: MDH (blue),PQQ (yellow), calcium ion (bright green), pMMO (green), , three copper ions (orange), and the zinc ion (grey).

This information aids the ongoing investigation by Bearden (Teaching Fellow) to determine if the two enzymes can have close contact to facilitate the methane to methanol oxidation reactions. In trying to orient MDH and pMMO in search of the configuration most likely used in nature, the complementary shapes of the enzymes suggest that they do interact and their active sites are not too far apart to make oxidation of methane to methanol in pMMO and methanol to formaldehyde in MDH a concurrent and regulatory process. Further simulations are needed to establish a trend in determining the most stable configuration of the MDH/pMMO interaction.

Incorporation of Research into the Classroom

As part of the RET initiative, participants had education goals along with their research goals. Culligan's education goal was to design and prepare learning materials and "in vivo" demonstrations of integrated research and educational activities for the students in her class. Culligan also collaborated with Bearden and Mainardi to create a learning module to take back to her biology classes.

From her learning module, Culligan took molecular modeling into her classroom and allowed her students to build molecular groups (atom by atom) including amines, ester, and hydroxyl groups using available software. She utilized the visualization aspects of the modeling software to present the concepts of condensation, hydrolysis, and nitrogen base bonding. The students also used ball and stick components from a purchased chemistry kit to physically build the purines (adenine and thymine) and pyrimidines (cytosine and guanine) structures found in DNA. This activity was constructed to confirm that adenine and thymine have two hydrogen bonds and cytosine and guanine have three hydrogen bonds, a benchmark in the Louisiana biology curriculum.

A second learning module Culligan created used molecular modeling to depict the different processes involved in cellular respiration. Students were engaged in the building of the molecular groups involved in the different steps of respiration (i.e. glycolysis, pyruvic acid breakdown, citric acid (or Krebs) cycle, and the electron transport chain). Students used the visualization aspects of the modeling software to present the concepts of hydrogen ions transporting and the use of an electron transport chain. The students were assessed in a carousel activity incorporating the information retrieved from the molecular bonding interactions, through computation and ball and stick models.

Conclusions

Collaboration aids all parties involved. The knowledge base of the Mainardi (research mentor) research group was increased through the results found by Culligan, a ninth grade biology teacher. During the six weeks of work, a very short period in the academic research community, Culligan was able to produce a sound baseline for further research. Her results aided in directing Bearden (GK-12 teaching fellow) to a region of particular interest in the goal of determining the active site of pMMO enzyme. Her research had a significant impact in suggesting the role of the Zinc ion in pMMO, which will be confirmed in future experiments performed by Bearden. Culligan, herself was exposed to molecular modeling techniques as she explored the interaction between MDH and pMMO enzymes. The learning experience provided by the NERO RET program expanded her knowledge of ongoing research in the Nanotechnology field, gained professional development tools, and created a module to implement in her classroom to convey the concepts of molecular modeling into her classroom. The partnership gave the members a better appreciation for the opportunities for collaboration available at the university and K-12 levels of education that provide direct channels for research to be integrated into the K-12 classrooms.

Acknowledgements

We gratefully acknowledge the financial support from the National Science Foundation (NSF) under the CAREER CTS-0449046, RET(0602029) and GK-12(0638730) grants. Support for computational resources for both software and hardware through the Louisiana Board of Regents, contract LEQSF(2007-08)-ENH-TR-46 and the National Science Foundation grant number NSF/IMR DMR-0414903 are also thankfully acknowledged.

We also appreciate the direction and constant assistance provided by Dr. David K. Mills and Ms. Linda Ramsey of Louisiana Tech University, Principal Investigators of the NSF NERO-RET and NSF-GK-12 programs that financially supported Tanya Culligan's summer experiences and Katherine K. Bearden's doctoral work.

References

1. F. Keppler, J.T.G. Hamilton, M. Bras, and T. Rockmann, *Methane emissions from terrestrial plants under aerobic conditions*, Nature, **439**. 2006. p. 187-191.
2. I.E. Galbally and W. Kirstine, *The production of methanol by flowering plants and the global cycle of methanol*, J. Atmos. Chem., **43**. 2002. p. 195-229.
3. A. Guenther, *The contribution of reactive carbon emissions from vegetation to the carbon balance of terrestrial ecosystems*, Chemosphere, **49**. 2002. p. 837-844.
4. S.K. Dubey, P. Padmanabhan, H.J. Purohit, and S.N. Upadhyay, *Tracking of methanotrophs and their diversity in paddy soil: A molecular approach*, CURRENT SCIENCE, **85**(1). 2003. p. 92-95.
5. R. Whittenbury and H. Dalton, *The methylotrophic bacteria*, in *The prokaryotes*, M. P. Starr, et al., Editors. 1981, Springer-Verlag: Heidelberg. p. 894-902.
6. S.K. Dubey, A.S.K. Sinha, and J.S. Singh, *Differential inhibition of CH4 oxidation in bare, bulk and rhizosphere soils of dryland rice field by nitrogen fertilizers*, Basic and Applied Ecology, **3**. 2002. p. 347-355.
7. B. Gilbert and P. Frenzel, *Rice roots and CH4 oxidation: the activity of bacteria, their distribution and the microenvironment*, Soil Biology and Biochemistry, **30**. 1998. p. 1903-1916.
8. K. Smith, K. Dobbie, B. Ball, L. Bakken, B. Sitaula, and S. Hansen, *Oxidation of atmospheric methane in Nothern European soils, comparison with other ecosystems, and uncertainties in the global territirial sink*, Global Change Biol., **6**. 2000. p. 791-803.
9. R.S. Hanson and T.E. Hanson, *Methanotrophic bacteria*, Microbiol. Rev., **60**. 1996. p. 439-471.
10. T.J. Smith and H. Dalton, *Biocatalysis by methane monooxygenase and its application to the petroleum industry*, in *Petroleum Biotechnology, Developments and Perspectives*, R. Vazquez-Duhalt and R. Quentero-Ramirez, Editors. 2004, Elsevier Science, BV: Amsterdam, The Netherlands. p. 177-192.
11. S.H. Stanley, S.D. Prior, D.J. Leak, and H. Dalton, *Copper stress underlies the fundamental change in intracellular location of methane monooxygenase in methane-oxidising organisms: studies in batch and continuous cultures*, Biotechnol. Lett., **5**. 1983. p. 487-492.
12. J.A. Vorholt, *Cofactor-dependent pathways of formaldehyde oxidation in methylotrophic bacteria*, Archives of Microbiology, **178**(4). 2002.
13. R.L. Lieberman and A.C. Rosenzweig, *Crystal structure of a membrane-bound metalloenzyme that catalyses the biological oxidation of methane*, Nature, **434**. 2005. p. 177-182.
14. C.A. Brantner, C.C. Remsen, H.A. Owen, L.A. Buchholz, and M.L.P. Collins, *Intracellular localization of the particulate methane monooxygenase and methanol dehydrogenase in Methylomicrobium album BGS*, Arch. Microbiol., **178**. 2002. p. 59-64.

15. N. Myronova, A. Kitmitto, R.F. Collins, A. Miyaji, and H. Dalton, *Three-Dimensional Structure Determination of a Protein Supercomplex That Oxidizes Methane to Formaldehyde in Methylococcus capsulatus (Bath)*, Biochemistry, **45**. 2006. p. 11905-11914.
16. M. Ghosh, C. Anthony, K. Harlas, M.G. Goodwin, and C.C.F. Blake, *The Refined Structure of the Quinoprotein Methanol Dehydrogenase from Methylobacterium Extorquens at 1.94 Å*, Structure (London), **3**. 1995. p. 1771-1787.
17. P.R. Afolabi, M. F., K. Amaratunga, O. Majekodunmi, S.L. Dales, R. Gill, D. Thompson, J.B. Cooper, S.P. Wood, P.M. Goodwin, and C. Anthony, *Site-Directed Mutagenesis and X-Ray Crystallography of the PQQ-Containing Quinoprotein Methanol Dehydrogenase and Its Electron Acceptor, Cytochrome C_L*, Biochem., **40**. 2001. p. 9799-9809.
18. S. White, G. Boyd, F.S. Mathews, Z.X. Xia, W.W. Dai, Y.S. Zhang, and V.L. Davidson, *The Active Site Structure of Calcium Containing Methanol Dehydrogenase*, Biochemistry, **32**. 1993. p. 12955-12958.
19. Z.X. Xia, Y.N. He, W.W. Dai, S. White, G. Boyd, and F.S. Mathews, *Detailed Active Site Configuration of a New Crystal Form of Methanol Dehydrogenase from Methylophilus W3A1 at 1.9 Å Resolution*, Biochem., **38**. 1999. p. 1214-1220.
20. Z.X. Xia, W.W. Dai, J.P. Xiong, Z.P. Hao, V.L. Davidson, S. White, and F.S. Mathews, *The Three-dimensional Structures of Methanol Dehydrogenase from Two Methylotrophic Bacteria at 2.6-A Resolution**, J. Biol. Chem., **267**. 1992. p. 22289-22297.
21. A.R. Leach, *Molecular Modelling Principles and Applications*. 1999, Harlow: Pearson Education Limited.
22. Accelrys, *Materials Studio*. 2006.
23. R. Balasubramanian and A.C. Rosenzweig, *Structural and Mechanistic Insights into Methane Oxidation by Particulate Methane Monooxygenase*, Accounts of Chemical Research, **40**(7). 2007. p. 573-580.
24. R.L. Lieberman, K.C. Kondapalli, D.B. Shrestha, A.S. Hakemian, S.M. Smith, J. Telser, J. Kuzelka, R. Gupta, A.S. Borovik, S.J. Lippard, B.M. Hoffman, A.C. Rosenzweig, and T.L. Stemmler, *Characterization of the Particulate Methane Monooxygenase Metal Centers in Multiple Redox States by X-ray Absorption Spectroscopy*, Inorganic Chemistry, **45**(20). 2006. p. 8372-8381.

230c The Polymer Envoys Program: A Study on the Effectiveness of An Innovative Research Program for Urban High School Students

LaRuth C. McAfee, Center for Layered Polymeric Systems, Case Western Reserve University, 2100 Adelbert Rd., Cleveland, OH 44106

In the United States certain ethnic groups are underrepresented in science, technology, engineering, and math (STEM) fields. National Science Foundation and Census Bureau data show that, while the population of Blacks and Hispanics in the United States is 12% and 13%, respectively, in STEM undergraduate programs these groups only account for 8.5% and 7.2% of the student population. This lack of minorities in STEM undergraduate programs also impacts the potential graduate student STEM population, where Blacks and Hispanics compose 5% and 4.8% of the populations. While these numbers have increased slightly over the past decade, they are still significantly lower than where they could be. An issue that makes this under-representation more significant is that the number of American jobs in STEM fields is growing at a rate much faster than the production of new STEM degrees. Therefore, it is important to consider how these underrepresented groups can position themselves to fill such positions. Many of these jobs will require college-level degrees, and students often must have an interest and show promise in math and science while in primary and secondary school to allow themselves the opportunity to pursue college degrees in the STEM disciplines. A better educated American STEM workforce is also advocated by reports such as "Rising Above the Gathering Storm" and federal initiatives such as the American Competitiveness Initiative.

To increase minority high school students' interest in STEM, the Polymer Envoys Program (PEP), an innovative outreach program, has been developed and is being piloted at Case Western Reserve University (CWRU) in partnership with the Cleveland Metropolitan School District (CMSD). This program features a two-year research internship where urban high school students are matched with a graduate student mentor, and participate in STEM research during their junior year, the summer between their junior and senior years, and their senior year. As a way to give back to their community and recruit future program participants, students also create presentations to teach younger students and the general community about polymer concepts.

A study is being used to determine the short- and long-term effectiveness of PEP. The researchers hypothesize that such an experience will increase a student's confidence in her or his technical abilities, and increase the interest that the student has in pursuing a STEM degree. The following questions are being addressed: (1) Are the high school selection and student selection processes appropriate to meet program goals? (2) How are a student's perception of and interest in STEM affected by the research experience? (3) How will the experience affect the student's perception of her or his technical abilities? (4) How will this program affect the number of urban students who go to college, and who pursue STEM degrees once there? An examination of question 4 has recently begun and limited data are available, but data on questions 1-3 will be presented. Baseline and end-of-term surveys are used to determine the impact of PEP on participants. These surveys ask students for academic and attitudinal information so that changes over time may be tracked using multiple-choice questions with Likert-like scales (1=strongly agree, 3=neutral, 5=strongly disagree). Short answer questions also ask for the students' opinions on their aspirations, and expectations or feedback on research and program activities.

CMSD mathematics and science coordinators in the district's central administration initially chose the schools involved in PEP. This process was revised recently to allow all interested CMSD high schools to submit proposals if they wanted to participate in the program, and this new school selection process was found to work very well because schools better understand their expectations and have committed to fulfill these expectations. Inversely, current schools that did not feel they could meet program expectations had an opportunity to opt-out of the program. Additionally, based on demographics in the new schools selected, it is expected that the high school student participants will become more diverse than what was achieved in the first classes of students.

Student evaluations indicate very positive results in the initial phase of PEP implementation. While many students initially indicated minimal experience with research and technical communication, in less than two years all have made significant improvements. This change was self-reported on surveys (responses of 3.6 initially, to 2.3 after one year in program), and observed in the oral and written communication assignments completed by the students. Additionally, students' academic and career aspirations have been affected through their experiences in the program. While most students initially expressed interest in becoming scientists and engineers and felt that they were capable of being in such fields (initial response of 1.9), through the program some became more interested in STEM careers (response of 1) and others became less interested (response of 2.8). While the program aims to excite students toward these fields, aiding a student in realizing that this is not her or his area of interest is equally important. Related to longer-term outcomes of PEP, two students in the initial class have been admitted to and will matriculate at CWRU, with an intention to major in engineering. Additional positive program outcomes include one student who started taking classes at a community college based on the encouragement of her graduate student mentor and her greater understanding of opportunities to better prepare for college, as well as another student who will have her research data included in a technical paper.

Based on these findings to date, three key recommendations are given for similar programs. First, schools need to have a choice on whether they are involved in this type of program to ensure necessary support for students and program activities. Additionally, students should begin in a program of this type in this type of school district as high school juniors. Finally, multiple levels of support (high school teachers, university graduate students, university faculty/staff, etc.) are necessary for students and the program to be successful.

Further studies of the program will include tests to determine how effective recently implemented technical content at program meetings is in supporting the students in their research projects and furthering their interest in STEM careers. Additionally, as possible, the long-term impact of the program will be studied to determine how many of the students eventually pursue graduate studies in STEM.

Financial support from the NSF Science and Technology Center for Layered Polymeric Systems (Grant DMR 0423914) is gratefully acknowledged.

230d K-5 STEM Enrichment Activities with Guilford County Schools and North Carolina a&t State University

Kenneth L. Roberts, Mechanical and Chemical Engineering, North Carolina A&T State University, 1601 E. Market St., 627 McNair Hall, Greensboro, NC 27411-0001 and Sophia R. Roberts, Dr. David D. Jones Spanish Immersion Elementary School, 502 South Street, Greensboro, NC 27406.

The development of increased numbers of future engineers and scientists in the United States, especially from impacted rural and urban communities, will require strong partnerships between academia, K-12 institutions, local and state government, corporate partners and community-based organizations. For students from economically-disadvantaged backgrounds, the use of tangible rewards such as candy, inexpensive toys, and fast food treats, for positive behavior reinforcement can be quite an effective motivator for student achievement. The establishment of professional, personal relationships between students and faculty which inspire trust show success in student achievement. These professional relationships can be initiated and maintained via such preplanned activities as classroom visits by faculty and students, enrichment efforts (i.e., tutoring, reading, etc.), and field trips.

The historical challenges in closing the academic achievement gap from current educational research appears to be more influenced by psychological and sociological phenomena as opposed to intellectual, economic, or technological factors. The most important tool for the promotion of student achievement success in underrepresented and also economically-disadvantaged groups had been noted to be the project of the expect high academic success without the presence of stereotypical attitudes and also without the use of emotionally and psychologically stressful consequence-based student learning evaluation techniques. In this work, the use of science and mathematics-based activities to educate students from impacted elementary schools in the Guilford County Schools will be discussed.

230e Using Lego^{TM} Robotics for K-12 Engineering Outreach

Bill B. Elmore, Dave C. Swalm School of Chemical Engineering, Mississippi State University, MS 9595, Mississippi State, MS 39762 and Emma Seiler, Bagley College of Engineering, Mississippi State University, McCain Hall, Mississippi State, MS 39762.

The LEGO® NXT robotics system is being used as a tool for engaging middle and high school students in brief, single-session workshops as a means of conveying opportunities in engineering education and as a "conversation starter". The author is beginning his third year of using the LEGO system in this context. Anecdotally, students, teachers, and school administrators, alike, are very enthusiastic about the potential for this robotics system to significantly impact learning in their schools. The program objectives have grown beyond the initial focus of using this robotics system as a recruiting tool. A broader objective now focuses on obtaining funding to place these robotics kits in strategic schools throughout Mississippi. Through this effort, the author seeks to establish relationships with teachers, teaming with them to make curricular enhancements—incorporating (or strengthening) project-based learning as a means of preparing students for pursuing STEM majors and ultimately STEM-related careers.

The use of the LEGOTM "Robotics Invention System" serves as a means for increasing student interest in STEM topics. Benefits include several complementary factors including 1) employing a platform familiar to many students (i.e. LEGO building components); 2) "seamless" integration of hardware and software systems; and, 3) minimal startup delays as a result of an easily understood symbolic software package and hardware components that facilitate "rapid prototyping" of devices. These features have allowed us to easily introduce the LEGO robotics system as the focal point in this particular aspect of our K-12 student recruiting system. Program activities to date will be presented, including observations of this program's strengths and weaknesses, and goals for future activities.

The author's first encounter with LEGO robotics was as a coach of a First LEGO League (FLL) middle school robotics team in 2004, sponsored by the College of Engineering and Science at Louisiana Tech University. Coaching FLL for a second year and in the inaugural year of BEST Robotics hosted by Mississippi State University convinced him of the utility for robotics to convey STEM concepts and principles for any branch of engineering and science. The author has teamed with Emma Seiler (K-12 Outreach Coordinator for the Bagley College of Engineering)

Now using the LEGO robotics in his freshman chemical engineering course, a strategy is being developed for sending MSU chemical engineering students out to schools as part of our recruiting program. This combination of using LEGO robotics with the help of college student recruiting teams visiting K-12 schools brings a "service learning" component to the program—benefiting not only the K-12 schools we visit, but adding value to the undergraduate engineering experience in line with our chemical engineering program objectives and outcomes.

Recognizing the diversity of interests and skills sets we encounter at the middle school level, we have built our discussion around general themes, seeking to appeal to all students interested in STEM fields of study and practice. Particularly among elementary and middle-school student groups, we have emphasized the stream of coursework necessary to enter university study in preparation for these careers.

230f Process Technology Summer Institutes for K-12 and Community College Educators and Students

Steve R. Duke[1], Gopal Krishnagopalan[1], R. Dale Smith[2], Ralph Benefiel[3], and Charles Shepherd[4]. (1) Chemical Engineering, Auburn University, 212 Ross Hall, Auburn University, AL 36849-5127, (2) Alabama Southern Community College, P.O. Box 2000, Thomasville, AL 36784, (3) Director, Process Technology, Lower Columbia College, PO Box 3010, Longview, WA 98632, (4) National Network for Pulp and Paper Technology Training, Alabama Southern Community College, P O Box 2000, Thomasville, AL 36784

Summer institutes and workshops were conducted in 2005 through 2008 for high school students and K-12 and community college educators as part of recruiting strategies to bring people into chemical and paper technology and engineering degrees and industries. The institutes and workshops were conducted through a partnership between Auburn University, Alabama Southern Community College (Thomasville AL), and Lower Columbia College (Longview WA). Faculty and industry partners from these institutions developed and delivered the workshops as part of the National Network for Pulp and Paper Technology Training (npt)2, a National Science Foundation Advanced Technology Education Center.

This talk will focus particularly on the importance of where the institutes and workshops are held as to their effectiveness and recruiting potential. The summer institutes are generally held on community college or university campuses to allow access to laboratory facilities and classrooms. In 2007 and 2008, pilot workshops were also held at off-campus sites that were near industrial regions. Summer institutes and workshops held on the college campuses included large pilot-scale experiments conducted in technology training or engineering laboratory facilities and small-scale experiments that can be integrated into the classroom. The talk will address effectiveness of the workshops comparing assessments and best practices of those on and off campus, as well as compare the number and demographics of the applicants and participants.

Web Page: www.npt2.org

305a Introduction

Robert Wellek, The National Science Foundation, Division of Chemical, Bioengineering, Environmental and Transport Systems, 4201 Wilson Boulevard, Arlington, VA 22230

The NSF workshop session will be introduced.

305b NSF CBET Overview and Other NSF Programs

John McGrath, Division of Chemical, Bioengineering, Environmental and Transport Systems, National Science Foundation, Arlington, VA 22230

The Division Director will present an overview of the CBET Division and other relevant NSF programs.

305c Highlights of CBET Cluster on Chemical, Biochemical & Biotechnology Systems

John Regalbuto, Division of Chemical, Bioengineering, Environmental & Transport Systems, National Science Foundation, 4201 Wilson Boulevard, Arlington, VA 22230

An overview and highlights of the CBET cluster on Chemical, Biochemical and Biotechnology Systems will be presented.

305d Highlights of CBET Cluster on Transport and Thermal Fluids Phenomena

Phillip R. Westmoreland, Division of Chemical, Bioengineering, Environmental and Transport Systems, National Science Foundation, Arlington, VA 22230

Highlights and an overview of the CBET cluster on Transport and Thermal Fluids Phenomena will be presented

305e Highlights of CBET Cluster on Environmental Engineering & Sustainability

Trung Van Nguyen, The National Science Foundation, Division of Chemical, Bioengineering, Environmental and Transport Systems (ENG/CBET), 4201 Wilson Boulevard, Arlington, VA 22230

The presentation will describe program aspects of the CBET cluster on Environmental Engineering and Sustainability

305f Highlights of CBET Cluster on Biomedical Engineering and Engineering Healthcare

Semahat S. Demir, National Science Foundation, 4201 Wilson Boulevard, Arlington, VA 22230

Highlights of the CBET cluster on Biomedical Engineering and Engineering Healthcare will be presented.

305g Interactive Question and Answer Session with NSF Staff

Robert Wellek, The National Science Foundation, Division of Chemical, Bioengineering, Environmental and Transport Systems, 4201 Wilson Boulevard, Arlington, VA 22230

Attendees will be able to ask questions of the NSF Staff present at the Workshop.

Mathematics Taught in Chemical Engineering, What material should be covered? What should be the complexity?

Benito Serrano*, Juan Jose Mejia, Victor Javier Cruz, Alfonso Talavera, Jesus Moreira.

Universidad Autónoma de Zacatecas, Programa de Ingeniería Química, Unidad Academica de Ciencias Quimicas, Campus Siglo XXI, Edificio 6, Carretera a Guadalajara, Km. 6, Ejido la Escondida, Zacatecas Zac., 98160, Tel. 52-492-925-6690-Ext 6135, Fax 52-492-921-3990, beniserra@prodigy.net.mx

Abstract.

This material will discuss the contents of mathematics that should be taught in the undergraduate programs of chemical engineering in the country Mexico and eventually in other countries, what material should be covered, what is the complexity and where the boundary begins with the mathematics taught in graduate programs. Several exercises are presented and solved for students in the undergraduate program of chemical engineering in Universidad Autonoma de Zacatecas, Mexico.

The solved problems are the following: 1) A model of laminar flow through a lubricated tube. This is one problem of momentum transport more complicated than that for a tube filled with only one fluid; therefore it leads to the construction of one set of differential equations requiring the use of continuity boundary conditions for determining the velocity profiles of both fluids. 2) Slow flow over a solid sphere. It requests the analysis of momentum transport and more complicated procedures than those traditionally taught in the undergraduate programs. The deduction of Stokes law is performed. 3) Transient flow inside a cylindrical tube. It demands the solution of the partial differential equation that describes the starting of a liquid flow, due a pressure difference. The Sturm Liouville problem is solved, the variable separation method is used, and one of the differential equations is solved using Bessel functions because it has variable coefficients. The velocity profiles changing with time are found. 4) No stationary heat conduction in a cylindrical metallic bar. The partial differential equation describing the heating of a metallic bar in one of its ends is solved, with different boundary conditions, using the weighted residual methods and mainly, the orthogonal collocation method. 5.- Potential flow on a cylindrical tube. The partial differential equation is solved using the method of variable separation and the Euler equation to find the angular component of velocity.

Once the ability of the undergraduate students to solve this kind of problems is demonstrated, an optional undergraduate course is proposed about applied mathematics, including, among others, the indicated themes above. It is considered that the knowledge, dominion and familiarity with mathematics, will guide and motivate the students to pursue graduated studies and to do scientific research and the level of mathematics in the chemical engineering program should be indicated by those used in the main nucleus of subjects that distinguishes this career with respect to others.

Introduction.

With the goal of passing some subjects in the chemical engineering program in the UAZ – Zacatecas - Mexico, and to fulfill some requirements to graduate, so called Self Formation Seminars, several undergraduate students choose to solve problems related with Transport Phenomena and applied mathematics, involving mathematics more complicated that those traditionally taught in the undergraduate courses.

The advisor of these students, intrigued about the success of these actions, decided to investigate what are the contents of the programs of mathematics in the undergraduate programs in chemical engineering in other universities, in order to realize whether this is or not a risky task.

A survey was performed about the courses of mathematics taught in the undergraduate programs of chemical engineering of several universities of Mexico, USA and one of Canada. According to the collected information about that sample of universities, it was found that the common courses and topics are: Differential and integral calculus, with one or several variables, linear algebra, vectorial calculus, statistics and probability, differential equations of first, second and n-th order, linear, with constant coefficients, homogeneous and non homogeneous.

There were some differences among the programs in the most advanced courses from one university to another, and they indicate the border with the graduate courses. Some courses included and others no, the following topics: differential equations with variable coefficients, series of powers, systems of differential equations of first order, partial differential equations solved with the method of variable separation, etc. Then this set of topics is a non definite frontier between the undergraduate and graduate courses.

The focus of the courses of mathematics in chemical engineering.

Generally, all the undergraduate programs of chemical engineering have the following structure: Basic area, basic professional area, professional area and complementary area.

The basic area comprises the fundamental and scientific subjects for all the engineering, such as mathematics, chemistry, physics, physical – chemistry, etc. All of these subjects are common to all engineering careers. Basic professional area includes subjects such as transport phenomena, thermodynamics, chemical thermodynamics, etc, and these topics introduce to the student the fundamentals of chemical engineering, while the professional area contains chemical reactors, unit operations, etc. This last set of topics is the nucleus of the main subjects of chemical engineering career, which make the chemical engineering different from the other engineering.

Mathematics plays a very important role in this career and the way it is taught and its contents can be an important factor to give the student or not an adequate analytical capacity in solving problems during the career and later in the professional development, graduate studies and etc. In addition, students have a fear of mathematics and the instructor has to deal with it. This is the motivation to prepare this material.

All of these topics in mathematics can be taught with all mathematical rigor or just as mechanical applications without really understand the theorems and theory involved in the possibility of finding solutions of the equations in a region of the dominion, the uniqueness of the solution, etc. The collected documents in the survey did not allow us to see the focus of the programs, and although some of them included a lot of topics, it is possible to teach them using the mathematical rigor or just the mechanical applications of procedures or recipes. We believe one or other extremes are not convenient in chemical engineering. There are opinions that the mathematics in chemical engineering career should be taught by mathematicians, with the mathematical rigor, and taught as in the schools of mathematics. This focus has the risk of giving to the student a lot of specific and specialized information which is not supposed to be learned and be used in the career of chemical engineering, to distract the attention just in mathematics and to address his interest and concentration just in one area or subject, to deviate the students from the main objective of the career, and eventually they do not learn all the mathematical rigor, and just to pass the subject, and few or nothing of applications are illustrated, and the connection with chemical engineering is not done. Something similar could happen for example with the chemistry taught in chemical engineering. Eventually the professor feels like to teach it as if the career were of chemistry. Also the professor of physics would like to teach and concentrate in physics, forgetting the main objectives of the career of chemical engineering.

On the other hand, to teach mathematics just to let the student to learn how to use the formulas to calculate derivatives, integrals, how to apply the recipe of steps to solve a differential equations, etc, will deny the student to acquire a scientific background, necessary to get an analytical capacity and knowledge, useful to have a good development as chemical engineer, as a graduate student, as a scientific researcher, etc.

We consider that the mathematics needed by the chemical engineer should be a combination of both focuses and must be indicated by the objectives of the career. The nucleus of professional subjects of the career and those of the basic professional area, dictates the level and type of requested mathematics. These sets of subjects are referred to the topics that do distinct the chemical engineering from the other engineering and those that orient the student to the chemical engineering and contains: Balances of mass and energy, transport phenomena, thermodynamics, chemical kinetics, physical and chemical equilibrium, unit operations, chemical reactors, dynamics and control of processes, etc. The study of these subjects asks the analysis of physical or chemical processes in order to establish the mathematical model, to solve it and to interpret the solution, and to do changes in some variables and to understand the different behavior for each case, to analyze a system thermodynamically, to construct an equilibrium diagram, etc.

The analytical capacity can be reached once the student has good formation in mathematics, due it is an important analytical tool and the osseous skeleton of science, and he or she is not going to learn mathematics just using formulas without knowing how they were deducted. Then it is important to teach them something about the demonstration of easy theorems, deduction of formulas etc, because this formation is going to let them to use the mathematical tools to analyze the problems corresponding to other subjects such as transport phenomena, chemical reactors, unit operations, etc.

Once the under graduate students acquire this knowledge, they will be able to understand the deductions, and to realize analytical procedures and physical analysis. Of course, a graduate course should include more amounts of knowledge and with more deepness, due he or she is supposed to solve more complicated problems, than those solved by undergraduate student. Anyway, even in graduate courses for chemical engineering, they should hold the same focus, related with the applied mathematics and not with just mathematics, unless the interest of the student is to concentrate in a mathematical problem, for example a numerical method to solve a very complicated partial differential equation representing a physical problem with a lot of restrictions.

Border between graduate and undergraduate courses.

A typical graduate course of mathematics in chemical engineering include among other specific objectives: The acquisition of more deep knowledge of mathematics about matrices, eigenvalues, etc, the construction of mathematical models about a physical system, the solution of partial differential equations, using the Sturm Liouville theory, methods to discrete the solution of one equation, application of numerical methods to specific problems, etc. In other words, the graduate courses should provide the student the mathematical tools needed to do the thesis, and according to the thesis topic the needs can change.

Therefore, the border between the under graduate and graduate courses in mathematics is not clear and definite, and sometimes, several topics are common in both programs. The following topics belong to this common border: ordinary differential equations of variable coefficients, systems of ordinary first order differential equations, eigen values and eigen functions, auxiliary tools necessary to solve partial differential equations that represent the model of a physical phenomena, the separation method to solve partial differential equations, etc. Sometimes and in some schools, these topics are either taught in undergraduate and graduate courses.

Normally, in undergraduate courses the studied physical problems are related with convective problems and chemical reactions, which lead to ordinary differential equations and in the graduate courses, the studied physical problems (diffusion or diffusion + convection + chemical reaction) lead to partial differential equations or complicated systems of ordinary differential equations of first order, and because of that, the majority of the undergraduate courses have as last topic the ordinary differential equations of second order or higher with constant coefficients.

So far, in summary, it can be said that it is necessary to delimit the contents and focus of all the subjects taught in the undergraduate program of chemical engineer, to fulfill the general objective of the career. Consider besides that the certifying organisms of the chemical engineering programs request a number of courses and hours for each area and subject and it is not possible to dedicate an excess of hours just to mathematics, to chemistry, and etc.

The adequate use of mathematics in the establishment and solution of mathematical models will provide the undergraduate student a solid basis and to further develop with confidence in the graduate studies, to feel like to pursue graduate studies, etc. We believe however, that eventually the focus of rigorous mathematics is not necessary for the general formation of chemical engineer, unless the student is specifically interested in a deep analysis of a topic, a specific problem etc., where it is necessary to know under what conditions it is possible to solve a solution, the part of the dominion where the solution is stiff, the theorems that guarantee the uniqueness of the solution, etc. This knowledge may be acquired in a graduate course or in an optional undergraduate course, about selected topics of mathematics.

A proposal for a special undergraduate course for applied mathematics.

For those students interested in learning more mathematics, and to understand better its application in the problems rose from chemical engineering, it is possible to offer them a special course that will allow them to have more training and therefore more success when they study the graduate program.

The suggested title for the course is: Applied Mathematics in Selected Topics.

The general objective is to use specific physical problems, to analyze them to build the mathematical model, to solve it and to study all the mathematics behind the problem, in order to familiarize the student with the applied mathematics.

The proposal of contents is:

1) A model of laminar flow through a lubricated tube.

This problem involves the topics of formulation of mathematical model in transport phenomena, related with momentum transport more complicated than that for a tube filled with only one fluid and without lubrication; due it leads to the construction of one set of ordinary differential equations requiring the use of continuity boundary conditions for determining the velocity profiles of both fluids. (1)

2) Slow flow over a solid sphere. It requests the analysis of momentum transport and more complicated algebraic procedures than those traditionally taught in the undergraduate programs, it leads to the deduction of Stokes law. (1, 2, 5)

3) Transient flow inside a cylindrical tube. It asks the solution of the partial differential equation that describes the starting of a liquid flow caused by pressure difference. The Sturm Liouville problem is solved, the variable separation method is used, and one of the differential equations is solved using Bessel functions because of the variable coefficients. The velocity profiles changing with time are found. (1, 2)

4) No stationary heat conduction in a cylindrical metallic bar. The partial differential equation describing the heating of a metallic bar in one of its ends, with different boundary conditions, is solved using the orthogonal collocation method that belongs to the weighted residual methods (4, 6)

5. - Potential flow on a cylindrical tube. The partial differential equation is solved using the method of variable separation and the Euler equation to find the angular component of velocity. (3, 7)

6. - Multiple steady states of homogeneous and heterogeneous chemical reactors. (8)

7. - Liquid – Vapor Equilibrium. Construction of data for equilibrium diagrams using thermodynamic relations (9).

Some illustration about the procedures and solutions is provided for two problems, worked by the students of UAZ. In this case, it is more important to give comments about the experiences of teaching these problems to the undergraduate students than to illustrate with detail the calculations and procedures, which can be found in the consulted books. Actually, the algebraic details are given as a guide to do comments.

Then the statement of the problem, some details about the algebra and some commentaries are given.

Problem One. - A model of laminar flow through a lubricated tube.

Statement: Important pressure drops are required to force viscous fluids to pass through tubes with small diameter. In this example we analyze a proposal to reduce the requested pressure by creating a thin layer of liquid lubricant on the internal surface of the tube. We assume that the thickness of the lubricant layer is constant and independent of the axial position and time. (1).

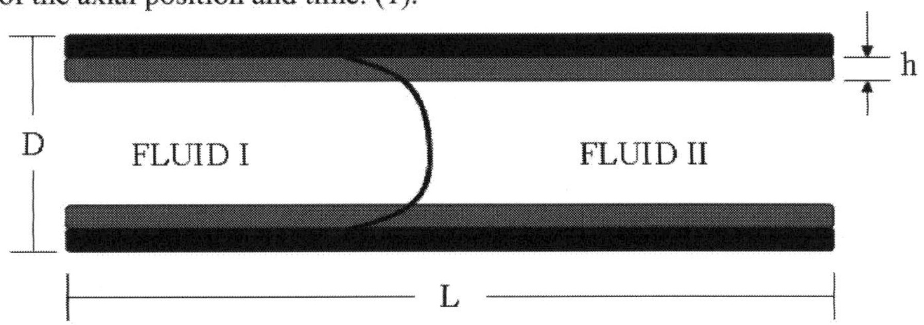

Figure 1. - Laminar flow of fluid I through a lubricated tube. Flow II is the lubricant with constant thickness h (1).

The solution for the velocity profile in fluid I is:

$$v_z^I = \frac{C^I}{4\mu^I} r^2 + F^I , \ 0 \le r \le \frac{D}{2} - h$$

Comment: Until now, to obtain the presented solution is easy and illustrated in undergraduate courses of transport phenomena, concretely, the problem of calculating the parabolic velocity profile for laminar viscous flow inside an empty circular tube. Then, the difficulty stars if the fluid contains a lubricating layer, where it is necessary to determine the expression for both fluids, and to achieve this goal, it is necessary to use additional boundary conditions, as it can be seen below.

The differential equation describing the change of velocity inside the lubricant layer, fluid II is:

$$\mu^{II}\frac{dv_z^{II}}{dr}=\frac{C_r^{II}}{2}+\frac{E^{II}}{r}\text{, in the region: }\frac{D}{2}-h\le r\le\frac{D}{2}$$

The solution is:

$$v_z^{II}=\frac{C^{II}r^2}{4\mu^{II}}r+\frac{E^{II}}{\mu^{II}}\ln r+F^{II}$$

Comment: To get this new differential equation for the lubricant layer, it is necessary to do an analysis in a differential volume element placed in the lubricant, or to analyze the Navier Stokes Equation.

There are five unknown constants, $\left(C^I, F^I, C^{II}, F^{II}, E^{II}\right)$ which will be determined using additional boundary conditions observed in the physical system.
So far we know that:

$$C^I=\frac{\Delta P^I}{L}\text{ and }C^{II}=\frac{\Delta P^{II}}{L}$$

The pressure drop is the same for fluids I and II, then:

$$C^I=C^{II}=\frac{\Delta P}{L}$$

Fluid II satisfies non slip condition because it is in touch with the tube wall.

$$v_z^{II}=0\ ,\ r=R=D/2$$

Substituting:

$$0=\frac{C^{II}R^2}{4\mu^{II}}+\frac{E^{II}}{\mu^{II}}\ln R+F^{II}.$$

Also, a no slip condition is satisfied in the interface between the two fluids.

$$v_z^{II}\big|_{R-h}=\frac{C^{II}(R-h)^2}{4\mu^{II}}+\frac{E^{II}}{\mu^{II}}\ln(R-h)+F^{II}=v_z^I\big|_{R-h}=\frac{C^I}{4\mu^I}(R-h)^2+F^I$$

Other boundary condition is the continuity of the shear stress in the interface between the fluids.

$$\mu^{II}\frac{dv_z^{II}}{dr}=\mu^I\frac{dv_z^I}{dr}\quad\text{at }r=R-h$$

$$\left(\frac{C^{II}r}{2}+\frac{E^{II}}{r}\right)\Bigg|_{R-h}=\left(\frac{C^Ir}{2}\right)\Bigg|_{R-h}$$

Now we have five unknowns and five equations. After algebra we get:

$$C^I=C^{II}=\frac{\Delta P}{L}$$

$$E^{II}=0$$

$$F^{II}=-\frac{\Delta P}{L}\frac{R^2}{4\mu^{II}}$$

$$F^I = -\frac{(R-h)^2}{4\mu^{II}} - \frac{R^2}{4\mu^{II}} - \frac{(R-h)^2}{4\mu^I}$$

Using these expressions we get the velocity profiles for both fluids.

$$\frac{v_z^I}{\Phi} = \left[(1-\eta)^2 + M\eta(2-\eta)\right] - s^2$$

$$\frac{v_z^{II}}{\Phi} = M\left(1-s^2\right)$$

Where:

$$\Phi = \frac{-CR^2}{4\mu^I} = \frac{\Delta PR^2}{4\mu^I L}, \; s = \frac{r}{R}, \; \eta = \frac{h}{R}, \; M = \frac{\mu^I}{\mu^{II}}$$

Comment: Finally the dimensionless velocity profiles for both fluids were obtained. However note that this procedure requested to observe additional boundary conditions, to use them, to get the profile expressions asking the student to analyze and understand the problem to get the additional boundary conditions, to face them with a long algebraic procedure, and finally to attain the velocity profiles. Now the question is, and what is this for? Why to get this?. It is very important to illustrate the student what is the utility of all the obtained results to avoid frustration.

Now we use this information to evaluate the possibility of increasing the velocity of fluid I through the tube using the lubricant layer.
Volumetric flow velocity:

$$Q^I = \int_0^{R-h} 2\pi r v_z^I dr$$

$$\frac{Q^I}{2\pi R^2 \Phi} = \frac{(1-\eta)^4}{4} + \frac{M\eta(2-\eta)(1-\eta)^2}{2}$$

The fluid velocity without lubricant is obtained with η = 0, (h = 0).
Then:

$$\Psi \equiv \frac{Q^I}{Q^I(\eta = 0)} = (1-\eta)^4 + 2M\eta(2-\eta)(1-\eta)^2$$

Figure 2 reports the plot of this last equation. Only for M > 1, when the lubricant is less viscous than the fluid I, there is an improvement in the fluid velocity. When η tends to one, even for a big value of M, the flow is reduced, since as the lubricant layer increases its thickness, and the area available for the flow is reduced.

Comment: To understand figure 2 will conclude the learning of the student about this problem the effort to reach these results, and avoid the frustration about to have performed a lot of algebra, without understanding the objective and the conclusions deducted from the results.

Problem Two: Transient flow inside a cylindrical tube

Statement: Consider a tube connected in the bottom of a big tank, and both are filled with water. The transversal section of the tube is circular. At time t < 0, the fluid is in repose and there is no flow. Suddenly, at t = 0, a constant pressure difference is applied in the ends of the tube, and the fluid starts to move from the tank to the tube and we want to estimate the necessary time in order the fluid reaches the new steady state. (1, 2)

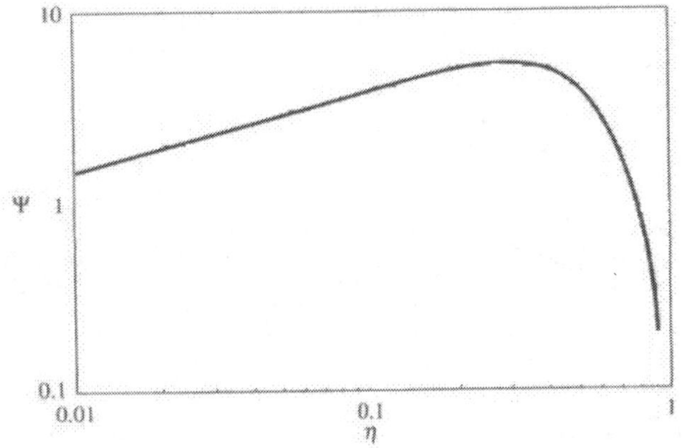

Figure 2.- Effect of a lubricant layer on the flow through a tube.

We assume laminar flow, a very long tube; the axial velocity is the only component of velocity. Using the continuity and Navier Stokes equations, it is possible to find the mathematical model describing this process.

$$\rho \frac{\partial u_z}{\partial t} = -\frac{\Delta P}{\Delta z} + \mu \frac{1}{r}\frac{\partial}{\partial r}\left(r \frac{\partial u_z}{\partial r} \right)$$

Boundary conditions:

$$u_z = 0 \quad \text{at} \quad r = R \text{ and } \frac{du_z}{dr} = 0, \, at \; r = 0$$

Initial condition:

$$u_z = 0 \; at \; t = 0$$

Using the following dimensionless variables:

$$s = \frac{r}{R} \qquad \Phi = \frac{u_z}{u_z^{\max}} \qquad \tau = \frac{\mu \cdot t}{\rho R^2}$$

We get the mathematical model without dimensions.

$$\frac{\partial \Phi}{\partial \tau} = \frac{1}{s}\frac{\partial}{\partial s}\left(s \frac{\partial \Phi}{\partial s} \right)$$

$$\Phi = 0 \quad \text{at} \quad s = 1, \quad \frac{\partial \Phi}{\partial s} = 0 \quad \text{along} \quad s = 0$$

$$\Phi = -\left(1 - s^2\right) \quad \tau = 0$$

This is one example of the Sturm Liouville problem.

Comment: To reach this point demands the understanding of the student about the physical problem, the capacity to use the Navier Stokes equation, and the capacity to adimensionalize equations, and to understand how to propose the dimensionless variables.

To find the transient solution, we use the method of variable separation, proposing the following form of the solution:

$\Phi(s,\tau) = S\cdot(s)\cdot T(\tau) = ST$

Inserting it in the differential equation, we get the following system of differential equations:

$$\frac{dT}{d\tau} = -\alpha^2 T\,, \qquad \frac{d^2S}{ds} + \frac{1}{s}\cdot\frac{dS}{ds} + \alpha^2 S = 0$$

The solution of the first equation is:

$$T = e^{-\alpha^2 t}\cdot A$$

Comment: To reach this point, it is important to understand that the problem solution has two parts, one stationary and one dynamic. The variable separation method is used to reduce the original partial equation in one system of two differential ordinary equations.

The second equation is the Bessel type. A proposed solution is:

$$S(s) = c_1 J_0(\alpha\cdot s) + c_2 Y_0(\alpha\cdot s)$$

$$J_0(x) = \sum_{n=0}^{\infty}\frac{(-1)^n}{(n!)^2}\left(\frac{x}{2}\right)^{2n} = 1 + \frac{x^2}{4} + \frac{x^4}{64} + \frac{x^6}{2304} + \ldots\ldots \qquad J_0(0) = 1$$

$$Y_0(\alpha^2 s) = \frac{2}{\pi}\left[\left(\gamma + \ln\left(\frac{\alpha^2 s}{2}\right)\right)J_0(\alpha^2 s) + \sum_{m=1}^{\infty}\frac{(-1)^{m+1}H_m}{2^{2m}(m!)^2}(\alpha^2 s)^{2m}\right], \quad \alpha^2 s > 0$$

Then,

$$\Phi_t(s,\tau) = \sum_{n=1}^{\infty} B_m\cdot e^{-\alpha_n^2\cdot\tau}\cdot J_0(\alpha_n\cdot S) \qquad \tau = 0, \quad \Phi = -(1-s^2)$$

$$-(1-s^2) = \sum_{n=1}^{\infty} B_m\cdot(1)\cdot J_0(\alpha_m\cdot S), \text{ to get } B_m$$

$$\int_0^1 J_0(\alpha_m\cdot s)(1-s^2)sds = = \sum_{n=1}^{\infty} B_m\int_0^1 J_0(\alpha_m\cdot s)J_0(\alpha_m\cdot s)sds$$

$$B_m = \frac{8}{\alpha_m^4[J_1(\alpha_m)]}$$

$$\Phi = -8\sum_{n=1}^{\infty}\frac{J_0(\alpha_n s)}{\alpha_n^3 J_1(\alpha_n)}\exp(-\alpha_n^2\cdot\tau)$$

This equation is plotted in figure 3, showing the velocity profiles for different times. It is observed that at 140 seconds the second steady state has been reached, and the steady parabolic profile is attained by the fluid inside the tube.

Conclusions.

1.- The undergraduate students in chemical engineering have the capacity of solving more complicated problems than those traditionally taught in the undergraduate courses, if they want and are motivated, as it is demonstrated with the worked problems by the students of UAZ – Mexico.

2.- It is possible to offer a special course of applied mathematics to the undergraduate students with specific interests, with the main objective of going through the physical phenomena and all the involved mathematics.

3.- The undergraduate courses of mathematics in chemical engineering must maintain an equilibrium between the usage of the mechanical procedures and the mathematical rigor.

4.- The border between the under graduate and graduate courses of mathematics in chemical engineering is indicated by the objective of the career. Generally the tools for solving partial differential equations and others are the main content of graduate courses.

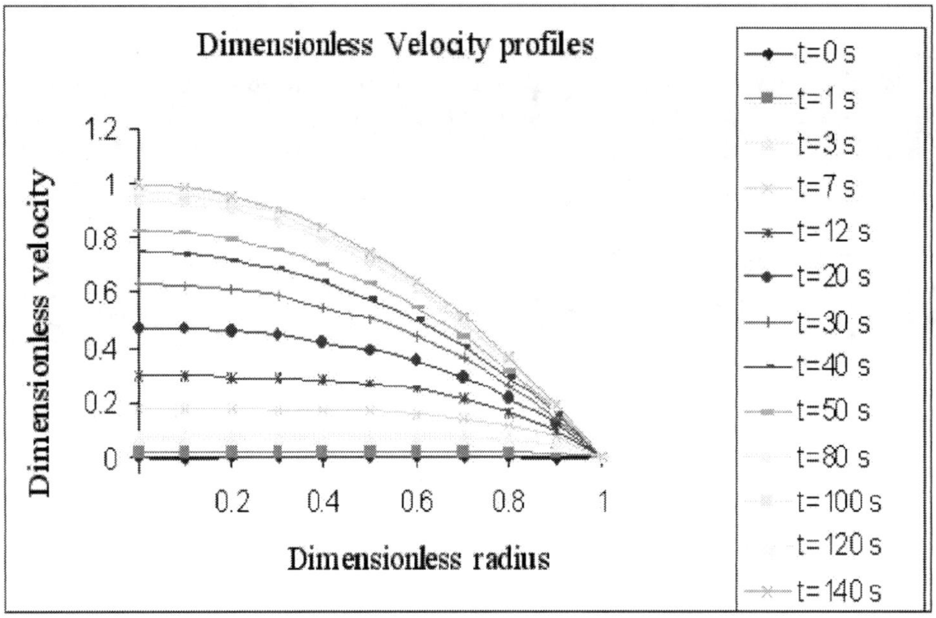

Figure 3.- Dimensionless radial velocity profiles at different times.

5.- According to the experience obtained, students that solved these problems are successful graduate students, which indicates that the dominion of mathematics prepare better the undergraduate students to pursue graduate studies.

References.

1.- Middleman Stanley, An Introduction to Fluid Mechanics, John Wiley and Sons Inc, 1998.

2.- Bird R. B., Stewart W. E., Lightfoot E. N., Transport Phenomena, Second Edition, John Wiley and Sons Inc., 2002.

3.- Welty J. R., Wicks C. E., Wilson R. E and Rorres G. L., Fundamentals of Momentum , Heat and Mass Transfer, John Wiley and Sons, Inc, Fifth Edition, 2008.

4.- O′ Neil Peter, Advanced Engineering Mathematics, Thomson, Sixth Edition, 2007.

5.- Bruce R. Munson, Donald F. Young, Theodore H. Okiishi, Fundamentals of Fluid Mechanics, Fifth Edition, John Willey and Sons, Inc. 2006.

6.- Finlayson Bruce A. The Method of Weighted Residuals and Variational Principles, with Application in Fluid Mexhanics, Heat and Mass Transfer, Academic Press, 1972.

7.- Merle C. Potter and David C. Wiggert, Mechanis of Fluids, Third Edition, Brooks/Cole Thomson Learning, 2002.

8.- H. Scott Fogler, Elements of Chemical Engineering, Second Edition, Perentice Hall, 1992

9.- Stanley I. Sandler, Chemical and Engineering Thermodynamics, Third Edition, John Wiley and Sons, 1999.

324b Why We Should Teach Product Design

Soemantri Widagdo, Display and Graphics Business, 3M Company, 3M Center, Building 236-1B-17, St. Paul, MN 55144

In the last few decades of the 20th century, chemical engineering graduates became increasingly involved in the development of new consumer products. The shift away from traditional process design activities, which dominate the manufacture of basic chemicals, coupled with the lack of education in product design, have left new chemical engineering graduates inadequately prepared for the challenge. To counter this, their employers have included them on new product-development teams, providing on-the-job training, through exposure to the art of developing new product as practiced by experienced engineers.

This presentation will address the necessity for teaching product design to chemical engineering students in the 21st century. The era of globalization and informatics has generated many forces that transformed the competition landscape in the newly developed global market. Speed to market is no longer a competitive advantage but rather a necessary condition to survive in the market place. The author will present a series of world trends and compelling needs from the perspective of an industrial practitioner, as well as challenges for the chemical engineering education community.

324c Computational Chemical Engineering: Process Simulators, Math Solvers, Spreadsheets and Computer Programming Languages

Eric L. Maase, Chemical Engineering Department, Lafayette College, Acopian Engineering Center Rm 268, 740 High St., Easton, PA 18042

The role of computers in chemical engineering endeavors has grown enormously and shows little sign of slowing. Computer skills in chemical engineering are often taught as 'soft skills' that are left to students to learn as self-taught subjects or instruction as short, intensive hands-on modules within specific courses.

After outlining and connecting all the essential computer skills chemical engineers require we then may subsequently identify the role that 'computer thinking' relates to other learning and skills that students might gain with more considered and perhaps more formal instructional experiences in computational chemical engineering.

324d Role of Scaling and up-Scaling In the Che Engineering Curriculum

Pedro E. Arce and Mario Oyanader. Chemical Engineering, Tennessee Technological University, Prescott Hall, Room 214 , 1020 Stadium Drive, Box 5013, Cookeville, TN 37083

Concepts of scaling and up-scaling are critically important for the future professional engineer. These concepts have been introduced, traditionally, into the curriculum via a relatively simple approach and, mostly, for linear-based processes. Situations that involved non-linear sources and scales in the micro and nano levels require a more fundamental-based approach. In addition, most the ChE curricula expose the students to the up-scaling and scaling concepts during the last part of the curriculum making the learning process stiff instead of working progressively with them. In this contribution, the authors will discuss the different aspects involved in scaling and up-scaling and illustrates the discussion with examples taken from real-world problems. In addition, they will discuss the importance of identifying a learning process (as opposed to hide it) that allows students to incorporate such concepts in a rational and efficient way into the curriculum. The authors will use examples from transport and reaction, biological, and biomedical areas to illustrate the process.

324e Development of a Chemical Engineering Comprehensive Concept Inventory

Jeff Csernica[1], Margot A.-S. Vigeant[2], Timothy M. Raymond[1], and Michael J. Prince[2]. (1) Bucknell University, Chemical Engineering Department, Lewisburg, PA 17837, (2) Chemical Engineering, Bucknell University, Dana Engineering Building, Lewisburg, PA 17837

In concert with consideration of what should be taught to chemical engineers, we should consider how we assess how much they have learned. Concept inventories are multiple choice tests aimed at assessing the level of conceptual understanding in a subject area, rather than recall of specific facts or equations. Tests of this nature are well established in physics, where the Force Concept Inventory (Hestenes et al, 1992) is a valid and reliable assessment of students' understanding of Newtonian physics concepts and has been used successfully to promote more effective teaching in physics. Concept inventories are of interest because they can be used to measure and demonstrate improvements in student conceptual understanding after a course, a novel educational intervention, or an entire curriculum. This last use is of potential interest as a direct assessment for ABET curricular evaluation.

While concept inventories exist or are in development for many aspects of the chemical engineering curriculum (materials science, statistics, transport phenomena, thermodynamics), there is no widely available comprehensive concept inventory assessing student understanding of the entire chemical engineering curriculum. At least two such tests are being tested, one at Auburn University (Eden, 2008, personal communication) and one at Bucknell University. This talk details the development of the test at Bucknell University, and shares initial results from the its pilot use in 2008. Graduating students achieved a passing average in every concept area and we plan to continue development and use of this instrument. Observations on the process of test development and limitations on interpretation of test results will also be shared.

324f Do We Really Have to Teach Team Work, Conflict Resolution, Leadership, Etc, Etc, Etc...?

Laureano Jiménez-Esteller and Gonzalo Guillén-Gosálbez. Department of Chemical Engineering, University Rovira i Virgili, Av. Països Catalans, 26, Tarragona, 43002, Spain

An educational model has been designed and implemented at the School of Chemical Engineering (ETSEQ) at Tarragona, Spain, to enable ChE students to acquire and integrate technical and scientific knowledge through the simultaneous and gradual development of competencies encompassing social and management skills. This model is based on the large-scale deployment of a project-based cooperative learning approach throughout the ChE curriculum. This extensive deployment can only be effective if it is supported by experts in change management and the systematic development of student teams that, in turn, requires that individual students develop key social and management skills.

An educational module has been developed and implemented at the ETSEQ (University Rovira i Virgili, Spain) to facilitate the transition of first-year ChE students into a comprehensive project-based learning environment. All first-year students participate in the first-year integrated design project with fourth-year ChE students acting as project managers for first-year project teams. The eighthour module is taught during the first two weeks of the first term and is structured to take into account the factors needed for students to understand and commit themselves to change. The first-year students' evaluation shows that the module helps them: (1) to identify what they need to accomplish to gain future employment as chemical engineers; (2) to understand what an integrated design project consists of and what the benefits of teamwork are, and (3) to realise that the integrated project and the related teamwork are great opportunities to acquire competencies that are essential in today's workplace.

The expertise in team development and change management methodologies, gained from nearly a decade of implementation work, has complemented the practice of the ETSEQ with experiential learning methodologies. A set of external training interventions has been designed to support the development of competencies by students. In the new educational system student teams grow from leader-directed teams in the first semester of the first academic year to self-directed or empowered teams in the fifth year of the curriculum. In this empowerment journey, fourth-year students play a key role as they act as facilitative leaders of first-year and second-year project teams, adjusting their facilitative leadership role according to the team development stage. The core of the competency-based educational model is client orientation. The need to satisfy clients and to adapt to their changing needs triggers the development of competencies related to the transformation of the individual students (versatility, entrepreneurship and innovation, systemic thinking, etc.), of the organization (facilitative leadership, teamwork and cooperation), and of the institution (organizational development and performance, and organizational leadership). Preliminary results show that student attendance has increased, that drop out has decreased, that more professors act as facilitators in the classroom, and that active-oriented and student-centered educational methodologies are increasingly applied.

324g Teaching Process Safety in a Laboratory Setting

John C. Prindle, Chemical & Biomolecular Engineering, Tulane University, 300 Boggs Building, New Orleans, LA 70118

The experimental composition of a conventional chemical engineering program's Unit Operations Lab is primarily comprised of the standard heat and mass transfer equipment as well as fluid flow. As departments have begun to embrace and incorporate the biological aspects of the discipline into their undergraduate curriculum, several experiments related to this area have been developed as well. However, only a few programs around the country have the necessary lab equipment to introduce the basic concepts of process safety. Given the importance placed upon this area by the chemical industry, more effort should be expended to introduce the concepts of relief system design and the experimental techniques employed to obtain the design data.

Design of a relief system involves the consideration of several fields: single and two-phase fluid flow; heat transfer; and possibly reaction kinetics. Hence, the introduction of an experiment related to relief system design would not only introduce the students to concepts important to this area, but would also challenge their use of the fundamental concepts in combination. Introduction of equipment used by industry professionals to obtain design data would also give students valuable experience in its use.

We have recently introduced a new experiment using the Advanced Reactive System Screening Tool (ARSST) available from Fauske and Associates, Inc. The ARSST is one of many tools available for screening chemically reactive systems to measure the potential for thermal runaway. Its predecessor, the Reactive System Screening Tool, was developed to address needs outlined by the Design Institute for Emergency Relief Systems (DIERS) program. Data obtained from the unit can be directly applied to the design of relief systems on large equipment. By using the ARSST, we have found that the chemical system studied can be readily changed each semester with minimal effort. There are ample examples of reactive chemical mixtures available from the literature. Some of these systems exhibit remarkable adiabatic thermal decomposition profiles, which capture student interest when they are asked for quantitative characterization.

An overview of the experimental methodology students are asked to follow along with results from different reactive chemical mixtures will be presented.

334a Does Faculty Research Improve Undergraduate Teaching?

Michael J. Prince, Chemical Engineering, Bucknell University, Department of Chemical Engineering, Bucknell University, Lewisburg, PA 17837, Richard M. Felder, Chemical and Biomolecular Engineering, North Carolina State University, Cary, NC 27518, and Rebecca Brent, Education Designs, Inc., Raleigh, NC.

Academicians have been arguing for decades about whether or not faculty research supports undergraduate instruction. Those who say it does—a group that includes most administrators and faculty members—cite many ways in which research can enrich teaching, while those on the other side cite numerous studies that have consistently failed to show a measurable linkage between the two activities. This article proposes that the two sides are debating different propositions: whether research can support teaching in principle and whether it has been shown to do so in practice. The article reviews the literature on the current state of the research-teaching nexus and then examines three specific strategies for integrating teaching and scholarship: bringing research into the classroom, involving undergraduates in research projects, and broadening the definition of scholarship beyond frontier disciplinary research. Finally, ways are suggested to better realize the potential synergies between faculty research and undergraduate education.

Design Course for Micropower Generation Devices

Alexander Mitsos
Aachen Institute for Advanced Study in Computational Engineering Science, RWTH Aachen, Germany

Abstract

A project-based design course is developed for man-portable power generation via microfabricated fuel cell systems. Targeted audience are undergraduate chemical/process engineering students in their final year. The course covers 6 weeks, with three hours of lectures per week. Two alternative projects are developed, one focusing on selection of alternatives and one focusing on process optimization. The course integrates product and process design.

Keywords

man-portable power, fuel cell system, design course, MEMS

1 Introduction

The chemical engineering education is undergoing changes, with more focus on emerging areas in molecular chemistry and biology, product design, as well as micro- and nanotechnology. On the other hand, design courses are still considered the capstone of an undergraduate chemical engineering program. This article describes a recently developed course for the Department of Chemical Engineering at the Massachusetts Institute of Technology (MIT) and the Aachen Institute for advanced study in Computational Engineering Science (AICES) at the RWTH Aachen. The course considers the design of microfabricated fuel cell systems for man-portable power generation.

Efficient alternatives to batteries for man-portable power generation are necessitated by the ever-increasing use of portable electric and electronic devices. The desired power level is in the order of 0.5 to 50W. There are several reasons for replacing batteries. Despite their high cost and large life-cycle environmental impact batteries have relatively low gravimetric (Wh/kg) and volumetric (Wh/l) energy density. State-of-the-art rechargeable batteries reach only a few hundred Wh/l and Wh/kg. Battery performance has significantly improved over the last decades, but it is believed that the upper limit on performance is being approached, because the list of potential materials is being depleted. A promising alternative is to use common fuels/chemicals such as hydrocarbons or alcohols as an energy source.

There is significant research activity in the area of microchemical systems, see, e.g., [1]. Chemical units such as reactors, separators and fuel cells with feature sizes in the submillimeter range have been considered for a variety of applications, due to their advantages compared to macroscale processes, such as the increased heat and mass transfer rates [2]. The replacement of batteries for electronic devices requires man-portable systems and therefore the use of microfabrication technologies is plausible since a minimal device size is desired.

There is great military [3] and civilian interest in developing battery alternatives based on common fuels/chemicals such as butane. As a consequence, a lot of research projects have been undertaken in academia and industry, see, e.g., [4, 5, 6] for reviews. While there are well-established microchemical courses with emphasis on microfabrication, the author is not aware of any course with emphasis on process synthesis, process design or optimization. Such a course is proposed herein; in addition to covering technological aspects of exciting topics (microchemical systems, fuel cells) it combines process and product design. This is important in view of recent trends for product-oriented design, [7, 8, 9, 10, 11, 12]. The course developed is based on several research publications of the Process Systems Engineering Laboratory at MIT, including [13, 14, 15, 16, 17, 18, 19, 20, 21]. In the remainder of the article, first the contents of the lectures are described in Section 2, and then the project tasks are summarized in Section 3. The article is concluded by the skills gained by students.

2 Lecture Contents

The course duration is six weeks, with three hours of lectures per week. No textbook is available for the course, but the material covered in [6] is the primary reference. Other useful references are books on microchemical systems, design and thermodynamics, such as [12, 22, 23, 24, 25, 26]. Approximately one week of lectures is reserved for software tutorials and discussions of issues raised by the students during the project execution. The remaining five weeks are devoted to five topics, namely the introduction & motivation, aspects of fuel cells, process synthesis, selection of alternatives, and process optimization. These topics are summarized in the following.

2.1 Introduction, Motivation & Project Description

The first week of lectures is devoted to a description of the project as well as an introduction. These lectures are intended to give the students the big picture of the project and help them understand the goals of their tasks. First, the need for micropower generation is motivated. This is done by comparing the trends in power consumption by portable electric and electronics to the performance characteristics of batteries. Pricing and performance of batteries is discussed, along with their environmental impact. A common critique to fuel cell-based systems for micropower generation are that they are deemed too dangerous. To put these claims into perspective the safety issues of batteries (fire, explosions, etc.) are discussed and demonstrated by pictures and movies.

The next step in the introduction is the definition of the key metrics for man-portable power generation devices, namely the gravimetric and volumetric energy densities

$$e_{grav}^{sys} = \frac{\tau_{mission} PW}{M^{sys}}, \qquad e_{vol}^{sys} = \frac{\tau_{mission} PW}{V^{sys}},$$

where the mission duration $\tau_{mission}$ [h] is the time between refueling or recharging, PW [W] is the power output (assumed constant for simplicity), M^{sys} [kg] is the mass of the system, and V^{sys} [l] is the volume of the system. These metrics are typically the objectives to be maximized by the process synthesis design and operation. In the case that the mission duration is very long and the device miniaturized, the size of the system is dominated by the fuel cartridge, in which case the simpler metrics of *fuel energy density* can be used:

$$e_{grav}^{fuel} = \frac{PW}{3600 \sum_i MW_i N_{i,in}}, \qquad e_{vol}^{fuel} = \frac{PW}{3600 \sum_i MV_i N_{i,in}},$$

where $N_{i,in}$ [mol/s] is the inlet molar flowrate of species i, MW_i [kg/mol] is the molecular weight of species i, MV_i [l/mol] is the molar volume of species i at storage conditions, 3600 is the conversion factor from hours to seconds, and the summation is taken over all stored fuels and oxidants.

In man-portable power generation the most important advantage of microfabrication is a device miniaturization. Microfabrication techniques are outside the scope of the course. On the other hand, various examples from microchemical systems are analyzed with emphasis on entire systems as opposed to components. The importance of physical phenomena at the microscale are analyzed and compared to the macroscale; for instance, it is shown that viscous forces dominate over inertial forces and that heat transfer (and losses) are much more important than in the macroscale. Various alternatives for man-portable power are summarized, such as microturbines [27] and devices based on man-power [28].

A common critique of micropower generation devices, and particularly of high temperature systems, is that they pose safety threats and generate a lot of heat. These concerns are analyzed via back-of-the-envelope calculations. It is argued that these concerns are partially true and partially misconceptions resulting from macro-scale experience.

2.2 Fuel Cell Working Principles and Types

Both the batteries and the fuel cell systems studied, i.e., the product to be replaced and the proposal for replacement, rely on electrochemical reactions. Electrochemistry is covered in some undergraduate curricula, but not in sufficient detail for performing and understanding the project tasks. Therefore, the principles of fuel cells are briefly summarized, along with a repetition of the relevant concepts from reactor engineering and thermodynamics. Then, the thermodynamic limits of fuel cell performance are analyzed and compared to heat engines.

Several fuel cells technologies have been proposed over the last decades. Some of the fuel cell types have a potential for scale down, such as solid-oxide fuel cells (SOFCs), polymer-electrolyte membrane fuel cells (PEMFCs) operating with hydrogen, direct methanol fuel cell (DMFC), proton ceramic fuel cells (PCFC), and membrane-less fuel cells, e.g., [29, 30, 31]. Miniaturization has been performed for some of the fuel cell types, often with the use of microfabrication technologies. These fuel cell types are analyzed with an emphasis on advantages, disadvantages and operating characteristics.

2.3 Conceptual Process Design at the Macro Scale

Process synthesis at the macroscale is typically included in an undergraduate curriculum. In the proposed course a brief summary of the techniques and methodologies is given, with emphasis on superstructure-based approaches [25]. This is deemed helpful for the students to be able to compare the challenges with the selection of alternatives at the microscale. For instance, the discussion of heat exchanger network synthesis, demonstrates that at the microscale the challenges are very different: no utility streams are available, and the operating conditions of various components are not independent from each other due to the pronounced heat transfer. In addition, having this short summary allows students from different disciplines to attend the course. The lectures also briefly discuss some of the mathematical and algorithmic background used in conceptual process design. The emphasis is on the material which is relevant to the project tasks.

2.4 Selection of Alternatives

A major challenge in the system design of micropower generation processes is the selection of alternatives, in particular which fuel to use for power and/or heat generation, what fuel cell type to select, whether a fuel reforming path should be followed and how heat integration should be performed. This selection of alternatives at the microscale is analogous in principle to macroscale process synthesis. Moreover, some of the mathematical techniques used in macroscale process synthesis can also be used for the selection of alternatives. However, there are several major differences, including different objectives and constraints and the fact that the unit operation paradigm must be replaced by that of highly integrated components in a system [32]. An additional challenge is the early stage of technology development.

The lectures describe the large number of alternative processes arising from the large choice of fuels, fuel reforming reactions and fuel cells. The advantages and disadvantages are discussed and a system-level approach for modeling is detailed [13, 14]. This modeling approach is then used in one of the projects offered, see Section 3.1. The advantage of this methodology is that the most promising alternative(s) can be selected without detailed knowledge about the technological details, such as the catalysts used or the reactor configuration. The disadvantage is that some parameters, which in principle can be calculated, are viewed as input parameters, e.g., the fuel conversion in the reforming reactor for a given operating temperature and residence time.

2.5 Optimization of a Given Process Alternative

Once a promising alternative has been chosen, the design and operation can be optimized via models of intermediate fidelity [15, 16, 17, 18, 19]. The spatial discretization results in problems with (partial) differential-algebraic equations. The models employ spatial discretization when necessary and are based on first-principle models. As a consequence they are predictive and can be used to find the optimal sizing of units (reactor, fuel cell, etc.) as well as operating variables (voltage, temperature, flow rates, etc.). It should be noted that the development of such a model takes significant effort and requires knowledge of kinetic rates.

For the optimization of design and operation algorithms from mathematical programming with differential-algebraic equations (DAEs) embedded can be used. These techniques are briefly described in the lectures along with techniques for the simulation of DAE systems. However, the state-of-the-art in dynamic optimization is such that the use requires significant mathematical background and computational experience, and is deemed limitedly suitable for an undergraduate class in chemical engineering. Instead, in the project (Section 3.2) the optimization is based on a simulation approach, in which the students must specify the degrees of freedom. To simplify the problem some variables (such as the operating temperature and voltage) are prespecified. On the other hand, to give some experience in the use of advanced methods, the simpler problem of parameter estimation is given as a subtask to be solved with an optimization algorithm.

3 Description of Projects

Two alternative projects are offered. The recommendation is to offer these in alternate years. Offering both projects in parallel (to different groups of students) is also possible, however complicates logistic considerations significantly, since the lectures that are directly relevant to the project should be given first. A third alternative would be to assign both projects, and extend the course duration.

3.1 Selection of Alternatives

Two main processes are considered, see Figure 1. Both are based on a solid-oxide fuel cell (SOFC); one of them uses NH_3 as the fuel for power generation while the other uses C_4H_{10}.

All units are modeled using stoichiometric reactors, i.e., a fixed conversion is assumed for each reaction. All units are microfabricated on a single silicon chip; as a consequence they share the operating temperature of $T = 1000$K. The entire process operates at ambient pressure. The gas phase is assumed to be ideal. A power production of 10W is requested. The enthalpy of the inlet streams is calculated at ambient conditions and the gaseous phase. For the outlet streams a temperature of $T_{out} = 600$K is assumed, based on heat recovery.

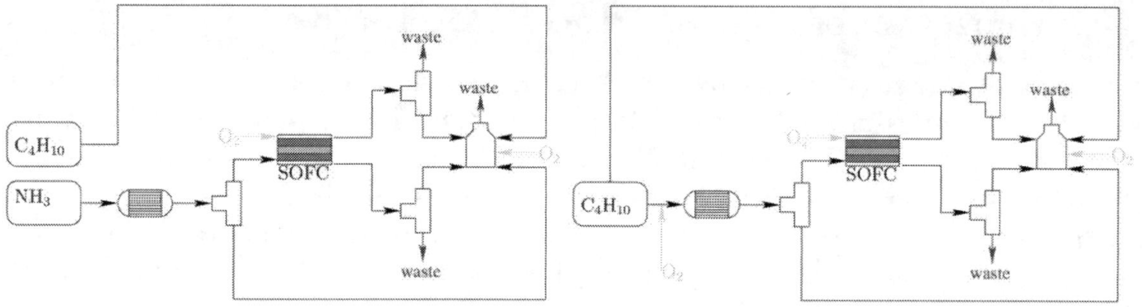

Figure 1: Process flowsheets for project on selection of alternatives.

The models for the processes are given to the students as a JACOBIAN® [33] input file. The students must perform additional calculations, such as the calculation of energy density based on the calculated flowrates. For these calculations the students have the choice of using JACOBIAN® or a software tool of their choice.

3.1.1 Project Tasks

The first task is to optimize the NH_3 and C_4H_{10} based processes. The operational variables are the flowrates of fuel and the splitfractions in the 3 splitters. The flowrates of air are a direct consequence of the fuel flowrates and a specified stoichiometric ratio. The objectives are to maximize the volumetric and gravimetric energy density; the device mass and volume can be ignored, but the fuel cartridges must be accounted for.

The second task is to compare the optimized processes with a conceptual process based on methane, stored at ambient temperature. An overall efficiency (power produced divided by chemical energy consumed) of 50% is assumed. The main challenge is to calculate the required cartridge thickness and volume as a function of pressure for various container types, e.g., plastic, or steel.

The third task is to compare the optimized processes with a process based on a H_2 generator, such as a hydride. The goal of this task is to identify the storage properties (hydrogen volume % and density) required to match the best process in terms of both gravimetric and volumetric energy density. To do so, an overall efficiency of 70% is assumed.

3.2 Optimization of NH_3-Based Process

The task of the project team is to optimize a micropower generation device for the production of $PW = 10$W. A fixed process is considered based on NH_3 cracking to H_2 and electrochemical oxidation of the produced H_2 in a solid-oxide fuel cell (SOFC). The device comprises two parallel lines, namely the NH_3 line for power generation and the C_4H_{10} line for heat generation, see Figure 2. These two lines are not independent, because they are microfabricated in a single silicon chip; as a consequence they share the operating temperature of 1000K. The entire process operates at ambient pressure. The gas phase is assumed to be ideal.

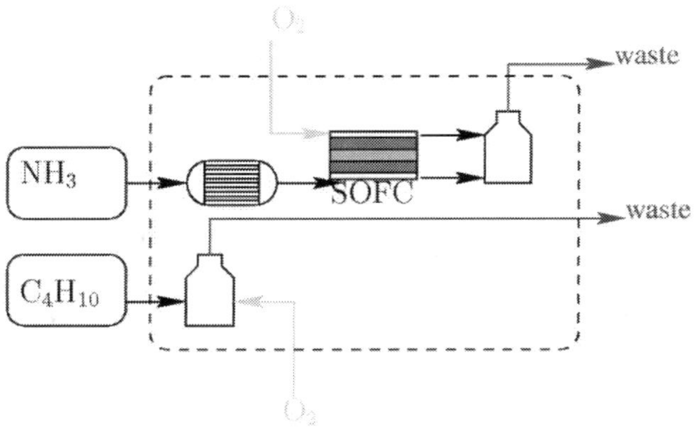

Figure 2: Process flowsheet for project on process optimization.

3.2.1 Project Tasks

The first project task is to determine appropriate values for the constants in the kinetic rate of NH_3 cracking, by fitting to a set of experimental values. The students are given a postulated kinetic mechanism along with experimental data of conversion as a function of residence time for four different temperatures. The kinetic mechanism has two adjustable parameters and the data contain random error. The students must extend an example provided to them. This task is relatively simple, thanks to the estimation capabilities of JACOBIAN®.

The main task of the project is to maximize the energy density of the device; this is done by optimizing the volumes of the device components and the flowrates of fuel and air. There are four *design variables*, namely the volumes of the reactor, SOFC hydrogen burner and butane burner. In addition, there are also four *operational variables*, namely the feed flowrates of fuel (NH_3 and C_4H_{10}) and air (to the SOFC and to the butane burner). The temperature and voltage have been fixed. The optimization is a challenging task, in which the students can only succeed if they employ a systematic procedure for varying the variables.

The final task is to analyze potential improvements to the process. This analysis includes the comparison of the chosen process configuration with alternatives, such as using stored oxygen and having a fresh air stream to the burner instead of using the cathode effluents. The students are also asked to comment on the effect of increasing or decreasing the temperature and the voltage. The process relies heavily on catalysts, and not surprisingly the performance of catalysts significantly affects the overall process performance; the students are asked to identify which component is the most important to optimize (reactor, burner or fuel cell). Finally, the students are asked to explain how a doubling of the desired power demand level will affect the process design and operation.

4 Conclusions

A new course on the design of microfabricated fuel cell systems is offered for chemical engineering students. The course is project-based and spans over six weeks. The theoretical material needed for a successful project execution is covered in three hour-long lectures per week. The students learn several skills through the lectures and project. Likely the most important skill is learning how to work in a team, as in any course based on group-projects. The most important technical skills are process and product design, and in particular their interaction. The students have a chance of integrating the knowledge acquired in their preparatory classes, especially thermodynamics and reactor engineering. Finally, the students are familiarized with the exciting technologies of fuel cells and microchemical systems.

So far the course is targeted to the graduating class of chemical engineering. It can also be used as a seminar-style course when the project is omitted. A potential extension would be to aim at interdisciplinary class. In particular it would be interesting to consider teaching joint classes in chemical, mechanical, material and electrical engineering.

Acknowledgments

I am indebted to Professor Paul I. Barton for his research guidance and for providing me the opportunity to develop this course.
Financial support from the Deutsche Forschungsgemeinschaft (German Research Association) through grant GSC 111 is gratefully acknowledged. The development of this class was sponsored in part by the Department of Chemical Engineering, Massachusetts Institute of Technology.

References

[1] V. Hessel and H. Löwe. Mikroverfahrenstechnik: Komponenten - Anlagenkonzeption-Anwenderakzeptanz - Teil 1. *Chemie Ingenieur Technik*, 74(1-2):17–37, 2002.

[2] K. F. Jensen. Microreaction engineering – Is small better? *Chemical Engineering Science*, 56(2):293–303, 2001.

[3] National Research Council Committee of Soldier Power/Energy Systems. *Meeting the Energy Needs of Future Warriors*. National Academy Press, Washington, D.C., 2004.

[4] J. D. Holladay, Y. Wang, and E. Jones. Review of developments in portable hydrogen production using microreactor technology. *Chemical Reviews*, 104(10):4767–4789, 2004.

[5] H. L. Maynard and J. P. Meyers. Miniature fuel cells for portable power: Design considerations and challenges. *Journal of Vacuum Science Technologies*, 20(4):1287–1297, 2002.

[6] A. Mitsos and P. I. Barton, editors. *Microfabricated Power Generation Devices: Design and Technology.* In Press: Wiley-VCH, June 2008.

[7] G. D. Moggridge and E. L. Cussler. An introduction to chemical product design. *Chemical Engineering Research and Design*, 78(A1):5–11, 2000.

[8] E. L. Cussler and J. Wei. Chemical product engineering. *AIChE Journal*, 49(5):1072–1075, 2003.

[9] J. Wei. Molecular structure and property: Product engineering. *Industrial and Engineering Chemistry Research*, 41(8):1917–1919, 2002.

[10] A. W. Westerberg and E. Subrahmanian. Product design. *Computers and Chemical Engineering*, 24:959–966, 2000.

[11] K. Wintermantel. Process and product engineering. *Trans IChemE*, 77(A), 1999.

[12] E. L. Cassler and G. D. Moggridge. *Chemical Product Design.* Cambridge University Press, New York, 2001.

[13] A. Mitsos, Ignasi Palou-Rivera, and P. I. Barton. Alternatives for micropower generation processes. *Industrial and Engineering Chemistry Research*, 43(1):74–84, 2004.

[14] A. Mitsos, Michael M. Hencke, and P. I. Barton. Product engineering for man-portable power generation based on fuel cells. *AIChE Journal*, 51(8):2199–2219, 2005.

[15] B. Chachuat, A. Mitsos, and P. I. Barton. Optimal design and steady-state operation of micro power generation employing fuel cells. *Chemical Engineering Science*, 60(16):4535–4556, 2005.

[16] B. Chachuat, A. Mitsos, and P. I. Barton. Optimal start-up of micro power generation processes employing fuel cells. Cincinnati, OH, 30th October–4th November, 2005. AIChE Annual Meeting.

[17] P. I. Barton, A. Mitsos, and B. Chachuat. Optimal start-up of micro power generation processes. In C. Puigjaner and A. Espuña, editors, *Computer Aided Chemical Engineering*, volume 20B, pages 1093–1098. Elsevier, ESCAPE 15, Barcelona, Spain, 29th May–1st June, 2005.

[18] B. Chachuat, A. Mitsos, and P. I. Barton. Optimal design and transient operation of micro power generation employing fuel cells. *Submitted to: Optimal Control Applications and Methods*, April 15, 2008.

[19] M. Yunt, B. Chachuat, A. Mitsos, and P. I. Barton. Designing man-portable power generation systems for varying power demand. *AIChE Journal*, 54(5):1254 – 1269, 2008.

[20] A. Mitsos, B. Chachuat, and P. I. Barton. What is the design objective for portable power generation: Efficiency or energy density? *Journal of Power Sources*, 164(2):678–687, 2007.

[21] A. Mitsos, B. Chachuat, and P. I. Barton. Methodology for the design of man-portable power generation devices. *Industrial and Engineering Chemistry Research*, 46(22):7164–7176, 2007.

[22] W. D. Seider, J. D. Seader, and D. R. Lewin. *Product & Process Design Principles.* John Wiley & Sons, New York, 2nd edition, 2004.

[23] J. M. Douglas. *Conceptual Design of Chemical Processes.* McGraw-Hill, New York, 1988.

[24] J. M. Smith and H. C. Van Ness. *Introduction to Chemical Engineering Thermodynamics.* McGraw-Hill, fourth edition, 1987.

[25] L. T. Biegler, I. E. Grossmann, and A. W. Westerberg. *Systematic Methods of Chemical Process Design.* Prentice Hall, New Jersey, 1997.

[26] V. Hessel, S. Hardt, H. Löwe, A. Müller, and G. Kolb. *Chemical Micro Process Engineering.* Wiley-VCH, Weinheim, Germany, 2005.

[27] A. H. Epstein and S. D. Senturia. Macro power from micro machinery. *Science*, 276(5316):1211–1211, 1997.

[28] L. C. Rome, L. Flynn, E. M. Goldman, and T. D. Yoo. Generating electricity while walking with loads. *Science*, 309(5741):1725–1728, 2005.

[29] K. J. Green, R. Slee, and J. B. Lakeman. The development of lightweight, ambient air breathing, tubular PEM fuel cell. *Journal of New Materials for Electrochemical Systems*, 5:1–7, 2002.

[30] N. M. Sammes, R. J. Boersma, and G. A. Tompsett. Micro-SOFC system using butane fuel. *Solid State Ionics*, 135:487–491, 2000.

[31] Z. P. Shao, S. M. Haile, J. Ahn, P. D. Ronney, Z. L. Zhan, and S. A. Barnett. A thermally self-sustained micro solid oxide fuel-cell stack with high power density. *Nature*, 435(9):795–798, 2005.

[32] A. Mitsos and P. I. Barton. *Microfabricated Power Generation Devices: Design and Technology*, chapter Selection of Alternatives and Process Design. In Press: Wiley-VCH, June 2008.

[33] Numerica technology. www.numericatech.com.

Units of Chemical Engineering Operations

Subbarao Duvvuri *and Hilmi Mukhtar,*
Department of Chemical Engineering, Universiti Teknologi PETRONAS, Bandar Seri Iskandar, Tronoh, 31750, Malaysia.

Abstract

Research creates rationale to interconnect existing bits of information which can help effective teaching while opening up new challenges for creative development. Research over the last several decades has expanded the knowledge base exponentially though it has not percolated into the curricula.

Major thrust of chemical engineering education is to teach design and development of process equipment for performing unit operations and reactions. Process equipment are essentially multiphase contactors (and separators) with momentum, heat and mass transfer with or with out reactions between the different phases. Understanding multiphase flow through wide variety of process equipment is needed for rational design. It is interesting to note that all the chemical engineering operations and reactions are carried out in just four types of "units". A course is proposed to organize the vast body of knowledge in terms of "units" of chemical engineering operations.

1 Introduction

Processing of the raw materials to new materials needed by society is the story of human civilization. First step that triggered the story could be controlled combustion of wood / biomass to harness thermal energy. Creativity of mankind has developed many materials of need such as food, clothing, habitat, security/defense, cleanliness and health care, pharmaceuticals, entertainment etc – the list is ever expanding. These developments triggered developments in the fields of transport, communications, security, defense etc and the feed back from these fields has prompted development of more and more new materials.

Chemistry of the materials and their transformations is the basis of development of chemical technology to produce materials needed by society in large quantities. Chemical engineering translates the chemistry into an economically viable chemical technology

Chemical Engineering

Chemistry ⟶ **Chemical Technology**

As newer materials get discovered to satisfy future needs, chemical engineering has to invent, innovate, develop and design process equipment to make the new technologies possible. It is necessary to bring the essence of ever expanding volume of knowledge on the design and development of process equipment in a compact rational format to the next generation to prepare them for future challenges.

2 Unit Operations, Reactions and Transport Phenomena

About a hundred years back, educators recognized the need for engineers who are capable of understanding, operating and developing the chemical industry for processing raw materials to useful products. The task was immense as there were very many chemical industries producing too many products. There were many ways of organizing the material. Arthur D. Little in 1915 observed: “Any chemical process on what ever scale conducted may be resolved into a coordinated series of what may be termed "unit actions" as pulverizing, mixing, heating, drying, roasting, absorbing, condensing, lixiviating, precipitating, crystallizing, filtering, dissolving, electrolyzing, evaporating and so on. The number of these basic unit operations is not large and relatively few of them are involved in any particular process”. Since then, Chemical Engineering education evolved through the concept of unit operations. Concepts of thermodynamic equilibrium and rate processes have shaped the framework for the equipment design of unit operations.

It is interesting to note that some how reactions were not considered as a unit operation though Little (1915) did mention roasting and precipitation also as unit actions. Groggins (1930) organized chemical reactions in terms of “unit processes” such as oxidation, nitration chlorination, sulfonation etc. based on the principles of chemistry. Concept of ideal reactors, such as stirred tank and plug flow reactors, shaped the development of Reactor Engineering. In most applications, reactants are all not in a single phase and challenges of interphase and intraphase mixing of reacting species dominated the developments in reactor engineering. Unit operations, unit processes and reactors add upto quite a number!

In 1960s, methodology of transport phenomena has brought out the physics of the rate processes at micro level in terms of fluid mechanics, heat transfer and mass transfer. Theories of Heat and Mass Transfer and Analogies have brought out the importance of fluid mechanics in understanding the rate processes. Translation of the knowledge of processes at micro level to understand the performance of process equipment at macro level is a formidable challenge as flow through process equipment is multiphase in nature.

One of the major objectives of Chemical Engineering Education is to equip young aspirants with the capability to choose/develop, design, operate, simulate and optimize process equipment with multiphase flow such as reactors, separators, heat exchangers and interconnecting pipes for various technologies. Hence focus should be on the science of process equipment than on the science of the process. An attempt in this direction is presented.

3 A fresh look: UNITS of Chemical Engineering Operations

A typical chemical industry can be expressed in terms of a "Flow Sheet" depicting the flow of materials through various process equipments, such as reactors, separators, heat exchangers etc. The reactors transform raw materials to products and as the reactor conversions are rarely total, the reactants in the out stream are separated from the products for recycle.

Most of the raw materials available in the nature except for air, natural gas, water and petroleum are in solid phase. In reactors, the raw materials in different phases need to be brought together at molecular level for the reactions to take place. After the reactions are complete, separation of the molecules which are in mixed state is not easy. The strategy of separating pure molecules from a mixture is to see that they partition preferentially from one phase to a different phase by diffusion. Later the molecules are recovered from the solvent by phase change techniques such as evaporation and crystallization. Separation of phases is comparatively simpler than separating the molecules directly. For these reasons, more than one phase need to flow through the separators.

In the separators, different phases are brought into intimate contact for achieving heat and mass transfer followed by separation of the phases; in the reactors different phases are brought into intimate contact for achieving heat and mass transfer with chemical reactions followed by separation of the phases. Application of external force is necessary to bring the different phases together as well as to separate them. Gravitational force is utilized with advantage for dispersion of one phase into another as well as for separation of the phases. Thus, both reactors and separators are essentially multiphase contactors (followed by multiphase physical separation) and the same "units" can be used for carrying out reactions or separations. It is interesting to note that, from process equipment point of view, all the chemical engineering operations are performed in just four basic "units" namely

Pipes

Packed Beds

Suspended Beds

Stirred Tanks.

To develop coherent focus in teaching process equipment design, it is proposed that multiphase flow through "units" of process equipment be organized as a new course. This type of course can rationalize many thumb rules used in the process equipment design. An attempt in this direction is presented in this paper.

4 Multiphase Flow in the UNITS

Single phase fluid mechanics provides information on three essential parameters viz.

- Pumping power required for the fluid to flow through the process equipment
- Heat and mass transfer coefficients across interfaces and
- Residence time (distribution) of molecules in the process equipment.

In the last several decades, fluid mechanics of multiphase contact in various process equipments have been extensively studied. Parameters such as

- pressure drop across the equipment,
- phase holdups,
- size of dispersions,
- area of contact,
- volumetric heat and mass transfer coefficients and
- residence time distributions of each phase

have been measured. In view of the dispersion of phases and complexity of flows, difficulties in the assumption and definition of continuum properties, internal circulatory flows, turbulence and equipment geometries, the traditional approach based on micro balances through Navier Stokes equations are not realistic. In recent times CFD is opening up new vistas in analyzing multiphase flow through process equipment though computational times are prohibitive.

Out of the haze, methodologies to use these information together for modeling performance of various process equipment are emerging. One dimensional meso scale conceptual approach is still of great educational value. The concepts of friction factor/drag coefficient/dispersion number are useful in describing the multi phase fluid mechanics to estimate phase hold ups, size of dispersions, area of contact and residence time distributions. Penetration theory can be adopted to estimate heat and mass transfer coefficients.

5 Course content

Contents of a course with detailed profile of each chapter is given. Synthesis of the concepts for one dimensional steady state performance modeling of reactors and separators is encouraged.

Contents

Ch1 Introduction

- Chemistry, Chemical Technology, Flow Sheets, Unit Operations, Reactors, Transport Phenomena,
- Basic Units in which Separations and Reactions can take place,
- Multiphase Fluid Mechanics, issues of importance
- Size distribution of dispersion, Phase holdups, interfacial area of contact,
- heat and mass transfer coefficients,
- degree of mixing in each phase, Tanks in series models,
- Performance modeling

Part I Pipes

Ch2 Single Phase Flow

- Fluid Mechanics, Bernoulli Equation,
- Laminar flow,Shell balances, Navier Stokes Equation,
- Turbulence, friction factors
- Power for pumping the fluid,
- Analogies and heat & mass transfer coefficients, Mixing, RTD, Tanks In series model, Axial dispersion model, Performance model for reactors

Ch3 Two Phase flow

- Pseudo homogeneous Model, Viscosity of complex mixtures, non Newtonian fluids
- Separated Flow Model, Lokhart-Martinelli correlation
- Drift Flux Model, Drag and buoyancy force on a dispersed phase, drag coefficient for a particle, drop and bubble, terminal velocity and slip velocity. effect of presence of other particles on slip velocity
- Mechanistic Models

Part II Packed Beds

Ch4 Single Phase Flow

- Ergun's Equation, tortuosity factor
- Drag coefficient on a particle in a packed bed of particles

Effect of particle size and bed voidage on pressure drop across the bed
Interphase heat and mass transfer coefficients
RTD of fluid

Ch5 Evolution of Packings
- 3D packings
  - Spherical, cylindrical particles, Trilobes
  - Bed Voidage
  - Surface to volume mean particle size
  - Surface area of particles per bed volume
- 2D packings
  - Raschig rings,
  - bed voidage, importance of wall thickness,
  - Ceramic, plastic and metal packings
  - surface to volume mean particle size, surface area of particles per bed volume
  - Berl Saddles, Pall Rings, IMTP packing, intellox saddles etc
  - Structured packing made of sheets
- 1D packings
  - Springs and helical coils
  - Rods and holes, monolith reactors

Ch6 Two Phase Flow through packed Beds
- Absorbers and Trickle Bed Reactors
  - Liquid distributors and channeling
  - Liquid holdup, flooding
  - Area of particles, gas-liquid interfacial area
  - Wetting efficiency
  - Two phase flow pressure drop
  - Volumetric mass transfer coefficients
  - RTD of liquid phase and gas phase

Part III Suspended Beds

Ch7 Bubble and Drop Formation
- From single orifice or nozzle
- Models of Davidson and coworkers
- Model of Kumar and Kuloor
- Bubble and drop formation from multiorifice distributors, design of distributors

Ch8 Bubble Columns,
- Homogeneous and heterogeneous/churn turbulent flow regime

Bubble size distribution, Liquid circulation in bubble columns, bubble velocity, gas holdup, gas-liquid interfacial area, volumetric mass transfer coefficient, liquid backmixing, gas backmixing

Packed bubble columns

Slurry Bubble Columns

Ch9 Fluidization

Minimum Fluidization velocity

Particulate fluidization, Richardson- Zaki correlation, slip velocity

Aggregative Fluidization

Geldart Particle Classification

Ch10 Bubbling Fluidized Beds

Two Phase theory of fluidization, Bubbles,

Size of bubbles at the Grid, Grid Design

Bubble rise velocity, coalescence and growth along the axial direction

Bubble holdup

Bubble to Dense phase mass transfer coefficients

Fast and slow bubbles, Cloud formation and bubble to emulsion phase mass transfer coefficients

Bubble Wakes and particle turnover in the bed

Emulsion phase gas velocity

Bubble assemblage Model for catalytic reactors, non catalytic reactions, physical operations

Cause of low conversions

Entrainment, TDH, Elutriation

Ch11 Turbulent, Fast Fluidization and Dilute Transport

Risers, Elutriation and circulating fluidized beds

Bubble less fluidization, slip velocities, particle aggregation in the form of clusters,

Cluster models for axial voidage variation, wall heat transfer and gas to particle mass transfer coefficients

Particle and gas back mixing

Reactor Models

Downers,and Emerging technologies

Part IV Stirred Tanks

Ch12 Mixing of liquids

Options of Equipment, tanks, impellers, baffles

Geometric, kinetic and dynamic similarity
Power number as a function of Reynolds Number
Liquid circulation and mixing times
Jacket/coil wall heat transfer coefficients

Mixing of two Phases
Gas-Liquid Mixing, Gas flow number, flooding, size of bubbles, bubble holdup, volumetric mass transfer coefficients
Liquid-Liquid mixing, droplet size distribution, liquid holdup, volumetric mass transfer coefficients
Liquid-solid mixing, critical stirrer speed, mass transfer coefficients

Mixing of Three Phases
Suspension of particles and simultaneous dispersion of gas, transfer processes

6 Concluding Remarks

The approach being developed in this elective course tries to include the results of research being pursued in the chemical engineering field for the last several decades. An attempt is made to present the various factors together for synthesizing performance models to explain the behavior of reactors and separators. It is expected that the subject matter evolves over next decade. There are many other topics which are of great importance and constraint of time dictates the extent of coverage. However, once the general approach to multiphase flow is delineated, the students will be able to figure out more advanced topics.

Acknowledgements

The authors wish to record their deep sense of appreciation for the financial support by Universiti Teknologi PETRONAS.

References

1. Little A.D., Report on the corporation of M.I.T. as quoted in Silver anniversary Volume, A.I.Ch.E. ,1933, p.7
2. "Unit Operations of Chemical Engineering", McCabe W.L., Smith J.C. and Harriot. P., 5th ed., McGraw Hill Chemical Engineering series, 1993
3. "Principles of Unit Operations", Foust. A.S., Wenzel. L.A., Clump. C.W., Maus. L. and Andersen. L.B. 2nd ed. , John Wiley & Sons, 1990.
4. "Unit Operations", Brown. G.G., New York, Wiley,1950

5. "Unit Processes in Organic Synthesis", P.H.Groggins, 5th ed., McGraw Hill, 1984.
6. "Transport Phenomena", Bird. R.B., Stewart W.E. and Lightfoot. E.N. John Wiley & Sons, 1960
7. "Chemical Reaction Engineering", Levenspiel. O., John Wiley & Sons, 1999
8. "One Dimensional Two Phase Flow", Wallis. G.B.,McGraw – Hill Inc. NY, 1969
9. "Fluidization Engineering", Kunii. D. and Levenspiel. O., Butterworth-Heinemann, Boston, 1991

From cold cataracts to an undergraduate laboratory

Daniel Forciniti
Chemical and Biological Engineering Department
Missouri University of Science and Technology

Introduction

I was an undergraduate student at the School of Sciences of the University of Buenos Aires in 1978 when I took a biochemistry class as part of my degree requirements. I had the fortune of taking the laboratory portion of that class with one assistant professor who was a member of the Campomar Institute, a research institute headed by Prof. Leloir who was a Nobel price in Medicine. I did not know exactly how our laboratory experiences (mostly isolation of DNA from rats whose reproductive organs were surgically removed) fitted into the large scheme of Prof. Leloir's research. However, I did had the feeling that I was doing something important --a "real life experience" in my undergraduate career that I will never forget.

Most of us try to incorporate in our classes pieces of our research projects but our success, sometimes, is limited. The engineering sciences have advanced so much that it is increasingly difficult to incorporate even small elements of our research projects into sophomore, junior or even senior classes. To make matters worse, we are often involved in research projects barely connected to what we like to call Engineering Fundamentals. The fact that the Chemical Engineering Curriculum of Missouri University of Science and Technology (formerly the University of Missouri-Rolla) includes a Biochemical Engineering Emphasis Program has facilitated the incorporation of research projects into my classes. Although this incorporation would have been quite more difficult in a traditional Chemical Engineering program, it is apparent to me that some (if not all) of the projects described in this paper can be incorporated into traditional unit operations laboratories.

Sometimes it is only possible to include small pieces of large research projects because of their complexity. I found that it is important to be somehow aggressive on this approach and above all, do not underestimate the students. The students' reactions may be mixed. Some of my students have used the expression "research laboratory" in a positive way while some others use it as a critique, mostly referring to the amount of hours that they are expected to spend in the lab.

Most of the experiments incorporated into our Bioseparation Laboratory were once either a main research project in our group or a source of raw material for more fundamental research projects. The two types of experiments have been received differently by different types of students. Laboratory experiences based on research projects aimed at the purification of a particular protein have been well-received by those students more inclined to work in the processing sector. Research-rooted projects have been, in general, better received by students inclined to pursue further studies (either graduate school or medical school). Still, it has been an excellent approach to attract students into our graduate program as many realize that to do research can be fun.

From the Research Group to the Trenches: A brief story.

Some of the projects that we have introduced into our undergraduate laboratory are: the isolation of transgenic antibodies from corn meal, the precipitation of antibodies by non-ionic polymers, the production and purification of alcohol dehydrogenase from yeast, the synthesis, purification and characterization of diblock peptides, and the fractionation of a family of proteins present in mammalian lenses. A few of these projects are described below.

Many years ago, our research group started investigating cold cataracts. In cold cataracts, mammalian lenses undergo a liquid-liquid phase transition as their temperature is lower below a critical temperature, which depends on the species and on the age of the specimen. As a result of that phase transition, a protein rich and a protein lean phases coexist. One of the proteins present in the lens (γ-crystallins) undergoes a phase transition similar to that of the whole lens. There is an interest in the study of solutions of crystallins because research in that direction is expected to suggest mechanisms for pathogenic or traumatic cataracts.

The α, β, and γ-crystallin subfamilies comprise most of the lens proteins in mammals. The α-crystallin subfamily consists of highly polydispersed proteins that constitute -40% of the protein content of the lens. The molecular weight of the α-subclass is approximately one million Da. β-crystallins are also polydispersed proteins and have a molecular weight ranging from 50,000 to 300,000 Da. γ-crystallins are a heterogeneous group of low molecular weight proteins, 30,000 Da, with subtle differences in their amino acid compositions. γ-crystallins can be divided into five subfamilies (γ-IV , γ-III, γ-II, γ-I, and γ-S). The fractionation of this subfamily into its components and their characterization by neutron and light scattering has been the focused of our work for quite some time.

The purification of crystallins was incorporated into a Bioseparations Laboratory, which is a mandatory class for those students in a Biochemical Engineering Emphasis Program that we started nearly 20 years ago. A number of different separation techniques are used to isolate γ-crystallins from whole calf lenses. A sequence of steps to purify the γ-fraction that is used routinely in our laboratory is the following (Petitt et al., 1997): 1. Cell Disruption; 2. Centrifugation; 3. Size exclusion chromatography ; 4. Dialysis; 5. Ion exchange chromatography; 6. Dialysis; and 7. Freeze Drying. The students characterize the preparation, explore new separation conditions (for example, new salt gradients in the ion-exchange step or different lengths of size exclusion columns) and study the proteins using static and dynamic light scattering. Good samples that are not used by the students are recycled into our research group. This laboratory requires some specialized equipment. A bench top or a floor centrifuge able to accommodate 50 mL tubes, a low pressure chromatography system, and a spectrophotometer are necessary. The freeze drier is, somehow, optional. The low pressure chromatography system can be replaced by some of its individual components like a flow cell spectrophotometer and a fraction collector. The pump of the chromatography system can be replaced by flow under gravity using a two-compartment gradient former. The most expensive materials are the chromatography gels (Sephadex G-75 and Sephadex C-50). We usually recycle gels from one year to the next without major problems. The low pressure columns are quite cheap in the sizes that we utilize (either 1.4 or 2.5 cm in diameter columns). Expendables are cheap. The lenses may be bought from some specialized vendor from whom it is possible to buy lenses of various ages for the students to explore variations in the composition of the lenses as the animal ages or they can be gotten from the local slaughter house by free.

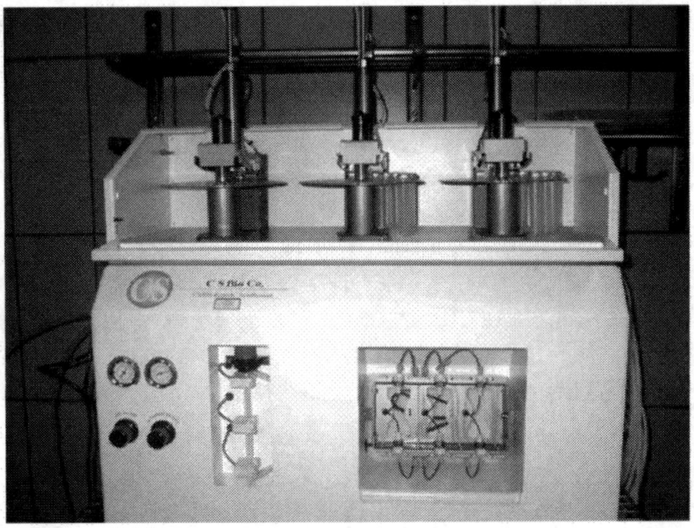

Figure 1. Solid state synthesizer

About ten years ago we began studying the behavior of plasma proteins at solid/liquid interfaces. Some time after, our interest switched to the study of short peptides at solid/liquid interfaces. These studies slowly evolved into an interest on the behavior of amyloid-forming peptides at solid/liquid interfaces. A few years ago, we acquired a solid-state synthesizer and immediately after we incorporated the synthesis of short peptides and their purification into the Bioseparations Laboratory. Every time that the synthesis is successful, the students are producing a tailored peptide that we can use to study, for example, its behavior at solid surfaces having diverse chemistries.

Solid state peptide synthesis is a powerful technique to prepare tailored peptides and even entire proteins (Merrifield, 1963). The fact that Prof. Merrifield won a Nobel Prize because of this work seems to have a positive effect on the students involved in this project.

The students learn how to synthesize a peptide using a solid state synthesizer (Figure 1), how to isolate the peptide by precipitation and how to identify (and purify further) the product by reverse phase chromatography. The basic pieces of equipment are a solid state synthesizer, a high speed centrifuge, a HPLC instrument and a freeze drier. The students are asked to explore different separation protocols in the reverse phase chromatography isolation of the target peptide. In addition, the students use a dynamic light scattering instrument to determine the thickness of the peptide adsorbed layer on poly(styrene) latex with different surface chemistries. This particular laboratory is relatively expensive because high pressure chromatography systems able to process hundreds of milligrams of peptides and the solid state peptide synthesizer are normally not found in a regular unit operations laboratory. It is possible to synthesize some simple peptides by wet chemistry with the consequent savings in equipment. However, the nature of the chemicals needed and the number of steps needed would make such a laboratory very difficult to pursue. The study of the behavior of these peptides at solid/liquid interfaces requires a particle sizing instrument. We use in our laboratory a backscattering dynamic light scattering instrument from Brookhaven that has proven to be quite rugged, it is very easy to use and it is reasonable cheap.

A few years back we began exploring the isolation of human antibodies expressed in corn. The main objective of our work was to find suitable alternatives to the use of a protein A column, by far the most expensive element in the purification of antibodies. The transgenic production of antibodies in plants needs to be followed by a robust separation process. Otherwise, the cost advantages associated with the production of the antibody in plants are lost.

We have developed a new process by which human antibodies expressed in corn are isolated to high purity and yield using aqueous two-phase extraction (Lee and Forciniti, 2005). We have found that one or two extraction steps, where the target antibody concentrates in the bottom phase or top phase, followed by a second extraction step, where the target antibody precipitates at the interface, yields the best results. The optimum purification protocol consists of the following steps: 1) Extraction of the antibody (and contaminating proteins) from the corn meal using a NaCl solution. 2) Addition of PEG and a salt to a concentration high enough to induce the formation of two liquid phases at equilibrium. 3) Removal and disposal of the upper phase (PEG-rich). 4) Addition of PEG to the bottom phase to create a second ATPS. 5) Recovering of the antibody from the new liquid/liquid interface. 6) Removal of the excess salt by diafiltration 7) polishing of the product by protein A chromatography.

The students in this laboratory extract the antibody from corn meal and then study its partitioning behavior in a variety of aqueous two-phase systems. This particular project is well-suited for the statistical design of experiments. The equipment needed is quite standard. In addition to basic laboratory equipment, the students will need a rotary mixer, a spectrophotometer, a HPLC system and a centrifuge. The raw material in our case is corn meal containing a particular human antibody. Of course, most programs will not have access to that raw material. One option is to use commercial corn meal spiked with a commercial IgG. Another option is to spike the corn meal with hemoglobin rather than an antibody. Hemoglobin from pig is quite cheap and besides its presence may be detected by its absorbance at 450 nm. Thus, it is possible to measure total protein by a colorimetric assay and hemoglobin concentrations by UV spectroscopy. In addition, the students need PEG of various molecular weights and various salts. The cost of this laboratory varies depending on the approach. If the experiments are done with corn meal spiked with an antibody, the main cost is a protein A column and the corresponding chromatography hardware. If the experiments are done with corn meal spiked with hemoglobin the cost is very low. The aqueous two-phase systems by themselves are very cheap, particularly if dextran is replaced with a salt like phosphate or citrate. Most labs will have a centrifuge and a spectrophotometer. Even in the absence of a centrifuge, the systems can be allowed to separate under gravity. Details of this project are described in the following section as an example.

Example: Isolation of Antibodies from Corn Meal

When macromolecules, such as proteins, are dissolved in a two-phase system, they selectively distribute between the phases. A partition or distribution coefficient is defined by,

$$K = \frac{C_t}{C_b} \qquad (1)$$

where C_t and C_b are the concentrations of the protein in the top and bottom phase, respectively. The partition coefficient depends on the pH of the phases, temperature, type and concentration of salt added, and type and concentration of the polymers used as phase forming species. The fact that the partition coefficient is a function of so many variables makes this project particularly suited for optimization studies. For example, the students may choose three molecular weights of PEG, three different salts (varying in their chaotropic properties) and three pHs (near, above and below the isoelectric point of the protein) A three levels full factorial experimental design will consist then, of 27 experiments. Because duplication is a necessity in this type of experiments students need to prepare, sample and analyze 57 extraction experiments.

Materials

Polyethylene glycol of various molecular weights (for example, 3,500, 8,000 and 20,000), dextran 70,000, acetate and phosphate buffers spanning three pHs, sodium chloride, lithium chloride, potassium chloride, pig hemoglobin, and Bradford reagent.

Equipment

The following equipment is needed: A spectrophotometer or a plate reader, a centrifuge (optional), magnetic stirrers, hot plates, balances, a pH-meter, micropipettes, a rotary shaker (optional), and an ultrafiltration cell (optional).

Methods

"Extraction" of antibody from corn meal

1. Extract 1 g of corn meal with 10 ml of 150 mM NaCl for 8 to 12 hrs at 4 ^{0}C with stirring.
2. Remove the solid particles by centrifugation at 11,000 rpm for 1 hr at 4 ^{0}C.
3. Filter the supernatant through a filter paper
4. Filter again through a 0.45 μm membrane.
5. Add 1 mg/ml pig hemoglobin to this extract.
6. Determine total protein using the Bradford test.

Preparation of Stock Solutions

A) Dextran (30% *w/w*)

1. Weigh 30 g of dextran into a bottle
2. Dissolve the dextran in 30 g of deionized water
3. Mix the above, it will make a paste
4. Add the remaining 40 g of deionized water
5. Heat the solution up to 95° C to facilitate dissolution of the polymer

B) PEG (50% *w/w)*

1. Weigh 25 g of PEG into a bottle
2. Dissolve the PEG in 25 g of deionized water

Preparing the Phase Systems

1. Shake the stock solutions
2. Place a 15 ml centrifuge tube on a balance
3. Weigh out the stock solutions into the tube in order of their increasing densities
4. Add enough buffer (blanks) or buffer plus 1 g of corn meal extract to complete 10 g.
5. Mix the contents of the test tube thoroughly, first by hand, and then in a rotary shaker (20 minutes)
6. Centrifuge the tubes for 15 minutes at 1500 x g to allow the phases to separate
7. Sample the phases as described below
8. The pH in each phase is measured with a microelectrode directly on the undiluted phases. Because of the high viscosity of the phases, the pH measurements must be done over a relatively long period of time.

Sampling and analyzing the phases

1. Carefully pipet out 2 g of the top phase
2. Carefully pipet 2 g of the bottom phase
3. Leave the separated phases resting and stir again. Inspect the solution to detect density differences along the axial direction of the test tube.
4. Read absorbance of each phase sampled at 450 nm.
5. Perform the Bradford assay for each phase.
6. If readings are out of range then dilute the samples with buffer (can be done by volume)
7. Calculate the partition coefficient of PIG hemoglobin and the overall partition coefficient for each sample. Partition coefficient is reported as K = $(\text{Absorbance sample} - \text{Absorbance blank})_{top} / (\text{Absorbance sample} - \text{Absorbance blank})_{bottom}$

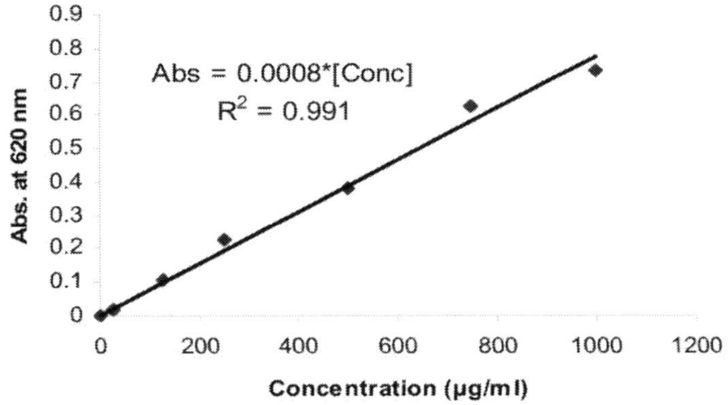

Figure 2. Bradford Standard Curve with BSA.

Bradford Assay Standard. The students create a standard curve for the Bradford Assay using standard solutions of BSA. A microplate procedure is ideal but, of course, the assay may be performed in a regular spectrophotometer. Figure 2 shows a typical standard curve. The assay is simple but careful (and polished) techniques are needed to obtain this type of results.

Results

The students process the raw adsorption data from the Bradford test and from absorbance at 450 nm. Proper blanks need to be discounted and dilution factors must be accounted for. The absorbance values are converted into concentrations using the appropriate calibration curves and those values plus the values of the volumes of each phase (using well-established correlations for densities of PEG and Dx solutions) are used to calculate recovery. Concentration values from Bradford (total protein) and absorbance at 450 nm (hemoglobin) are used to calculate purity and purification fold. Finally, the partition coefficients (total protein and target protein) are calculated and reported.

Figure 3 is a contour plot of the partition coefficient versus pH and PEG molecular weight. We found this presentation of the data is very illustrative.

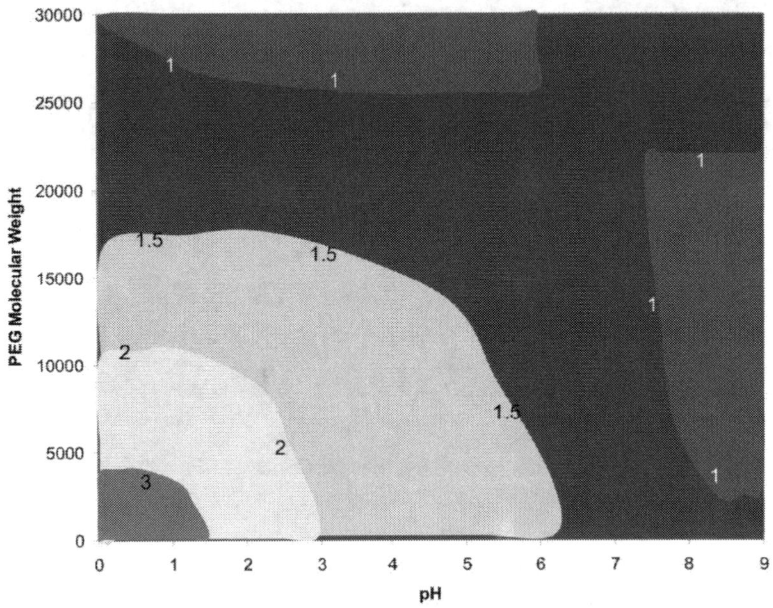

Figure 3. Contour plot of the partition coefficient (total protein) vs. PEG molecular

The partition coefficient data may be regressed to find a correlation between the partitioning coefficient with pH and PEG molecular weight for all salt types. One possible correlation is,

$$K = a + b[pH] + c[pH]^{-2} + d[MW] + e[MW]^{-1} + f\frac{[pH]}{[MW]} + g\frac{[pH]^2}{[MW]} + h\frac{[MW]}{[pH]} + i\frac{[MW]^2}{[pH]} + j\ln[pH] + k\ln[MW] \quad (2)$$

where MW is the molecular weight of PEG, pH is the pH, and "a" through "k" are adjustable parameters. Calculated and experimental partition coefficients are plotted in Figure 4.

Conclusions

Pieces of research projects have been successfully incorporated into the undergraduate curriculum. The students benefit because they are exposed to state of the art techniques, using equipment bought with research funds. In exchange, they contribute to a project by, for example, optimizing a particular separation or by producing materials that are fed into our research group.

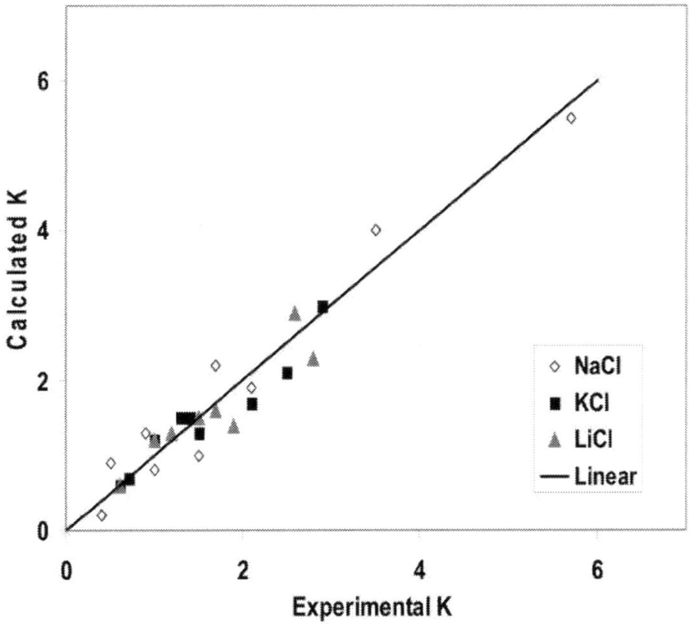

Figure 4. Calculated Partition Coefficient Versus Experimentally

References

R.B. Merrifield. 1963. Solid phase peptide synthesis. I. The synthesis of a tetrapeptide. J. Am. Chem. Soc., 85: 2149-2154.

P. Petitt, M. Edwards, and D. Forciniti. 1997. A Small Angle Neutron Scattering Study of Proteins at Their Isoelectric Point. European Journal of Biochemistry 243: 415-421 (1997).

J.-W. Lee and D. Forciniti. 2005. Purification of human antibodies by using liquid/liquid extraction. 7th World Congress of Chemical Engineering, Glasgow, Scotland, July 2005. 84799/1-84799/10.

334e Integrating Pharmaceutical Engineering Research in Graduate, Undergraduate, and Pre-College Educational Activities

Piero M. Armenante, Otto H. York Department of Chemical Engineering, New Jersey Institute of Technology, 323 King Blvd., Newark, NJ 07102-1982 and Henrik Pedersen, Chemical and Biochemical Engineering, Rutgers University, 98 Brett Road, Piscataway, NJ 08854-8058.

New Jersey Institute of Technology and Rutgers University are both in New Jersey, a state in which the pharmaceutical industry has a dominating presence. Both schools now have graduate-level pharmaceutical engineering programs that contribute to address the educational needs of the industry, and they are both members of the NSF-sponsored, four-university (Rutgers U. (lead), NJIT, Purdue U., and UPR-Mayagüez) Engineering Research Center for Structured Organic Particulate Systems (C-SOPS). As a result, these institutions have developed additional educational tools to promote the incorporation of C-SOPS research advances in pharmaceutical engineering and in engineered structured organic particulate systems into the existing engineering curriculum at both the undergraduate and graduate levels, as well as in high-school education. One of most significant outcome of these activities has been the development by C-SOPS researchers of specific graduate-level modular educational units ("modules") on specific topics of relevance to pharmaceutical engineering. In addition, C-SOPS researchers have been involved in a number of other initiatives aimed at training undergraduate students through the REU program (Research Experience for Undergraduates), high-school teachers involved in the NJIT's RET program (Research Experience for Teachers), high-school students through the New Jersey Governor's School for Engineering at Rutgers University, and graduate students involved in C-SOPS-based research and in several IGERT programs (Integrative Graduate Education and Research Training Program). In all cases, C-SOPS researchers have been actively engaged in bringing to the classroom the information and experience gained through their research efforts. This presentation briefly describes the whole range of the C-SOPS-based educational initiatives, showing how the current research activities have permeated the educational tools that were developed as a result.

Incorporating Microdevice Research into a Split-Level Elective Course for Undergraduate and Graduate Students

Adrienne R. Minerick
Dave C. Swalm School of Chemical Engineering
Mississippi State University
Mississippi State, MS 39762

Abstract:
It is widely touted that the use of research ideas can create excitement for learning in the classroom. Microscale research is a challenging area within which to do this because microscopes are required to observe most phenomena in microdevices. This paper describes the development of an Analytical Microdevice Technology (AMT) course for undergraduates and graduate students at Mississippi State University. The approaches used in this course were designed to overcome the challenges with directly observing microscale phenomena. The primary course goal was to get the students familiar with small-scale technology with a focus on biomedical diagnostic applications. The course covered both theoretical and experimental advances in the realm of chemical, mechanical, optical and biological analysis. This was accomplished through four activities throughout the semester. One day per week was a dedicated lecture day where the professor came with a structured set of material, in class activities, videos, etc. to provide a foundation of knowledge for the students. A second class day was dedicated to a Survivor Game modeled after J. Newell's 2005 article [1]. The third class day was comprised of student presentations and discussion of technical articles and current news articles on Analytical Microdevice Technology. The fourth activity was a semester-long open-ended concept development project. This activity included progress reports every two weeks where these intermittent reports built to a fully developed concept well grounded in the research literature and featuring a novel approach or device for a biological analysis. This paper will describe both the course content, its close influence with research, and conclude with results of student assessment of the four learning tools.

Introduction:

A multidisciplinary, split graduate level and senior undergraduate level course was taught on Quantitative Analysis and Electrokinetics in Microdevices, an active area of research for the professor and author. The course covered both theoretical and experimental advances in the realm of biomedical diagnostic applications. Course content is given in the adjacent table with the reference texts listed below the topic area by author [1-12].

The objective of the course was to provide undergraduate and graduate students with a background in analytical microdevice technology. Surveys of news and corresponding technical articles were intended to empower students with familiarity, skills and knowledge to envision microdevice applications, and apply this in research or in future job pursuits. Upon completion of the course, the student were to have command of the following topics and skills:

- Review of Micro / Nano technology news and critique of corresponding technical publications
- Micro and nano scale forces
- Materials and methods for device fabrication
- Existing and future detection tools
- Concept plan of a fully integrated device

Students were engaged in four main learning activities during the semester: Lectures, a Survivor Game, Article Presentations, and a large Concept Development Project. The lecture content was largely governed by the content given in the adjacent table. The Survivor game was modeled after J. Newell's Survivor: Classroom [13]. Questions for the game were pulled from each of the other three activities. Students were encouraged to actively read the literature as a learning tool and as a supplement to text information provided in class by weekly article discussions in class. Lastly, the students were arranged into four teams each having a graduate student leader. Each team worked together to develop a concept project and write it up into a journal article by the end of the semester. Each of these activities is discussed in turn below.

Two class tours were conducted. The first was of the microfabrication facilities (photolithography, electroplating, and epitaxy) at SemiSouth, Inc., a start-up company from Mississippi State University. The second tour was of MSU's Life Sciences Biotech Institute, a multi-user genomics and proteomics facility.

AMT Course Topics

1. Intermolecular and Surface Forces
   Israelachvili, Probstein [1,2]
   - Intermolecular interactions
   - Interparticle forces
   - Polar (izable) molecules
   - Electrostatic forces
2. Microdevice Designs & Considerations
   Literature, Rathore [3-5]
   - Shape & Materials
   - Fabrication techniques & Lab Tour
   - Sample Injection / Mixing
3. Pressure Driven Flow (micron length scales)
   Bird, Fournier, Truskey, Literature [6-8]
   - COMSOL Mulitphysics DEMo
4. Linear Electrokinetics
   Rathore, Li [5,9]
   - Electrophoresis
   - Electroosmotic Flow
   - Lab Tour: Life Science Biotechnology Institute
5. Nonlinear Electrokinetics
   Morgan, Delgado Literature, [10,11]
   - Dielectrophoresis
   - Magnetophoresis
   - Traveling wave DEP
6. Detection and Quantification of Analytes
   Webster [12]
   - UV, flow cytometry, fluorescence

Activity 1: Lectures

Lectures were held each Monday in order to provide the students a well-organized foundation in the physics of materials at the micro and nanoscales as well as fundamental knowledge of the optical and electronic tools utilized in microdevices. The topics covered are outlined in the adjacent list.

Lectures were of traditional format with content written on a whiteboard in a sequential fashion. Exercises and short class activities were interwoven into the lecture. Due to the open atmosphere of the classroom (likely facilitated by the other class activities), students regularly asked questions. When topics related to biochemistry or biology were included in the lecture, a Biochemistry graduate student would frequently add insight and instruction beyond what the professor presented. It was an excellent learning opportunity for the students and professor alike.

Activity 2: Micro-Technology Survivor

A game of Survivor: Classroom was conducted each week and based upon J. Newell's adaptation of this popular TV show to Mass and Energy Balances [13]. Two surveys were conducted during the semester to gauge student perception of this as a learning tool. The first was conducted after the first Survivor game day and was focused primarily for students to provide feedback on questions and logistics of conducting the game in class. The second was conducted halfway through the semester. In both surveys, the students were asked to respond to two questions on a 5-point Lickert scale from Strongly Agree down to Strongly Disagree. The responses are compiled in Table 1 and demonstrate positive student feedback to the game. Interestingly, the students do recognize that the fun factor is a little greater than the learning factor. There was a slight increase by mid-semester in the number of students who felt they were learning from the Survivor game.

Table 1: Student Assessment of Survivor game merits

	Q1: I learn a great deal from the Survivor Problems		Q2: I had fun playing Survivor in class.	
	Begin	Mid	Begin	Mid
Strongly Agree	4	6	10	9
Agree	8	6	3	4
Neutral	2	2	1	1
Disagree	-	-	-	-
Strongly Disagree	-	-	-	-

With the feedback from the first survey, a number of modifications were made to the rules of the game. Original rules are available in J. Newell's article [13], AMT class rules are summarized below.

- The class is broken into "tribes," which sit together in groups to solve knowledge / reasoning, calculations, and design-based questions. Students are permitted to use any resource (notebooks, computers, phones, etc.).
- The first tribe to have an answer raises their hand and their answer is compared against the professor's. If necessary, additional time is provided. The tribe with the correct (or most optimal solution) is immune from losing a member. A representative from the successful tribe works through their solution so that the other teams may consider their solution strategies or novelty of their concept designs.

- At the end of each round, non-immune tribes lose a member according to (sequence repeats in order): a) tribe members vote off a member of their own tribe, b) tribe with immunity votes off a member of other tribes, c) one member of each tribe is eliminated by drawing a name out of a hat.
- The tribe with fewer members is immune from losing a member the first round. Graduate students are immune from being voted off for the first 3 questions.
- Students who have been eliminated in any round join the Peanut Gallery. They are also given the task of designing and solving a problem to be used in later rounds. Each eliminated student will need to produce 1 problem for every gaming session. If a student does not turn in their problem by the start of the next game, they will lose 2 points. If a student question is used in the game, they will earn 1 point.
- Students in the Peanut Gallery can also work on the Survivor questions independently. If they "win" a round, they can choose to join any team they like (and earn 2 points). The team they join then earns immunity and the loss of a member of the other tribes proceeds as above for that round.
- Tribes will be revamped / reformed once a single player has been crowned Grand Champion. Successful students will be rewarded with bonus points on their final according to a) every player who survives to the second day of play gets 3 points, b) every member of the champion's tribe gets 2 points, c) the Grand Champion earns an additional 5 points, the points are additive, so the Grand Champion will earn 10 points (or more!).

Interestingly, the ability of members of the Peanut Gallery to answer questions independently to earn back a position on any team helped immensely with keeping those students who had been voted off engaged in the game and learning.

Activity 3: Current Micro / Nano News & Archival Journal Articles Discussions

This section of the class was modeled after the author's research group Journal Clubs [14]. Article discussions were intended to be a practice forum where one student would lead the class in a discussion of an article of his / her choosing. Undergraduate students were allowed to pull from popular news. However, graduate students could scan the science / technology news, but had to secure the archival article and present from that. When the student provided the article in advance, it was posted on webCT, but the other students were not required to read articles in advance. The professor did keep track of questions asked and those who participated in the discussion in order to give participation points, but did not control the progression of the discussion.

Given the number of students in class, each student presented once every 3 weeks. The students were told that a significant part of engineering research is oral and written communication with tangible depth and conveyance of understanding. As a result, emphasis was placed on the clarity, organization, and understandability of the student's presentation. It was strongly encouraged to present at the level of the audience such that an involved discussion could develop from the article. Students were allowed to use any method of delivery they preferred.

Activity 4: Semester Concept Development Project

The skills that serve graduates of engineering programs best are not all learned from textbooks. Tangible skills are harder to teach, but are essential to prepare students to be productive, technical, members of society include problem solving skills, information filtering skills, and logic skills. The traditional classroom does not focus on these skills nor does it usually provide individual practice linking unique concepts together. This semester long Concept Development Project was a concerted effort to strategically develop these skills in the students enrolled in this course.

The assignment for the project teams was a large, open-ended, concept development project. The students were integrally involved in deciding the small-scale technology that they wanted to pursue which would help address an important biomedical application using either micro or nanotechnologies. The concept was to build from both theoretical and experimental reported technologies in the realm of chemical, mechanical, optical and biological analysis. Their resulting virtual microtechnology was to be a novel extension of published work. Upon completion of the project, each team member was expected to be able to conduct an extended discussion of the following topics and skills related to their project:

- Review of pertinent technology from peer-reviewed publications
- Micro and nano scale forces acting within their micro / nano device
- Materials and methods utilized in their conceptual device (and why chosen)
- Why and how their project was novel from existing detection tools

The output of the effort was to be a concept plan of a fully integrated biomedical technology. The concept was to be articulated in an archival journal paper and presented via a team oral presentation. Progress reports were due throughout the semester and were designed to roughly build the sections of the final archival journal article. The reports were a) description of proposed novel analytical microtechnology, b) complete literature review on the scientific premises of the proposed analytical microtechnology (> 10 references, fully discussed), c) prototype drawing and accompanying description of the analytical microtechnology, d) final device design and a first draft of complete final report, and e) final archival journal article.

Assessment:

Student's preferences for each class activity were assessed via a short survey in the final day of the course. The 11 question survey was approved by MSU's IRB and the students present all signed consent forms for their data to be included. The survey was designed to test the hypothesis that the class activity that student's most prefer was influenced by their learning style.

The premise was that everyone has learning style preferences [15,16] which can be measured on a sliding scale between two extremes in four stages of learning: processing, perception, input modality, and understanding [17]. During processing, students favor either **active (ACT)** or **reflective (REF)** learning by introspectively thinking about material. In perception, **sensing (SEN)** learners focus on external input while **intuitive (INT)** learners focus on internal contemplation. Input modality preferences scale between **visual (VIS)** and **verbal (VRB)**. Lastly, learners can achieve understanding in **sequential (SEQ)**, linear logical steps or **globally (GLO)** where information is pieced together into a big picture.

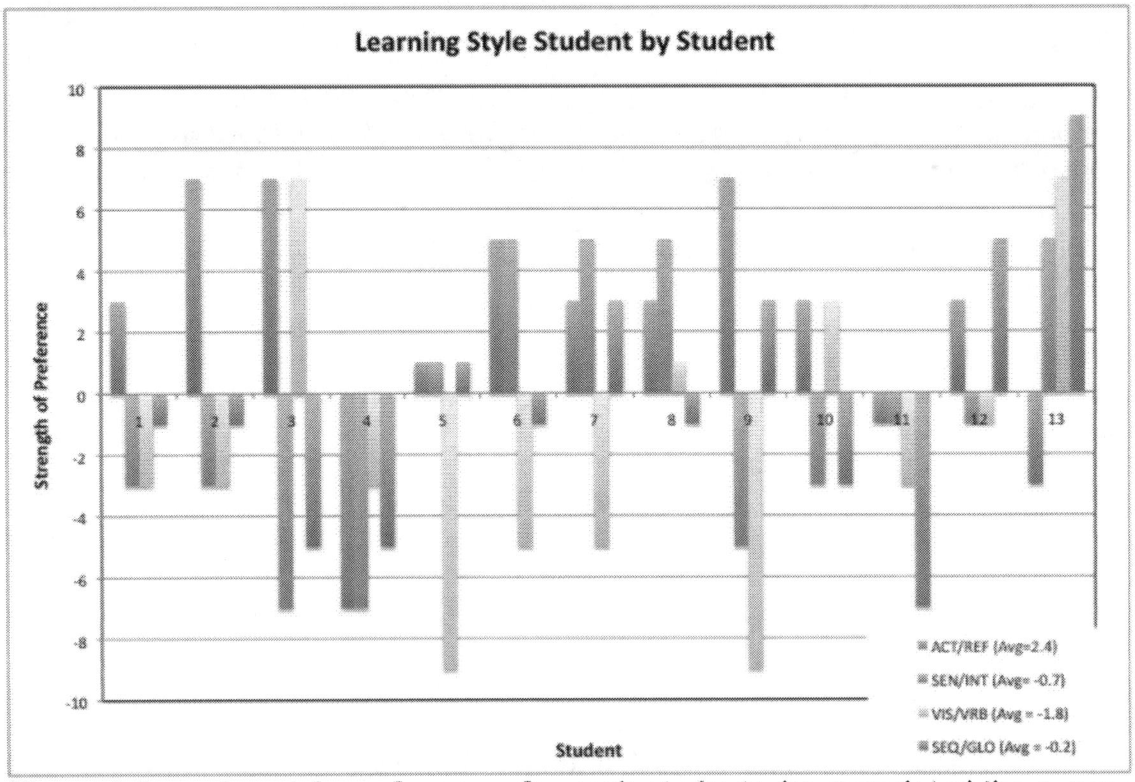

Figure 1: Learning style preferences for each student who completed the survey. Negative values indicate a preference towards the first of the two modalities. Non-weighted averages are provided in the legend labels.

To test this hypothesis, students were directed to the Soloman-Felder online learning styles inventory and asked to complete the 44 questions [18]. They then self-reported their scores on the written survey and proceeded to answer the remaining survey questions. In figures 1 through 3 reporting the student's learning style, the four stages of learning are reported as ACT / REF with a negative number indicating preference towards active learning while a positive number indicates preference towards reflective learning. The same nomenclature is used for SEN / INT, VIS / VRB, and SEQ / GLO.

Overall student learning preferences are included in Figure 1 and are slanted towards reflective (REF) with an average of 2.4, slightly sensory (SEN) with an average of -0.7, more visual (VIS) than verbal (VRB) with an average preference of -1.8, and nearly balanced between sequential (SEQ) and global (GLO). The SEQ / GLO average is slightly misleading as 8 of the 13 students were sequential learners, but did not show as great of a preference in this direction. Given that global learners historically have not gravitated into engineering, this is an interesting distribution of students in this course.

The first question asked students to rank their most favorite to least favorite learning activity in this class. The activities were: lecture, Survivor game, article discussions, and concept project. The most popular activity was the article discussions, which is interesting because this was the activity that students demonstrated the most apprehension over when the concept was introduced. While some students voiced dread over their turn to present, the remaining students became more engaged in the discussions as the semester progressed. Figure 2 cross correlates the student's preferred class activity with their composite learning style (style preference added together). The stronger active learners were skewed toward Survivor as the preferred activity (N=4 students) while the weakly active learners and reflective learners preferred the Article discussions. Overall, sensory learners preferred Article discussions while intuitive learners preferred Survivor. The stronger visual learners had a slight preference towards Survivor than for Article discussions. Overall, global learners preferred Article discussions while sequential learners preferred Survivor.

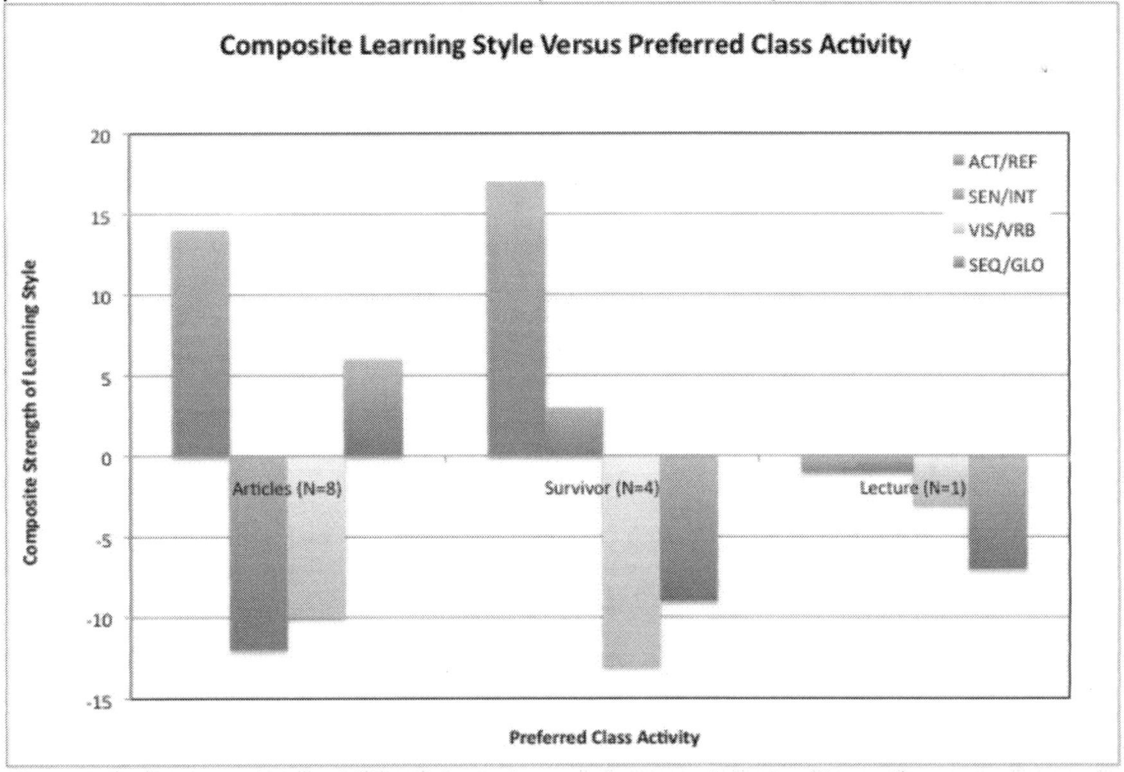

Figure 2: Composite (additive) learning style correlated with preference toward class activities.

One individual chose lectures as the favorite class activity and their strongest preference was toward sequential. Lectures tended to be quite linear in concept progression, while the article topics were random due to student's freedom of choice to obtain an article and the Survivor game questions were purposefully randomized.

Students were also asked to score how much they learned during each activity on a scale from 1 to 10 with 10 being "learned a great deal" and 1 being "learned very little". As demonstrated in Table 2, the students felt they learned the most from their Concept Development Project followed by their favorite activity, Article discussions. The student's learned the least from the Survivor game, but it should be noted that the average score here is still greater than average (5).

Table 2: Student self-rated learning in each of four class activities.

	Lecture	Survivor	Articles	Project
Average	7.7	6.5	8.2	8.5
Standard Deviation	1.5	1.3	1.5	1.3

This critique of their own learning was cross-correlated with their learning style and is shown in Figure 3. Trend lines are added to give an overall sense of which type of learner learned more. For example, active learners felt they learned more from the project, article, and lecture than reflective learners. However, reflective learners felt they learned more from Survivor than the active learners, which is surprising given the learning mechanism these modality suggests. The preferences in one direction or another are slight and given the population size, the numbers can be skewed by the response of a single individual. Only slight preferences are shown when comparing sensory and intuitive learners. Other notable trends include that visual learners felt they learned more than verbal learners from the lectures. In all cases, all types of learners rated Survivor lowest as a learning tool.

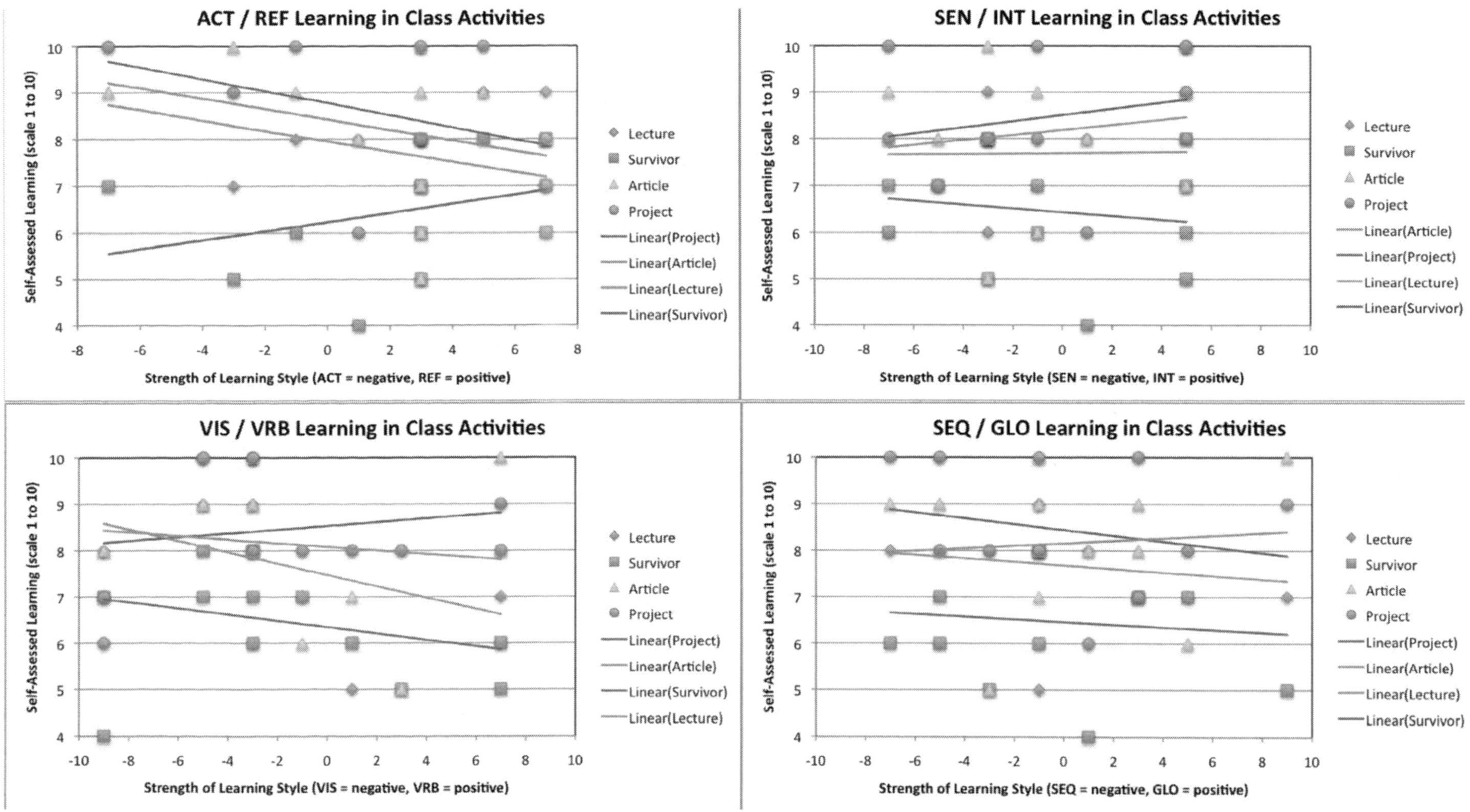

Figure 3: Student ratings of learning value of the four class activities (Lecture, Survivor, Article discussions, Concept development Project) cross-correlated with their learning style preference.

Students were also asked if there a particular combination of learning activities that was instrumental in helping them learn. The two most common combinations mentioned (by 4 of 13 in each case) was Lecture / Survivor and Article / Project.

The original Survivor paper [13] discussed a model where students had four primary types of motivation. The original work describing this was Biggs and Moore in 1993 [19]. They summarized the four types as: "**Intrinsic** – learning because of natural curiosity or interest in the activity itself; **Social** – learning to please the professor or peers; **Achievement** – learning to enhance position relative to others; and **Instrumental** – learning to gain rewards beyond the activity itself (better grades, increased likelihood of getting a high-paying job, etc.)" Students were asked to rate themselves against these four motivators on a scale from 1 to 10 with 10 being "very motivated by this" and 1 being "not motivated by this at all". On average, the students felt they were very intrinsically motivated (8.8 ± 2), fairly neutrally socially motivated (5.0 ± 2.2), a little more strongly motivated by achievement (6.5 ± 1.9), and even stronger instrumentally motivated (7.8 ± 1.5).

Conclusions:

A research focused special topics course on Analytical Microdevice Technology was taught for the first time in Spring 2008 with four different class learning activities. The activities were lectures, a Survivor game, Article discussions, and a concept development project conducted in teams. While the students showed a strong preference toward the article discussions, most felt they learned the most from the projects.

Correlations with student learning styles were compiled and demonstrated that while students have a preference for certain activities, learning is possible with all activities as demonstrated by quantitative learning rankings greater than neutral for all activities. Overall, the students rated themselves as intrinsically and instrumentally motivated. Grades were de-emphasized in this course, yet student participation in all except one student was greater than the professor has seen in core chemical engineering courses she has taught. The population size is rather small (13 completed the surveys), yet interesting trends suggest learning merits in each of the four class activities.

References:

1. Israelachvili, J., *Intermolecular and Surface Forces: Second Edition*, Academic Press, San Diego, 79,82,102, 1992.
2. Probstein, R.F. Physicochemical Hydrodynamics: An Introduction (Second Edition), Wiley-Interscience Publication, New York, 1994.
3. Rathore, A.S, and A. Guttman, Electrokinetic Phenomena, Marcel Dekker, New York, 2004.

4. Hill, P. J. "Introducing Technical Articles to Graduate Students Through Active Learning" presented at the Southeast Regional ASEE meeting, Chattanooga, TN, April 2005.
5. Micro / Nano Magazine. Sponsored by Nano Science and Technology Institute (NSTI). http://www.rdmag.com/micronano.aspx Accessed 18 July 2006.
6. Bird, R.B., W.E. Stewart, and E.N. Lightfoot, Transport Phenomena (Second Edition), Wiley, New York, 2002.
7. Fournier, R.L. Basic Transport Phenomena in Biomedical Engineering, Taylor & Francis, Philiadelphia, 1999.
8. Truskey, G.A., F. Yuan, and D.F. Katz, Transport Phenomena in Biological Systems, Pearson Prentice Hall, Upper Saddle River, NJ, 2004.
9. Li, D., Electrokinetics in Microfluidics in Interfacial Science and Technology V. 2, Elsevier, Amsterdam, 2004.
10. Morgan, H., and N.G. Green, AC Electrokinetics: colloids and nanoparticles in Microtechnologies and Microsystems Series, Research Studies Press, Baldock, 2003.
11. Delgado, A.V. Interfacial Electrokinetics and Electrophoresis in Surfactant Science Series V. 106, Marcel Dekker, New York, 2002.
12. Webster, J.G., Bioinstrumentation, John Wiley & Sons, 2004.
13. Newell, J.A. "Survivor: Classroom. A method of active learning that addresses four types of student motivation." *Chemical Engineering Education*, 39(3), 228-231, 2005.
14. Minerick, A.R. "Journal Club: A Forum to Encourage Graduate and Undergraduate Research Students to Critically Review the Literature." New Engineering Educators Division - American Society of Engineering Education Proceedings, June 2006.
15. Felder, R.M. and L.K. Silverman, *Engr. Education,* 78(7): 674-681, 1988.
16. Felder, R.M. *ASEE Prism,* 6(4): 18-23, 1996.
17. National Effective Teaching Institute (NETI), offered in cooperation with American Society of Engineering Education, 2005, http://www.ncsu.edu/felder-public/NETI.html, Accessed 19 July 2005.
18. Soloman, B.A. and R.M. Felder, "Soloman-Felder Learning Styles Questionnaire," http://www.engr.ncsu.edu/learningstyles/ilsweb.html Accessed 22 Sept. 2008.
19. Biggs, J., and P.J. Moore, The Process of Learning, Prentice Hall, Englewood Cliffs, NJ (1993).

366a Introduction to the NSF Workshop II

William B. Krantz, Chemical and Biomolecular Engineering, National University of Singapore, Room E5-02-11, Block E5, 4 Engineering Drive 4, Singapore 117576, Singapore

The second session of the NSF workshop will be introduced.

366b Proposal Writing Tutorial

John Regalbuto, Division of Chemical, Bioengineering, Environmental & Transport Systems, National Science Foundation, 4201 Wilson Boulevard, Arlington, VA 22230 and **James M. Lee**, Washington State University, PO Box 642710, Pullman, WA 99164-2710.

A proposal writing tutorial will be presented by both coauthors.

366c Introduction and Overview of the Format for Breakout Sessions

Robert Wellek, The National Science Foundation, Division of Chemical, Bioengineering, Environmental and Transport Systems, 4201 Wilson Boulevard, Arlington, VA 22230

The format for the breakout session will be described

366d Interactive Breakout Panels Coordinated by NSF Program Directors

Robert Wellek, The National Science Foundation, Division of Chemical, Bioengineering, Environmental and Transport Systems, 4201 Wilson Boulevard, Arlington, VA 22230

There will be breakout groups and discussion led by NSF Program Directors. Attendees can discuss topics specific to their areas of interest

406a Panel Discussion: Building a Successful Relationship with Your Dean and Upper University Administration

Eduardo D. Glandt, School of Engineering and Applied Science, University of Pennsylvania, 220 S. 33rd Street, Philadelphia, PA 19104-6391, **Michael F. Malone**, College of Engineering, University of Massachusetts, Amherst, MA 01002, Kirk H. Schulz, Mississippi State University, Box 9595, Mississippi State, MS 39762, Dianne Dorland, College of Engineering, Rowan University, 201 Mullica Hill Road, Glassboro, NJ 08028, and Harvey Stenger, College of Engineering, University of Buffalo, 412 Bonner Hall, Buffalo, NY 14260.

This will be a panel discussion on how to build a successful relationship with your dean and upper university administration.

406b Abet Update

Jeffrey J. Siirola, Eastman Research Division, Eastman Chemical Company, PO Box 1972, 200 South Wilcox Drive, Kingsport, TN 37662-5150

This presentation will give ABET updates

What's New at the Petroleum Research Fund?

Chris Hollinsed, Ph.D., Director

The Petroleum Research Fund was founded in 1944 to support "advanced scientific education and fundamental research in the "petroleum field"". The work may include "any field of pure science which ... may afford a basis for subsequent research directly connected with the petroleum field." The "petroleum field" as defined by the fund includes: (1) exploration for, and the production, transportation and refining of, petroleum, petroleum products and natural gas, and (2) the production and refining of substitutes for petroleum and petroleum products from natural gas, coal, shale, tar sands and like materials.

The fund was started as trust which was to receive the income of United Oil Products (UOP) in 1944. Unfortunately, UOP had no income in the early years of its association with the Fund and as a result, no funds were available to fund research until 1954, 10 years after the creation of the Fund. By 1956, it was clear that the company, UOP could not be a viable operation without access to its own income and in 1959, the original trust document was changed to allow the sale of UOP and the conversion of the company to a diversified portfolio of securities. The value of the Fund at this point in time was $72.5 million. These were held in trust by the Guaranty Trust Company, later Morgan Guaranty. In the late 1990s, the question was raised as to whether or not 5% of the corpus of the fund could be used instead of being restricted to the income from the trust. The Attorney General of the State of New York decided at the time that the trust could not be changed but could be dissolved under its own terms and reconstituted in Washington, DC. As a result, in 2001 the Fund was transferred to the American Chemical Society.

The Fund has committed over $500 million to the community in support of research over its lifetime. Known primarily for support of new faculty and undergraduate institutions, the Fund commits between $23 and $25 million per year in support of its mandated mission. The majority of the money goes to established investigators at research institutions.

Over 19,000 articles in ACS journals have been published acknowledging PRF support. The Fund is well represented in other publications as well. Twenty four grantees have received Nobel prizes; several of these had PRF grants prior to receiving their prize which had titles similar to their Nobel Prize citation. Many grantees have gone on to become members of the National Academies and each year the majority of the award winners at the ACS meeting have at one time or another been funded by PRF.

In 2005 the current Director of the Fund, reviewed the history, the programs and the governance structure responsible for guiding the fund. It was clear at that time that the Fund was not aligned with norms in the philanthropic community. In particular, the Fund had limited strategic oversight. The Fund fell under the auspices of the Board of Directors of the American Chemical Society which delegated oversight to the Board

Standing Committee on Grants & Awards. While certainly serious in exerting their responsibility to govern the Fund and its programs, the committee was not able to devote sufficient time to developing a comprehensive strategic plan for the Fund. Secondly, evaluation plans and practices were missing. Modern philanthropic practice requires clear definition of programs in terms of the goals the program is attempting to achieve and clear measurements of success. The Fund had in place an absolutely first rate process for selecting grantees from the pool of applicants, however there was no mechanism in place for determining the efficacy of the programs or for evaluating whether some pilot programs which had been created should be continued or terminated. Certainly there was no process for understanding whether or not the Fund was supporting the "right' or 'best' activities through its programs.

In early 2006, the Petroleum Research Fund Advisory Board and the Board Standing Committee on Grants & Awards were asked to authorize a comprehensive review of the Fund. A consultant with considerable experience in strategic planning and change management for philanthropic organizations was brought in to work with the Director and the PRF staff to determine the operational baseline, the aspirational goals of the fund and to help develop a path to a strategic plan which would be focused on those goals.

The Board of Directors and the selected Advisory Board members were interviewed in addition, several community stakeholders and representatives of counterpart funding agencies were interviewed as well. It soon became very clear that the community had three clear goals for the Fund:

- Support of young faculty members at research and undergraduate institutions.
- Support for undergraduate institutions through the engagement of undergraduate students in scientific research.
- Grants to research institutions should have a clear and significant research impact.

A Steering Committee consisting of members of the Grants & Awards Committee and the PRF Advisory Board was assembled to guide the programs of the Fund toward a direction consistent with the above three goals and to develop a governance structure appropriate for such a large research fund.

The Steering Committee developed a vision statement for the Fund which was later modified in cooperation with the Planning Committee of the PRF Advisory Board. That statement is:

> The ACS Petroleum Research Fund will support innovative fundamental research, advanced scientific education, and the careers of scientists, to aid in significantly increasing the world's energy options.

In order to determine the community's response to the change, a workshop was held seeking community input. The starting point for this work was the already develop goals listed above and the vision statement. The External Review Workshop was held on January. The workshop reinforced the goals as outlined and in addition, added some ideas which were not anticipated. In particular, it was suggested that multidisciplinary, multi-investigator grants be created, a portion of the fund should be set aside to fund high risk, high impact research and that the visibility of the fund should be enhanced through the creation of a "signature" event. In terms of staff operations, it was suggested that a fund be set aside for the Director to use at his discretion, that PRF Program Staff be granted more decision making authority and that *ad hoc* panels be used for proposal review to relieve pressure on the reviewer community.

The information from the goals, the development of the vision and the feedback from the workshop provided a basis for a revised program proposal for the fund. The revised set of programs for the Fund was developed by the Steering Committee. These were fleshed out and refined by staff and presented first to the Advisory Board and with their recommendation to the Board Standing Committee on Grants & Awards for approval and authorization.

The resulting programs are as follows:

- New Investigator Grants

For scientists and engineers who are within the first three years of their first academic appointment. One program is for new investigators at Ph.D. granting institutions, a second is for new investigators at primarily undergraduate institutions.

- New Directions Grants

For scientists and engineers who have very limited or even no preliminary results for a new research project they wish to pursue, especially in an area that they have not worked in before. It is expected that the PRF funded preliminary results would be used to seek continuation funding from other agencies.

- Undergraduate Research Grants

For scientists and engineers with established programs of research at primarily undergraduate institutions.

The Steering Committee sought and received approval of these programs from the Committee on Grants & Awards, however the committee also wanted to address the issues of high impact and multidisciplinary research. As a result, the committee established working groups for three additional programs. The groups studied and developed additional programs which were presented for recommendation to the Advisory Board and subsequently for approval to the Committee on Grants & Awards.

These programs are as follows:

- Exceptional Innovation Grants

To stimulate exceptionally innovative fundamental research in the petroleum and energy and related fields; to provide "seed" funding for potentially high reward research; to provide a venue for testing novel concepts and/or addressing key energy related science gaps; to populate the front end of the innovation pipeline with new technical concepts and leads.

- Strategic and Multidisciplinary Research Grants

Designed to address key technology barriers that, if overcome, would have a high impact; use of a fundamental research approach which may be higher risk, require longer term solutions and be both novel and non-incremental; specifically address those barriers that require a multi-disciplinary approach.

- Fellowships for Increasing Representation of Underrepresented Minority or Women Scientists and Engineers in Academia.

Fellowships or other support directly to potential faculty aimed at increasing the representation of underrepresented minority or women faculty in academia. The grants are aimed at empowering and educating these communities as the transition is made from graduate school to post-doctoral research to first academic appointment.

The results of a nationwide survey on Chemical Engineering faculty salaries will be presented along with other survey results which includes graduate student assistantships, post-doctoral fellow salaries, and starting assistant professor salary offers. Regional and national averages will be compared to similar results from past surveys.

406e Integration and Segregation: Mixing Curricular Reform with Disciplinary Research

Joseph J. McCarthy, Chemical Engineering, University of Pittsburgh, 1249 Benedum Hall, University of Pittsburgh, Pittsburgh, PA 15261

Enhancing the educational process for training engineers is a recurring topic in Chemical Engineering whose momentum periodically waxes and wanes. This is due, in part, to the perceived gulf between educational and disciplinary research. While there are a number of institutions or individuals that are well known for "educational research", I am often struck by the creativity and passion for educational innovation evidenced by "regular" faculty. In this talk, I will briefly describe how I have managed to serve as PI of a 6 year NSF-funded $1.5M curriculum reform project aimed at vertical integration of ChE, while concurrently supervising 7 disciplinary projects funded from NSF, DOE, and ACS aimed primarily at mixing and segregation of granular materials. The goal of the talk is to highlight best practices for succeeding in both educational and disciplinary research in an effort to inspire some of the "regular" faculty to increase the scope of their impressive efforts.

406f Building Relationships with Alumni and Fundraising

Christopher Roberts, Auburn University, 212 Ross Hall, Auburn, AL 36849

This presentation will share successful strategies for building relationships with alumni and fundraising

406g Eroding Engineering Education: a Profession Under Threat

Colin S. Howat, Chemical and Petroleum Engineering, University of Kansas, 1530 W. 15th Street, Room 4132, Lawrence, KS 66045-7609

The Emperor has no clothes! We are not who we think we are.

A profession requires that students are taught by those well-versed in practice, that there is a body of knowledge that identifies the profession, that life-long learning is required to master the knowledge, there is a passion to serve society and that the profession is self-regulated through its code of ethics and governing body. But, chemical engineering education is controlled by closed tribes who hire the 'Learned Ones' to fulfill other goals. The training and background of these 'Learned Ones' has markedly changed.

Engineering education, the training of our future chemical engineering practioneers - those who control hazards, impact society to a larger extent than any other profession and advance our technical culture, is increasingly in the hands of people who have never practiced and controlled by non-engineers. But, the university dictates:

• Talents of who is hired;

• Evaluation of advancement within the tribe;

• Measures of stature; and,

• Effort spent on education.

New faculty members are hired for their research potential, not their industrial experience at the undergraduate practice level. Promotion pays lip service to teaching while counting publications, research dollars and Ph.D.'s supervised. The literature is replete with statistics that the stature within in a university is measured by how little the individual has to do with undergraduates. Assistant professors truly must publish or perish.

Textbooks are written for faculty by faculty. They do not provide perspective. New faculty members have an increasingly narrow perspective of engineering as their education requires increasing specialization. Textbooks are written for faculty by faculty. They do not provide perspective. We only have to look at the development of the new paradigm (G. G. Brown, to BSL to a Committee) or to the completely flexible ABET standards where we can define who we are because we aren't who we were to see that the 'Learned Ones' have changed. But, have the elements of practice changed?

Our undergraduates move into an environment where they must exercise their problem, creative, integrative, critical and life-long learning skills in an environment where answers are not black or white, right or wrong. Engineers no longer stay with the same company for a life-time. This places increasing responsibility on the 'Learned Ones' to socialize engineers into the requirements of professional practice.

Engineering education serves the profession, not the universities. We must have a passion for education. We need re-establish:

• Teaching by practitioners;

• Focusing on the core practice oriented knowledge;

• Focusing on practice oriented skill development;

• Developing ethical reasoning.

We cannot rely solely on our capstone courses to fulfill these goals.

412a Modernizing the Content and Pedagogy of a Classic Textbook

Mark T. Swihart, Chemical and Biological Engineering, The University at Buffalo (SUNY), 303 Furnas Hall, Buffalo, NY 14260-4200 and Hendrick C. Van Ness, Chemical and Biological Engineering, Rensselaer Polytechnic Institute, 110 8th St., Troy, NY 12180.

The authors are currently in the midst of a substantial revision of Introduction to Chemical Engineering Thermodynamics, by Smith, Van Ness, and Abbott. This presentation will discuss the challenges involved in updating the style and content of a long-successful text without losing the essential character of the book or dramatically disrupting the practices and syllabi of current users. Changes that have been made, planned or considered will be presented, along with demonstrations of a variety of newly available online resources for instructors using the book. Key questions and issues include (1) how best to broaden the perspective of the text beyond the chemical process industry, (2) how to support flexible use of computer tools and a variety of software packages, (3) how, and to what extent, to incorporate molecular concepts, and (4) what changes in style or format may enhance pedagogy without compromising the character of the text. Most importantly, feedback on proposed changes, and further suggestions will be solicited.

412b Events That Shaped the Chemical Industry

Rebecca K. Toghiani, Dave C. Swalm School of Chemical Engineering, Mississippi State University, Mississippi State, MS 39762

At Mississippi State University, the Center for Teaching and Learning has established a First Year Seminar series. This seminar series has established learning outcomes that include a student being able to: 1) argue issues specific to their discipline; 2) distinguish their discipline within the academic community and 3) think critically about topics within their discipline. A one-hour seminar entitled "Events that Shaped the Chemical Industry" was proposed and is being taught during the fall semester to 25 freshman chemical engineering students. The last century is ripe with events that significantly impacted the chemical industry, the practice of chemical engineering, and the education of chemical engineers. In this seminar, these freshmen are exposed to the rich history of achievement and accomplishment in chemical engineering as well as the challenges faced by the industry in light of environmental events such as Love Canal and Times Beach and economic events such as the 1973 oil embargo. The seeds of their education are planted in how the chemical industry responded to these challenges. To understand where the chemical industry is heading during the coming decades of their careers, students need to be knowledgeable of the events that have shaped the industry.

412c Practical Implementation of Data Acquisition to Enhance Problem-Based Learning in Chemical Engineering Labs

Keith L. Levien, School of Chemical, Biological and Environmental Engineering, Oregon State University, 103 Gleeson Hall, Corvallis, OR 97331-2702

In Winter 2002, ten LEGO® Team Challenge kits, based around the RCX microprocessor, were introduced in lab sessions in CHE 102 (Introduction to Engineering Problem Solving and Computations) for data acquisition utilizing light, rotation and temperature sensors. The software package ROBOLAB®, developed at Tufts University based on the LabVIEW package by National Instruments, was chosen as the programming base for those experiments. An integrated course project was developed based on the NASA Mars Rover project which was being planned for a 2003 launch. The class has seen continued growth throughout the 7 years that we have utilized the LEGO-based labs. Based on the success of student engagement in the project and course, students in the Bioengineering program were added to the class in winter 2003. First-year students in the Environmental Engineering program were added in winter 2007. The teamwork design project has grown to 40 teams of 3 students working in three lab sections. This class included 43 women, one of the largest enrollments of women engineering students in a required course in the College of Engineering.

While the RCX microprocessor, developed and released by LEGO® in 1998, has proven to be very reliable and robust, a newer more powerful microprocessor called the NXT with many enhanced capabilities was released in 2007. An additional attraction to the continued use of the LEGO® platform is the development of interfaces to the NXT which allow direct use of a wide variety of new, lower cost sensors to provide process and physiological measurements. Our experiences with the RCX and first-year student retention will be discussed along with our plan to utilize the improved NXT capabilities and sensors in our second-year process data analysis lab and third-year transport lab.

THE TRIALS AND TRIBULATIONS OF IMPLEMENTING A PROBLEM BASED LEARNING APPROACH IN THE FRESHMAN YEAR OF UNDERGRADUATE CHEMICAL ENGINEERING

Dr. Diane Rossiter and Dr. Catherine A. Biggs
Department of Chemical and Process Engineering, The University of Sheffield, Sheffield, UK
Abstract Number 132415

Abstract

Chemical Process Principles I (CPE1002) is a core and compulsory course taken by approximately 70 students in the freshman year of a Chemical Engineering undergraduate degree at The University of Sheffield, UK. The primary learning objective of this module is to develop the students' knowledge and understanding of material balances. Here we will present the progression of improvements in CPE1002 over the past 3 to 4 years, including the adoption of a problem based learning approach.

New aspects introduced into the course were group work, industrial visits and assignments incorporating industrially relevant processes. The course was extended to cover key personal and transferable skills such as group work, communication skills, independent and self-directed learning and peer assessment, without losing any of the key technical material. In addition, online resources were provided via WebCT Vista (managed learning environment) [1] to enable the students to self-assess their weaknesses and strengths in the core chemical engineering principles and practice these skills so that they were able to come to the group work more prepared.

Through the changes in the course, we were able to expose the students to the way process engineers work, think and communicate their ideas as well as introducing them to the processing/chemicals industry. These changes transformed the course from a problem module with the class generally struggling to grasp the key concepts to a successful module with positive student feedback.

Introduction

The core learning content of the Chemical Process Principles I course looks at material balances for single unit operations, such as mixers, splitters, separators and reactors, and then various combinations of these single units in series. This allows for purges and recycles to also be considered. A lot of our students were finding the learning content challenging and were struggling to make the connection between the mathematical manipulations required and what a practicing Chemical Engineer does. As a result of this feedback, it was decided to adopt a different style of course delivery to try and engage the students more effectively with the learning content and its application.

The new course delivery mode selected was a type of Problem Based Learning (PBL). Our primary reason for this choice of change in module delivery was that our department had good links with Chemical Engineering at the University of Queensland, in particular Professor Paul Lant. They had had considerable success with their students adopting a project centred curriculum [2] and as part of his sabbatical at Sheffield, Professor Lant was willing to assist with its introduction within our course. At Queensland, this approach has been implemented programme-wide whereas we were looking initially to introduce it for one course. Professor Lant visited in the academic year 2005/06 and brought all his learning resources relating to his syllabus for material and energy balances. This was the start of a major transformation for our course and change within our department.

This extended abstract will provide details of the chronological developments of the course including the issues that were addressed; feedback from the students and lessons learnt by staff. Then, some conclusions will be provided.

Course Developments

The developments have occurred over a number of years with the initial transition being from traditional didactic delivery in the academic year 2003/04 (see Table 1), when students numbers were relatively low, to two-hour weekly "problem solving" workshops with a one-hour weekly supporting lecture from 2005/06 onwards. The idea was to shift the emphasis from the lecturer being the "sage on the stage to the guide on the side" [3] with the students being given tasks that required them to seek out information from the resources and supports provided, and construct their own knowledge.

Looking at the review by Newman [4], this aligns with his definition of Problem Based Learning i.e. "a conception of learning as an integrated process of cognitive, metacognitive and personal development". This concept has been introduced successfully into many professional fields of study including engineering. (For anyone wishing to development a PBL course this review provides very good guidance on curriculum design.)

Table 1. Chronological Course Delivery Developments

Academic Year	**2003/04**	**2004/05**	**2005/06**	**2006/07**	**2007/08**
Student Numbers (SN)	28	55	66	67	70
Lectures (per week)	2 hours	2 hours	1 hour	1 hour	1 hour
Tutorials	2 per term 3 hours each Large group	3 per term 3 hours each Small group	weekly 2 hours each PBL-style	weekly 2 hours each PBL-style	weekly 2 hours each PBL-style
Assessment	80% exam 20% coursework	75% exam 25% coursework	50% exam 50% assignments and test	50% exam 50% assignments and test	50% exam 40% assignments 10% online test

From Table 1, it can be seen that the evolution of the course has coincided with an almost 3 fold increase in student numbers from 2003 to 2007. Traditionally, the University of Sheffield has been fortunate to attract very able students to its Chemical Engineering programmes who exhibited higher order learning skills. This type of student was able to cope with limited tutorial support (pre- 2003/04). However as various changes have occurred such as the British government's widening participation agenda for higher education and changes to the 16-18 curricula in science and mathematics [5], although the entry criteria have remained high, the students entering into Year 1 appear in several cases to lack the higher order learning skills required for independent learning. Hence, the change of the course delivery to include weekly PBL-style tutorials has provided more scaffolding of the students' learning.

"Traditional" Problem Based Learning Approach (2005-2007)

The key to successful PBL approach is the use of authentic problems to engage the students whilst developing their core technical skills. Professor Lant provided several case studies relating to industrial processes and the students were required to work in groups to carry out the weekly assigned formative tasks. We also introduced an industrial visit to a pulp and paper mill and summative assignments incorporating industrially relevant processes. The course content was extended to cover key personal and transferable skills such as group work, communication skills, independent and self-directed learning and peer assessment, without losing any of the original key technical material! Students were also encouraged to do weekly homework problems to support their learning.

When the changes were first introduced in 2005/06, feedback from the students was extremely positive. However, 23% of the students failed the summative assessment compared with 16% the year before! A review was carried out by Biggs [6] and showed that those students who didn't engage with the PBL group work were not engaging overall with the whole programme. Suggesting it wasn't necessarily the PBL style of delivery that had impacted but their general lack of motivation and involvement.

However the following year 2006/07, two additional changes were made: (1) emphasis on the need to engage with the group PBL tasks and assignments to ensure success in the course and (2) the weekly homework was marked so that the students got regular formative feedback. This had the desired effect of improving the overall summative assessment results (only 2% failed the course).

The introduction of weekly homework marking and the size of the class meant 2 members of academic staff and 2 teaching assistants were now involved with the course delivery and marking compared with the usual 1 member of academic staff on other traditionally delivered courses within the department. This has had an impact on the running cost of the course and it is not practical for all courses to be delivered in this way without a major increase in the departmental teaching budget which is unlikely to occur in the short term.

Also accommodating up to 70 students at the same time in a learning and teaching space suitable for a PBL-style tutorial proved a challenge. (Due to timetabling constraints it was not possible to split the cohort into say two smaller groups.) Although the University of Sheffield has had a major investment into Inquiry Based Learning (IBL) flexible learning and teaching spaces (see CiLASS [7] website) the largest collaboratory only accommodates up to 48 students. Alternative venues had to be sought which had a flat floor with moveable tables. These proved to be very limited in number. We also wanted to be able to present information using Powerpoint and/or overhead projector slides so it was also essential that the students could see the wall from any where in the space where this information was to be projected. These issues may seem trivial but for PBL-style tutorials to be successful it is very important that the students are able to "huddle" in their groups and exchange ideas and information. This is the essence of the communication and collaboration for PBL to take place.

The new course delivery style of PBL was deemed a "success" by the end of 2006/07 and a new phase of developments started. Initially, it had been envisaged that further developments would involve rolling out the same course format across other courses. However, the combined impact of the running costs and infrastructure limitations meant that this type of widespread roll out was not an option. However, lessons learnt from this course on how to deliver a PBL-style tutorial have been transferred to other small group teaching such as being carried out in the process design strand of teaching throughout Year 1 to 3 (beyond the scope of this paper). So the developments continued to focus on the same course with internal learning and teaching grant from CiLASS.

Blended Learning PBL Approach (2007-onwards)

Despite good student feedback on the whole, for the PBL-style of delivery, there was evidence from the peer evaluation data that some students were being "carried" by the members of their groups. There was a danger that because the summative assessment was 50% group assignments and 50% individual examination that students could succeed on the course but not have the core technical skills required for later courses in their undergraduate programme. These concerns needed to be addressed but constraints on staff time and increasing students numbers (see Table 1) meant that providing additional remedial group tutorials and/or one-to-one support were not practical options to support weaker students. It was necessary to find an alternative approach that would "blend" with the PBL approach.

The strategy adopted was to incorporate online formative quizzes that the students could use to self-test their strengths and weaknesses prior to participating in the PBL group tasks and assignments. We wanted the students to get instant feedback which was possible using computer aided assessment tool. These quizzes cover five core technical skills:

(1) unit conversions using the unity brackets approach;
(2) mass to mole conversions;
(3) calculations and definitions relating to material balances;
(4) material balance calculations without reactions; and
(5) material balances with reactions.

They were delivered within WebCT Vista [1] managed learning environment although some of the development work was carried out within a 3rd party tool for creating online assessments called Respondus [8]. (For detailed discussion of the development of the online quizzes see Rossiter and Biggs [9].)

As part of the CiLASS funded learning and teaching project during 2007-08, a detailed evaluation study was carried out to assess the impact of the developments on students, academic staff, their department, the University and the wider community. This evaluation study provided a rich source of data. The data was collected via focus group with students, interview with academic project staff and questionnaire to all students involved with the course. Some of this data is discussed in the following sections.

What did the students say?

The students were asked in the questionnaire and the focus group about all aspects of the course including the PBL-style tutorials and their use of the online quizzes. Here are some comments made:

Table 2. Extract from student feedback at focus group[1]

Interviewer: Do you think the module (CPE1002) changed the way you approach learning or problem solving activities?
Student: "In the first semester, we were learning things and **I could see why we were learning** them because I could see how to apply them. This semester, we're learning a lot, but I'm not always sure why. I wish we had more practical group assignments throughout the course."
Interviewer: If you had your choice between the group format and what you're getting now (predominantly lectures), which would you prefer?
Student: "Last semester, I really liked visiting the paper mill site. Now we know about industry." Student: "You **actually felt like an engineer** when you were doing the group assignments. Now **I just feel like a student**, learning a lot of things"

These comments suggest that by working on the PBL group tasks and assignments the students were seeing the connection between chemical engineering practice and what they were learning on the course. This was reinforced by comments in the personal statements that they included with their PBL group assignments and also the questionnaire data[2] where:

- 93% of respondents found the PBL activities enjoyable and motivating.
- 98% of respondents found their experience of PBL helped them to at least some extent to develop confidence and skills in working collaboratively. (Team working is an essential skill for chemical engineers.)
- 96% of respondents found their experience of PBL helped them to at least some extent to develop confidence and skills in problem solving. (Problem solving is an essential skill for chemical engineers.)

In relation to the online quizzes, the students were equally positive i.e. 81.5% of students who responded said quizzes had helped to some extent in developing their technical core skills. The rest either didn't use the quizzes, didn't know or didn't respond.

The immediate feedback from the quizzes was also found to be beneficial:

[1] Eight students were present at the focus group. – Interviewer R. Petrulis (CiLASS) on 3rd March 2008

[2] Questionnaire, End of Semester 1 Feedback by Year 1 students, 15th Feb 2008, 54 responses out of 69.

Student: "They (Quizzes) were beneficial because of the immediate feedback. With the homework, it took a week or so. The online quizzes also referred you to the book and page for more information. If you just get a grade, that's meaningless."
Student: "The online self test quizzes gave INSTANT feedback. So if you didn't get the question right, I understood why and did not do the same mistake again. The quizzes were unlimited and this helped me practice."

What do staff say?

Overall, blending online formative quizzes with offline PBL-style tutorials has proved successful for this course. This has helped to provide a mechanism with instant feedback for the less able students to get help in developing their core technical skills in preparation for the PBL group tasks and assignments. Thus, helping all students to more actively engage in the group work.

For the first time in the academic year 2007-08, the examination for the course was held at the end of Semester 2 whereas the lectures and tutorials were held in Semester 1. This meant that the online quizzes were also available throughout the year to assist students with their revision prior to the examination. Providing these online quizzes did involve significant development time however it has been offset by a reduction in remedial one-to-one support being required. Some time was also gained through some of the coursework (an online test worth 10%) being automatically marked online. Also, the homework sheets were modified as some of the questions formed the basis of the online quizzes. Hence, this led to some reduction in the weekly homework marking load.

The major benefit though has been the observed awakening of the students when taking ownership for their own learning. The students are clearly more motivated when engaging in PBL group work than passively listening to lectures. The PBL-style tutorial sessions have a "buzz" about them!

Conclusions

There have been several major challenges to be addressed relating to this first year chemical engineering course such as: increased student numbers in some cases with reduced independent learning ability, constraints on provision of additional academic staff, and adequate learning spaces for Problem Based Learning style tutorials.

Creative solutions have had to be found to deal with these challenges and this paper has outlined the blended learning approach adopted for course delivery and support. The initial provision of the PBL content and training from Professor Lant of the University of Queensland was invaluable for starting a major process of change within our department. Back in 2005, it was envisaged that this change would lead to major programme change in favour of PBL style approach but this has not occurred to the extent originally imagined. Although, the process design strand of the programmes have been influenced by the PBL approach as we, the authors, have both been involved in design project supervision. However, an unexpected outcome has occurred in that an alternative mechanism for face-to-face support of the weaker students has been developed through the provision of the online formative quizzes so taking the course towards a so-called blended learning approach i.e. providing both online and offline learning resources and supports.

References

1. WebCT Vista (2008), Blackboard Learning System – Vista Enterprise License, E-learning platform. . [Accessed 14 September 2008]. Available from: http://www.blackboard.com/products/Academic_Suite/Learning_System/vista.htm
2. Crosthwaite, C., I. Cameron, P. Lant and J. Litster (2006), "Balancing curriculum processes and content in a project centred curriculum: in pursuit of graduate attributes", *Education for Chemical Engineers Trans. IChemE, Part D*, March, pp. 1-10.
3. King, Alison (1993), From the sage on the stage to the guide on the side. Abstract of Journal article in College Teaching, Vol. 41. [Accessed 15 September 2008]. Available from: http://www.questia.com/googleScholar.qst?docId=94305197
4. Newman, Mark (2008), "Problem based Learning," in *Higher Education Academy Imaginative Curriculum Guide*. [Accessed 14 September 2008]. Available from: http://www.heacademy.ac.uk/resources/detail/id362_Imaginative_Curriculum_Guide_Problem_Based_Learning
5. Government H.M. (2005), "14-19 Education and Skills", White Paper. [Accessed 15 September 2008]. Available from: http://www.dcsf.gov.uk/publications/14-19educationandskills/
6. Biggs, Catherine (2006), *CPE112 Chemical Process Principles I – A Review*, The University of Sheffield, Department of Chemical and Process Engineering, Sheffield S1 3JD, March, Internal Report.
7. CiLASS (2008), CiLASS spaces for learning and teaching, The University of Sheffield, Centre of Excellence for Inquiry-based Learning in the Arts and Social Sciences. [Accessed 15 September 2008]. Available from: http://www.shef.ac.uk/cilass/learning-spaces
8. Respondus (2008), Respondus version 3.5 – Assessment Tool for Learning Systems, Respondus, Inc.. http://www.respondus.com/products/respondus.shtml [Accessed 15 September 2008].
9. Rossiter, Diane and Catherine.A. Biggs, (2008), Development of Online Quizzes to Support Problem-based Learning in Chemical Engineering, In Proceedings *Third International Blended Learning Conference* 2008, University of Hertfordshire, UK, 18-19 June.

Acknowledgements

The authors are grateful to: Professor Paul Lant of Chemical Engineering at the University of Queensland for his inspiration; and the Leverhulme Trust for funding his sabbatical to Sheffield. The authors also wish to acknowledge the Centre of Excellence for Inquiry-based Learning in the Arts and Social Sciences (CiLASS) at the University of Sheffield for project funding and support to carry out some of this educational development work.

Ideas to Consider for Chemical Engineering Educators Teaching a New "Old" Course: Freshman and Sophomore Level Courses

Jason M. Keith, Department of Chemical Engineering, Michigan Technological University
David L. Silverstein, Department of Chemical Engineering, University of Kentucky
Donald P. Visco, Department of Chemical Engineering, Tennessee Technological University

Abstract

So, you are going to teach a core chemical engineering course next term that you have not taught before. It's time to come up with some new ideas to revolutionize that core course in ways that will amaze students and maximize learning, right? Or perhaps the maxim about "an hour in the library is worth a month in the laboratory" might be meaningful in the context of teaching. This paper summarizes the authors' selection of the most effective, innovative approaches reported recently in the literature or discussed at previous conferences for lower-division core courses in chemical engineering, as presented at the 2007 ASEE Summer School for Chemical Engineering Faculty. The challenges associated with particular courses and solutions successfully applied to address those challenges will also be described. Courses covered in this paper include introductory courses for freshmen, material and energy balances, fluid mechanics, introductory thermodynamics, and separations.

Objectives and Motivation

Although teaching is a critical mission of any college or university, today's faculty members are increasingly becoming involved in other scholarly activities. Thus, when teaching a new course, developing a good set of instructional materials can be a challenging, time-consuming task. In this paper we provide a review of some of what we consider the best practices in engineering education, applied to the following courses: freshmen chemical engineering, material and energy balances, fluid mechanics, introductory thermodynamics, and separations. Note that a companion paper which covers the upper-level undergraduate classes in the chemical engineering curriculum is planned for the following year.

The format used for each course is:

- Brief description of typical course content
- Discussion about novel and successful methods used
- Listing of "toughest concepts" for the students (and how to address them)

We note that most of this material was originally presented by the authors at the 2007 ASEE Chemical Engineering Division Summer School in Pullman,

WA[1]. This work was originally published (and also presented) at the 2008 ASEE Annual Meeting [2] as paper number #AC 2008-1147.

Freshman Chemical Engineering Courses

Depending on the school, this course is either a "stand-alone" introduction to chemical engineering or is part of a college-wide introductory course (with a portion devoted to chemical engineering). Ironically, many chemical engineering educators may never have taken this course.

A major goal of the course, as it is a freshman course, should be to cultivate student interest in engineering[3] and motivate students to pursue an engineering career. This course can have a wide variety of formats, depending upon the number of credits and objectives of the course for a particular institution. For example, Brigham Young University has a three-credit course which introduces (via an integrated design problem) all of the aspects of the chemical engineering curriculum[4], while Tennessee Technological University has a one-credit course that focuses more on hands-on experiments and information exchange[5]. Whatever the course, it is important for a department to identify why they have introduced or are teaching such a freshman course and whether (via specific assessment) the goals and objectives of the class are being met, from both the faculty and student standpoint.

In the rest of this section, we briefly highlight (as a resource) some of the novel work available on freshman courses in chemical engineering.

Some best practices that we have used (or discovered) for this course are:

- The use of freshman design projects:
  - Design and economic analysis of a controlled-release nitrogen fertilizer plant[6]
  - Design, build, and test an evaporative cooler[7]
  - Design and build a pilot-scale water treatment plant[8]
  - Analyze and design sneakers with better material prperties[9]
- Introduce in-class, hands-on experiments:
  - Melting chocolate and coating cookies[10]
  - Electrophoresis and brewing with microreactors[11]
  - Heat transfer scaling with hot dogs[5]
  - Human respiration process[12]

One overlooked concept in designing this course is to consider the needs of the student *from the student perspective*. Recently, the University of Pittsburgh asked their freshman engineering students to conduct a survey of other first-term freshman engineering students on topics the students felt were important[13]. While the results of the surveys are interesting in their own right, the most useful result is the *types* of surveys the students developed. The top ten types of surveys were as follows:

1. Getting enough sleep?
2. Has high school prepared you for college?
3. Do you feel safe on campus?
4. Any new romantic relationships?
5. Is partying getting in the way of schoolwork?
6. Exercise more or less than in high school?
7. Homesick?
8. Campus food options?
9. Susceptible to doing drugs / alcohol now?
10. Confidence in time management skills?

It is noted that there is nothing about a student's major listed in the top ten. Thus, a freshman engineering course requires a balance between what an instructor knows (or thinks) that a student needs, and what the students think they need. Therefore, while a freshman chemical engineering course must (obviously) contain information about the field of chemical engineering, it should also find ways to address non-chemical engineering related issues as well. Here, ample use of guest speakers in Counseling Services or similar offices on campus should be explored.

In addition to what has been discussed above, other ideas in freshman chemical engineering courses exist as well. Roberts discusses a course that focuses on, among other areas, communication skills[14]. Worcester Polytechnic Institute looks to mix writing with first-year engineering in a course taught shared by a ChE faculty and Writing faculty member[14]. Vanderbilt University describes a course where students are introduced to chemical engineering by "using examples from cutting-edge research to illustrate fundamental concepts"[16]. At Youngstown State University, they are demonstrating combustion principles to chemical engineering (and non-chemical engineering) students using a potato cannon[17].

Trouble spots for this course include:

- Most students do not know what chemical engineers do – one idea is to have teams of like minded students investigate where Chemical Engineers work in a particular field. Each team will present this information to the rest of the class at the end of the semester. Also, The Sloan Career Cornerstone Center[18] has short "Day in the Life" interviews of various young chemical engineers in a wide variety of industries that is quite informative at emphasizing the diversity of career options accessible for B.S. chemical engineering graduates.
- Most students only have a vague idea as to why they are taking math – one idea is to have upperclassmen come into the class and tell them how they are using math in their courses. In fact, using upperclassman as much as possible during the semester is a good idea as it indoctrinates the students easier into the program.

- Many students struggle with the transition from high school – one idea is to use upper-class peer mentors or speakers from on-campus who can discuss student-relevant issues. Having students conduct their own surveys, as discussed in a previous section of this work, might identify the most important issues for your students.

Material and Energy Balances

This course may also be called the “Stoichiometry” or “Process Principles” course by faculty. Students may refer to it as “The Cut Course,” or by even less flattering names. This course poses a unique challenge in many chemical engineering curricula since it requires students to think at a higher level than in previous courses. A typical course will cover: units and dimensions, properties, measurements, phase equilibria, material balances, energy balances (nonreactive and reactive systems), and combined mass and energy balances. The course should prepare students to apply conservation laws to process simulation as the first source of modeling equations. The course is the foundation for the rest of the curriculum—it is all about planting seeds for the future!

Trouble spots for this course include:

- Reluctance to show work. Students should be required from the start to show clean, detailed solutions even on the easiest problems assigned earlier in the class. Significant point deductions for deviations early in the course help train students to clearly communicate with their problem solving[19].
- Reluctance to apply rigorous methods to simple problems. The grader must pay attention to the method and not just the final answer. Requiring students to start from the general material balance even on problems that can be solved intuitively will aid students in solving more complex problems later in the course.
- Misunderstandings about density / specific volume and g_c. Repetition, drills, quizzes, and clear examples help to clear up some of these common misunderstandings. Warning students that these can be challenging issues may help a few pay more attention. Keeping a reference page at the beginning of their notebook or in the cover of the textbook with notes on these and other key subjects can also help.
- Trouble with thermodynamic diagrams. Students will not grasp these diagrams without working with them. One approach is using online interactive tutorials. Another effective approach is to bring copies of charts (even if they are in the text) for students to use in working problems either with the instructor, or better still, in small groups. They will only learn how to use these charts if they practice using the diagrams.
- Reluctance to apply rigorous methods to simple problems. Yes, this problem is significant enough to mention twice.
- Integration of “old” material into subsequent chapters. Student are going to tend to compartmentalize knowledge from each chapter (or each

homework assignment, each exam, etc.) and not internalize the concepts into their problem-solving repertoire. Blending lectures in a manner that bridges the chapter divide, using problems that draw extensively on previous topics, and even giving quizzes on material covered earlier in the course can help develop anchors to key elements in a course as they move on to new topics.

Some best practices and useful tools that we have used (or discovered) for this course are:

- Emphasize importance of communication in problem solving[19]. Requiring students to submit a solution or two that meets corporate standards can be a useful exercise in developing students' communication skills. Overuse of such a requirement can distract from the problem-solving objectives, so use sparingly.
- Teaching by analogy[20]. Using simple analogies for explaining confusing topics like mass / mole fractions, steady-state, specific volume, saturated air, and others can help students grasp topics that might elude them from lecture and reading alone. Analogies provide a link between what the student already knows and what you are trying to teach them.
- Mass and energy balances on the human body[21]. In this module students are asked to measure flows and compositions using a medical gas analyzer while exercising and at rest. They then apply several ChE fundamental principles (ideal gas law, partial pressure, stoichiometry, relative humidity, heat of reaction, work, efficiency, and process simulation) to analyze their results.
- Starting the unit operations early in the curriculum[22]. The equipment is already in the laboratory, so why not use it within the material and energy balance course? This allows for introduction of measurement, application of conservation laws, and an introduction of the fundamentals of design. Any time students can apply knowledge to a real task, they will learn better.
- Incorporating programming with templates[23]. Programming is an effective way of teaching students numerical methods. The problem with programming is that it often has significant overhead (input/output, user interface, etc.) that has nothing to do with the objectives of an assignment. Using templates, or "almost finished" programs lacking only the numerical method code, enables students to focus on implementing the numerical method and concentrate on the learning objectives for the assignment.
- Student-centered teaching[24-26]. This reference provides a host of suggestions for the material and energy balance course, including: developing a well-structured team approach to homework, posting homework answers (but not solutions), giving open book exams, and developing clear objectives and exam study guides to aid in student learning.
- Psychrometric chart applet[27]. This applet allows the user to calculate properties of humidified air, and helps students understand how to use the

psychrometric chart. It also frees up valuable lecture time when assigned to students to study on their own and then assessed through in-class active learning exercises.

- Richard Felder's Resources in Science and Engineering Education Website[28]. This is a popular site containing a link to the stoichiometry course taught by the textbook[29] author. The site also contains links to Excel tutorials[30]. Furthermore, there are many links to information on using active learning in your courses.
- Graph paper website[31]. Assuming you still expect students to learn fundamentals of graphing like use of logarithmic axes, these papers will come in handy.

Fluid Mechanics

Fluid mechanics has an interesting history within Chemical Engineering programs[32]. It developed from steam and gas technology for industrial chemistry and chemical engineering needs. From this evolved Unit Operations, which helped make chemical engineering a unique field. Meanwhile, fundamental studies in fluid mechanics were quite popular (and remain so) in the literature. This research work became integrated into the chemical engineering curriculum mostly due to the *Transport Phenomena* text[33].

Trouble spots for this course include:

- Students may possess weak math skills. Instructors can develop handouts to step students through difficult solution processes (such as solving differential equations). Have them practice with in-class problems and homework before testing them.
- Difficulty in connecting highly theoretical content to real industrial applications – if there is an internet connected computer and projector in the classroom, instructors can use online and/or laboratory demonstrations to make a strong connection. This connection can also help students with their following classes.
- Students often do not know order-of-magnitude values for pressure drops, velocities, Reynolds numbers, etc. The teacher can provide them with general values on a handout they can paste in the front of their textbook
- Students struggle with when to eliminate terms in the governing equations. If they are provided with handouts to step them through difficult solution processes (such as solving differential equations), they will be prepared for more advanced homework and exam questions.

One major advantage of teaching a course in fluid mechanics is the visualization that could be easily brought into this course. Some best practices that we have used (or discovered) for achieving this in the fluids course are:

- Ford's paper on "Water Day"[34] developed several observation stations so that students can visualize continuity, the Bernoulli equation, conservation of linear momentum, the vena contracta effect, and relative and absolute velocities.

- Incorporate high school outreach into the course
  - Using pressure conepts[35]
  - Using a tank-tube viscometer experiment[36]
- Use unit operations and/or research laboratories
  - Unique experiments have been developed by Fan[37] who discusses flow surrounding a bubble, two phase theory, flow segregation, phenomena of bubble wake dynamics, and computational fluid dynamics of particulate systems).
  - Particle technology is a field which offers a large number of simple experiments that can be brought into the classroom[38]. These include wet powder systems (single-particle settling, hindered and lamella settling, sedimentation and flocculation, interparticle force effects on colloidal suspension rheology, wetting behavior of dry powders, and granulation coalescence behavior) and dry particle systems (hopper flow, consolidation effects of powder flow, particle dilation, wall friction, segregation during hopper flow, vibrational segregation, fluidization, and flow improvement due to powder agglomeration). There is also a CD[39] and website[40] available with additional powder technology education information.
  - Golter et al.[41] have developed a methodology to teach students fluid mechanics and heat transfer inductively. Many of their modules are see-through to aid in visualization. These include Reynolds dye/flow-through clear pipe, pressure drop through fittings and valves, flowmeters (Venturi, orifice, and Pitot tube), extended surface heat exchangers, kettle boiler / steam condenser, 1-2 shell and tube heat exchanger, fluidized bed (compressed air through sand), and a double-pipe heat exchanger.
  - Wright et al.[42] introduced bioseparations through a three-part laboratory experiment. This includes bed expansion characterization under fluidization conditions, tracer studies, and protein adsorption studies.
  - Other experimental unit operations that could be demonstrated include agitation and aeration[43], solid/liquid and liquid/liquid mixing[44], and compressible flow analysis[45].
- Use fluid mechanics videos from the web
  - Most notably is the "Fluid Mechanics" video series starring Prof. Hunter Rouse of the University of Iowa. These videos are available online at the Iowa website[46]. General topics include the introduction to the study of fluid motion; experimental principles of flows; characteristics of the laminar and turbulent flows; fluid motion in a gravitational field; form drag, lift, and propulsion; and effects of fluid compressibility.
  - There is also the "National Committee for Fluid Mechanics Film Series[47]" with sample topics: aerodynamic generation of a sound,

cavitation, channel flow of a compressible fluid, deformation of a continuous media, Eulerian Lagrangian description, and flow instabilities.

- Use commercially available software
  - Computational Fluid Dynamics (CFD) case studies in the fluids course[48] and for fluid-particle flow[49]
  - COMSOL modules for fluid dynamics and heat and mass transfer applied to fuel cells[50]
  - Use of Mathematica[51] to analyze non-Newtonian flow systems

Introductory Thermodynamics

This course is normally the first of two thermodynamics courses where fundamental thermodynamics concepts are introduced (1st and 2nd Law of Thermodynamics) while solution properties are normally not discussed. Processes and equipment are emphasized, including various thermodynamic cycles and the analysis of their components (turbines, compressors, throttling valves, etc.) The course enrollment can also contain non chemical engineering students, so the instructor must also be aware of issues that mechanical or civil engineers may encounter in their careers.

German Physicist Arnold Sommerfeld said it best when discussing the topic of thermodynamics:

> *Thermodynamics is a funny subject. The first time you go through it, you don't understand it at all. The second time you go through it, you think you understand it, except for one or two small points. The third time you go through it, you know you don't understand it, but by that time you are so used to it, it doesn't bother you any more*

The subject of thermodynamics can be confusing due to a number of issues, but most notably the lack of an intuitive feel for certain integral concepts, such as entropy, internal energy, fugacity, chemical potential, etc. Recently one of us observed, in research involving student-prepared study guides, that entropy and the 2nd Law of Thermodynamics are the most confusing topics. In fact, students did not put much information, if at all, on their study guides for these two topics not because they were comfortably with them, but because they had a poor understanding of the topic. This manifested itself in exam scores on problems with these concepts[52].

One way to connect this concept to students is through unique, non-lecture methods. Kyle discusses the mystique of entropy, applied to a wide range of fields including cosmology, time, life, and art[53]. Muller integrates 2nd law concepts into common life experiences and economic theories[54]. Foley presents a view of entropy as a quality of energy degraded[55]. There are also newer thermodynamic terms that are gaining in popularity, including exergy (maximum work done by a system which brings it into equilibrium with a reservoir) and emergy (the cost of a process or product in solar energy equivalents).

Another problem that students face with thermodynamics is the strong importance placed on the use of differential calculus concepts. While students have normally been exposed to all of these concepts in their Calculus sequence, the act of placing it in a thermodynamic context often proves a significant barrier. Working with F=F(x,y) is, seemingly, different than working with P=P(T,v). Accordingly, the thermodynamics instructor has two options. The first involves re-teaching the fundamental concepts of differential, partial derivatives, meaning of integrals, etc. within the thermodynamics course. The second is to work with the people who are teaching students these math concepts, which are Mathematics Faculty members. If chemical engineering (or any engineering) faculty were to work with Calculus instructors to provide context to some of the math they are learning, this would mitigate the need for the remedial work when they arrive in the classes that depend on this knowledge.

Other new ideas associated with this course include:

- Incorporation of biological concepts in addition to traditional chemical engineering examples. For example, Haynie[56] describe the irreversible increase in entropy involved in how a grasshopper jumps. Additional problems are available in this area as part of the Bioengineering Educational Materials Bank[57].
- Development of a Personalized Class Binder[58] which requires students to put class notes, handouts, in-class problems, quizzes, exams, and homework into a binder. The binder is graded at various points during the semester. Students are also required to rewrite the notes neatly for inclusion in the binder and to show reworked exams, quizzes and homework. Finally, the binder will include brief biographies of the scientists mentioned in the course which goes towards humanizing the subject matter.
- Creative Expression Day, where students make visible posters that can be placed above the chalkboard which contains various concepts or formulas important for the course. Students can then easily "view" this information during the whole semester.
- Extensive use of NIST WebBook for data to perform any of a number of comparisons of involving polar and nonpolar substances[59].

Do note that many articles in the journal *Chemical Engineering Education* have been written on thermodynamics problems, especially in the "Class and Home Problems" section. Some notable ones include a powerful example on energy consumption relating the 2nd Law by Fan and co-workers[60], an open-ended design estimation problem from Lombardo[61], and the description of an experimental vapor-liquid equilibria laboratory at the University of Delaware[62].

Trouble spots for this course include:

- Second law of thermodynamics. One idea is to use the statistical nature of entropy as an introduction as well as the work of Foley[55] and Fan[60].
- Translating concepts of mathematics into this course. Rather than assume knowledge of differentials, partial derivatives, etc., spend some time to remind students of these concepts.

Equilibrium-Staged Separations

This course typically combines steady state material and energy balances with phase equilibrium to form the student's first experience with equipment design. Students apply equilibrium relationships to the design of staged separations equipment. Typical operations include flashes, cascades, absorption, stripping, binary distillation, and extraction. This course may also cover rate-based processes such as membranes, adsorption, and ion exchange.

Graphical methods are used to learn conceptual relationships and for order-of-magnitude design. Analytical methods are then used as rigorous design tools and provide a foundation for simulation.

Trouble spots for this course can include:

- Reluctance to show work; Reluctance to apply rigorous methods to simple problems; Trouble with thermodynamic diagrams. These are problems encountered in earlier courses and have been discussed in the Material and Energy Balances portion of the paper.
- Looking for "answers" instead of trends. Students often fail to see that the point of solving model equations (outside of homework and exams) is not to find a particular number. Models are always approximations or subject to other forms of error. The real value of models is in simulation to determine answers to questions like "What happens if my flow rates vary +/- 50%" or "What would be the effect of a malfunctioning thermocouple?"
- Expecting rigor in graphical approximate solutions. You will need to constantly remind and reinforce the fact that assumptions are being made throughout the course. Some of the assumptions may not be significant (equimolar counter diffusion for a binary distillation with similar substances) or may change the character of the entire separation (use of inappropriate thermodynamic models).
- Disconnect between theory and simulators. If students do not learn how to use a process simulator for separations as they learn theory, they will have difficulty reconciling the terminology used in their text and the input fields in the simulator. Fostering that connection throughout the course makes use of simulators more effective.

Some best practices that we have used (or discovered) for this course are:

- Ask the experts. Sometimes we do not teach the courses for which we have the most relevant experience. Both Chemical Engineering Progress[63] and Chemical Engineering Magazine[64] routinely publish relevant articles on separations applications on a regular basis. They are written at a level that students can often understand better than their textbook.

- Bring in the history of the field[65]. Separations have been performed for millennia. The earliest recorded use of distillation dates back to 50 B.C.; it was used in the 12th century for ethanol processing; and in the 16th century was widely used for perfumes, vinegars, and oils. Occasionally interrupting terribly interesting technical lectures with historical anecdotes can renew student interest in a lecture while giving them perspective on their current course of study.
- Use literature from industrial suppliers[66]. Many manufacturers and distributors of industrial equipment have useful applications papers describing not only their equipment in particular but general concepts as well. A web search will easily find vendor articles like "Factors Affecting Distillation Column Operation", "Evaporator Handbook" and "Liquid-Liquid Coalescer Design Manual". These are also written at a very accessible technical level.
- Wankat's "Why, What How?" approach. Establish why you're teaching something (economics, core of chemical engineering), what exactly you're teaching (equilibrium staged separations), and then teach it using best pedagogical practices (lecture with simulation labs, inductively structuring the course, using both graphical and then analytical methods, and then reinforcing with laboratory exercises and design projects)[67]. This process should lead to a deeper understanding of the subject.
- Levels of understanding[68]. Dahm combines Wankat's approach with Haile's Special Hierarchy of understanding to give a specific possible formulation of the levels of understanding in teaching separations.
- Separations using spreadsheets[69]. Working with students to develop an analytical approach to graphical separations on a spreadsheet forces a connection between the graphical methodology and the theoretical underpinnings. Automating shortcut separations develops an understanding of what is required to be known in what order.
- Use of commercial simulation[70]. Use of commercial simulators in the classroom enables a range of inductive exercises to be incorporated into a course. Instead of performing time-consuming laboratory exercises (which do have an esteemed place in the course) to explore a piece of equipment, experiments can be performed virtually with the simulator, enabling students to observe results and draw conclusions. When the theory is later discussed, students have a framework of understanding whereby they can assimilate the salient points of the discussion.

Use of Active Learning

Studies have shown[71-75] that students typically learn best in an active mode; however, engineering is usually taught as lectures. The use of active learning is underscored in teaching textbooks[71-72] and those intended for the new professor[73] as well as in numerous conference proceedings and engineering education archival publications and conference proceedings. A good listing of references are presented by Smith[74] and by Dyrud[75].

A great deal of information on improving student-teacher interaction through active learning is presented at the National Effective Teaching Institute (NETI)[76] and the Excellence in Engineering Education (ExcEEd)[77] workshops. One former attendee and active learning advocate is Ken Reid who highlighted the positive experiences in his classroom[78], and summarized simple ways that faculty can increase active and collaborative learning in their lectures and within the laboratory[79].

Improving student motivation may also improve learning, as was recently illustrated by Newell who developed a game based on the reality television show "Survivor" within a material and energy balance course[80]. Newell referenced the student motivation classifications of Biggs and Moore[81]:

1. Intrinsic – learning because of a desire to learn
2. Social – learning to please others
3. Achievement – learning to enhance one's position
4. Instrumental – learning to gain long-term rewards

- Think-pair-share – think for 1-2 minutes, talk with neighbor for 1-2 minutes, then share answers with the rest of the class)
- Poll the audience – with a show of hands, colored notecards, or clickers
- Minute paper – the students write down 1-2 ways to do something, then the instructor solicits answers from the students. This is also a good way to get anonymous feedback on the course content, what the "muddiest" point of a lecture is, etc.
- Engineering Education articles from Rich Felder[28] – this site highlights recent teaching methods that have been proven to improve student learning
- Use of Quiz Shows - Within the chemical engineering education literature, a popular way to use active learning within the classroom is through quiz shows such as "Jeopardy" or "Trivial Pursuit"[82], "Hollywood Squares"[83], and professor-created games such as "Green Square Manufacturing,"[84] "True Blue Titanium Game,"[85], "Chemical Engineering Balderdash,"[85] and the "Transport Cup."[86]. Although these games usually only address the knowledge or comprehension component of Bloom's taxonomy[87], these games certainly address the social and achievement components of Biggs

and Moore. Newell[80] found that the "Survivor" game addressed all four motivation categories and improved student learning.

Conclusions

This paper has described some of the best practices for use in the following lower level chemical engineering courses: freshman chemical engineering, material and energy balances, fluid mechanics, introductory thermodynamics, and separations. For copies of the presentation slides from the Summer School, contact one of the authors.

Bibliography

1. D. L. Silverstein, D. P. Visco, and J. M. Keith, "New Ideas for Old Courses: Lower Division," presented at 2007 ASEE Summer School, Pullman, WA.
2. J. M.Keith, D. L. Silverstein, and D. P. Visco, "Ideas to Consider for Chemical Engineering Educators Teaching a New "Old" Course: Freshman and Sophomore Level Courses," Proceedings of the American Society for Engineering Education, 2008.
3. E. Seymor and N. Hewitt, *Talking about Leaving: Why Undergraduates Leave the Sciences*, Westview Press, Boulder, CO, 1997.
4. K. Solen and J. Harb, "An Introductory ChE Course for First-Year Students", Chem Eng. Ed., **32** (1), 52 (1998).
5. D. Visco and P. Arce, "A Freshman Course in Chemical Engineering: Merging First-Year Experiences with Discipline-Specific Needs" Proceedings of the American Society for Engineering Education, 2006.
6. S. G. Sauer, "Freshman Design in Chemical Engineering at Rose-Hulman Institute of Technology" Chem. Eng. Ed., **38** (3), 222 (2004)
7. C. Coronella, "Project-Based Learning in a First-year Chemical Engineering Course: Evaporative Cooling", Proceedings of the American Society for Engineering Education, 2006.
8. A. Barritt, et. al.. "A Freshman Design Experience: Multidisciplinary Design of a Potable Water Treatment Plant", Chem Eng. Ed., **39** (4), 296 (2005)
9. M. Vigeant and R. Moore, "Sneakers as a First Step in Chemical Engineering, Proceedings of the American Society for Engineering Education, 2006.
10. K. A. Hollar, M. Savelski and S. Farrell, "Guilt-Free Chocolate: Introducing Freshman to Chemical Engineering" Proceedings of the American Society for Engineering Education, 2002.
11. A. R. Minerick and K. H. Schulz, "Freshman Chemical Engineering Experiment: Charged up on Electrophoresis & Brewing with Bioreactors" Proceedings of the American Society for Engineering Education, 2005.
12. S. Farrell, M. J. Savelski, R. Hesketh, "Energy Balances on the Human Body", Chem Eng. Ed., **39** (1), 30 (2005)
13. D. Budny, A. Allen and J. Quarcoo, "What Do Our Students Think is Important During Freshman Year?" Proceedings of the American Society for Engineering Education, 2007.
14. S. Roberts, "A Successful 'Introduction to ChE' First-Semester Course", Chem. Eng. Ed., **39** (3), 222 (2005)
15. L. Lebduska and D. DiBiasio "Mixing Writing With First-ear Engineering", Chem. Eng. Ed., **37** (4), 248 (2003).
16. F. Bowman, et. al. "Frontiers of Chemical Engineering", Chem. Eng. Ed., **37** (1), 24 (2003).
17. H. Pierson and D. Price. "The Potato Cannon", Chem. Eng. Ed., **39** (2), 156 (2005)
18. Sloan Career Cornerstone Center, available online at: http://www.careercornerstone.org/chemeng/profiles/chemengprofiles.htm, accessed February 2008.
19. J. Anthony, "Elements of Calculation Style", *Chemical Engineering Progress*, pp.50-55, Nov. 2001.
20. M. J. Fernandez-Torres,,"Analogies: Those Little Tricks That Help Students to Understand Basic Concepts in Chemical Engineering", *Chem. Eng. Ed.*, **39**(4), 302(2005).

21. S. Farrell, R. Hesketh, M. Savelski, "A Respiration Experiment to Introduce ChE Principles", *Chem. Eng. Ed.*, **38**(3), 182(2004).
22. V. Young, "Using Your Unit Operations Laboratory", Proceedings of the Annual Meeting of American Society for Engineering Education, June 1999.
23. D. L. Silverstein, "Increasing Time Spent of Course Objectives by Using Computer Programming to Teach Numerical Methods", *Chem. Eng. Ed.*, **37**(3), 214(2003).
24. L. Bullard and R. M. Felder, "A Student-Centered Approach to the Stoichiometry Course", Proceedings of the Annual Meeting of American Society for Engineering Education, June 2007.
25. L. Bullard and R. M. Felder, "A Student-centered Approach To Teaching Material And Energy Balances 1. Course Design", *Chem. Eng. Ed.*, **41**(2), 93(2007).
26. L. Bullard and R. M. Felder, "A Student-centered Approach To Teaching Material And Energy Balances 2. Course Delivery, Assessment", *Chem. Eng. Ed.*, **41**(3), 167(2007).
27. K. Biasca, Web page, https://www.uwsp.edu/it/tlrn/lo-scos/LOs/Natural_Resources/Paper_Science/ Biasca/currentpages/table.htm, accessed February 2008.
28. Richard Felder's Resources in Science and Engineering Education, Web page, http://www4.ncsu.edu/unity/lockers/users/f/felder/public/RMF.html, accessed February 2008
29. R. M. Felder and R. W. Rousseau, *Elementary Principles of Chemical Processes*, Wiley, New York. 5th edition, 2005.
30. Software Tutorials, Web page, http://www4.ncsu.edu/unity/lockers/users/f/felder/public/tutorials.html, accessed February 2008.
31. Free Online Graph Paper/ Grid Paper PDFs, Web Page, http://incompetech.com/graphpaper/, accessed February 2008.
32. MIT Frontiers of Chemical Engineering Education, website available online: http://web.mit.edu/che-curriculum/statements/RCA_NSF_ChE_Frontiers_Overview.pdf, accessed February 2008.
33. R. B. Bird, W. E. Stewart, E. N. Lightfoot, Transport Phenomena, Wiley, New York, 1960 (1st edition).
34. L. Ford, "Water Day: An Experiential Lecture for Fluid Mechanics," Chemical Engineering Education, **37**, 170 (2003).
35. J. M. Ross and T. M. Bayles, "Incorporating High School Outreach into ChE Courses," *Chemical Engineering Education*, **37**, 184 (2003).
36. T. M. Floyd-Smith, K. C. Kwon, J. A. Burmester, F. F. Dale, N. Vahdat, and P. Jones, "Demonstration and Assessment of a Simple Viscosity Experiment for High School Science Classes," *Chemical Engineering Education*, **40**, 211 (2006).
37. L.-S. Fan, "Particle Dynamics in Fluidization and Fluid-Particle Systems. Part 2: Teaching Examples," *Chemical Engineering Education*, **34**, 128 (2000).
38. S. Iveson and G. V. Franks, "Particle Technology Demonstrations for the Classroom and Laboratory", *Chem. Eng. Ed.*, **37**,274 (2003).
39. M. Rhodes and A. Zahkari, "Laboratory Demonstrations in Particle Technology," CD, Monash University, Australia (1998).
40. Educational Resources for Particle Technology website, available online at: http://www.erpt.org/, accessed February 2008.
41. P. B. Golter, B. J. Van Wie, P. V. Scuderi, T. W. Henderson, R. M. Dueben, G. R. Brown, and W. J. Thomson, "Combining Modern Learning Pedagogies in Fluid Mechanics and Heat Transfer," *Chemical Engineering Education*, **39**, 280 (2005).
42. P. R. Wright, X. Liu, B. J. Glasser "A Fluidized Bed Adsorption Laboratory Experiment," *Chemical Engineering Education*, **38**, 14 (2004).
43. A. Badino, P. I. F. De Almeida, A. J. G. Cruz, "Agitation and Aeration: An Automated Didactic Experiment," *Chemical Engineering Education*, **38**, 100 (2004).
44. S. Barar Pour, G. Benoit Norca, L. Fradette, R. Legros, P. A. Tanguy, "Solid-Liquid and Liquid-Liquid Mixing Laboratory for Chemical Engineering Undergraduates," *Chemical Engineering Education*, **41**, 101 (2007).

45. S. E. Forrester, A. V. Nguyen, G. M. Evans, P. M. Machniewski, "Compressible Flow Analysis: Discharging Vessels," *Chemical Engineering Education*, **38**, 190 (2004).
46. IIHR Hydroscience and Engineering Laboratory at the University of Iowa, online at: http://www.engineering.uiowa.edu/fluidslab/referenc/processes.html, accessed February 2008.
47. National Committee for Fluid Mechanics Film Series, available online at: http://web.mit.edu/fluids/www/Shapiro/ncfmf.html, accessed February 2008.
48. J. Sinclair Curtis and R. O. Fox, Computational Fluid Dynamics, presented at 2007 AIChE-ASEE Summer School, Pullman, WA.
49. J. Sinclair, "CFD Case Studies in Fluid-Particle Flow," **32**, (1998).
50. J. M. Keith, F. Morrison, J. A. King, "Finite Element Modules for Enhancing Undergraduate Transport Courses, Applications to Fuel Cell Fundamentals," Proceedings of the American Society for Engineering Education, 2006.
51. H. Binous, "Introducing Non-Newtonian Fluid Mechanics Computations with Mathematica in the Undergraduate Curriculum," *Chemical Engineering Education*, **41**, 59 (2007).
52. D. P. Visco, Jr., S. Swaminathan, L. Zagumny and H. Anthony.., "Evaluation of Student Constructed Study Guides", ASEE Annual Meeting and Exposition Proceedings, 2007.
53. B. G. Kyle, "The Mystique of Entropy", Chem. Eng. Ed. **22** (2) 92 (1988)
54. E. Muller, "Human Societies: A Curious Application of Thermodynamics", Chem. Eng. Ed., 32 (3) 230 (1998).
55. A. Foley, "Escape from Carnot: A New Way to Introduce the Mysterious Property, Entropy", Proceedings of the American Society for Engineering Education , 2007
56. D. T. Haynie, Biological Thermodynamics, Cambridge University Press, 2001.
57. Bioengineering Educational Materials Bank website, available online at: http://www.engr.sjsu.edu/~bioemb/index.php, accessed February 2008.
58. R. Rawlings, S. Allen, and P. Arce, The Class Binder: A Powerful Enhancer of Active and Collaborative Learning Environments Proceedings of the American Society for Engineering Education – Southeast Regional Section Proceedings, 2005.
59. NIST Webbook, available online at: http://webbook.nist.gov/chemistry/fluid/, accessed February 2008.
60. L. Fan, T. Zhang and J. Schlup "Energy Consumption vs. Energy Requirement", Chem. Eng. Ed., **40** (2), 132 (2006)
61. S. Lombardo, "An 'Open-Ended Estimation' Design Project for Thermodynamics Students", Chem. Eng. Ed., **34** (2), 154 (2000)
62. S. Wrenn, et. al., "Vapor-Liquid Equilibria in the Undergraduate Laboratory", Chem. Eng. Ed., **34** (1), 74 (2000)
63. i.e., D. O'Brien, M. Schultz, "Ask the Experts: Distillation", *Chem. Eng. Prog.*, **100**(2), 14(2004).
64. i.e., T. Yanagi, "Inside a Trayed Distillation Column", Chem. Eng., 122(November 1990).
65. The Chemical Heritage Foundation, Web page, http://www.chemheritage.org, accessed February 2008.
66. CAChE TRC, Separations Processes, Web page, http://www.che.utexas.edu/cache/trc/t_separate.html, accessed February 2008.
67. P. Wankat, "Teaching Separations: Why, What, When, and How", *Chem. Eng. Ed.*, **35**(3), 168(2001).
68. K. Dahm, "Process Simulation and McCabe-Thiele Modeling: Specific Roles in the Learning Process", *Chem. Eng. Ed.*, **37**(2), 132(2003).
69. J. Jevric, M. Fayed, "Shortcut Distillation Calculations via Spreadsheets", *Chem. Eng. Prog.*, **100**(12), 60(2002).
70. P. Wankat, "Using a Commercial Simulator to Teach Sorption Separations", *Chem. Eng. Ed.*,**40**(3), 165(2006).
71. W. J. McKeachie, *McKeachie's Teaching Tips: Strategies, Research, and Theory for College and University Teachers*, Houghton Mifflin Compnay, Boston (1999).
72. A. Brinkley, B. Dessants, M. Flamm, C. Fleming, C. Forcey, E. Rothschild, *The Chicago Handbook for Teachers: A Practical Guide to the College Classroom*, The University of Chicago Press, Chicago (1999).

73. P. C. Wankat, *The Effective, Efficient Professor: Teaching, Scholarship, and Service*, Allyn and Bacon Publishers, Boston (2002).
74. K. A. Smith, "Cooperative Learning," Proceedings of the American Society for Engineering Education Annual Meeting (1998).
75. M. A. Dyrud, "Getting a Grip on Groups," Proceedings of the American Society for Engineering Education Annual Meeting (1999).
76. National Effective Teaching Institute website, http://www.ncsu.edu/felder-public/NETI.html, accessed February 2008.
77. Excellence in Engineering Education website, http://www.asce.org/exceed/, accessed February 2008.
78. K. J. Reid, "More Effective Teaching: One Year After the National Effective Teaching Institute," Proceedings of the Frontiers in Education Conference (1999).
79. K. J. Reid, "Implementing Active and Collaborative Techniques: Lectures, Labs, Grading, and More," Proceedings of the American Society for Engineering Education Annual Conference (2002).
80. J. A. Newell, "Survivor: A Method for Active Learning in the Classroom that Addresses Student Motivation," Proceedings of the American Society for Engineering Education (2004).
81. J. Biggs and P. J. Moore, *The Process of Learning*, Prentice-Hall, Englewood Cliffs, NJ (1993).
82. P. C. Wankat and F. C. Oreovicz, *Teaching Engineering*, McGraw-Hill, New York (1993).
83. J. A. Newell, "Hollywood Squares: An Alternative to Pop Quizzes," Proceedings of the American Institute for Chemical Engineers Annual Meeting (1999).
84. S. S. Fenton and J. M. Fenton, "The Green Square Manufacturing Game: Demonstrating Environmentally Sound Manufacturing Principles, *Chemical Engineering Education*, **33 (2)**, 166 (1999).
85. Educational Materials for grades 13-16, Professor Suzy Fenton, http://www.engr.uconn.edu/~suzy/13-16_educational_materials.htm ,accessed February 2008.
86. J M. Keith, "The Transport Cup," Proceedings of the American Society for Engineering Education (2005).
87.Bloom, B.S. (Ed.) Taxonomy of educational objectives: The classification of educational goals: Handbook I, cognitive domain. New York: Longmans, Green (1956).

412f Enhancement of Bioengineering Course with Results from Biophyics Research

Richard L. Long, New Mexico State University, Chemical Engineering Department, P.O. Box 30001, MSC 3805, Las Cruces, NM 88003-8001

An elective topic area course can be enhanced by including results of faculty research in the course. This allows students to see the boundary of current knowledge and offers them the chance to consider what future developments are likely, possible and needed. At NMSU we offer an introductory bioengineering course to both graduate and undergraduate students that has sufficiently limited prerequisites that it can be appreciated by both groups. An effort is made to show current challenges in the field and one way to do this is to incorporate current faculty research. Both challenges in bioprocess and biomedical engineering are addressed. Applications involving biophysics allows the students to see in a very basic way how mass, momentum, energy, and entropy balances along with principles of kinetics and transport processes can be employed in addressing biological problems. This is an effective way to open the eyes of engineers to living systems.

Weblab in Chemical Engineering as a Tool for Cooperative Learning within a Global Environment: a report of experiments among Brazilian Universities connected through KyaTera

Galo A.C. Le Roux[1], Giann B. Reis[2], Charles D. F. de Jesus[2], Roberto de C. Giordano[2], Antonio J. G. Cruz[2], Luiz V. Loureiro[2], Paulo F. Moreira Jr.[2], Galo A. C. Le Roux[2], Claudio A.O. Nascimento[2] (1)Chemical Engineering Department, Polytechnic School of the University of São Paulo, SP, Brazil, (2)Chemical Engineering Department, Universidade Federal de São Carlos, SP, Brazil

1. The KyaTera Project

KyaTera is a project financed by FAPESP (São Paulo State´s Agency for Research Development) that aims to develop high speed internet applications for research and educational purposes. It is based on an optical high speed packet network interconnecting a number of laboratories, research institutes and universities in São Paulo state focusing on the study, development and usage of technology and applications in advanced Internet.

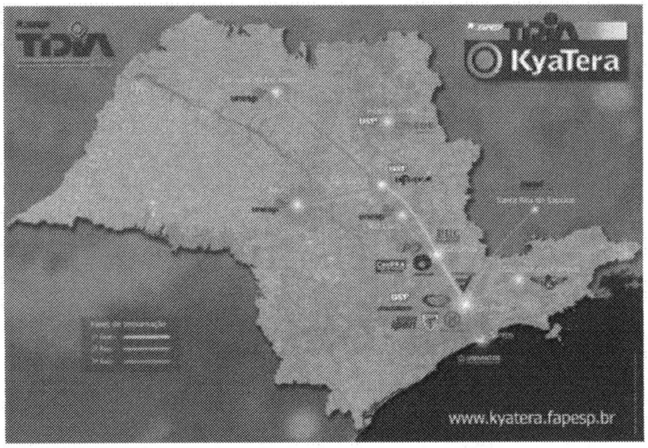

Figure 1. Optical network connecting research laboratories in the state of São Paulo, Brazil.

2. Cooperative Weblabs in Chemical and Biochemical Process Engineering

The term Weblab was employed to name real - not virtual - experiments remotely operated via Internet. This kind of experiments has been proposed and implemented worldwide by some major universities. In this article we present a unique approach for Weblabs: to offer intercultural experiences to students, while enhancing their communication skills.

Cooperative Weblab experiments must be performed by two groups of undergraduate students, in two different locations and aims to partially fulfill the lack of interaction among students in this level of engineering education. The Cooperative Weblab cluster intends to develop a set of experiments for undergraduate students that should be performed by "mixed teams". The experimental setups are physically placed in laboratories in São Paulo, São Carlos, Ribeirão Preto and Campinas, four cities in São Paulo state, Brazil.

The Laboratory of Development and Automation of Biochemical Process (LaDABio) of the Chemical Engineering Department of the Federal University of São Carlos (DEQ/UFSCar) and the Center for Chemical Systems Engineering (CESQ) of the Chemical Engineering Department of the Polytechnic School of the University of São Paulo (DEQ/EPUSP) are involved in this project. These groups are 255 km distant and they are among the best chemical engineering courses in Brazil.

Collaboration is achieved by gathering participants of the different institutions into working groups that are asked to simultaneously solve a technical problem, for which an experiment is available. This procedure emulates challenges that will frequently take place in their future professional lives. The Weblab also fosters the learning of Chemical Engineering instrumental concepts. To achieve these goals, students of different departments are required to work cooperatively.

3. Cooperative Weblabs

In this work two Cooperative Weblabs, one in São Carlos and the other one in São Paulo, were employed.

3.1 WebLab for Mass Transfer Experiments at Federal University of São Carlos

During the aerobic cultivation of microorganisms or cells in tank bioreactors, the level of dissolved oxygen must be kept high enough for the organisms to thrive. Thus, it is important for the education of chemical engineers to handle the fundamentals of mass transfer herein involved, and to get familiar with techniques that assess rates of oxygen transfer from the gas phase into the liquid culture medium as well. A scheme of the WebLab for mass transfer experiments is presented in Figure 2.

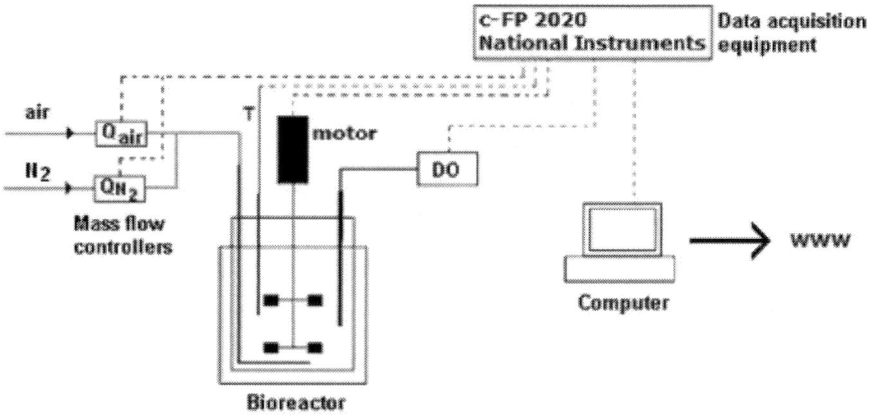

Figure 2. Experimental setup used for mass transfer experiments.

The dissolved oxygen is removed from the liquid phase by bubbling nitrogen into the medium. After reaching a zero oxygen concentration, the nitrogen flow is stopped and air flow is started. The dissolved oxygen (DO) in the liquid phase is measured by an electrode probe. The mass transfer coefficient is represented by the parameter k_La that is estimated by fitting a model for the change of the DO to the experimental data. The experiment intends to calculate k_La values at different operating conditions of air flow rate and stirrer speed employing the gassing-out method (SHULER and KARGI, 2002; BLANCH and CLARK, 1997).

The bioreactor is an aerated and agitated tank reactor. Gas (air and nitrogen) under pressure is supplied to the sparger (a ring with holes) located inside the reactor and above of the impeller. The system is constituted by a stirrer with two Rushton impellers. This impeller is typically a disc with 6 to 8 blades designed to pump fluid into radial direction. The WebLab was build employing National Instruments hardware for data acquisition and LabVIEW software as the supervisory system.

Figure 3 shows the main screen of the WebLab for mass transfer experiments. Through this screen users can choose the experiment operating conditions (air flow rate and stirrer speed).

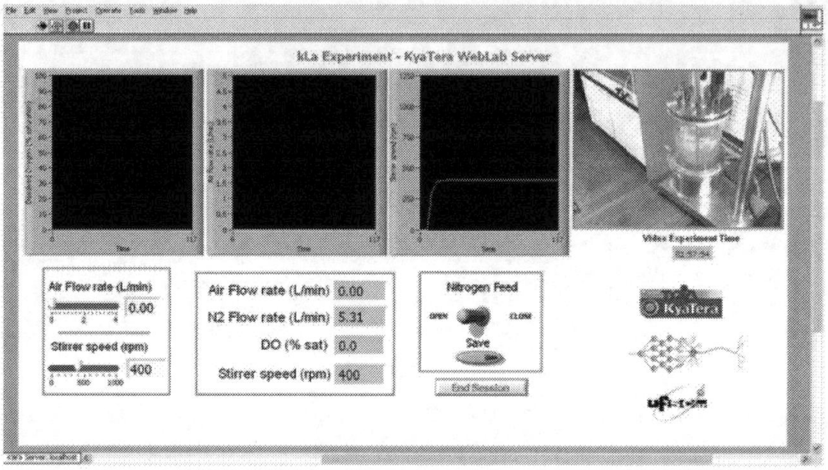

Figure 3. WebLab for mass transfer experiments' main screen.

3.2 Weblab for Reactor Temperature Control at University of São Paulo

The reactor is a jacketed glass vessel with capacity of 4.5 L. Inside the reactor there is a stainless steel coil connected to a 3.000 W thermal bath, a temperature sensor (PT100) and a stirrer (Heidolph). The jacket is connected to cold water (room temperature). Both the streams, from the thermal bath and cold water have their flow controlled by electropneumatic valves. An electrical heater is inserted inside the reactor, in order to simulate various kinds of reaction without risks. The electrical power to the resistor is controlled by a power module.

The temperature of the reactor can be controlled by manipulating the hot water flow through the coil or the cold water flow through the jacket. In figure 4 a screen shot of the temperature control experiment is presented. This screen includes a live camera for visualizing the experiment.

4. Performing the Experiments

In order to perform the experiments, two students in São Carlos and two in São Paulo are gathered into a group. An instructor supervises each group of two students. The students are left free to decide what measurements and experiments are to be performed and how to manage the experiments (who is in charge of what). Students interact using video conference software. Video streaming is very effective because KyaTera network supplies a performing communication capability. This gives the students an improved perception of the reality.

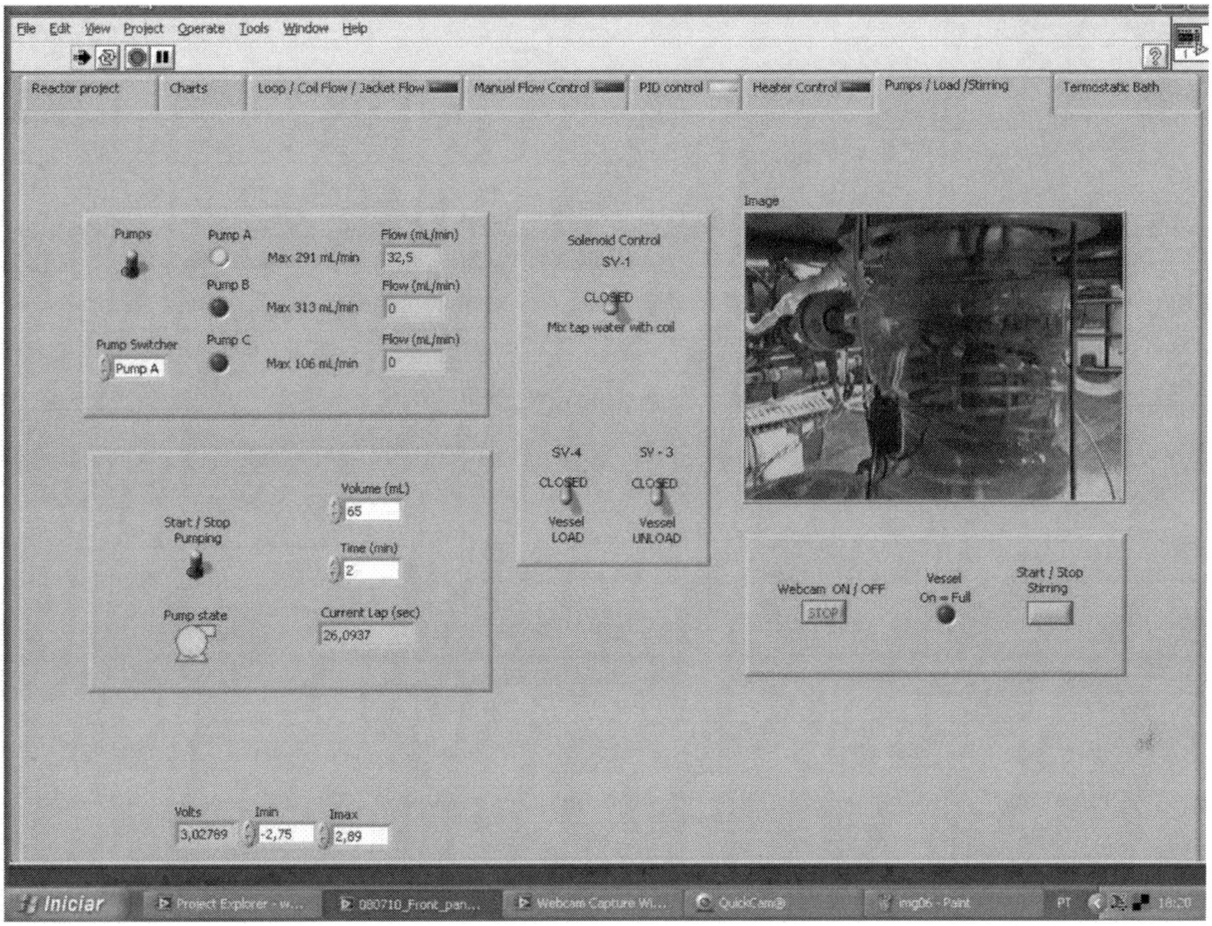

Figure 4. Weblab Reactor Temperature Control' Screen shot.

The students are encouraged to interact. All along the experiments, the students exchange information and opinions about the phenomena that take place. In the mass transfer experiment, the main concerns are about the quality of the mixing, the size of the bubbles and the various problems that arise during the experiment (bubbles blocking the tip of the probe, high coalescence phenomena, and conditions of inefficient mixing). In the temperature control experiment, a step change on the coil flow is performed in order to obtain the response curve. The students must perform calculations together, during the experiment, in order to apply the Cohen-Coon procedure (Seborg et al. 1989) to obtain the PID tuning parameters. The set of PID tuning parameters is implemented by closing the control loop and is tested by different methods, e.g., set point changes and disturbance rejection tests. They are left free to implement other disturbances with the immerged resistor in the reactor.

During the experiment, a high level of interaction takes place between the students, who keep talking and discussing all long the experiment. After the experimental session, the students must process the data collected together in order to produce a unique report for the group. The students are invited to employ KyaTera network to communicate and prepare the final report.

6. Conclusions

Weblab experiments offer a collaborative learning experience to the participants because they are faced with the hurdles of remote team working. Cultural reality certainly does not differ much for a 250 km distance, but the views on which the topics are taught are certainly different. This is an interesting issue, because these experiments also provide new experiences to the instructors involved who perceive how the colleague teach the same phenomena to her/his students.

The overall opinion of students is extremely positive and encouraging. Cooperative Weblabs experiments are valuable educational tools in a world with an increasing interaction. An effort is necessary in order to fit this kind of activity in the curricula of Chemical Engineering courses, because it requires the agreement of two or more institutions. Also some practical difficulties must be faced and overcome such as the difference in time zones and calendars. Nevertheless the impact of these issues is unimportant if compared with the potential improvement that this tool can bring to the chemical engineering education.

References

BLANCH, H. W. and CLARK, D. S. *Biochemical Engineering.* Marcel Dekker, Inc., New York, 1997.

SEBORG, D.A., EDGAR, T.H., MELLICHAMP, D.A., Process dynamics and control, John Willey & Sons, New York, 1989;

SHULER, M. L. and KARGI, F. *Bioprocess Engineering: Basic Concepts.* 2nd edition, Prentice Hall International Series, Upper Saddle River, NJ, 2002.

Acknowledgments

Authors would like to thank FAPESP for its financial support.

514b Global Chemical Engineering Workforce Statistics

Tom R. Marrero, University of Missouri-Columbia, 1313 Fieldcrest, Columbia, MO 65203

The purpose of this study is to assess the current number of chemical engineers graduating and the demand for chemical engineers by industries in various nations throughout the world. The statistics and information were obtained from international institutions, government agencies, engineering societies, and recent journal articles. The results are sorted into five major areas: North America, South America, Asia, Africa and Mid-East, and Europe. Results include number of undergraduate and graduate students completing BS and PhD degrees; plus descriptions of typical curriculum. Recent additions to curriculum are also described for example, sustainable process design. Estimates are available for chemical engineers needed by various geographic areas and industries. The study indicates imbalances between supply and demand of chemical engineers; the flow of chemical engineers from one nation to another is affected by political constraints.

514c Are We Preparing Our Students for a Flat World?

Nada M. Assaf-Anid, Chemical Engineering, Manhattan College, Manhattan College Parkway, Riverdale, NY 10471, William F. Zucker, Mobility Access Business Division, Alcatel-Lucent, Gordon Silverman, School of Engineering, Manhattan College, Manhattan College Parkway, Riverdale, NY 10471, and Arthur T. Andrews, Global Pharmaceutical Development, Johnson & Johnson, 1000 Rt. 202, Raritan, NJ 08869.

This paper will attempt to answer three main questions: what qualities do employers at multi-national industries value in our students? Are our engineering students prepared for the global market? What can be done to better prepare them and help them compete in a flat world? The paper will argue that qualities sought after the most besides a thorough grounding and understanding of science, math, and engineering fundamentals, are excellent communication and interpersonal skills combined with the ability to work on teams and understand complex assignments and the subtleties of adapting to different environments. While our current engineering students may not have these qualities, engineering curricula could incorporate them without loss of any educational value. Strategies could be the inclusion of enrichment components to our students' education like language and culture immersion or technical cooperative programs and internships abroad. In addition, courses in social sciences and the humanities could emphasize world civilizations, global issues, and successful international business and management practices. Extra-curricular activities could also provide opportunities for innovation, intellectual curiosity, implementation of a vision, and goal-oriented leadership skills.

STUDENTS AND GRADUATES EXCHANGE PROGRAMMES IN THE CHEMICAL ENGINEERING DEPARTMENT OF THE UNIVERSITY OF COIMBRA IN PORTUGAL

M.G. Rasteiro,
Chemical Engineering Department, Coimbra University, Coimbra, Portugal
E-mail: mgr@eq.uc.pt

Abstract

Undergraduate exchange programs have been active in Europe for quite some time and, as a consequence, the Portuguese university students have been profiting from those programs since long. In the Chemical Engineering Department of the University of Coimbra we have been engaged in those programmes, mainly with European universities, for at least 15 years. In this paper we will report that experience, giving notice of its importance to the education of our chemical engineering students. Additionally, we will discuss also the importance to the education of our chemical engineers of the contact with the incoming students from those programs, spending a period of their studies in the University of Coimbra. Notice will also be given to the more recent mobility programs involving industrial partners which allow our young Chemical Engineers a training period in a European company outside his home country or, still, short training periods abroad to undergraduate students. Finally, the new trends and targets at the EU level regarding the mobility of the young people, namely engineering students and young graduates will also be addressed in this paper.

Introduction

Undergraduate exchange programs have been active in Europe for quite some time and, as a consequence, the Portuguese university students have been profiting from those programs since long ago. In the Chemical Engineering Department of the University of Coimbra we have been engaged in those programmes, mainly with European universities, for at least 15 years.

These exchanges have been sponsored by the European Union, first under the **Erasmus Program** (since mid 1980s) and later, since 1995, under the **Socrates/Erasmus Program** which has got a broader scope since it includes also exchange of staff members, development of joint degrees (Erasmus Mundus Program) and, more recently, mobility traineeship in industry of undergraduate students and also mobility for life long learning and students exchange at the secondary school level.

With the implementation of the Bologna Process in The European Union (EU) Universities, the emphasis put in mobility is still stronger. A recent report from a group of experts gathered by EU (1) recommends that by 2020 at least 50% of all young people aged 16-29 in the EU should be offered the opportunity to engage in some form of cross-border mobility.

It is believed that besides the positive effects learning mobility can have on the personal development of young people, there is a more general dimension to it. Mobility does not only help exchange and circulate ideas; it can also break down barriers between people and groups and add to a sense of European citizenship. Thus, this forum of experts calls for further strategic action to address barriers to mobility and to engage in a new partnership all the stakeholders of the EU Member States: governments, education institutions, business and civil society.

In this paper we will report the experience of having undergraduate students or young graduates from the Chemical Engineering Department of the University of Coimbra involved in mobility programs, giving notice of its importance to the education of our chemical engineering students, not only because it gives them access to chemical engineering fields not covered in our university, but mainly because it allows them an international experience, which we repute very important for their employability and future attitude in their profession as chemical engineers.

The Experience of the Chemical Engineering Department of the University of Coimbra

The Chemical Engineering Department of the University of Coimbra (ChemEng/UC) has been involved in students exchange at the undergraduate level, in the framework of EU programs, for at least 15 years. In the beginning the mobility was just from the University of Coimbra to other European Universities, namely to the University of Leeds, UK, University of Delft and University of Eindhoven, the Netherlands, University of Murcia, Spain, etc. Moreover, the number of outgoing students every year was small, being mainly final year students developing one semester of their final project in a foreign university. In fact, in the earlier years of this experience, and till very recently (only two years ago) the students had to obtain equivalence from the University of Coimbra for each course taken abroad, which required approval of the professor responsible for the equivalent course in the University of Coimbra.

Lately, several changes have been occurring at the two levels: outgoing and incoming students. Referring to the outgoing students and following the adoption of the Bologna Process, the students go out with a one or two semesters complete program, prepared with the help of the Erasmus coordinator in the ChemEng/UC. Thereafter they get equivalence from the University of Coimbra for the corresponding semester/year rather than course by course. This is expected to increase the number of outgoing students every year compared to what was happening previously, which, in fact, is already occurring. During the last ten years and till 2006/07 we have been sending abroad an average of two students per semester. In the school years 2007/08 and 2008/09, we have been having abroad four students per semester. Moreover, these students are no longer students belonging only to the final year of their degree (5th year), as used to happen traditionally, but are now students from the 3rd, 4th and 5th years of our integrated MSc degree, according to the Bologna nomenclature, though it is true that the 5th year students are still the majority of the candidates.

Referring to the incoming students we started receiving undergraduate students from other European Universities in 2001/02, in the beginning only from Spanish Universities. The number of incoming students has also been increasing and, currently, we receive an average of four foreign students per semester from European Universities of the most different countries: UK, Austria, Romania, Italy, Spain, the Netherlands, etc.

A similar trend as that observed regarding the outgoing students can be observed for the incoming ones. Initially those students would come just to develop their final year project in the University of Coimbra during one semester. Nowadays, they are already coming with a one or two semesters program, which means that they will take a set of pre-defined courses.

One of the limitations that still prevents more foreign students to get engaged in a full semester of the Chemical Engineering curriculum in Coimbra is the language barrier, since courses are still taught in Portuguese. Two strategies are being pursued to help those students: first they enroll in an intensive Portuguese course provided by the University itself, two or four weeks before the beginning of the Chemical Engineering classes; secondly, when the teacher is aware of the presence of non-Portuguese speaking students in the class he volunteers to teach in English.

However, a more radical change is envisaged in the very near future: teaching in English at least for the second cycle of the Bologna format (4th and 5th year of the integrated MSc in our degree). In fact, the language is still by far the main barrier to receiving more foreigner students in mobility in the Chemical Engineering Department of UC. One of the events aimed at making the integration of the foreign students easier and better, which started also in the school year 2007/08, is the organization of a welcome week for the mobility students of the whole University, organized by the international relations office.

The opinion of all our students that have been involved in mobility programs during the last ten years has always been very positive, with the recognition by them of the advantages of contacting different realities both at the university level and in general at the social level. They also recognize the advantages for their mastering of the English language and enhancement of personal autonomy. A small percentage of those students have even been involved more than once in mobility programs: in an intermediate year and in the final year of the degree for the final year project, or even after graduation as will be described later.

Also very positive is the opinion of our students to the contact with the incoming students. Usually these students are integrated in mixed teams including also Portuguese students. Again, they stress the benefits for the Portuguese students of this interchange of experiences with students coming from different cultures, as well as the benefits to the mastering of the English language. It must be stressed that our local Portuguese Chemical Engineering students are a big help to the integration of the foreign students that come to ChemEng/UC through mobility programs.

A more recent experience is with the mobility of incoming students from Brazil. A recent agreement between the University of Coimbra and Brazil (a Portuguese speaking country) is enabling Brazilian students, from a set of pre-determined universities, to be involved in mobility programs with the University of Coimbra, at the undergraduate level, since, at the graduate level (MSc and PhD) that mobility has been going on since long ago. In the Chemical Engineering Department of the University of Coimbra we have received the first group of Brazilian students in the school year 2006/07 (coming from the Universities of Campinas, S. Paulo and Santa Catarina). We have been receiving 2-3 students per year since then, either for one or two semesters. Although they report the experience as positive, there are still changes to introduce, namely learning from the Socrates/Erasmus Program. In spite of these students coming through an agreement between the University of Coimbra and their home University, when they go back they sometimes face problems with the recognition of the courses they performed in Portugal, because some of the home professors still requiring a complete agreement, course by course, between the programs in Portugal and in Brazil. An effort has to be put in preparing at home, and through the dialogue between the staff members responsible for the exchange program in each university, a complete curriculum for the semester or school year that will be recognized as a whole by the partner university without a need for recognition course by course.

It must also be stressed that we have a large and long experience of receiving students, in the Chemical Engineering Department of UC, from the African Portuguese speaking countries but, in this case, enrolled for the whole degree and not on mobility. The number of these students we accept every year has also been increasing.

More recently, during the last five years, we have been involved, simultaneously, in another exchange program, funded also by the European Community, this time directed to graduates, in our case graduated chemical engineers: the Leonardo da Vinci Program. This program allows us to send our recent graduates, during the year after their graduation, to European companies or research institutes adhering to the program, and thus give to our engineers access to a first international professional experience.

We have been sending an average of 6 young graduates every year (on average we graduate 35-40 Chemical Engineers per year) to European companies in UK, the Netherlands, Germany, Italy, Spain, etc. for training periods around six months. All the chemical engineers involved report very positive experiences. At the end of the training period there is also a report supplied by the company and, so far, the evaluation by the companies has also been very positive. In three situations the young graduates have even been offered a position in the company and one of them (a young female) accepted it and stayed with Procter & Gamble. So, it can be said that the experience is proving fruitful for both sides.

This mobility program relies a lot on the contacts the staff members (Professors and Researchers) have got with companies outside Portugal, in Europe. The staff members supply their contacts to the international relations office which then establishes the agreement with the company. The graduate receives a monthly allowance, fund for the travel costs, and is entitled to insurance and a paid language course in the language of the country. The company itself is obliged to assist the graduate in finding an accommodation.

These periods of training abroad constitute a plus in the CV of the Chemical Engineering graduates from the University of Coimbra, being quite valued by the Portuguese employers.

Additionally, and starting this school year, a new program, also under the Leonardo da Vinci funding scheme, has been launched. This time the program aims at supporting training experiences of undergraduate students, namely chemical engineering students, in European companies, for periods of two to three months, during the summer vacations. We are also institutionally engaged in that program though, so far, these training periods will be considered as extra-curricular work, which will, nevertheless, be included in their official diploma supplement, a document the university has to add to each graduate certificate describing his or her entire education course in the university. The possibility of considering these international summer trainings as curricular work is being discussed. Since this will be the first year in which students have got the possibility to get engaged in this program, it is impossible, at this moment, to make any kind of assessment.

To finalize we must recognize that in spite of the long experience described above, of involvement of the Chemical Engineering Department of UC in mobility programs (several types), the number of students that get involved, at least once during their degree, in those programs, is still unsatisfactory (less than 10% of the graduates). The same is felt all around EU where the average is of 6% of University students per year in whatever type of mobility action. So, the general feeling is that a lot has still to be done to promote mobility in education.

Conclusions

The well known practice of the European Universities to engage their students in mobility programs has been increasing over the years, namely motivated by EU programs (Socrates/Erasmus and Leonardo da Vinci).

The Chemical Engineering department of the University of Coimbra has adopted that practice since long ago, starting firstly only with outgoing students and, more recently, with incoming students.

The latest trends of these mobility programs are, nowadays, towards the involvement also of the industrial sector, enabling, in this way, that either young graduates or undergraduate students can spend training periods in industrial or services companies in European countries other than their home country.

These practices have been rated very positively by the students or engineers involved and also, in the last case, by the companies collaborating with the universities.

Moreover, the Portuguese students were also very positive about the exchange of experiences with the incoming students visiting the ChemEng Department of UC.

Some lessons have been learned over the years, the main one being that the student exchange program must be prepared and agreed to in advance, between the two universities involved, so that there is no problem with the recognition of the courses attended in the host university. The recognition should be granted to either the semester or the school year rather than to individual courses, in line with the recommendations of the Bologna Process.

Another lesson is that language can be a barrier for mobility. Thus, in the Portuguese case it is advisable that courses, at least second cycle courses, be taught in English in order to attract more incoming students.

To conclude, it is generally agreed that besides the positive effects learning mobility can have on the personal development of the students, there is a more general dimension to it. Mobility can help exchange and circulate ideas as well as on intercultural know how, breaking down barriers between people and groups.

Thus, at the European level, with the aim to best equip young people with skills and competences needed in the future, in a competitive and knowledge based society, the target is to have, in 2020, 50% of all young people aged 16-29 being offered the opportunity to engage, at some point, in some form of cross-border mobility. To achieve this goal it is necessary to promote more strongly mobility programs, increasing the synergies between the different programs available at the moment in EU.

Moreover, mobility with non-European countries is also recommended, as already is happening at the research level, namely with other Portuguese speaking countries which, so far, have only been supplying incoming students.

References

1. "Making Learning and Mobility an Opportunity for All", report of the high level expert forum on mobility, coordinator Maria Joao Rodrigues, EU, June 2008.

THE NIGERIAN SOCIETY OF CHEMICAL ENGINEERS: CHALLENGES AND OPPORTUNITIES FACING CHEMICAL ENGINEERS IN NIGERIA.

Alexander O. Ogedegbe; President, NSChE, Lagos, Nigeria.

Preamble

Let me begin by expressing my profound gratitude to the Board and President of the A.I.Ch.E for the honor done to me and the Nigerian Society of Chemical Engineers, by inviting us to participate in the celebrations of your Centennial anniversary in the founding City of Philadelphia, Pennsylvania.

The Nigerian Society of Chemical Engineers (NSChE) was formed by a small group of graduate chemical engineers in Lagos in 1968. They met regularly at the BP (British Petroleum) House in Lagos, from 1968 until it was formally incorporated on the 3rd of December 1981. NSChE has held 37 consecutive annual conferences in the last 40 years.

On a personal level, I recall with some nostalgia and satisfaction my membership of the A.I.Ch.E, first as a student at University of Iowa, Iowa City in 1966; and later as a graduate member from 1967 to 1975 before I returned home to Nigeria in December 1969. I attended my first A.I.Ch.E Conference in Los Angeles California in 1975. At that time, I was on a special assignment with Union Oil of California (UOC) in Brea, Orange County, CA. I was a process engineer with the Nigerian National Oil Corporation (NNOC) which had licensed the (UOC) Unifiner process for hydro treating petroleum distillates in crude oil refining.

Introduction

Before addressing the substantive issues stated in the title of this paper, I consider it of utmost importance that I present the country called Nigeria to you in a brief but comprehensive manner. With the understanding of its geography, a bit of demography, natural resources and political economy, I hope to make it easier for you to appreciate the main issues addressed in their proper context.

Nigeria is a multi-ethnic and multi-cultural country situated on the West African Coastal region known as the Gulf of Guinea, between the Republics of Benin and Cameroon. Geographical Coordinates are between 4^0 and 14^0 latitude north and 2.30^0 and 14.30^0 longitude east. Zone standard time is GMT+1

Area

Total: 973.770 sq km; Land: 910,770 sq km, Water: 13,000 sq km
The land area is about 1.5 times the size the state of Texas.

Land Boundaries:

As shown in the **map of Nigeria** below, the perimeter or total length is about 4,047 km, with a coastal length along the Atlantic Ocean of 853 km.
Border countries are: **Republic of Benin** 773km of common border to the west; **Republic of Cameroon**, 1690km of common border to the east; **Republic of Chad**, 87km of common boundary to the northeast and **Republic of Niger** 1497km of common border to the north.

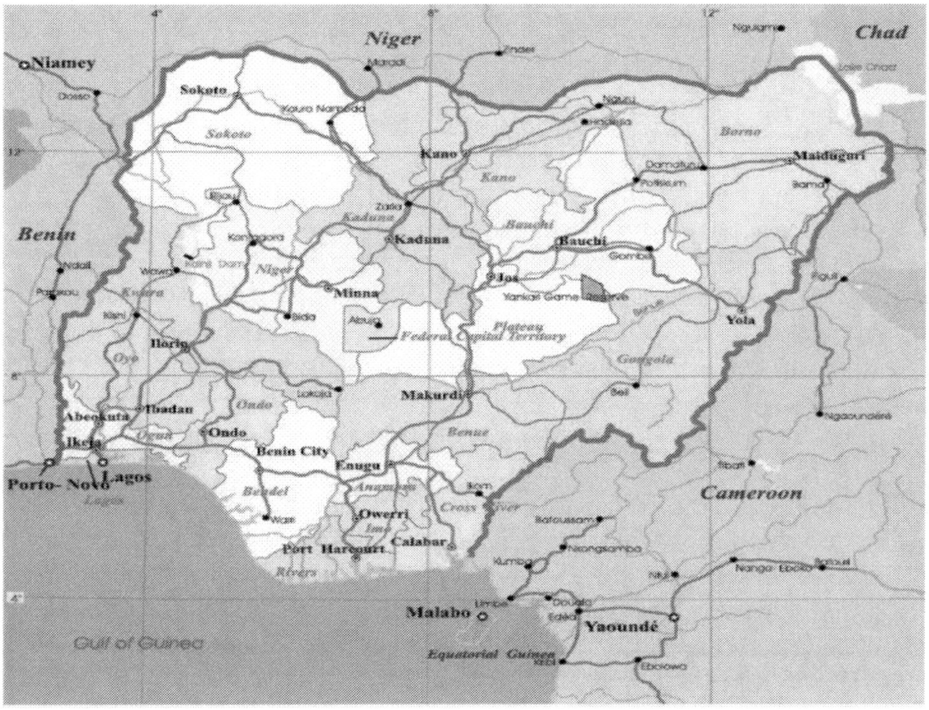

Maritime exclusive economic zone:

Nigeria's traditional territorial limit within the Atlantic Ocean was 12 nautical miles. This zone includes a portion of the Continental shelf with water depths up to 200 meters. The internationally recognised exclusive economic zone extends 200 nautical miles into the sea, with water depths varying from 2600m to 3000 meters. The vast oil and gas offshore (proven) reserves consisting of about 18 billion barrels of oil and 80 trillion standard cubic feet of gas have been discovered in this area. These petroleum reserves offshore represent about fifty percent of the current national total reserves.

Climate

The Niger Delta coastal areas are equatorial; while the tropical region extends northerly to the middle section of the country and the arid region is in the extreme north. Temperatures average between 18^0C - 35^0C in the North and 23^0C - 35^0C in the South. There are two main seasons: rainy or wet and dry seasons of equal length of six months approximately; from April to October for the wet season and from October to April in the following year, for the dry season. There is practically no rainfall during the dry season.

Mineral Resources

Petroleum oil and natural gas deposits, oil tar sands, tin and columbite, iron ore, coal, limestone, lead; zinc and gold are available in commercial quantities. These mineral deposits except the oil and gas are geographically and naturally spread throughout the country, from north to south and east to west. Commercial quantities of oil and gas deposits have been found only in the southern lands, swampy Niger delta and offshore regions. There are some striking similarities between the large natural deposits of petroleum oil and gas of the Niger River Delta in Nigeria and the Mississippi River Delta in the state of Texas. These similarities, according to petroleum geologists, are based on the natural processes of the formation of petroleum many millions of years ago, from plankton deposited from rivers flowing into the seas.

Land use

Arable land consists of 33% as permanent crops: 3%, permanent pastures, about 44% of arable land still exists as forest and woodland about 12%. Irrigated land is about 12000 sq km. Natural hazards are periodic droughts in the far northern areas and soil erosion in the southern areas, especially in the south eastern parts.

Environmental Issues

Pollution, soil degradation, erosion and rapid deforestation, desertification in the northern areas, recent droughts in the northern areas severely affected agricultural activities.

Population

The last census figures (2003) indicate a population of about 140 million people. Growth rate is about 3%.The population is largely rural. There are 13 cities with population in excess of 500,000 inhabitants. These include the Lagos metropolis, also the commercial and financial capital with an estimated 12million; Kano, 8million; Abuja, the official capital, 600,000 people respectively. Nigeria has the largest population in Africa. Every one out of four Africans is a Nigerian. There are 60 Nations in Africa.

The main Nigerian ethnic groups are Hausa, Yoruba, and Ibo. In addition there are more than 250 minor ethnic groups, some numbering fewer than 50,000 people. Other larger groups include the Edo, Fulani, Gwari, Ibibio, Idoma, Igala, Ijaw, Jukun, Kanuri, Nupe, Tiv, Urhobo, Kalabari and Itsekiri

There are three main languages widely spoken throughout the country. However, there are as many spoken languages as there are ethnic groups. The official language is English.

Economic Indicators/factors

GDP: real growth rate increased from 31/2 – 7% during the last 8years.
GDP- per capita: purchasing power parity decreased to $930 from $1300 in the same period. This could be due to a combination of several factors including continuous increases in international oil prices, higher local inflation, higher living costs and low productivity. All these factors have characterised the Nigerian economy which has been highly dependent (more than 80%) on its oil earnings.

Agriculture:	28%
Industries:	53%
Services:	19%

Inflation rate: average 9 – 12% per annum
Labor force is about 45 million. Unemployment is estimated conservatively at about 40% nationally. There is a very high and varied informal employment sector which makes it difficult to ascertain any measurements.

Oil and Gas reserves are 37.9 billion barrels Oil and 179 trillion cubic feet of Gas respectively. Daily oil production peaked at 2.5mb/d in 2005, before the gradual decline to the current 1.5 – 2mb/d due to the crisis in the oil producing areas of the Niger Delta region. Nigeria is a member of OPEC (Oil Producing and Exporting Countries) and is the 6th largest producer of oil in the group. The oil- and- gas- rich Niger Delta region covers an area of 70,000 sq km, which is roughly 7% of the total area of the country, and contains about half of all of the oil and gas reserves referred to above. The other half is situated in the shallow and deep offshore areas in the waters of the Atlantic Ocean.

Industries:

Oil and gas exploration and production, refineries, petrochemicals, coal, tin and columbite mining, agro-processing (palm oil, peanut, cotton, rubber, wood, hides and skins), textiles, cement and other construction materials.
Agriculture: cash crops are cocoa, peanuts, rubber, oil palm and cotton; food crops are maize, sorghum, cassava, millet, yams, rice, plantain, Soya, tomatoes, onions; fruits are pineapple, oranges, grapefruits, bananas and Vegetables.

Infrastructures

Highways: 35,105 km
Railways: 3557km
Waterways: 8575km
Road network: Just over 16000km of intercity, paved roads, including dual carriage express trunks.
Pipelines: Crude oil 2100km; petroleum products 5200km; natural gas 1000km.
Electric Power generation: 5000mw installed, current production is about 3OOOmw (30percent of the national requirement).
Estimated requirement is 10, 000mw.
Estimated local production of petroleum fuels/lubricating oils is less than 50% of consumption.
Piped potable water production supplies less than 70 % of the homes in urban areas and non existent in many rural towns and villages.
Communication: This is probably the most developed industrial system in the country. There are over 40million users on the telephone networks in the country. Most users are on the Global System of mobile communications (GSM).
Ports and Harbours:
There are seven deep water ports in Lagos (2), Port Harcourt, Calabar, Onne, Sapele and Warri. Naturally these are all located along the coast of the Atlantic Ocean. Imports and exports of all dry cargo pass through these ports. In addition there are petroleum oil and gas terminals at Forcados, Escravos, Brass, Bonny, and Qua Iboe. These specially constructed seaport facilities are used exclusively for export of crude oils and LNG (Liquefied Natural Gas) to the world markets.

Government

The form of government operated in the Federal Republic of Nigeria is very similar to the USA. The president heads the country and the Federal Government. The government of each state is headed by the governor. There are 36 states plus a Federal Capital Territory, FCT. There is a tertiary level of elected officials consisting of 774 local governments intended to bring development to the grassroots level. All three levels are enshrined in the Nigerian Constitution. The current constitution formulated in 1999 is the third after the country won its independence from the British colonial masters in October 1960. The two previous Constitutions were abolished by military regimes which governed the country intermittently for 31years out of the 48 years since independence. The third republic started in 1999 after the last military government agreed to handover peacefully to democratically elected governments at the federal, state and local government levels.

Who is the Nigerian Chemical Engineer?

The first Nigerian chemical engineer graduated from the University of Manchester, United Kingdom in 1957.Since then another 12000 chemical engineering students have graduated from 14 Nigerian universities and colleges of technology. The University of Ife (now Obafemi Awolowo University), Ile-Ife, Osun State was the first to establish a faculty of chemical engineering in 1968.Many Nigerian chemical engineers like me were educated in foreign universities in the USA, UK, Canada, Australia, East and West European countries. It is estimated there are more than 4000 Nigerian chemical engineers who graduated from Nigerian and foreign universities now work and live abroad. Most of the foreign based Nigerian chemical engineers live in English speaking countries such as the USA, UK and Canada. There are 2123 currently registered members of the Nigerian Society of Chemical Engineers. This number excludes student members in the Nigerian universities and colleges of technology.

CHALLENGES FACED IN NIGERIA BY CHEMICAL ENGINEERS.

All Nigerians living at home today face many common challenges, irrespective of age, gender, ethnic origin, and level of education, rich and poor, profession or occupation. As would be expected, these common challenges affect each person or group in different ways and in varying degrees. To the extent that, the opportunities for individual or group aspirations and growth or even survival, can become seriously curtailed temporarily or permanently.

Therefore in discussing the challenges that Nigerian chemical engineers face I will generalise the issues and be more specific on the effects, in order to differentiate them from other Nigerians or groups.

Corruption and poor governance:

In my humble opinion, corruption and bad governance in public life is the greatest challenge faced by every Nigerian living at home at this time. The simple reason for this position is that corruption and poor governance of government institutions are so pervasive, entrenched, and have become so endemic that most attempts, by the Nigerian authorities and private citizens' groups to tackle corruption have so far, failed woefully.

One of the products of corruption in all three tiers of government is bad governance, which in turn is the result the wrong or unqualified people in leadership and other important positions. The effects on Nigerians and Nigerian chemical engineers manifest in such areas as: perennial decline in the quality of education (Nigerian universities are worse off than they were a decade ago due to poor funding and inability to recruit and retain the right personnel); the inability of most of the new chemical engineering graduates to secure employment easily in a potentially developing petroleum based economy. Especially when there is an acute shortage of engineers required for very important and profitable projects.

The projects in the oil and gas industry are very large multimillion dollar projects. They are not only profitable but also serve other critical functions to the society, such as stopping the flaring of billons of cubic feet of gas into the atmosphere (pollution and waste of valuable resource), For example, as many as 12 of such projects worth over US$ IO billion have been advertised in the local press repeatedly in the last four years, but have not been awarded. This is due to interminable delays, ultimately attributable to the federal government and its responsible agencies.

The timing of the approval of Federal Government's annual budget and the release of funds always affects the operations of all the major oil and gas companies operating in Nigeria. This is because by law, all such companies operate under a joint-venture partnership agreement with the Federal Government. The Federal Government has the majority share holding with each operating joint-venture and is represented by the Nigerian National Petroleum Corporation (NNPC).

Hundreds of Nigerian engineers who could have been employed on these projects are either unemployed now or are underutilised in the companies which employed them, in the hope of winning some work advertised in the delayed projects.

The Shrinking Industrial base:

Another serious problem chemical engineers are facing is the closure of well established manufacturing plants which employed many of them in the past. Following are three examples:

Thirty years ago the Federal Government after an extensive studies conducted at great cost, for a period of five years, established an indigenous steel industry for the country. This consisted of two very large steel complexes for the production of steel from iron ore at Ajaokuta and Warri: and three steel rolling mills at Katsina, Jos, and Oshogbo. These Steel plant complexes employed more than 9000 people. The total number of chemical engineers recruited, trained and worked in the plants was about three hundred. Due to the bureaucratic nature of the Nigerian Steel Authority which controlled the management of those plants, all of them technically went out of business. Most of the chemical engineers lost their jobs.

As a result of continuous international pressure from the International Monetary Fund (IMF) and the Nigerian public for many years, the Federal Government recently made unsuccessful attempts to sell what remains of the plants to private investors. These attempts failed because of corruption and the insincerity of those bureaucrats charged with the assignment.

Recently the two largest local manufacturers of tyres and rubber products in the country have been forced, by a combination of fiscal, economic and infrastructural problems to close down their businesses and factories in Nigeria. These international companies, Michelin and Dunlop, are still leaders in this industry in France and the United Kingdom respectfully. Until their withdrawal from Nigeria these two companies obtained their feedstock, latex rubber from local rubber plantations and carbon black produced from the petrochemical plant in the Warri Refinery complex. Several engineers including chemical engineers, who had worked in the plants for many years, lost their jobs.

The Federal Government's policy of privatisation (at the insistence of the IMF) of the downstream sector of the petroleum industry has been reversed. The main objective of the policy was for the Federal Government to divest from its monopoly of the refineries in particular, to make them operate more efficiently and competitively under competent private ownership and management. This reversal has prevented the much needed rehabilitation and expansion of the refineries. The potential new job opportunities and hopes for the rejuvenation of existing jobs for chemical engineers have been dashed.

Lack of adequate infrastructural facilities

From the description of the infrastructural facilities above, it is clear that all the statistics show large deficiencies in all basic infrastructures such as national electric power generation and distribution; good road networks, water supply and distribution, urban mass transit, transportation fuels

All these infrastructural deficiencies not only adversely affect the quality of life of the people but also businesses and production of goods and services. Most factories in the country suffer from high cost of producing their own power from diesel generators. Worse still diesel fuel is in short supply and very expensive, because the local refineries produce only a small percentage of the total national requirements. The balance has to be imported.

OPPORTUNITIES FACING CHEMICAL ENGINEERS:

Entrepreneurship:

Entrepreneurial ventures appear to be one of the few areas of endeavour relatively free from the apparent stranglehold of government. Opportunities abound for chemical engineers with entrepreneurial capabilities and tendencies in Nigeria because of the available natural resources, sizeable labour force and a large national market. However, success in any ventures will still depend on a number of other factors, including those infrastructural challenges identified above.

Cooperation with foreign organisations for training and collaboration in projects are essential prerequisites for successful start up projects in Nigeria. The NSChE encourages its members to avail themselves all the opportunities available through close ties with organisations local and foreign, such as A.I.Ch.E, and N.O.B.C.Ch.E etc. to develop its capabilities and commercial interests. This will be especially useful in the area of small-scale industrial projects for which the Nigerian financial sector appears very willing and ready to participate. The availability of funds even in these trying times for obtaining credit facilities appears to be assured.

Employment in local industries.

The opportunities for chemical engineers still abound in large numbers if and when the challenges described above are mitigated by proposed reforms or eventually solved. The NSChE still avails the Nigerian government authorities of its findings and suggestions for solutions to the problems of the economy, and the industries in particular. We continue do this in spite of the lack of appreciable improvements over the years. We use the use communiqués prepared during our annual conferences and personal contacts with important officials whom we invite to participate in our activities, to pass information and advice to the Federal and State governments.

CONCLUSION

I would like once again to thank the A.I.Ch.E for the unique opportunity I have had to present my country and my fellow Nigerian chemical engineers during such a historic occasion of the celebration of A.I.Ch.E Centennial. I wish to assure you that the cooperation between our two societies will be strengthened further for the benefit of all our members.

On behalf of the N.S.Ch.E, I will also thank the A.I.Ch.E for the opportunity extended to our members since 2007 to join A.I.Ch.E at discounted rates. This we know has opened up immense opportunities for our members to benefit in many ways, educationally and professionally.

Finally, in order to personally strengthen the ties between our two societies; I will rejoin the A.I.Ch.E and encourage my successors to do the same.

Thank you all.

CHEMICAL ENGINEER FOR TODAY AND TOMORROW THROUGH CURRICULAR REFORM

Álvaro Ramírez García
Ch E school, Universidad Industrial de Santander
Bucaramanga, Colombia

Introduction

This paper presents some major questions raised in the process of curricular reform of the Chemical Engineering school at Universidad industrial de Santander (UIS), in Bucaramanga, Colombia.

UIS was founded in 1948 and began institutional life offering three engineering programs: mechanical, electrical and chemical. Today is a very well recognized public university, with 22000 students in 50 undergraduate programs in engineering, natural science and mathematics, human science and health science. There are about 1400 students in graduate programs; 40 in three doctoral programs: physics, chemistry and chemical engineering. Total alumni are 48.000. The Ch E school offers programs in undergraduate, master and doctoral levels: 1.200 undergraduate, 17 master and 11 doctoral.

The former curriculum of Ch E at UIS (this is the program up to year 2005) had a strong industrial chemistry influence of the Spaniard school. German teachers who came to UIS in the fifties reinforced this tendency. The North-American influence began to be strong since the seventies. Consequently, curricular changes have gone in the direction of increasing unit operations, transport phenomena, reactor and process design for which it is also observable a decreasing tendency on chemistry and chemical process description. The chemical engineers from UIS have very good recognition in national industry and have been successful in graduate schools abroad.

The curricular reform process took more than a decade and involved students, teachers, alumni and foreign university visitors. The output could be considered as a changeable compromise between several tendencies. Actual program changes began in 2006. Some major questions to begin with are: being a leader and successful chemical engineering program, why should it be reformed? What to reform? In what directions?

The need for curricular reform

The former curriculum of Ch E program at UIS was successful in a social environment characterized by a closed national economy in which big chemical industry companies, many of them state owned, offered stable employment with high salaries and good social benefits and were confronted with scarce competitors in a protected market. Advances in chemical engineering as a discipline came in the direction of transport phenomena, use of computer calculations and materials science and engineering. Those advances were introduced to reinforce the core of the former, traditional curriculum.

The present and near future appear to be quite different. The local Chemical Engineering is faced to an environment in which employment is temporal, unstable and scarce; competition occurs in a global market for a country that is not in the main stream of the economy. Also relevant is the factor that the local established chemical process industry is not growing fast enough to accommodate the total offering of new chemical engineers coming out of the eleven ChE programs in the nation. But, there are plenty of technological challenges and potential raw materials to generate chemical process industries.

There are new frontiers in the first world Chemical Engineering science and technology, namely, bio, computational and nano engineering.

The former program of chemical engineering, of course, was not intended for these new frontiers neither for the emerging social conditions.

There are also internal or university requirements to be satisfied: a 5 years program with 50 courses at most; students should come to take Ch E. courses since the first semester when most of their courses are in science and general studies. Another important condition is that there is a growing demand of young people to study Chemical Engineering.

Since the former curriculum was fitted for an environment that does not exist any more, there is a need for curriculum reform to prepare a Chemical Engineer for today and tomorrow; one able to cope with both traditional and new frontiers of Chemical Engineering, because young Chemical Engineers will probably find a closed traditional environment as well as an open challenge to build a new Ch E career as discipline and profession.

<u>General purpose and characteristics</u>

The new curriculum is intended for the student to build strengths for generating, developing and managing traditional and non-traditional chemical process industries and technological services.

The new directions of the curriculum embraces new scopes: from atomistic to macro scale; steady and unsteady states; traditional and new frontiers of Chemical Engineering science and technology, business vision and a worldwide perspective of chemical process industry and technology.

Consequently, in the normal set of characteristics of a good chemical engineer should be included competitiveness, enterprise orientation and flexibility (universality). These three characteristics are intervolved and call for productivity, quality culture and environment care. A chemical engineer has to contribute to society in an interdisciplinary work environment from a technological perspective, that is, to do on the bases of knowledge.

The curricular reform

Accepted that a curriculum reform should be performed and given the above purposes and characteristics, a question arises on what to reform. The curriculum involves at least course program, pedagogy, assessment, environment and rules.

The traditional and the new chemical engineering program are supported on strong and modern formation in math and natural sciences including chemistry, modern physics and biology. The math component is rather strong: six courses on superior algebra, descriptive geometry and calculus and three courses on informatics and numerical analysis. The number of chemistry courses was diminished leaving only three; in physics the number of courses, three, remain the same, but contents in the third course were changed to include modern, twentieth century, physics fundamentals. Two new biology related courses appear, microbiology and bioprocesses.

The core of traditional chemical engineering has to be preserved because it contains the fundamentals for running and even developing the traditional chemical process industry still existing in the world and locally and also because it is the basis for developing new frontiers in the discipline and the profession. It includes stoichiometry, thermodynamics, transport phenomena, unit operations, reactor design, control and design of processes. According to trends in the market, there seems to be an increasing industrial interest in batch processes, which should be reflected in the academic program, including also, of course, continuous processes. The core of traditional chemical engineering is expanded to include nanotechnology and computational engineering.

To give attention to the social conditions already described, the program offers opportunity for formation on enterprise based on knowledge. Two courses are devoted to this purpose.

The proposed curriculum brings together contradictory tendencies: new subjects keeping the traditional ones and also giving time to the student to develop a worldwide vision, all of that in the same 5 years program; on the other hand the students are going to be taught about subjects that teachers never took at school; and also students are being formed in subjects that, currently, are not being applied in the local chemical process industry.

To cope with these different trends some strategies were adopted in regard to pedagogy, assessment, environment and rules:

In the course contents and assessment emphasis should be made on basic concepts rather than in detailed completeness; to develop a general and holistic view in contrast with an exhaustive purpose.

Repetitions of topics in different courses should be avoided. The student should be aware of the academic requisites for undertaking a given course. To help in previous preparation for a course, review material in the form of by computer interactive learning programs should be available for students. These programs are prepared for senior students as final graduate work under the direction of a teacher. Of course, this material is also valuable for students taking the respective course.

New courses are added to the program; some traditional courses are withdrawn and some others are readjusted. In this case, 22 courses were withdrawn and 13 new were added. Some of the withdrawn were lab curses that in fact were included in the basic courses, example, the basic chemistry lab now makes part of Chemistry I.

As a laboratory course there is only one that occupies half of the credits of one semester. This course is devoted to process design and control. The idea is that the student have a practice on a complete process, in such a way that the student faces a complex interdisciplinary situation instead of doing specific measurements in a, for example, transport phenomena lab course. However, those specific measurements should make part of the transport phenomena course that includes lab practices.

As indicated above, each core course includes laboratory practices designed by the teacher to emphasize basic concepts. It is advisable to familiarize the student with modern equipments such as differential calorimetry, gravimetric thermal analysis, rheology, microscopy or any other piece of equipment available in the school. This is the least implemented part of the curricular reform. Equipments are available, a new building facility is available, designed practices are not, yet. A necessary and still incomplete effort should be done to make available hardware and software to support core courses.

The introduction of teachers and students into new frontiers of chemical engineering has been done through research and academic mobility. Through supported research programs some thesis have been devoted to topics on computational engineering in collaboration with universities abroad. Also, final works of senior undergraduate students have been devoted to exploring topics on computational engineering and nano basic knowledge and technology.

Another important strategy is to take advantage of international collaboration agreements to establish some kind of a two-way exchange. UIS undergraduate students have gone as interns to Universities like Texas A&M (USA) and Campinas (Brazil) to make a lab and research practice that is valid in UIS as a final graduate work. Those students present in public his findings and experiences in both traditional and non-traditional fields of chemical engineering. This has given support to the underlining ideas of the curriculum reform. Some of these students, after graduation at UIS, went back to continue graduate studies and some of them probably will contribute to make a difference in the Colombian chemical engineering in years to come. Also these collaboration agreements have made possible the visit of Professors from Texas A&M and Campinas to UIS to give constructive criticism and very valuable ideas to make a better curricular reform. Some visiting professors have given short courses on fundamentals of computational engineering, nanotechnology, update thermodynamics and material science.

Discussion and conclusion

The proposed curricular reform has an inherent risk for new generation of chemical engineers who will meet a chemical process industry for which they are, at best, over prepared.

At first glance it seems to be a waste of time and of scare resources and a potential cause of frustration.

The confronting view is that engineers for today chemical process industry should have been formed yesterday and that today formation should be intended to support traditional chemical process industry, but also, to build a new Ch E career and profession.

That is why new frontiers of chemical engineering (bio, nano, computational), complete scale reference (nano and macro), complete process regime (steady and unsteady) and enterprise capacity are key points in the proposed curriculum reform.

References

C. Retamoso, A. Ramírez . Estado de desarrollo e inserción social de la Ingeniería Química en Colombia. La conformación de comunidades científicas en Colombia. Vol 3, Tomo II, Fonade, Misión de Ciencia y Tecnología. Bogotá, 1990

A Ramírez , C. Retamoso. A study on Chemical Engineering development state and social insertion in Colombia, Proceedings "The 1993 Centennial Conference of American Society for Engineering Education", Urbana, USA, 1993

K.R. Hall, Seeing the future: How to do it! Memorias XXIII Congreso Colombiano de Ingeniería Química, Manizales, 2005

Visión Colombia. Segundo centenario: 2019. Departamento de Planeación Nacional, Bogotá, 2005

Frontiers in Chemical Engineering Education, http://web.mit.edu/che-curriculum/

514g Student Learning through Engineers without Borders

Randy S. Lewis, Chemical Engineering, Brigham Young University, 350 Clyde Building, Provo, UT 84602 and W. Vincent Wilding, Department of Chemical Engineering, Brigham Young University, 350 Clyde Building, Provo, UT 84602.

Student chapters of Engineers Without Borders (EWB) are becoming increasingly popular in Engineering Colleges across the nation. The purpose of EWB is to improve the quality of life of developing communities throughout the world through the implementation of sustainable engineering projects, while training internationally responsible engineering students.

In 2006, an EWB student chapter and an associated project development course were implemented at xxx University. This combination (student chapter and course) is providing a rich environment for active and effective student learning.

The EWB chapter is a volunteer club that participates in local service projects, assesses opportunities for global projects, provides sustainability to projects previously implemented, organizes humanitarian-based informational meetings, and provides social opportunities for club members.

The course, entitled "Global Projects in Engineering and Technology," provides an opportunity for approximately 30 EWB chapter members to work in interdisciplinary teams on the development of global engineering projects that are implemented at the end of the academic year. Learning objectives for the course include

* Exhibit innovative problem solving skills as they apply their engineering training to global projects

* Demonstrate effective oral and written communication skills, particularly to individuals in countries in which projects are implemented.

* Demonstrate effective teamwork and leadership skills and an appreciation for other disciplines as they work together on interdisciplinary teams

* Develop an understanding of and appreciation for global aspects of engineering.

During the 2006-07 academic year, the course focused on the production of biodiesel from coconuts in the country of Tonga. The project was implemented during a two-week trip to Tonga with 26 students and two faculty. For the 2007-08 academic year, the course involved energy efficiency and water resource development in Salkantay, Peru. Thirty-two students participated in the course. Twenty students and three faculty recently completed a two-week trip to Peru in May 2008.

This work will discuss lessons learned from the startup of an EWB chapter, the development of the projects course, and the implementation of the project in the associated country. Particular emphasis is on student learning and assessment of that learning.

Preparing 21st Century Engineering Students through a Student-Initiated Interdisciplinary Service-Learning Project

Alexander Bick, Dana Lazarus, and Harvard College Engineers Without Borders. School of Engineering and Applied Sciences, Harvard University, 29 Oxford Street, Cambridge, MA 02138

Introduction

As traditional demarcations between academic disciplines fade, and the virtual distance between Boston and Bangalore approaches zero, the engineering educator faces new challenges and unique opportunities. Preparing chemical engineering students to operate in international collaborative efforts is of critical importance in our global society. Engineering students should be equally conversant with anthropologists as their fellow chemical engineers, and as fluent in Spanish as in MatLab.

The Accreditation Board for Engineering and Technology (ABET) agrees. In their 2008-2009 Criteria for Accrediting Engineering Programs document, Criterion 3 "Educational Outcomes" specifically states the following as engineering program outcomes required for all engineering students [1]:

3 (c) an ability to design a system, component, or process to meet desired needs within realistic constraints such as economic, environmental, social, political, ethical, health and safety, manufacturability, and sustainability
3 (d) an ability to function on multidisciplinary teams
3 (h) the broad education necessary to understand the impact of engineering solutions in a global, economic, environmental, and societal context

The Engineers without Borders (EWB) program is devoted to improving the global quality of life with a bottom-up approach. Individual EWB chapters partner with developing communities and together design and implement sustainable engineering projects. In doing so students learn to design and implement resource efficient projects in an international context. EWB contributes to meeting the United Nations Millennium Development Goals through capacity building in local communities [2].

Here we discuss findings of a two year old interdisciplinary project in student initiated sustainable design education. First we present an overview of our project. We then briefly describe how our project benefits from commonly observed advantages of working in an interdisciplinary service learning framework. We then examine specific deviations from the usual model that we believe are particularly advantageous to our project and students. We conclude by discussing ongoing challenges posed by highly interdisciplinary projects.

Harvard EWB Project Status

Under the suggestion of St. Jerome's Constanza Mission, a group of doctors from Boston Children's Hospital and South Shore Medical Center, we started a project to design a sustainable solution to a contaminated water supply in rural areas outside Constanza, Dominican Republic.

St Jerome's Mission first approached Harvard EWB because the members noticed that water-borne diseases are the cause for many of the illnesses they treat. They asked HCEWB to further investigate the situation to see where engineering design could help. Two trips were undertaken to Constanza in the 2007-2008 school year.

During the first trip in November, we spent most of our time visiting aqueducts, mapping the water cycle in the area, and building personal relationships with the community. Before returning to the US, we held a community meeting where we presented our findings.

In a second trip in May, eight members of our chapter, plus our professional mentor, traveled to Constanza. The community consists of about 1200 families. Coming into the trip, our primary contact in the village was the local nun, Sor Rosario. We were also in contact with Carlos Mena, the local INAPA representative, who is in charge of all the drinking water in the area. Before departure, we also made contact with two Peace Corps volunteers in the area who translated for us throughout the trip and who are willing to work with us remotely. We had knowledge coming in that there was a small-scale treatment facility not operating properly, and that there was not enough water supply. During the trip we split up into two groups. One group investigated the water treatment facility to obtain data and figure out why it was not functional, the other focused on ways to increase the water supply through either drilling wells or capturing more water at the source. We also did some health assessments on a house-by-house basis and water testing from the top of the system to the bottom.

During the second trip, we made many contacts in the area, including an engineer who helped build the treatment facility, local government officials, and a couple of families who had wells. We managed to obtain both topological maps of the area as well as plans of the treatment facility from INAPA Headquarters in Santo Domingo. We learned that water at the source of the system was already polluted with biological contaminants.

Additionally, the water treatment plant was not built correctly. In the facility, water first goes through a settling tank. However, there is a leak in the apparatus as the water enters the plant, and this allows some water to bypass the settling tanks. Following the settling tank, water passes through a sand filter. The filter has four layers: sand, gravel, stones, and large rocks. The total depth of the filter is 3.5 feet. However, the sand filter was built incorrectly and as a result gets clogged up very quickly, leading the locals to bypass it entirely. Following the sand filter, the plan calls for chlorination of the water, however this aspect of the design was never completed.

During the trip, we performed water tests at various points in the system, from the initial source to the homes at the end of the pipe. We tested for biological as well as chemical contaminants. Biological tests gave a count of Coliform and E.coli per 100 mL of water. Chemical tests gave values for nitrate and nitrite, chlorine, pH, hardness and alkalinity.

Water serving the community is taken from three sources in the mountains. We visited two of these sources. The first source is an open stream, the second flows under vegetation and is dammed for collection. These sources all come to the water treatment facility. The water then splits: part of it is stored in a holding tank and serves part of the community, the rest goes to another storage tank downstream and services the rest of the community. Because there is not enough water to go around, water supply in different parts of the community is intermittent.

Our water tests showed that water in homes further downstream from the treatment facility was had more biological contaminants than those closer to the treatment facility, suggesting that there is either sedimentation or leaks causing contamination in the pipes. In homes, people store water due to its intermittent supply. Some people treat their water with chlorine, which reduces the amount of contamination. Some families purchase bottled water, which we found to be free of any contamination. We also tested a couple of local wells, and found that some were significantly contaminated as well. The few wells we found in the area were drilled by hiring outsiders to come in for a costly fee.

During the trip, we obtained topographical maps as well as the original plans for the treatment facility. These maps also include the pipes that distribute water to the homes. We also took measurements of the entire treatment facility so we can compare it to the original drawings. We will use this information to make calculations about the water flowing through the facility and in the pipes. These next steps will be part of the work of our project teams this year. We also took some samples of the local soil back with us for analysis, as well as samples of the sand from the filter; these will be analyzed during the year.

We found that we could obtain many materials that we believe will be necessary for drilling wells and improving the treatment facility either in Constanza or at the capital on Santo Domingo, including piping, metal washers, etc. The local people were also very creative when it came to suggesting alternatives. Due to curiosity and desire for change, it was very easy to find volunteers to help us out, including people with engineering experience.

This year, our project will focus on two areas. The first is fixing the water treatment facility, since the infrastructure is already present. We want to 1) fix the sand filtration system, 2) install a chlorination system, and 3) create a flushing mechanism for the pipes. All of these initiatives will require community support and maintenance.

Secondly, even if we fix the water treatment facility, there will still not be enough water to supply the area. We are working on designing a method to drill wells at a low cost for families or groups of families. This task will increase the water supply and also ensure that the wells are maintained by the families we drill them for. Some families already have wells, so we know it is possible to reach the water table, and over the year we will be investigating 1) exploration techniques, soil analysis, and finding water; 2) perfecting the design of our drilling technique, and 3) filtration and storage of water for use.

The community reaction was varied. There is a group of people who are very concerned about the situation and are glad we came to help. These people are aware that their drinking water is unclean and want to do something about it. They are excited about our project and were very helpful during the trip. Some others were complacent or apathetic about their situation, but there is enough community interest such that when we return we will have the resources we need.

We are working with the local community in everything we do. With whatever we build, the local community will help us at every step and will be responsible for maintaining it after we leave. We will work through the Peace Corps volunteers, as well as Sor Rosario to coordinate our planning efforts with the people on the ground, so that they know what we are planning in every step of the process. We found a higher level of knowledge and innovation than we expected among the people, and expect that, when we return for implementation, they will be very helpful in obtaining materials and figuring out alternatives. We expect the members of the community to provide labor to help us with building.

EWB: Highly Interdisciplinary

EWB is a well received program as evidenced by its pervasiveness on engineering campuses with 153 established chapters out of 600 schools with ABET accredited engineering programs. [2] In fact EWB projects can be seen as meeting all 12 of the program objects listed in ABET requirement 3. [1] EWB programs often have three interdisciplinary aspects: the projects themselves, sustainable design, and the participants. [3] We will discuss briefly how this occurs using our project as an example.

The projects of EWB rarely fall within a single traditional engineering discipline. [4] A majority of EWB projects deal with water as, despite significant progress, access to clean drinking water remains one of the single largest health challenges in the developing world. [5] Water related projects commonly deal with increasing access to water supplies—where the quantity of water is insufficient to meet the needs of the local peoples or water purification projects where the water supplies are not fit for human consumption. To find water supplies; drill wells or build pipes to access water sources; pump the water from the ground and purify it to the point where it is fit for human consumption; test the water for biological contaminants involves a whole range of skills that cross traditional disciplines. Our project thus relies on the input and individual knowledge of a team of individuals with a variety of skills and classroom knowledge, making the activity highly interdisciplinary.

Sustainable engineering – a cornerstone of the EWB philosophy in general and our project in particular – requires an interdisciplinary approach. In brief, sustainable design charges engineers to create a solution that can be wholly created and maintained by local individuals with locally available materials using locally available tools and manpower. [6] This charge demands an interdisciplinary response. Although a mechanical engineer in our group designed a system for drilling based on tools, machines and techniques available in a university lab environment, he then looked for help from a material scientist to see whether the same drill bit could be produced from materials available in a local plumbing store.

In a looser sense of interdisciplinary—the integration of students and mentors of different years and stages of their careers adds to the interdisciplinary nature of the undertaking. Although an undergraduate student, graduate student and professional engineering mentor all thought of themselves as mechanical engineers, when it came to actually fixing the water filtration system, each had very different expertise and experience within mechanical engineering and consequently had different contributions to make. Thus the integration of multiple kinds of thinking that resulted from a number of years of experience was yet another interdisciplinary aspect of the project—commonly found in EWB projects.

Harvard EWB: Uniquely Interdisciplinary Aspects

In addition to the wonderfully interdisciplinary nature of the generic EWB experience, our EWB chapter has further benefited from further interdisciplinary interaction in two ways. First a large proportion of EWB undergraduates are non-engineers. Second we have had the unique opportunity of interfacing with non-engineer professional mentors.

The interdisciplinary environment created with the inclusion of non-engineering undergraduate students has uniquely benefited both engineering and non-engineering students alike. Approximately half of the Harvard EWB club is composed of engineering students, a third come from the natural science ranks and the remaining sixth are social science and humanities students. While engineering students are primarily drawn to EWB as an opportunity to apply what they learn in the classroom, students from non-engineering disciplines are drawn to EWB to experience what they do not learn in the classroom—namely design. Both groups are drawn to the idea of helping those in the developing world and the bottom up philosophy of change championed by EWB.

Engineers truly benefit from having natural scientists challenge them to develop an idea more mathematically before just hacking together a design—which is perhaps an engineer's first instinct. Natural science students benefit from actually building something that works—rare in the physics or biology classroom. Both groups of scientists benefit greatly in terms of communication. Explaining how to improve water flow rates through a sand filter to a History student is not easy, however in having to do so they better articulate their own ideas leading to overall better communication. Similarly, science and engineering students learn much when their grant proposals and write-ups are marked up by those with better verbal skills. The humanities students learn much about what engineering is all about and specifically how an engineer approaches a problem. These students describe how they are then able to incorporate their experience with EWB into their coursework. The uniquely interdisciplinary composition of the Harvard EWB project benefits all students concerned.

A second uniquely interdisciplinary aspect of EWB is cooperation with the medical doctors of the St. Jerome's Parish Constanza Mission. While an engineer might describe the heart as an electrically activated, two stroke pump with mean time to failure of two billion cycles, a doctor would describe the heart in terms of its structure and function in the cardiovascular system. As previously described, this group of doctors has been traveling to the community of Constanza twice a year for the past five years. Although they had made a significant contribution to the health of the local community, their efforts to prevent waterborne illness was not sustainable. They would distribute the appropriate medication and then returning six months later find that the same individuals needed the same medicine again. The doctors realized that in order to get to the root of the problem, they would have to improve the water quality of the community. Like working with humanities majors, interacting with the Doctors forced engineers in Harvard EWB to communicate without jargon or technical terms and to be very specific in our future request for information for as we learned quickly the way that doctors think about a problem and engineers is very different. This interdisciplinary interaction with highly educated experts in a different field was very informative.

Challenges Posed by Interdisciplinary Activities

Creating a functional interdisciplinary learning environment where all participants benefit is no easy feat. Although all participants outwardly appear to speak the same English language, in fact each group of participants speaks in a discipline specific tongue and is armed with discipline specific knowledge. [7] Although a student pursuing engineering, a natural science or a domain within the humanities approach problems differently, all are bound together with an overriding enthusiasm for the end result—helping those without access to clean drinking water. The project would likely fall apart without this strong individual motivation, which causes all participants to teach and all to learn. Since the project is student initiated and students run, all students are there because they want to be. The desire to succeed is far higher than for any engineering class problem set. A strong sense of camaraderie develops and slowly these challenges can be overcome.

Conclusion

Although the traditional engineering student course in design remains at the capstone of most engineering curriculums, interdisciplinary student-initiated service-learning programs like EWB can provide an alternative and less formal introduction to the world of engineering design. [8, 9] Students involved are highly motivated as they set their own objectives for themselves and their team. Additionally, there is a strong sense of helping others that drives the process forwards. Working in highly interdisciplinary groups on long term projects is highly advantageous to 21st century engineers as projects like EWB will likely define much of their future work experience. [4] Improving how engineers are able to relate and communicate with non-engineers is of upmost importance as well. [10] While a head-to-head comparison of the EWB approach to learn design in comparison to a more traditional engineering design course has not been undertaken and would be difficult to conduct due to significant variation between programs at different schools, it would be very interesting to qualitatively asses differences in student readiness for the capstone design course between those who participated in an EWB experience and those who had not. [11] We anticipate as the EWB students go on to careers in industry and academia, that they will report EWB as one of the more formative components of their engineering education.

Acknowledgements

The authors greatly appreciate the support of the Harvard School of Engineering and Applied Sciences Faculty and in particular faculty advisors Dr. Marie Dahleh, Dr. Rob Wood and Dr. Jim Rice. Additionally the assistance of Alison M. Forsyth and Dr. Howard A. Stone was greatly appreciated in the preparation of this manuscript.

References

[1] ABET, "CRITERIA FOR ACCREDITING ENGINEERING PROGRAMS 2008-9," Engineering Accreditation Commission, 2007.

[2] EWB-USA, "EWB-USA," 2008.

[3] B. Sukumaran, J. Chen, J. Mehta, D. Mirchandani, and K. Hollar, "A Sustained Effort for Educating Students about Sustainable Development," *CD) Proceedings, 2004 American Society for Engineering Education Conference,* 2004.

[4] B. Amadei, "PROGRAM IN ENGINEERING FOR DEVELOPING COMMUNITIES: VIEWING THE DEVELOPING WORLD AS THE CLASSROOM OF THE 21ST CENTURY," *FRONTIERS IN EDUCATION CONFERENCE,* vol. 2, 2003.

[5] S. A. Esrey and J. P. Habicht, "EPIDEMIOLOGIC EVIDENCE FOR HEALTH BENEFITS FROM IMPROVED WATER AND SANITATION IN DEVELOPING COUNTRIES 1," *Epidemiologic Reviews,* vol. 8, pp. 117-128, 1986.

[6] S. M. Lele, "Sustainable Development: A Critical Review," *World Development,* vol. 19, pp. 607-621, 1991.

[7] J. D. J. VanderSteen, "Humanitarian engineering in the engineering curriculum," in *Civil Engineering*. vol. Doctor of Philosophy Kingston, Ontario: Queen's University, 2008, p. 374.

[8] E. J. Coyle, L. H. Jamieson, and W. C. Oakes, "EPICS: Engineering Projects in Community Service," *International Journal of Engineering Education,* vol. 21, pp. 139-150, 2005.

[9] C. I. Helgesson, "Engineers Without Borders and Their Role in Humanitarian Relief," *IEEE ENGINEERING IN MEDICINE AND BIOLOGY MAGAZINE,* vol. 25, p. 32, 2006.

[10] D. Riley and A. Bloomgarden, "Learning and Service in Engineering and Global Development," *International Journal for Service Learning in Engineering,* vol. 2, pp. 48-59, 2006.

[11] P. R. Brenner, M. Schroeder, and G. Madey, "Student Engineers Reaching Out: Case studies in service learning and a survey of technical need," *Frontiers in education conference-global engineering: knowledge without borders, opportunities without passports, 2007. FIE'07. 37th annual,* pp. T2J-1, 2007.

516b A Novel Interdisciplinary Biotechnology Process Engineering Laboratory Course at the University of Massachusetts Amherst

Susan C. Roberts, Chemical Engineering, University of Massachusetts, Amherst, 686 North Pleasant Street, Amherst, MA 01003 and Louis A. Roberts, Biochemistry & Molecular Biology, University of Massachusetts, Amherst, Amherst, MA 01003.

The disciplines of chemistry and chemical engineering are continually incorporating biological systems into research and discovery in both academic and industrial settings. Combinatorial chemistry for the purpose of drug discovery, rational drug design and delivery, and metabolic engineering for production of pharmaceutically-relevant natural products are three specific examples of biologically-important applications of chemistry and chemical engineering methodologies. Conversely, biologists have rediscovered "chemical genetics," i.e., the enormous utility of chemical inhibitors to explore biomolecule function as an alternative to the classical genetic methods of gene mutation and transformation.

Research in academia and industry has begun to successfully integrate the life sciences and physical sciences and engineering. Undergraduate curricula must likewise evolve to promote similar interactions to adequately prepare students for professional careers at the interface of science and engineering. Consequently, faculty in the Departments of Biochemistry and Molecular Biology (BMB) and Chemical Engineering (ChE) at UMass-Amherst have been collaborating in curriculum development over the past several years in order for undergraduates in these majors to benefit from cross-fertilization amongst these disciplines. Successes to-date include expansion of the chemical engineering curriculum to include courses in biochemistry and cell and molecular biology and addition of new lecture courses in biochemical engineering and systems biology that service students from chemical engineering, biochemistry and chemistry.

Most recently, we have extended our interdisciplinary curricular efforts in the development of a novel laboratory course in Biotechnology Process Engineering, open to both life sciences and chemical engineering majors at UMass-Amherst. This course is truly interdisciplinary and comprehensively focuses on the laboratory skills necessary to bring a product to the marketplace in the pharmaceutical and biotechnology industries. Students work in interdisciplinary teams (two chemical engineers and two life scientists) to design a process to supply a high-value product. Molecular biology tools are used to engineer cells lines to express the protein of interest; cells are then cultivated in bioreactors with designed process strategies; the protein product is recovered from the fermentation broth and purified using a wide range of techniques including filtration and chromatography; and the final protein product is characterized in terms of protein quality and function. State-of-the-art techniques using high-throughput technologies are integrated throughout the course along with quantitation and statistical analyses. This is the first course at UMass Amherst designed specifically for undergraduates to gain laboratory experience at the important interface of engineering and the life sciences and we will focus this talk on course structure, design and implementation.

516c Chemical Engineering and the Development of the the Doctoral Portfolio Program at the University of Texas at Austin

Brian A. Korgel, University of Texas at Austin, Department of Chemical Engineering, Austin, TX 78712

The Doctoral Portfolio Program in Nanoscience and Nanotechnology is a graduate certificate program offered through the Graduate School's Doctoral Portfolio Program at The University of Texas at Austin (UT Austin). The Chemical Engineering Department played a vital role to the formation of the program and continues to play a significant leadership role. The program was approved by the Provost and began accepting admissions to the program in Fall, 2004. 49 PhD students have graduated from the program and currently there are 76 students enrolled in the program. The students are pursuing their PhD degree in one of eight different departments, including the Departments of Biomedical Engineering, Chemical Engineering, Chemistry and Biochemistry, Electrical and Computer Engineering, Engineering Mechanics, Materials Science and Engineering, Mechanical Engineering and Physics. The Portfolio students that have been enrolled in the program have been supervised by 47 different advisors! This presentation will describe the program, its successes and pitfalls, how it is supported at UT Austin and the role of the Chemical Engineering Department in its continued success.

516d Mems (Microelectromechanical Systems) – a Model System for Interdisciplinary Design

Joseph J. Biernacki[1], Christopher D. Wilson[2], Satish Mahajan[3], Glenn T. Cunningham[2], and Stephen A. Parke[3]. (1) Chemical Engineering, Tennessee Technological University, Department of Chemical Engineering, P.O. Box 5013, Cookeville, TN 38505, (2) Mechanical Engineering, Tennessee Technological University, Department of Chemical Engineering, P.O. Box 5013, Cookeville, TN 38505, (3) Electrical Engineering, Tennessee Technological University, Department of Chemical Engineering, P.O. Box 5013, Cookeville, TN 38505

For the past seven year, a team of faculty from Mechanical, Electrical and Chemical Engineering from Tennessee Technological University (TTU) have been offering a course called "Introduction to Microelectromechanical Systems (MEMS)." Having grown out of the integrated circuit (IC) revolution, MEMS fabrication shares similar elements with IC device production. Design of MEMS devices, however, are not limited to the domain of electrical engineering since they, by definition, incorporate elements of mechanical and chemical processing components. The TTU faculty team has explored a number of course concepts, incorporating true interdisciplinary activities with design as the centerpiece of the course. A survey of experiences illustrating the concepts of interdisciplinary training will be presented along with suggestions for an optimized course structure. The logistics of managing interdisciplinary student teams and teaching with a larger than typical number of co-instructors will also be discussed.

516e The Interdisciplinary Nature of Energy Applications. A Course In Carbon Capture and Sequestration

Jennifer Wilcox and Sally Benson. Energy Resources Engineering, Stanford University, 367 Panama Street, Green Earth Sciences 065, Stanford, CA 94305

World carbon dioxide (CO_2) emissions are projected to rise from 26.9 billion metric tons in 2004 to 33.9 in 2015 and 42.9 in 2030.[1] Concerns of global warming, increasing carbon emissions, and an ever-present demand for energy production, has led to an increase in students pursuing energy-related careers. To be prepared for such a career, a connection needs to be made between traditional fields such as engineering, fundamental sciences, economics, and policy so that the student can understand energy production and its impacts from both environmental and economic perspectives. The Department of Energy Resources Engineering at Stanford University is concerned with the design of processes associated with energy recovery and recently a course on CO_2 capture and sequestration has been designed. Bridging concepts within the fundamental aspects of physics and chemistry for understanding the effective separation of CO_2 from both flue and fuel gas streams with the processes involving its transport and potential long-term storage can be difficult without an interdisciplinary teaching approach. An underlying life cycle assessment concept to determine the overall cost of electricity based upon carbon separation and sequestration exists throughout the course to provide students with an understanding of the economic feasibility of such approaches to carbon emissions mitigations. The course is taught by instructors from multiple disciplines, ranging from materials science and mineral engineering to physical chemistry and chemical engineering.

The course is separated into two components, with the first focusing on CO_2 separation from both syngas and flue gas, for gasification and combustion processes, respectively. Traditional absorption separation techniques using monoethanolamine are investigated, along with more novel polymer-based and inorganic-based membrane processes. The second part of the course will address transportation of CO_2 in pipelines and sequestration in deep underground geological formations. Using a real-world sequestration project, we will learn about all of the aspects of CO_2 transportation and sequestration in geological formations. Options for geological sequestration in oil and gas reservoirs, deep unmineable coal beds and saline aquifer are compared and assessed.

[1]Energy Information Administration, www.eia.doe.gov, May 2007.

516f D4: Drug Design, Development, and Delivery. A New Capstone Course for Chem E's Interested In the Bio-Industries

Mark Prausnitz and **Andreas S. Bommarius**. Chemical and Biomolecular Engineering, Georgia Institute of Technology, 311 Ferst Drive, Atlanta, GA 30332-0100

While the pharmaceutical industry often focuses on drug discovery and drug sales, there are many more aspects to the process of developing successful pharmaceuticals. Within the industry, aspects such as designing a scalable route for small molecules, desgining a robust process for proteins or vaccines, as well as manufacuring and delivery have long been neglected but have garnered much attention recently. D4: Drug Design, Development, and Delivery is a new interdisciplinary course at Georgia Tech taught annually by the authors and a colleague from Chemistry/Biochemistry. Cross-listed in the Schools of Chemical & Biomolecular Engineering, Biomedical Engineering, and Chemistry/Biochemistry and open to graduate students and seniors, the course teaches an introduction to the industry and the drug development process as well as basics of each of the three D's, design, manufacuring, and delivery. The latter part of the course deals with case studies of actual and failed pharmaceutical products, emphasizing one of the three D's. Teams of 3 students from a variety of majors and standings (U/G) prepare a study and present it to the rest of the class, followed by discussion and a statement on broader impacts by the instructors. Examples of the case studies include the testosterone patch, dorzolamide against glaucoma, leuprolide extended release against cancer, and insulin manufacture and delivery.

Remote vs. Hands-on Laboratory Experiences: What Works and What Doesn't

David DiBiasio[1], Jim Henry[2], Marina Miletic[3], and William M. Clark[1]. [1]Chemical Engineering Department, Worcester Polytechnic Institute, 100 Institute Road, Worcester, MA 01609, [2]Chemical Engineering, University of Tennessee at Chattanooga, 615 McCallie Avenue, Chattanooga, TN 37403, [3]Department of Chemical and Biomolecular Engineering, University of Illinois Urbana-Champaign, 206 Roger Adams Lab, Urbana, IL 61801

Background

Increasing enrollments, reduced access to laboratory equipment, and limited resources place new pressures on departments wishing to maintain a significant undergraduate lab experience. One possible solution to the problem involves incorporating remotely-operated lab experiences.

Ideally, remote experiments can be conducted at any time from any place. They are particularly useful for students at universities where resources are severely limited and there is no access to any significant experimental equipment. Knowing how to optimize learning in a remote lab experiment is critical, as is knowing the limitations compared to hands-on. The literature indicates mixed results about student learning using remote experiments. And, there are only a few studies of chemical engineering systems. Our goal was to understand student learning and attitudes in one remotely-operated lab experience compared to hands-on.

At WPI distillation is taught in the sophomore year throughout a project-based spiral curriculum. All projects are team-based. We introduce basic concepts early in the sophomore year then revisit distillation throughout the year with successively more complex assignments and projects. This includes at least two lab experiences. The first experiment uses a batch column operated at total reflux to introduce students to multistage distillation including efficiency and energy balances at total reflux. A follow-on course explores pressure swing distillation without a lab component. Near the end of the year, the batch column operated at a constant external reflux ratio is used again in a lab project. This project engages students in process dynamics and challenges them to compare differences between theory and reality using the Rayleigh analysis.

Large enrollments forced us to teams of 8-10 people (not optimal) or to run experiments during times when safety and lab monitoring because serious issues. Simultaneously with struggling to deal with this problem, we discovered the work at UTC involving remotely accessible experiments. In fact, the UTC distillation column is physically nearly identical to the WPI column but much more flexible in its' operation. A partnership was quickly formed with the idea of probing learning differences between identical remote and hands-on experiments.

Methodology

We conducted a preliminary, pilot study that would inform a subsequent larger, more rigorous investigation. This spring we recruited 7 volunteer WPI teams from the cohort enrolled in the final sophomore year course to do the remote-only experiment using the UTC column. We also had 7 different WPI teams run the identical experiment locally using the WPI system.

Remote-only and hands-on-only teams conducted identical experiments with identical assignments and project reporting requirements. Evaluation had three components with direct and indirect assessments, giving us some degree of triangulation. Both cohorts completed surveys managed by a third partner: UIUC. Course instructors compared final reports from remote and hands-on teams. An in-class quiz compared individual learning for students in the remote cohort compared to hands-on cohort in a quantitative manner. Qualitative measures include evaluating student attitudes about the experience and assessing students' improvement in ABET outcomes such as: (b) – design, analysis and interpretation of data; (d) – functioning on multidisciplinary teams; (g) – effective remote communication; (h) - have broad education necessary to understand the impact of engineering solutions in a global, economic, and environmental societal context; and (k) - using techniques, skills, and modern engineering tools (such as computers and web interfaces) necessary for engineering practice.

Results

Logistics

From the instructors' standpoint the logistics associated with running 7 teams in less than two weeks through either the local lab or the remote were roughly identical. The support team at UTC had to be ready to prepare the column for new runs on a daily basis, yet remote teams still had 24/7 access. Likewise, at WPI the support team had to cool down, recharge and restart the column on a daily basis (a 3-4 hour operation) to prepare for the following day's team. At least one hands-on team had to run during a weekend to complete their work. It would have essentially impossible to run 14 teams through the local WPI lab during the time allotted for the course. Hence, accessibility to the remote column literally kept us from dropping the experiment completely and going to something like a simulation instead.

Student Attitudes

Surveys from both cohorts were distributed to understand student perceptions about their hands-on and local experiences. Response rates were 52% for the local cohort and 75% for the remote cohort. There appeared to be little or no differences overall between the cohorts. In fact, for one item (improving your ability regarding ABET outcome b) 29% of the remote cohort selected "substantially new" compared to 18% for the local group. On the negative side, local students complained about a minor malfunction with an analytical instrument while remote students complained about minor server problems. In summary, we were pleased with these results. Remote students in general liked the 24/7 aspect and did not feel as if they had lost out on a valuable hands-on opportunity.

Student Learning

All reports were graded by the WPI instructor using the same rubrics. We evaluated presentation quality, analysis of results, and demonstration of appropriate concepts. Out of a possible 100 points, the average remote cohort score was 79.9 and the average local cohort score was 84.9. We gave an individual, in-class quiz about basic batch distillation concepts. Results showed an average score of 4.6 for the remote students and 4.9 for the local students, out of a possible 8 points. These results indicate that students who did the hands-on experiment, on the average, performed slightly better than students who did the remote experiment. However, as a class each cohort's performance was acceptable. These results are promising because they indicate that appropriately designed and implemented remote experiences can provide student learning at levels near hands-on experiences.

We completed a content analysis of the reports and quiz results to probe differences between cohorts regarding specific learning topics. Despite the differences in quiz scores there were no systematic differences between cohorts. For one item that asked about typical column transients, both cohorts scored similarly, indicating approximately equal knowledge regarding dynamic column behavior.

There were common deficiencies that were independent of remote or hands-on cohorts. These included a perplexing confusion of thermodynamic efficiency with stage efficiency; the typical mistakes with energy balance calculations; and another typical confusion over quantitative comparison of experiment to Rayleigh analysis. Students expect exact comparisons yet reality is never that predictable. In too many cases, the report presentation quality was barely acceptable. This result is not unusual given the end-of-semester time demands. However, remote students had access to archived data that could be easily and effectively displayed yet they persisted in printing pages of Excel files instead. In summary, it was difficult to extract any patterns or systematic learning differences between the cohorts. Most of the lab report scoring difference can be attributed to one particularly poor remote team, that most likely would have done equally poorly had they been in the hands-on cohort.

Our lessons learned in this first study included: running 7 teams through the WPI lab is difficult and requires 2-4 support staff to handle properly, while running 7 teams through the UTC remote lab is less difficult and requires one support staff with oversight by a faculty member. Student teams got unnecessarily frustrated with normal kind of issues. For hands-on teams it was use of simple specific gravity measurement while for remote teams it was server issues and software downloads. In conclusion the remote lab experience has great promise and can provide an acceptable learning experience compared to hands-on. WPI may introduce a simplified remote experience earlier in the year to better prepare students for this project, and future students will use Aspen BatchSep® simulations together with the lab experience to potentially enhance the learning experience.

The University Power Plant: A Readily Accessible Remote Learning Platform

Kerry Dooley* and F. Carl Knopf, Dept. of Chemical Engineering, Louisiana State University, Baton Rouge, LA 70803

*presenting author, dooley@lsu.edu

Introduction

Combined heat and power (CHP) or cogeneration is critical to the economic viability of many industries, including fuels, chemicals, plastics, paper, and ferrous and nonferrous metals. U.S. investment in new CHP is expected to remain >$10 billion/year, and there are many ongoing projects in existing CHP systems resulting in substantial dollar savings through optimal energy management schemes.

CHP systems provide an excellent platform for students to study thermodynamics, certain unit operations and process design principles, optimization, data reconciliation, and emissions control. Even better, many Universities have ready (albeit remote) access to such a system on their own campus. At LSU, we are opening our newer (commissioned 2005, 20 MW, $20 MM) cogeneration system to the world of engineering education through real-time data delivery using formats which can communicate with virtually any control/data acquisition system and then broadcast compressed data via the net. Easy to use (from Excel) educational modules have been developed which make use of these data to teach the above-mentioned topics in classes ranging from introductory Sophomore-level to capstone design.

This remote educational platform also offers students the opportunity to make real-dollar energy management decisions on a complex yet easily visualized system, where students must consider tradeoffs in cost minimization and operability. It can therefore become a unique tool for engineering economics education.

Experimental

We purchased OSISoft PI® software to interface to and collect data from various control systems (JCI Metasys, ABB Symphony) of the LSU Cogeneration Facility. The interfaces make use of both OPC and ODBC data transfer protocols. Data are then archived using a PI Server (version 3.4) installed in a commercial PC, transferred to Excel via PI Excel Datalink. Data are now being linked to the web using a web server running PI's RTWebparts (version 2.0) software, which is deployed with MS SharePoint. User-friendly displays were created using Visio and PI Process Book.

Results

A Visio-created schematic of the entire system showing real-time data is given in Figure 1 below.

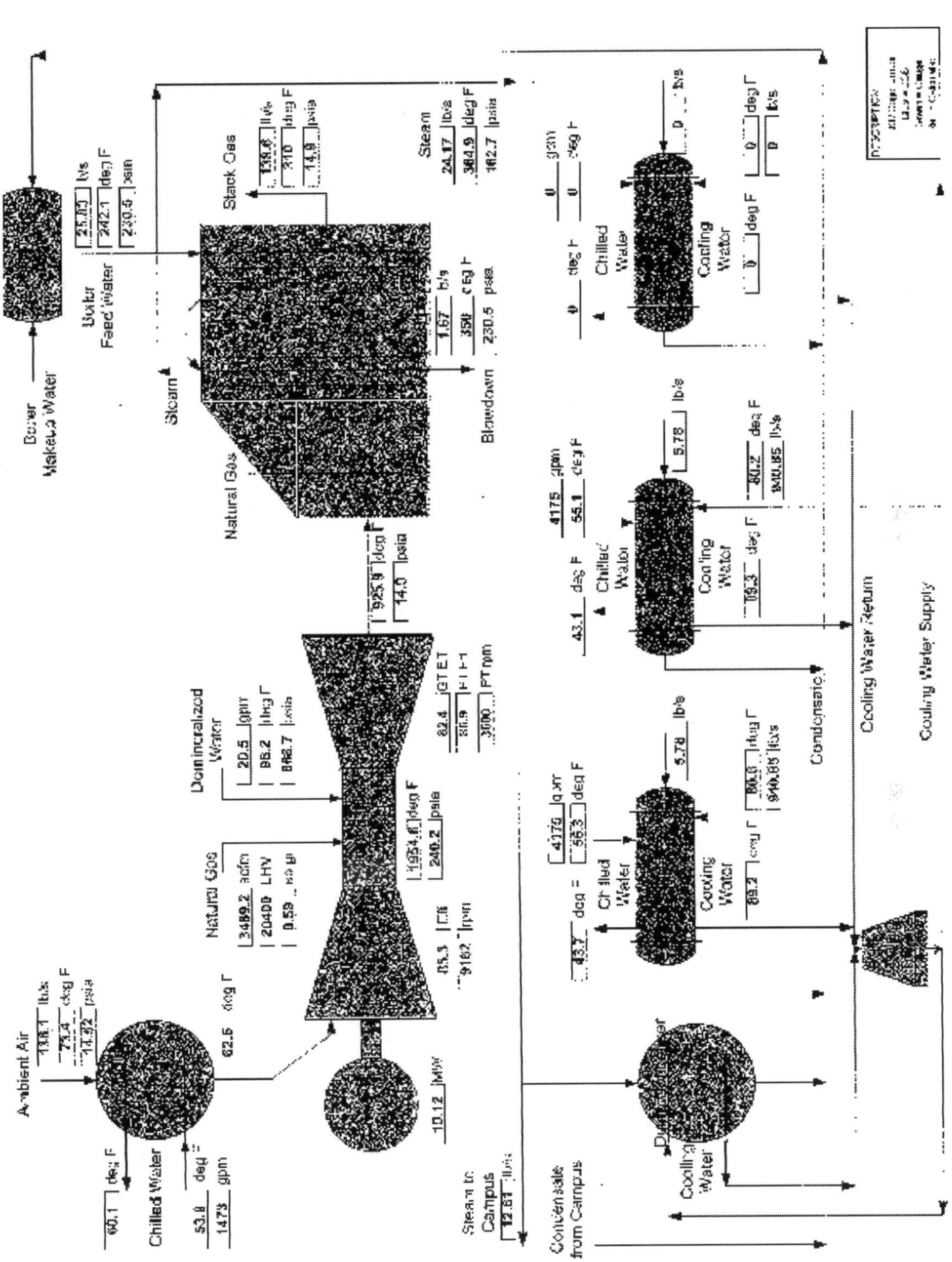

Figure 1. Schematic of LSU Cogeneration System

Educational modules completed so far include Excel-callable rigorous thermodynamics functions for all fluids (steam, water, air, methane, refrigerants, etc.) in the cogeneration system. This provides great flexibility for the students to solve cogen problems. These functions are based on the same open-source functions as used in the commercial GE code Gatecycle for cogen calculations.

1. Data Reconciliation / Gross Error Detection – these use real data from the cogen system to illustrate these concepts, and can be used in a numerical methods, lab or design course.

2. Design of Cogen System - here the non-detailed design of a complete cogeneration system is accomplished, at first using ideal gas and other standard assumptions, and then using rigorous thermodynamics. All is done using Excel so optimization and system improvement and alternatives can be quickly accomplished.

3.) Energy Dispatching / On-line Optimization - here a simplified version of the cogeneration system with data from all units for one-year are provided - students must fit data and then develop a plan to reduce energy costs. How utilities charge for electricity is explained, and then students must predict utility costs (easy to do once they understand how utilities calculate rates) and understand varying loads.

4.) Heat Recovery / Steam Generator (HRSG) – This key component of a cogen system is used to develop concepts in boiling heat transfer, unsteady-state heat transfer, and numerical solutions of ordinary and partial differential equations by finite differences. This module can be used in either Numerical Methods, Heat Transfer or Unit Operations classes.

For example, a typical problem might be **Determination of the HRSG Physical Design**. For the HRSG shown below assume the evaporator is 30 tubes wide and the economizer is 20 tubes wide. The tubes are: O.D. = 2 in..; wall thickness = 0.095 in.; length = 15ft.; tube spacing 4 in. square. Using a typical value for overall heat transfer coefficient based on the tube outside area of 12.3 Btu/hr-ft^2-^{o}F, determine the equivalent number of rows needed in the evaporator and economizer sections.

The equations are solved numerically by a simple finite difference method implemented in Excel. The grid design and resulting temperature profiles are shown below in Figure 2. The solution strategy and computation of the heat transfer coefficients are discussed by Knopf ("Energy Design and Optimization", manuscript in preparation, 2008). An Excel module showing the calculated temperature distribution for a few nodes of the economizer section is shown as Figure 3.

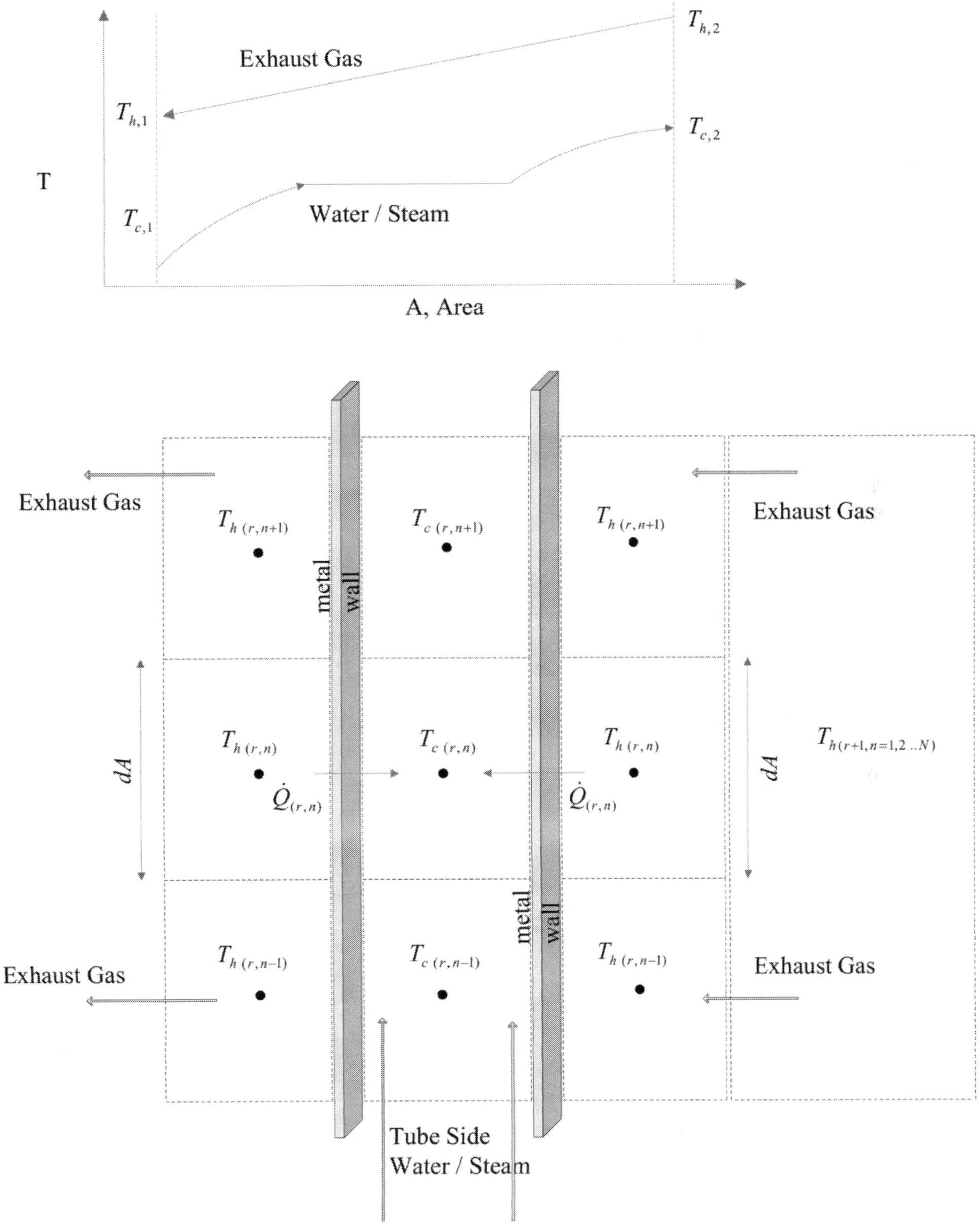

Figure 2. a.) temperature profile in the HRSG evaporator section b.) nomenclature for numerical analysis of a single tube in the HRSG (evaporator or economizer section).

Numerical Solution Economizer Tubes Temperature Profile

Tube Data — data input in green cells

Outside Diameter	2	in
Inside Diameter	1.81	in
Number of Tubes Wide	20	
Tube Spacing	4	in
Tube Length	15	ft
Number Tube Passes	200	
k_{wall}	25	btu/hr-ft-°F

Water Flow Data

$F_c = F_{water}$	111096	lb/hr
$T_{c(r=R,\ n=1)}$	77	°F
P_{water}	290.08	psia

Exhaust Flow Data

$F_h = F_{exhaust}$	840327.8	lb/hr
$G_h = G_{exhaust}$	16806.56	lb/ft²-hr
$T_{h(r=R,\ n=1)} \ldots T_{h(r=R,\ n=N)}$	251.524	°F

Finite Difference Data

dL	1	ft
number nodes / tube	15	

To Stack ← Exhaust Gas

	Tube Pass =	1			Tube Pass =	2	
Position	$U_{(r=1,\ n=1..N)}$	$T_{h(r=1,\ n=1..N)}$	$T_{c(r=1,\ n=1..N)}$	Position	$U_{(r=2,\ n=1..N)}$	$T_{h(r=2,\ n=1..N)}$	$T_{c(r=2,\ n=1..N)}$
1	20.21	251.52	77.00	1	20.24	254.09	81.96
2	20.21	251.52	77.33	2	20.24	254.09	82.29
3	20.22	251.52	77.67	3	20.25	254.09	82.62
4	20.22	251.52	78.00	4	20.25	254.09	82.95
5	20.23	251.52	78.33	5	20.25	254.09	83.28
6	20.23	251.52	78.67	6	20.26	254.09	83.60
7	20.23	251.52	79.00	7	20.26	254.09	83.93
8	20.24	251.52	79.33	8	20.27	254.09	84.26
9	20.24	251.52	79.66	9	20.27	254.09	84.58
10	20.25	251.52	79.99	10	20.27	254.09	84.91
11	20.25	251.52	80.32	11	20.28	254.09	85.23
12	20.25	251.52	80.65	12	20.28	254.09	85.56
13	20.26	251.52	80.98	13	20.29	254.09	85.88
14	20.26	251.52	81.30	14	20.29	254.09	86.21
15	20.27	251.52	81.63	15	20.29	254.09	86.53

Figure 3: Numerical Solution of Temperature Distribution for Economizer, Heat Recovery Steam Generator

Abstract #137613

In Class Experience Using the Web and a New Text for Teaching Undergraduate Mass and Heat Transfer

T. W. Fraser Russell, Chemical Engineering, University of Delaware, 150 Academy St, Colburn lab, Newark, DE 19716, **Anne S. Robinson**, Department of Chemical Engineering, University of Delaware, 150 Academy Street, Newark, DE 19716 and **Norman J. Wagner**, Department of Chemical Engineering, University of Delaware, 150 Academy St., Newark, DE 19716

In this presentation we describe a novel method for teaching undergraduate mass and heat transfer. The material is based on a new text (Mass and Heat Transfer, Cambridge Univ. Press, 2008) developed at the University of Delaware over the past 8 years. The text is organized into two parts and is a significant departure from the traditional courses in heat and mass transfer. The course and text is enriched by an interactive web site (www.mht.che.udel.edu) that can be used by students individually as well as by the instructor in class.

A competitive game on the web site requiring reactor model development and economic analysis allows us to effectively review reactor design. Student activities on the web site build upon this to develop interactive derivations of the key model equations. To illustrate the technically feasible design of mass contactors and heat exchangers, interactive role playing exercises are included.

Part I of the text we use in class analyzes and models laboratory, pilot, and commercial scale equipment for reaction, heat exchange and mass contact. The equipment is classified by the simplified fluid motions occurring in reactors, heat exchangers and mass contactors: mixed-mixed and mixed-plug tank-type equipment and plug flow cocurrent and countercurrent tubular equipment (Table 1). Part I clearly shows the critical role of experiment essential for evaluating parameters in the correlations for heat transfer and mass transfer presented in Part II. The model equations from Part I are applied to interactive technically feasible analysis and design on the web site.

Part II first examines molecular-level conduction and diffusion and relates this to the equipment-scale model equations. We next relate physical properties and detailed fluid motions to the equipment-scale analysis in Part I. State-of-the-art methods for estimation of fluid-fluid interfacial areas critical to the analysis and design of mass contactors are developed as a separate chapter. Finally detailed technically feasible designs for a heat exchanger, a countercurrent mass contactor, a staged liquid-liquid extraction unit and a fermentor are presented.

We teach the material in this manner so students first study equipment, which is of importance for planning and execution of experiments and for processing of material at the commercial scale. An additional benefit of this organization is that Part I and selected correlations from Part II can stand alone and can replace the traditional courses in fluid mechanics and heat and mass transfer. This allows new material to be introduced into a chemical engineering curriculum, for example, critical courses in the biological engineering and sciences. It could also serve as a stand alone course for non-chemical engineers to learn basics of transport phenomena and their application.

REACTORS SINGLE PHASE	REACTORS TWO PHASE	HEAT EXCHANGERS	MASS CONTACTORS
Single Control Volume	Two Control Volumes	Two Control Volumes	Two Control Volumes
Tank Type ***Mixed*** • **Batch** • **Semi- Batch** • **Continuous**	**Tank Type** *Mixed –Mixed* • Batch • Semi-Batch • Continuous *Mixed -Plug* • Semi-Batch • Continuous	**Tank Type** *Mixed –Mixed* • Batch • Semi-Batch • Continuous *Mixed -Plug* • Semi-Batch • Continuous	**Tank Type** *Mixed –Mixed* • Batch • Semi-batch • Continuous *Mixed -Plug* • Semi-Batch • Continuous
Tubular ***Plug Flow***	**Tubular** *Plug Flow* *Co Current*	**Tubular** *Plug Flow* *Co Current* *Counter Current*	**Tubular** *Plug Flow* *Co Current* *Counter Current*

TABLE 1
(reproduced from "Mass and Heat Transfer: Analysis of Mass Contactors and Heat Exchangers." Cambridge Univ. Press, 2008)

Abstract ID#: 137613
Password: 634495
Title: In Class Experience Using the Web and a New Text for Teaching Undergraduate Mass and Heat Transfer
Session Selection: 04007 Applications of Remote/distance Learning
Preferred Presentation Format: Oral
Lead Presenter's Email: twfr@udel.edu
Related Web Page: www.mht.che.udel.edu

Related to existing technology or advanced technology associated with the overall use and supply of energy or global climate change research? No

First author

Presenting Author
T. W. Fraser Russell
Allan P. Colburn Professor
Chemical Engineering
University of Delaware
150 Academy St
Colburn lab
Newark, DE 19716
Phone Number: 302-831-1714
Email: twfr@udel.edu

Second author

Anne S. Robinson
Department of Chemical Engineering
University of Delaware
150 Academy Street
Newark, DE 19716
Fax Number: 302-831-1048
Email: asr@udel.edu

Third author

Norman J. Wagner
Department of Chemical Engineering
University of Delaware
150 Academy St.
Newark, DE 19716
Phone Number: 302-831-8079
Email: wagnernj@udel.edu

523d Use of Distance Learning Tools as Part of a Long Distance Non-Traditional Industrial PhD

Jeffrey R. Seay, Department of Chemical and Materials Engineering, University of Kentucky, Paducah, KY 42002 and Mario R. Eden, Department of Chemical Engineering, Auburn University, 230 Ross Hall, Auburn University, AL 36849-5127.

Distance learning tools have gained increasing popularity in graduate engineering education. At many universities, students can earn coursework only Master of Engineering degrees without ever visiting campus. Traditionally though, these tools have been limited to the masters and undergraduate level. For students seeking a research based doctorate in one of the engineering disciplines, the traditional path, including a residency requirement has remained the only option. However, through the use of distance education tools, this traditional path does not have to be the only one.

This contribution will describe a unique, non-traditional approach taken by the Auburn University Department of Chemical Engineering to allow a student to utilize distance education tools to pursue a traditional PhD in a very non-traditional way. By utilizing video based coursework, and long distance collaboration tools, the student has been able to complete all the requirements for a doctorate, while only spending a minimal amount of time on campus, and continuing to maintain a career in industry.

This contribution will address the challenges of implementing a long distance, collaborative research project between industry and academia as part of an industrial Ph.D. dissertation. The research project is a collaboration between Evonik Degussa, Auburn University and the University of South Alabama. This work will outline the challenges of integrating all the groups into a cohesive project team and will address how the research group is structured, the challenges of organizing team members across multiple universities and industry groups in the United States and Germany and how the research is incorporated into a non-traditional, industrial Ph.D. dissertation for a distance education graduate student at Auburn University.

TriLab – a combined remote, virtual and hands-on laboratory as a novel reusable learning object (RLO) for Supporting Engineering Laboratory Education

M. Abdulwahed, Z. K. Nagy*

Chemical Engineering Department, Loughborough University, LE11 3TU Loughborough, UK

Introduction

Reusable Learning Object (RLO) is, in brief, a conceptual model of a shared vessel of teaching and learning content which is normally developed using IT tools. Attention have been paid recently towards using RLOs in educational processes for facilitating better learning and minimizing costs associated with improving teaching and learning. The importance of a laboratory experience in engineering education curricula has been emphasized in a large number of science and engineering education articles (Johnstone et al 2001; Hofestein et al 2004; Feisel et al 2002; Kirschener et al 1988; Ma et al 2006). Wankat observes that only 6% of the articles published in the Journal of Engineering Education from 1993-2002 had 'Laboratory' as a keyword (Wankat 2004). In response to the recent recommendations in the literature regarding the engineering laboratory education, we worked on developing an Online Laboratory Learning Object (OLLO) for enhancing the students laboratory experience, the OLLO was developed for the process control laboratory taught through various courses at the chemical engineering department of Loughborough University.

The Process Control Lab

The process control lab is a coherent part of the second year Instrumentation, Control and Industrial Practice module at the Chemical Engineering Department at Loughborough University. The experimental rig of the hands-on process control lab was designed to mimic a real surge tank system which is a typical chemical engineering process. The laboratory is a compulsory part of the module designed for undergraduate engineering master (MEng), bachelor (BEng), and bachelor in science (BSc) programmes in chemical engineering at Loughborough University. The lab aimed to introduce students to the principles of control engineering, such as the main components and instruments of a feedback loop, the concept of open-loop control, feedback control, proportional-integral-derivative (PID) control, and PID tuning. The hands-on laboratory consists of two 3 hours sessions, scheduled for two consequent weeks. In the first week the students were introduced to the elements of typical feedback loops such as sensors, actuators, controller, and process. The main objectives were of the first session are: calibration and hysteresis of the level sensor; Calibration, hysteresis, installed characteristics and relative resistance of the control valve. During the second week, students were introduced to control engineering concepts. The aim of the experiments in the second week is to help students appreciating the advantages of automatic control compared to manual operation, and to equip the students with qualitative evaluations of the differences among proportional (P), proportional-integral (PI), and proportional-integral-derivative (PID) controllers. In general, all students were asked to prepare for the lab by reading the lab manual before the session and by downloading the virtual laboratory software.

* Corresponding author (z.k.nagy@lboro.ac.uk)

TriLab and the Online Laboratory Learning Object (OLLO)

There has been recent shift in engineering education towards embracing constructivist pedagogy and experiential learning practices. There is more demand on supplementing the theoretical lectures in the engineering courses with authentic real applications, i.e. laboratory demonstrations. Most of the previous obstacles can be overcome by developing a self contained entity that includes instructions and information about the experimental rig, the hardware operation, the experiment purpose, brief background theory, simulation of the rig, and experimental procedure for the sake of learner-centered approach of conducting the experiment, and furthermore a remote operation capability of the hands-on lab for the sake of sharing the whole entity and the lab rig among different institutions which could result in dramatic drop of setting up new labs and will considerably enrich the engineering pedagogy by embedding new laboratory resources that would have been possible to access. We call such entity the "Online Laboratory Learning Object", or in brief OLLO. In definition, the Online Laboratory Learning Object (OLLO) is a learning object that is particularly designed for the laboratory pedagogy and is characterized by the following:

1- *It is a learning object which includes self contained learning content related to the hands-on laboratory experiment that enables learner-centered approach of learning the experiment.*
2- *It is incorporating remote operation of the physical hands-on lab rig, mainly through the internet.*
3- *It is designed to mimic a relevant hands-on laboratory when operated offline by using virtual instrumentations.*
4- *It is preferred that OLLO would include video transmission of the hands-on test rig.*

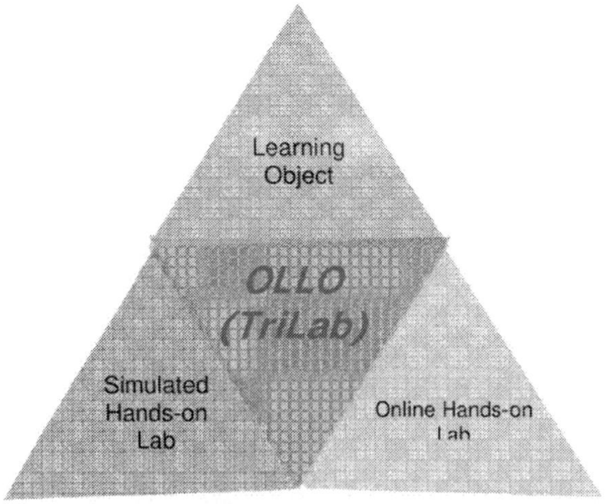

Figure 2. Conceptual model of the TriLab as an OLLO.

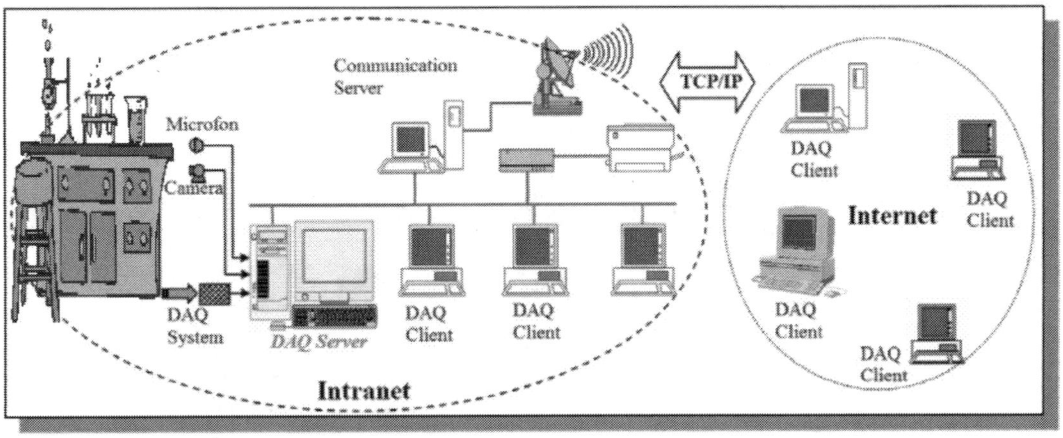

Figure 3. Core Communication Architecture of the OLLO.

The OLLO without remote operation of the physical instruments is a learning object LO only and is NOT an OLLO. Conceptual model of the OLLO is shown in Figure 2, the core communication architecture of an OLLO is shown in Figure 3. The benefits of developing OLLO's for as much labs as available within the institute is beyond the teaching and learning process, since collaborative research could be significantly fostered when such large database of OLLO's is easily accessible. For further enhancing the collaborative part of an OLLO, video conferencing and editing tools can be added on the top of the OLLO. Laboratories are often developed for providing the students with authentic real experience, if we restrict the OLLO for simulation only as the case of LO's, the developed laboratory learning object will loose the most important motivation behind labs tuition, which is realism.

We used LabView as major tool for developing an OLLO for the process control lab that is used for many courses tuition in the chemical engineering department. We developed a virtual simulated version of the laboratory, which allows student to perform all experiments in a simulation mode using an interface identical with the real operator interface in the lab. A remotely operated version of the lab was also been developed and used in the classroom to illustrate the theoretical concepts on real-life experiments. The OLLO allows remote operation and provides real-time video transmission for creating the feeling of telepresence. The DAQ hardware device we used is a NI USB-6000 series USB data acquisition (DAQ) device worth about 150£. Part of the process control lab OLLO GUI is shown in Figure 4. The developed OLLO has been deployed on the web for internal use only right now but a simulated version can be downloaded and installed as stand alone application from http://www-staff.lboro.ac.uk/~cgzkn/).

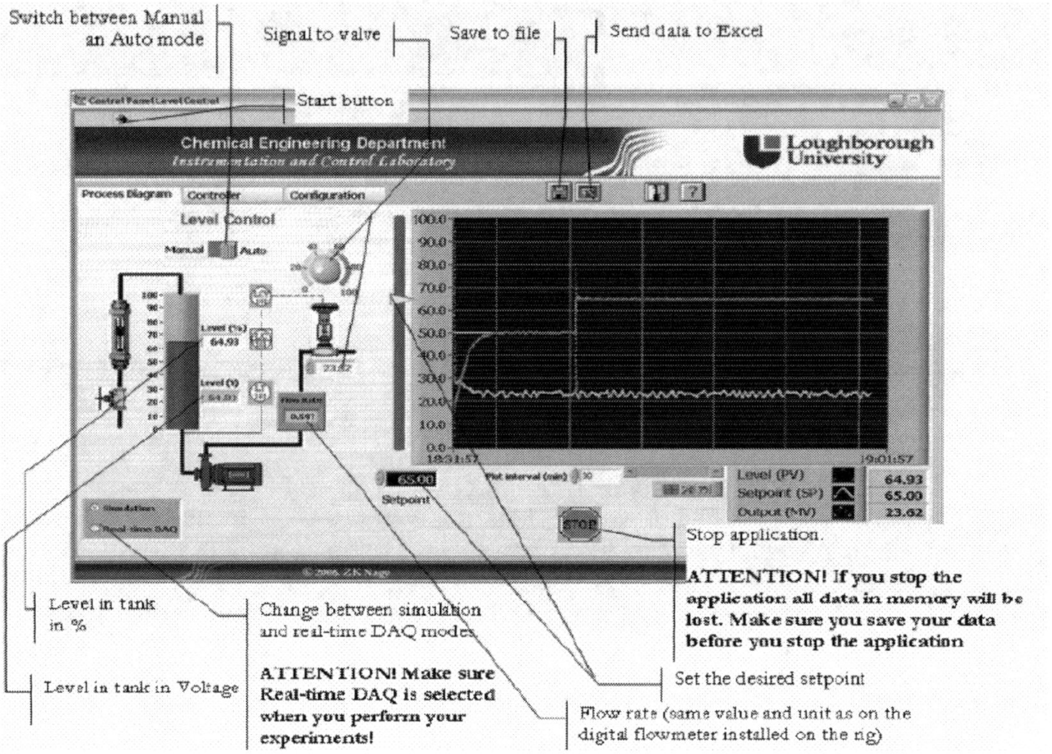

Figure 4. The Process Control Lab OLLO GUI

Educational Utilization of the Process Control Lab OLLO:

The P.C.L OLLO has been used in the master module "Advanced Computational Methods for Modelling and Analysis of Chemical Engineering Systems". The Module aims to introduce students into topics such as dynamic modelling, optimization, PID control, which are applied to chemical process. In this course. We used the P.C.L OLLO remote operation property in the course exam.

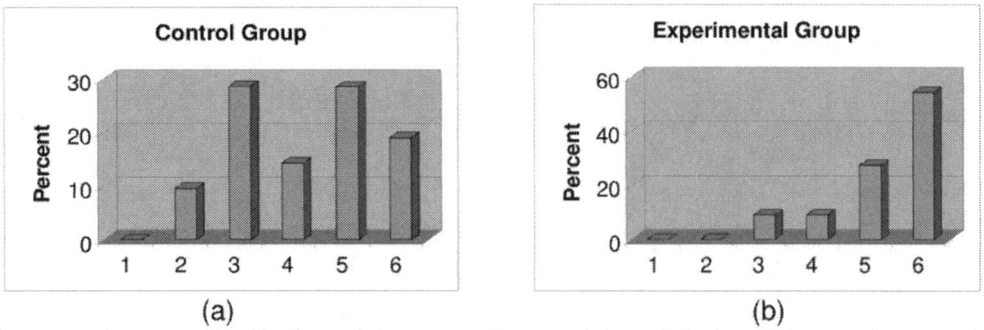

Figure 5. Answer distribution of the control group (a) and the experimental group (b)

The P.C.L OLLO was also used in a pilot study for the "Instrumentation and Control" course taught for second year students at the chemical engineering department,

Loughborough University. In this course, the OLLO was used in pre lab session for getting the students prepared before they conduct the hands-on experimentation. Some students done the preparation session with the OLLO and some did not. In surveying the students will of conducting extra experimental work, we found that those who did the preparation have got more motivation towards experimenting further ideas after the hands-on session facilitated by remote connection character of the P.C.L. OLLO. The responses of the two groups differed considerably; the average of the control group is 4.19/6 while the average of the experimental group is 5.27/6. Figure 6 shows the answers distribution of both groups (Control group are those who did not use the OLLO).

Furthermore, we found that the experimental group students have got higher conceptual understanding of the theory behind the experiment. We conducted post test quiz right after the lab for the students, the quiz contained many questions with emphasize on the conceptual and the hands-on experience that the students were expected to gain after conducting the lab. The means average in general was higher for most question of the experimental group students. The Hypothesis statistical test revealed a higher difference significance factor for the conceptual questions.

References

Feisel LD, and Peterson GD 2002. A Colloquy on Learning Objectives For Engineering Education Laboratories. ASEE Annual Conference & Exposition: Vive L'ingenieur!, 16-19 June.2002, Montreal Canada.

Feisel, L D, Rosa, A, J, "The role of the laboratory in undergraduate engineering education", J. Eng. Educ., 94, No 1, 2005, pp. 121-130.

Hofstein A, and Lunetta VN, 2004. The laboratory in science education: Foundations for the twenty-first century. Science education, 88, (1), 28-54.

Johnstone AH, and Al-Shuaili A, 2001. Learning in the laboratory; some thoughts from the literature. University Chemistry Education, 5, 42-51.

Kirschner PA, 1988. The Laboratory in Higher Science Education, Problems, Premises, and Objectives. Higher Education, 17, (1), 81-90.

Ma J, and Nickerson JV 2006. Hands-on, simulated, and remote laboratories: A comparative literature review. ACM Computer Survey , 38, (3), Article 7.

Wankat PC, 2004. Analysis of the first ten years of the Journal of Engineering education. Journal of Engineering Education, 93, (1), 13–21.

Development of Remotely-Controlled Fluidized Bed Reactor for Distance Learning

Dmitry A. Sladkovsky, Nikolay V. Kuzichkin, Nickolay V. Lisitsyn
The Resource Saving Technologies Department
(St. Petersburg State Institute of Technology)
26 Moskovskiy Prospekt - St. Petersburg - Russia

E-mail: slatech@mail.ru

1. Introduction

Remote laboratories are an increasingly popular innovation in engineering education and they offer an alternative learning environment to traditional laboratories. Such remote laboratory is developing in the Resource Saving Technologies department of Saint-Petersburg State Technological University. Distance educational complex called «ResTech Remote Lab (RTRL)» is based on a hands-on laboratory-scale process unit. RTRL provides a wide range of educational topics such as hydrodynamic and chemical reaction research, process control and optimization, adjustment of measuring and regulate equipment. Lab work is organized on the basis of internet technology which allows conducting remote experiments from any place without installing additional software on students' computers.

Fluidized bed reactor is the main process element (equipment) of the remote laboratory unit. Fluidized bed reactors are widely used in the chemical industry and are essential to the production of key commodity and specialty chemicals such as petroleum, polymers and pigments.

One of the most significant advantages of the laboratory process unit is that hydrodynamic condition of the catalyst bed is simulated by the physical model while chemical reaction in the reactor unit is simulated by the virtual model on the computer. The virtual model is real time calculated with the aid of the measured hydrodynamic bed condition. Therefore we can use any chemical reaction as an educational example of any chemical process.

2. Remote controlled process unit

The laboratory unit consists of a three-dimensional fluidized bed reactor and a two-dimensional (flat) reactor. An ordinary atmospheric air is used as a fluidized agent which at the same time is the only process feed. The laboratory unit is a completely safe process unit because in the catalyst bed there are no chemical reactions and no really high pressure or temperature occurs in the process. The structural scheme of the process unit and the photo of the remote laboratory are given in figure 1 below.

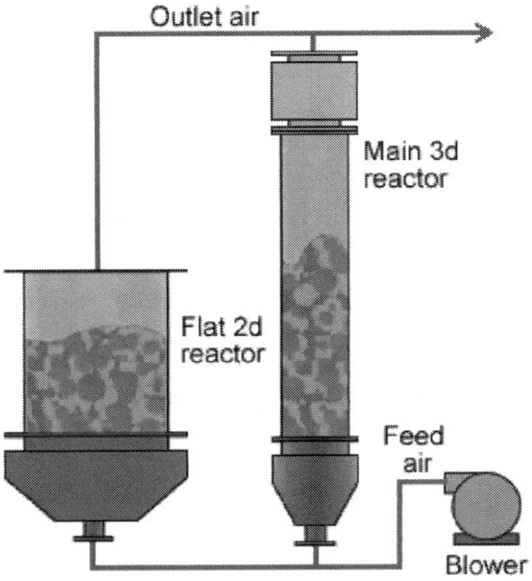

Fig. 1. Structural scheme of the process unit and the photo of the remote laboratory.

The laboratory unit is equipped with a pressure and flow measuring instrument, electrical phase inverter, high resolution video camera and TOSHIBA distributed control system (DCS). DCS has a feed-forward connection with general educational software RTRL for laboratory remote control and hydrodynamic bed condition research. The structure of the distance education complex is represented in the below-given figure 2.

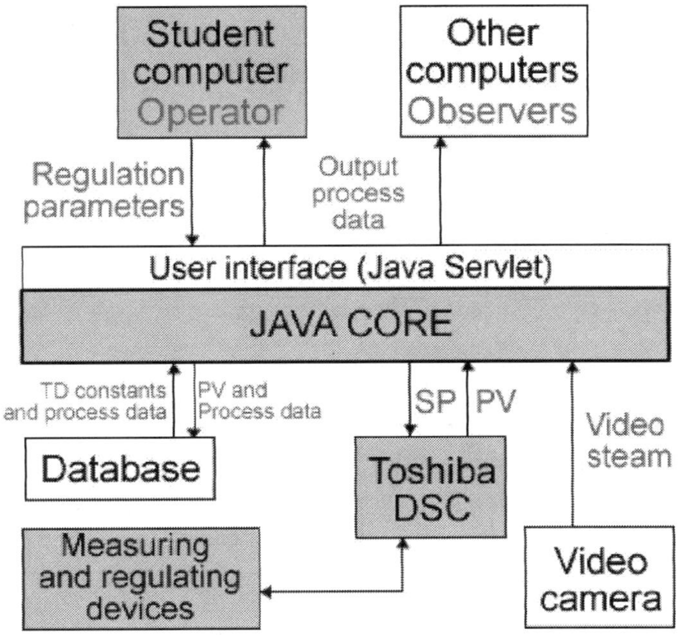

Fig. 2. RTRL structure.

3. Fluidized bed and chemical reaction modeling

Fluidized bed increasing catalyst process in the reaction zone by using internal surface of catalyst (diameter less then 0,04 in). Pore-diffusion resistance is practically reduced. At the same time the chemical reaction is progressing in isothermal condition.

The bubbles of different size are formed in the catalyst bed. Therefore some raw passes through the catalyst bed without physical contact with catalyst fraction. There is correlation between pressure pulsation and bubble size. If the average bubble size is known it is possible to calculate phase-boundary equilibrium coefficients. The bubbles size is researched on additional two-dimensional reactor. Size and distribution of the bubbles are calculated with the processing video frames which video camera takes. That provides to obtain correlation between pressure pulsation and bubbles size which are used in chemical reaction computer model. Therefore the chemical reactions in the main reactor are simulated on the base of actual hydrodynamic bed condition with the pressure pulsation measuring. Some examples of the bubble distribution in two-dimensional reactor are given on figure 3. Video capturing and real time processing are carried by the software developed on JAVA computer language (see Figure 1). Calculated data are used in chemical kinetic model.

Chemical reactions calculated on the base of measured pressure pulsation and gas velocity are in real time mode. It is possible to simulate any chemical reactions which kinetics are known. In the present time RTRL chemical reactions base consists of sulphur dioxide oxidation, butane dehydrogenation and methane oxidation. Chemical reactions are dynamically simulated in two phase model.

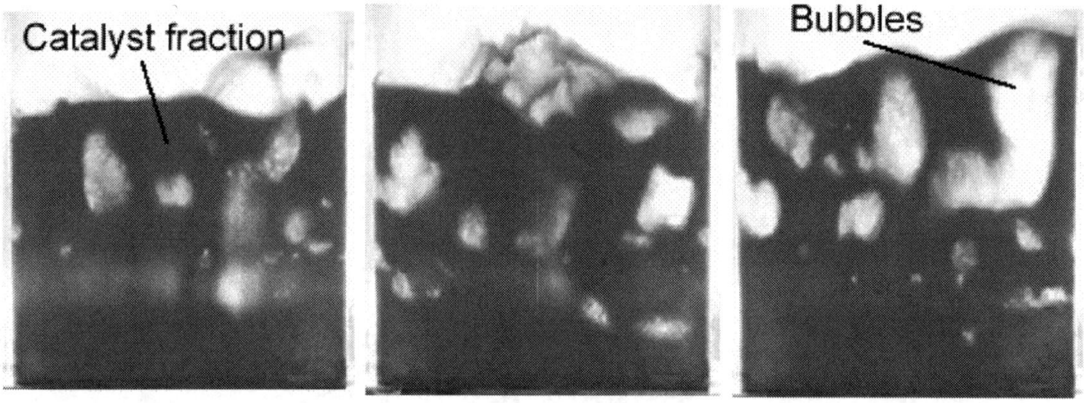

Fig. 3. Some examples of babbles distribution in two-dimensional reactor.

4. Remote control

Laboratory unit is equipped with TOSHIBA distributed control system. In the non distance control mode the regulators set points values are entered from the operator control panel and all the process is displayed on operator monitor. In remote control mode DCS receives the regulators set points values from the RTRL JAVA core. DCS exports present values and other measured parameters to the data base in the real time.

This data exchange provides feedback connection between internet server application (JAVA servlet) and conventional industrial DCS. Server application is internal part of general JAVA core. Students do not install special program for remote education, because RTRL software is developed on JAVA language. All remote lab work is carried out in internet browser (for example Internet Explorer, Opera, Firefox).

5. Distance education process

Remote education is organized entirely through the internet. In the same time only one user has operator function. Other connected users are observers. There are several types of lab work, which are developed on this laboratory unit.

Process startup and control work. First of all the students must research parameters connected with hydrodynamic bed condition (velocity of bed fluidization, distribution and size of bubbles). The next step is to reach technological process rate which provides to produce specify product flow and specify key component concentration in virtual model. The regulate parameters are feed flow ratio and heat exchange mode. The lab work report is automatically generated after the work in Excel. The report consists of PV-parameters trends, regulate parameters and summary table. The work is organized with an operator internet interface.

Fluidized bed hydrodynamic research. The work purpose is to find the correlation between pressure pulsation and bubble size for fluidized bed. The lab work carried out on the two-dimensional reactor using video researching of catalyst bed condition. RTRL program captures the video from video camera and automatically cuts it into frames. Bubble size is calculated by frames processing. Students carry out series of experiments with different pressure pulsation level and line gas velocity speed. The lab work report consists of source data for calculating the correlation out of RTRL program.

PID controller adjustment. The work purpose is to find the optimal regulate parameters (K, Ti). To adjust the line gas velocity regulator we use current frequency in blower electrical line as control parameters. To adjust the virtual feed temperature regulator we use heat exchanger bypass flow rate as control parameters.

6. Remote laboratory additional ability

We plan lab works connected with using of different nozzles. They decrease bubble size in fluidized bed and set better hydrodynamic condition for the chemical reaction. We also plan lab works to research interphase mass exchange coefficient by using gas tracer, for example helium.

532a Learning to Design Experiments - An Integrated Approach

Joseph J. Biernacki[1], Pedro E. Arce[2], Mario Oyanader[2], Ileana Carpen[2], and Donald P. Visco[2]. (1) Tennessee Technological University, Department of Chemical Engineering, P.O. Box 5013, Cookeville, TN 38505, (2) Chemical Engineering, Tennessee Technological University, Department of Chemical Engineering, P.O. Box 5013, Cookeville, TN 38505

Learning to design experiments cannot easily be achieved in one or two terms, rather, becoming an experimentalist takes time and practice for most students. The faculty of the Department of Chemical Engineering at Tennessee Technological University (TTU) have developed an integrated lab-lecture curriculum that introduces lab in the sophomore year, fuses together the lecture (theory and computational studies) with observation (experimentation and real-world experiences) and offers a more graduate progression in the learning experience. Guided inquiry begins as a sophomore, with increasing levels of independence expected, nurtured through various design related activities such as the "experimental prototype." The experience culminates with an open-ended capstone laboratory.

532b Use of a Generalized Rubric for Teaching Students to Design Experiments

Claire Komives, Chemical and Materials Engineering, San Jose State University, 1 Washington Sq., San Jose, CA 95192-0082

While traditional cookbook experiments provide opportunities to students for learning laboratory techniques and the operation of equipment, they are ineffective at teaching students to design and execute an experiment. Student-driven inquiry labs are becoming more common, but the logistics of enabling students free reign to implement a focused, unique experiment can be burdensome for faculty. As a way to enrich the biochemical engineering laboratory curriculum at SJSU, students were assigned the task of designing an experiment of their choice during two of the 5-hour class periods. Students were given a rubric for the preparation of a short proposal of the experiment they had selected, and once the proposal was reviewed and accepted, they could then carry out their experiment. Experiments in the lab ranged in creativity from mere variations of the regularly assigned modules to altogether different procedures. Together with a small group of faculty from different engineering disciplines and building on the first experience with the lab, a focused rubric was designed for the students and was tested in different labs in the college. The presentation will summarize the teaching and learning experience, logistics of lab organization, and a brief description of the rubric.

532c Simulation In the Unit Operations Laboratory: Comsol Multiphysics® Model of a CO_2 Absorber

William M. Clark, Yaminah Z. Jackson, and Giacomo P. Ferraro. Chemical Engineering Department, Worcester Polytechnic Institute, 100 Institute Road, Worcester, MA 01609

We are using COMSOL Multiphysics® to simulate the process equipment in our unit operations laboratory and testing the hypothesis that using this software to solve momentum, energy, and mass balances and to visualize the resulting velocity, pressure, temperature, and concentration profiles within the equipment will provide an improved learning experience. This presentation will focus on simulation of a gas absorption experiment.

The increasing concern over greenhouse gas emissions and global warming makes our CO_2 absorption experiment more relevant now than ever. Our old steel absorption tower had become clogged with rust and residue from years of use with sodium carbonate solution as absorbent, however. To continue offering a CO_2 absorption experience but to reduce cost and avoid column fouling in the future, we chose to use pure water as absorbent in our new smaller glass column packed with glass Raschig rings. Although using water does focus the lab on mass transfer concepts without the added complexity of reactions, the limited solubility of CO_2 in water makes it necessary to have accurate analysis of the gas phase and to work with concentrated gas streams to get good results. A Rosemount Analytical, Inc. infrared analyzer provides accurate and reliable gas phase analysis and although determination of mass transfer coefficients is more complicated for concentrated systems than for dilute systems we will show that modern computing environments like MATLAB® can be used to integrate complex functions and easily back out mass transfer coefficients from laboratory data.

We will also show that a simple simulation of our column using COMSOL Multiphysics® finite element software can be used to gain insight into mass transfer fundamentals. In our simple convection and diffusion model, CO_2 disappears from the gas phase and appears in the liquid phase at the same rate according to a "reaction" described by an overall mass transfer coefficient and appropriate driving force. The parametric solver feature of COMSOL allows straightforward determination of mass transfer coefficients that describe the experimental results and agree with those obtained by traditional analysis methods and published correlations. Using built-in plotting capabilities, the composition of the vapor and liquid phases and the driving force for mass transfer can be observed as a function of position in column. Further post processing features allow determination of fluxes and flow rates in and out of the column by boundary integration. A time dependent analysis can also be made to illustrate the column performance as a function of time. With this tool, students can perform "what if" analysis and design experiments and/or scaled up columns with ease.

A more detailed simulation of the absorption process gives even more insight into mass transfer fundamentals. In a more detailed convection and diffusion model, the column is assumed to contain a large number of glass rods with a thin layer of liquid flowing down surrounded by a gas stream flowing up. Equilibrium is assumed according to Henry's law at the gas liquid interface and realistic values are input for liquid and gas phase diffusion coefficients. The two film theory of mass transfer is illustrated by including a non-flowing layer of finite thickness on either side of the interface. By adjusting the effective diffusivity in each of these stagnant layers the experimental results can be reproduced by the model. Moreover, it can be shown that the effective diffusivities divided by the film thicknesses correspond to the individual gas and liquid side mass transfer coefficients that describe the process via traditional analysis methods. Using built-in plotting capabilities, students can plot the CO_2 concentration in mol/m^3 at any point in the column and observe the relative magnitude of the resistances to mass transfer given by the gas film, the interface, and the liquid film.

In future implementations of our unit operations laboratory course, we plan to evaluate any improvement in student learning and attitudes afforded by these simulations as we have already done with simulations of a heat exchanger, a gas permeation membrane, and a fluid flow experiment. We anticipate that use of these simulations will provide students with a clear understanding of the often confusing concepts of mass transfer fundamentals.

532d Using Computational Methods to Design Experiments

David A. Gallagher, CAChe Research, 13690 SW Otter Lane, Beaverton, OR 97008

The design of an experiment is typically limited by the experience and knowledge of the designer and the available equipment and chemicals. This presents a barrier to exploring breakthrough concepts or areas that are outside the experience of the designer or the practical limitations, and so limits creativity and learning.

The ability to run fast simulations of physical and chemical properties of materials on a standard laptop computer, can open up new avenues of exploration. The student can now quickly explore and test new ideas for feasibility before designing new experiments, and even run some experiments on the computer itself.

This talk will illustrate the use of some fast computational tools for property prediction (based on quantum chemistry, statistical thermodynamics, and QSAR, etc.) that can offer new insights and novel information for the student. Released from the constraints of their limited experience and the available chemicals and hardware, these new tools enable the students to more fully explore their own creativity, test out their own ideas, and gain valuable new insights.

Adding Equipment to make E-85 in the UO laboratory

Extended Abstract

Laboratory and design courses bring students closer to real world projects than in any other part of the engineering curriculum. However, most of our departments laboratory courses use well defined systems that have mathematically predictable outcomes, whereas, most of the industrial systems are much more complex. Graduates of our program sometimes had problems extending their education to these much more complex systems. Furthermore, while the experiments serve to reinforce classroom topics, none of our past Unit Operations Laboratory experiments connected students with the outside world or leant themselves to economic analysis. To address these concerns, the McKetta Department of Chemical & Bioprocess Engineering modified its existing methanol / water distillation experiment to produce E-85 grade fuel from grain. Additional equipment was purchased to enzymatically convert the starches to sugar (mash), ferment the sugar into alcohol, and dry the distilled ethanol. These enhancements cost well under $5000 and can produce over 1 gallon of E-85 per batch. The benefits include an increase in student's interest, the inclusion of a much more complex system into our laboratory mix and a direct connection to trends in modern society. In addition, this laboratory may help performance on outcomes associated with the departments learning objective on both "Social Awareness and Responsibility" and "Problem Solving Ability".

Introduction

Students in the Chemical Engineering degree program at Trine University are exposed to several different unit operations laboratory experiments. These experiments serve several functions including reinforcing concepts that the students have studied in class, introduction to real life technical problems and reinforcement of both written and verbal communication. However, a few weaknesses have been uncovered as part of the assessment system in regards to the laboratory curriculum.

- All of the experiments involved relatively simple well defined systems. While these simple systems are valuable in reinforcing concepts studied in class, they are not typical of our recent graduates work experiences. Typically the systems that the students must face upon graduation do not lend themselves to such well defined analysis. Feedback from employers and members of the Industrial Advisory Board indicated that recent graduates had difficulties at applying analytical reasoning to these more complex and less well defined systems. Furthermore these former students, when faced with complex systems frequently relied on custom and hear say and ignored applying any analytic or statistical principles.

- None of the experiments in the core curriculum involved fuels or were of concern to popular society. Because Trine University is located in the Corn Belt and students are certainly familiar with the push to "grow" more of our fuel. With the rising cost of energy, it seemed timely and appropriate to add a fuels component to the laboratory.

- With the exception of water, the previous experiments do not involve substances that the students are already familiar with. For instance, one of our experiments involves the extraction of propionic acid out of kerosene. While this is a nice demonstration of countercurrent multistage liquid - liquid extraction, students don't know why they would want to extract propionic acid out of kerosene. On the other hand, they know exactly why they would want to make ethanol. Having some familiarity with the products from the experiments helps the students make the connection to the "outside world".

- None of the laboratories involved a usable product and therefore the products from the laboratories were discarded. While these "products" are non-hazardous, this is certainly not the goals of "green engineering" and differs from the typical industrial situation. The ethanol product can be used for powering the basic utility vehicle that the Mechanical Engineering students work on. The spent grain from the mashing step of the process can be used for supplemental livestock feed. Yeast from the primary fermentation step can be used as the starter culture for subsequent fermentations. Molecular sieves can be regenerated using heat and vacuum.

- There was little emphasis in the laboratories on society and the source of raw materials or products. The raw materials for ethanol production are common commodities from well publicized international markets and are directly tied to the state of the world economy. This further ties the experiment to the real world. It is an opportunity to talk about such diverse topics as grain and future markets and the societal impact of using a food source for a fuel feed stock. The cost and efficiency of gathering the grain compared to other fluid sources of fuel andthe possibility of making ethanol from other sources such as cellulosic sources.

- Most of the laboratories dealt with substances that cannot be easily communicated with non-technical individuals. Prospective students and their parents typically did not relate to the previous experiment such as separating methanol from water or pipe friction. This same group becomes very interested when you bring up making fuel from grain sources.

- Most of the experiments were not concerned with economics and completing both the material and energy balances. By making a usable product like E-85, it is possible to calculate the analyze both the cost to make a gallon of the fuel but also how much energy it takes to get a unit energy of liquid fuel. Students can further analyze where they are using the most energy and suggest ways to reduce the energy usage and increase the efficiency of the conversion. Recovery for each of the processes can also be analyzed.

- While more biology has been added to both the required and elective curriculum, there was no biochemical engineering component to any of the unit operations laboratories. By adding, the mashing and fermentation steps to the UO lab we have added both enzymatic conversion and yeast fermentation.

- All of the previous unit operations were continuous in nature. No transient state or batch processes were studied. However many of the processes that are graduates faced were either transient of batch wise. The nature of the mashing, fermentation and adsorption is batch. In addition, the distillation column operation was switched from steady state operation to a batch process.

World class ethanol production facilities are currently under construction in the region so plant trips to these facilities enhance and re-enforce the conversion and separation processes. Students get to see how concepts and principles used in the laboratory are applied in world scale plants. Furthermore the plant trip becomes more meaningful.

Equipment:

Ethanol production can be divided into several discrete steps or unit operations. These steps include:

- Mashing - Conversion of starch into fermentable sugars
- Fermentation - Converting the fermentable sugars into ethyl alcohol
- Separation by distillation and drying of the ethanol using adsorption on molecular sieves.

The department purchased a 20 gallon mashing system. Since this equipment is typically purchased for private beer making the system is relatively inexpensive. The total cost for a largely automated and instrumented stainless steel system was just over $3000.

Fermentation was conducted in 5 gallon water jugs. These are very inexpensive and are easy to relocate and clean.

The department had a six stage, 6 inch diameter distillation column. This was run in continuous mode to separate methanol and water. As was mentioned previously, this column was utilized in batch mode to make the water ethanol separation. The composition of the distillate off the top of the column was close to the azeotropic composition.

The department purchased molecular sieves to dry the ethanol from the distillate composition to near 100% ethanol. The molecular sieves are re-generated by using a vacuum oven that the department had in storage.

Feedback from students indicated that a dramatic increase in student interest in the laboratory. Examination data indicate an increase in statistics and statistical reasoning. However these results need to be monitored to see if a long term shift has occurred. Plans are being made to assess if the ability of students to analyze complex systems and problem solving has been improved. Longer term measurements from employers and the Industrial advisory board have not been obtain but are planned.

In conclusion, with a relatively small expenditure we believe we have dramatically improved our Unit Operations laboratory experience. The current E-85 manufacturing takes raw grain and turns them into a liquid fuel. A dramatic increase in student interest in this laboratory has been observed.

Teaching a Bioseparations Laboratory: From training to applied research

Daniel Forciniti
Chemical and Biological Engineering Department
Missouri University of Science and Technology
Rolla MO 65409

Introduction

One of the problems in the teaching of many laboratories is the lack of direct participation of students in the experiments. This trend has become more acute as the instrumentation used in laboratories has become more and more expensive. Their lack of direct participation in the experiments has not precluded us to expect them to produce meaningful data in spite of their lack of experience. We start evaluating the students during their basic learning process increasing tensions in the classroom and leaving little room for errors.

A widespread practice in the teaching of laboratories is the generation of a laboratory guide that the student follows step by step. The guides are often quite comprehensive and they leave little room for improvisation. After completing their work, the students write a report following a pre-established template that often defies the minimum rules of good writing. At the end of the semester, the student moves happily into the next set of courses. After graduation, the student cannot function in a laboratory setting because she was never trained to function in one but rather to follow a series of commands. We strongly believe that such a deficiency needs to be addressed. Poor communication between the basic sciences and our discipline places that burden on us.

The students at the Chemical Engineering Department of Missouri University of Science and Technology (formerly the University of Missouri-Rolla) take a series of laboratory courses in the Chemistry, Biology and Chemical Engineering Departments. The Missouri S&T Chemical Engineering curriculum has a Biochemical Engineering Emphasis program in which the unit operations laboratories have been replaced by a Biochemical Separations and a Bioreactors Laboratories. In addition, the students take several Biology classes including molecular genetics, general biology and microbiology; two of them, at least, have laboratory sessions. I have been teaching the Bioseparations Laboratory for the last 15 years. It did not take me long to realize that the ability of the average student coming into my class (normally first semester seniors) to function in a laboratory setting is, at least, questionable. Most of them are not able to use a balance properly, do not know how to operate a micro-pipette, but more importantly, do not know how to "move" or function in a laboratory.

The difficulties that the students have functioning in a laboratory may be difficult to quantify but it is very easy to find a few examples here and there. Once, a young student came to see me in distress. There was some fear in her eyes. I sat down ready to listen some disastrous news. "I needed to use the small centrifuge but when I tried it did not work properly. The centrifuge was moving very fast along the bench; I was afraid it was going to fall down so I stopped the experiment", she said. I sighed in relief. Nothing has been broken and the students were in one piece. The case of the stubborn balance happened a few years after. A student was (trying) weighing potassium phosphate on a top load balance to prepare a buffer. I was in the vicinity when I heard: "This balance is definitely not working!!!". Oh Oh, I thought, here it goes my budget for the semester. As I asked him what was the problem he began demonstrating to me how he was adding a lot of material to the weighing dish but the numbers in the digital display did not change. When I noticed that he did not remove the balance cover I couldn't but laugh. More tragic (and costly) are the story of constant pH pH-meter caused by the smashing of the electrode against the bottom of a beaker and the story of the "broad" range micro-pipette (or how to measure 1.1 ml with a 100 to 1000 µL adjustable micro-pipette). These stories are all from my Bioseparations Lab and they were partly caused by what I like to call "engineer" hands. Rugged and fat hands that, without proper training, are good to disassemble a centrifugal pump but whose dexterity to handle micro-liters of material is questionable. Somehow, I feel sympathetic to my chemistry and biology colleagues who do not allow the students to use anything in their laboratories. Of course, their approach does not teach the students anything but, at least, it is cheap. Obviously, better approaches are necessary.

Our Approach

We decided, a few years back, to attempt to alleviate the poor laboratory training of the students by dividing our Bioseparations Laboratory into two parts: Training and Project. During the first weeks of the semester, the students learn the fundamentals of various separations techniques (mainly membrane filtration and chromatography) as well as ancillary techniques (for example, centrifugation, spectroscopy, use of pH and conductivity meters, gel electrophoresis, etc.). This portion of the semester consists of six to 8 short experiments (one to two weeks each) in which the students follow recipes and write technical reports in scientific journal format. There is not penalty for failure as long as they have been careful in their work. This training section requires close monitoring of the students. We have found that with the exceptions of Ph.D. candidates in the author's laboratory teaching assistants are not very useful in our laboratory. Therefore, many semesters I have taken the sole responsibility of teaching the class. Although the number of contact hours is considerable during the training portion of the lab that is somehow compensated during the project part in which the students work mostly alone.

A few laboratories that we use for training purposes are included in Table I. This list is not exhaustive as the type of experiments included in the training section is tailored to the type of projects planned for the second part of the semester. The students need to be trained in general laboratory practices but at the same time they must prepare themselves for the operations that they will use in a particular project. The only recurrent themes in the laboratories included in Table I are chromatography and ultrafiltration because of the undisputable importance of these operations in the bioprocessing sector. We have included in the table the main tasks involved in each laboratory as well as the overall goals of each training section.

A common goal of each training laboratory is to gain general laboratory skills. Table II summarizes the basic skills that the students learn in the training laboratories. Notice the redundancy of many of them. We have found that repetitive direct instructions are necessary for the students to perform properly. The training in the use of each piece is quite intensive.

For example, the students learn the working principles of a pH electrode, how to operate a pH meter and to calibrate it.

Table I. Examples of training laboratories		
Laboratory	Tasks	Goals
Gel Permeation Chromatography	Buffer Preparation Swelling a Gel Packing a Low Pressure Column Running a Low Pressure System Preparing a Calibration Curve	Learn how to pack a column. Learn how a low pressure chromatography system works. General Lab. skills
Ion Exchange Chromatography	Buffer Preparation Swelling a Gel Packing a Low Pressure Column Running a Low Pressure System	Learn how to pack a column. Learn how a low pressure chromatography system works. Learn how to make gradients. General Lab. skills
Partitioning	Preparation of Phase Systems Mixing Centrifugation Sampling the Phases Protein Assay Enzymatic Assay Calculation of Partition Coefficients Statistical analysis of the Results	Learn how to do a Liquid/liquid extraction. Learn enzyme kinetics. Learn protein quantification protocols. Learn statistical analysis of data. General Lab. Skills
Isoelectric Focusing	Preparation of Gels Running the Gels Developing the Gels Documenting the Gels	General Lab Skills. Photo-documentation. Electrophoresis
Cross-Flow Filtration	Preparing buffers and solutions Running a Cross-Flow Filtration Unit Statistical analysis of results Calculation of Mass Transfer Coef.	Learn about concentration polarization. Applied mass transfer principles to filtration. General Lab Skills
Dialysis	Preparation of Dialysis bag Performance of Dialysis Calculation of Diffusion Coefficients	Learn about diffusivity through porous media. General Lab skills.

Table II. Mapping between basic skills and training laboratories.

Laboratory	pH-meter	Balance	Conductivity meter	Micropipettes	Centrifuge	Spectrophotometer (Absorbance)	Spect. (Kinetics)
Gel Permeation	■	■		■		■	
Ion Exchange	■	■	■	■		■	
Partitioning	■	■		■	■	■	■
Ultrafiltration	■	■		■			
Isoelectric focusing		■		■			
Protein Precipitation	■	■		■	■	■	
Dialysis	■	■				■	

As an illustration, we describe below the experiments the students perform in the ultrafiltration laboratory. Ultrafiltration is a means of concentrating dilute biological solutions that are heat sensitive. The solution to be concentrated is added to a cell that contains a membrane that prevents large molecules from passing through. The membrane pore sizes range from 0.001 to 0.02 μm. Ultrafiltration can also be used to partially separate proteins whose size difference is relatively large. Pressure is applied to force liquid through the membrane until the volume has been reduced to the desired amount. A common problem with ultrafiltration is gel polarization, which is a build-up of the solute at the surface of the membrane which reduces the flux through the membrane.

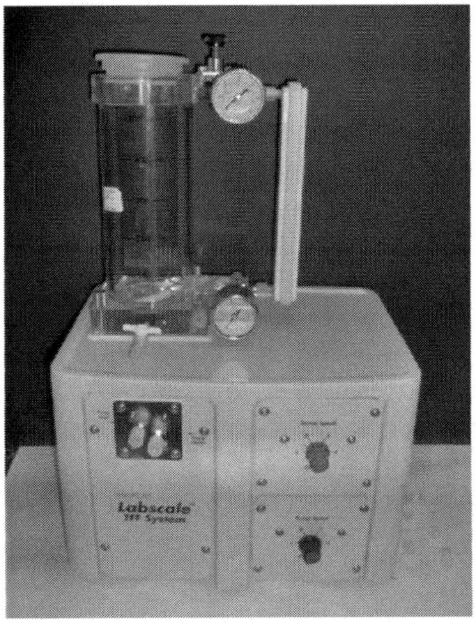

Figure 1. TFF system by Millipore

The learning objectives of this laboratory are: 1) To understand the functioning of a cross-flow ultrafiltration device; 2) To understand the concept of concentration polarization; and 3) To determine the mass transfer coefficient on the retenate side of the membrane (in the pressure independent regime). The students perform all the experiments in duplicate. The students are asked first to collect the necessary materials, which are Dextran (MW 150,000 or higher) and NaCl. Then, they prepare a saline solution (0.050 M NaCl) and dextran solutions at various concentrations in 0.050 M NaCl. The students are then introduced to the particular instrument. For example the Labscale TFF system by Millipore (Figure 1) equipped with a cross filtration membrane modulus with a molecular weight cut off of 100,000. The instructor describes the main components of the instrument, briefly indicates its capabilities and more importantly, he offers the students a copy of the manual and points out at the home page of the manufacturer.

The students then perform several runs in the instrument varying the concentration of dextran. Their objective is to identify the pressure independent regime. The laboratory is very time demanding because washing between runs takes approximately 90 minutes. After enough data has been collected they used the equation below to estimate the mass transfer coefficient (k_o) and the membrane concentration (C_{AW}),

$$J = k_0 \ln \frac{C_{Aw}}{C_{Ab}} \qquad (1)$$

where J is the mass flow-rate and C_{Ab} is the concentration of the dextran solution.

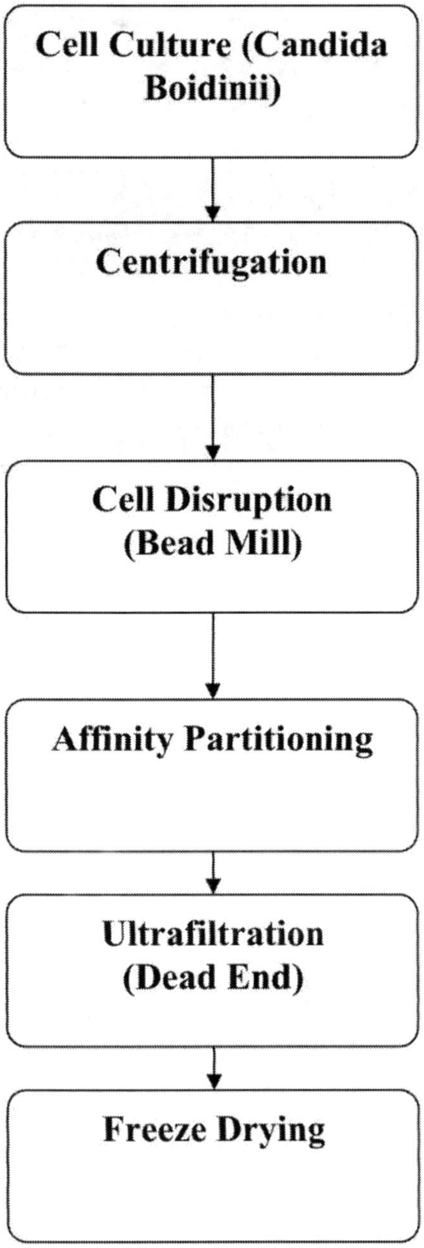

Figure 2. Flow-sheet for the isolation of alcohol dehydrogenase from *Candida boidinii.*

After they have completed the training part, they start a project that consists of a cascade of operations. Examples of projects that we have included in our laboratory are: 1) the isolation of human antibodies from transgenic corn (Lee and Forciniti, 2005); 2) the isolation of alcohol dehydrogenase from yeast (Waldorf et al., 1990); 3) the isolation of coagulation factors from human plasma (Weerasinghe et al., 1985); and 4) the fractionation proteins present in the lenses of mammals (Petitt et al., 1997). During this portion of the semester the participation of the instructor is minimal. The students are responsible to keep the supplies available, they take care of waste disposal, and their working hours are free. Their objective is to finish their project successfully. Because they have been trained before they start the project, it is their responsibility to repeat unsuccessful experiments in their own time. The projects are described to them but it is expected that the students would explore different operating conditions to determine the effect of those changes on yield and purity. The students in the class are divided into groups that pursue different projects. This is necessary because of equipment availability in our laboratory. One student in each group is chosen as the group leader who has the responsibility to distribute work and to report to the instructor. The acceptance of our approach by the students changes from year to year and it depends heavily on their quality.

Figures 2 to 4 show the flow-sheets for the isolation of alcohol dehydrogenase from *Candida boidinii*, the isolation of γ-crystallins from bovine lenses, and the purification of coagulation Factor XII from human plasma respectively.

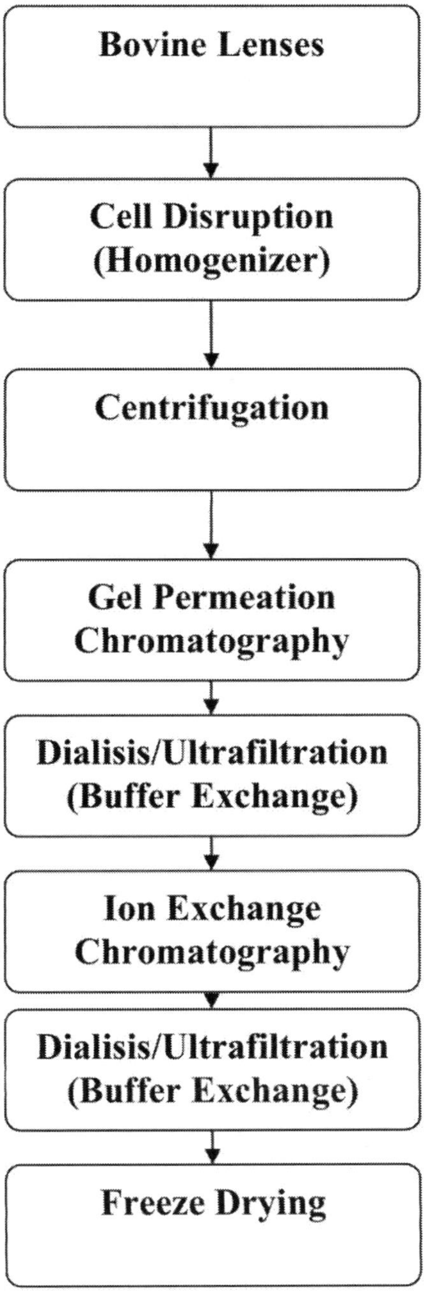

Figure 3. Flow-sheet for the isolation of γ-crystallins from calf lenses.

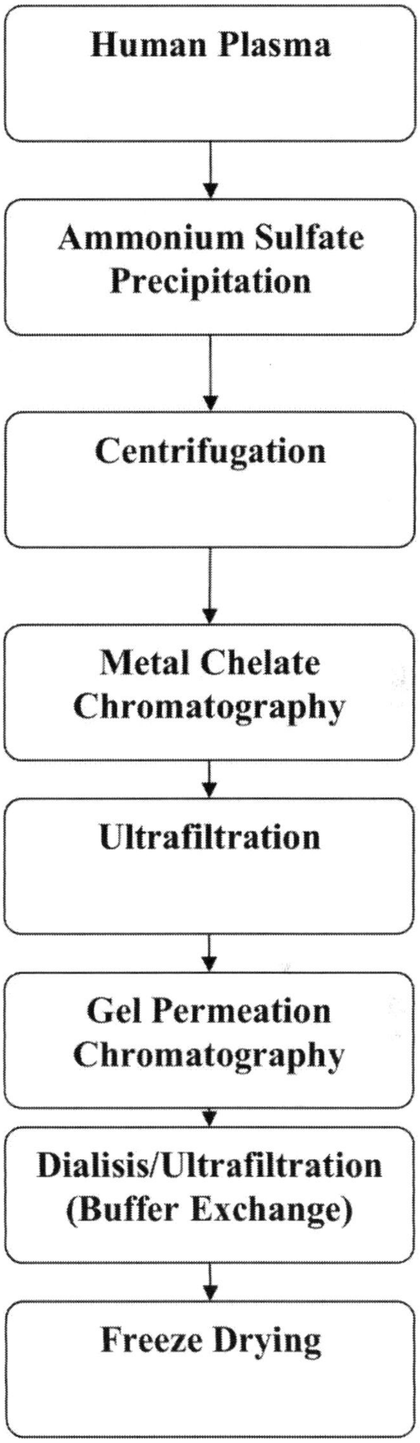

Figure 4. Flow-sheet for the isolation of Coagulation Factor XII from human

The genesis of each laboratory project is to be found in research projects that have been pursued in the author's research group. The projects are only vaguely described to the students (just a little bit more than the flow-sheets included in this manuscript). After the students have become familiar with the project (by reading a couple of key articles in the open literature), they discuss with the instructor the kind of experiments they want to conduct and what kind of conditions they would like to explore. For example, in the isolation of gamma crystallins from bovine lenses, the selection of the type of buffer gradient in the ion exchange chromatography step is key to obtain a good separation.

Laboratory Organization

The number of students in each group varies from year to year depending on the number of students in my course. During the training sessions groups are small, no larger than 3 students per group. During the project portion of the lab group sizes range from 2 (in years with few students) to 5 in years with a large student population. The expectations in each case also change. For example, students belonging to small groups of two students are expected to actively participate in each task of the project but they are not asked to explore a variety of operating conditions. Students belonging to large groups are normally divided into two or three subgroups with very specific tasks. For example, in the isolation of alcohol dehydrogenase project a large group of five students will be divided into three subgroups. A group of two will be in charge of cell grow and cell disruption. They explore how different metabolic stages of the yeast affect the production of the enzyme. They may also explore different milling conditions. A group of two students is in charge of the process separation development whereas the remaining student is in charge of all analytical work (determination of protein concentration and enzymatic activity in each step of the process). The group organizes its own tasks and the lab is run on "open door" policies (including nights and weekends). Graduate students working in the author's laboratory help in keeping the lab open at unusual hours and they provide the "adult" supervision that an undergraduate student should have.

References

J.-W. Lee and D. Forciniti. 2005. Purification of human antibodies by using liquid/liquid extraction. 7th World Congress of Chemical Engineering, Glasgow, Scotland, July 2005. 84799/1-84799/10.

Petitt, P., Edwards, M., and Forciniti, D. 1997. A Small Angle Neutron Scattering Study of Proteins at Their Isoelectric Point. *European Journal of Biochemistry* 243: 415-421.

Walsdorf, A., Forciniti, D. and Kula, M.-R. 1990. Investigation of Affinity Partition Chromatography Using Formate Dehydrogenase as a Model. *Journal of Chromatography*, 523:103-117.

Weerasinghe, K.M., Scully, M.F. , Kakkar, V.V. 1985. A rapid method for the isolation of coagulation factor XII from human plasma. Biochimica et Biophysica Acta, 839:57- 61.

AUTHOR INDEX

AUTHOR INDEX

AUTHOR INDEX

AUTHOR INDEX

AUTHOR INDEX

AUTHOR INDEX

AUTHOR INDEX

AUTHOR INDEX